Advances in Conservation Agriculture

Volume 2: Practice and Benefits

It is widely recognised that agriculture is a significant contributor to global warming and climate change. Agriculture needs to reduce its environmental impact and adapt to current climate change whilst still feeding a growing population, i.e. become more 'climate-smart'. Burleigh Dodds Science Publishing is playing its part in achieving this by bringing together key research on making the production of the world's most important crops and livestock products more sustainable.

Based on extensive research, our publications specifically target the challenge of climate-smart agriculture. In this way we are using 'smart publishing' to help achieve climate-smart agriculture.

Burleigh Dodds Science Publishing is an independent and innovative publisher delivering high quality customer-focused agricultural science content in both print and online formats for the academic and research communities. Our aim is to build a foundation of knowledge on which researchers can build to meet the challenge of climate-smart agriculture.

For more information about Burleigh Dodds Science Publishing simply call us on +44 (0) 1223 839365, email info@bdspublishing.com or alternatively please visit our website at www.bdspublishing.com.

Related titles:

Advances in Conservation Agriculture Volume 1: Systems and Science
Print (ISBN 978-1-78676-264-1); Online (ISBN 978-1-78676-266-5, 978-1-78676-267-2)

Assessing the environmental impact of agriculture
Print (ISBN 978-1-78676-228-3); Online (ISBN 978-1-78676-230-6, 978-1-78676-231-3)

Improving organic crop cultivation
Print (ISBN 978-1-78676-184-2); Online (ISBN 978-1-78676-186-6, 978-1-78676-187-3)

Managing soil health for sustainable agriculture Volume 1: Fundamentals
Print (ISBN 978-1-78676-188-0); Online (ISBN 978-1-78676-191-0, 978-1-78676-190-3)

Managing soil health for sustainable agriculture Volume 2: Monitoring and management
Print (ISBN 978-1-78676-192-7); Online (ISBN 978-1-78676-195-8, 978-1-78676-194-1)

Chapters are available individually from our online bookshop:
https://shop.bdspublishing.com

climate•SMART•publishing
IN AGRICULTURAL SCIENCE

BURLEIGH DODDS SERIES IN AGRICULTURAL SCIENCE
NUMBER 62

Advances in Conservation Agriculture

Volume 2: Practice and Benefits

Edited by Professor Amir Kassam, University of Reading, UK and Moderator, Global Conservation Agriculture Community of Practice (CA-CoP), FAO, Rome, Italy

burleigh dodds
SCIENCE PUBLISHING

Published by Burleigh Dodds Science Publishing Limited
82 High Street, Sawston, Cambridge CB22 3HJ, UK
www.bdspublishing.com

Burleigh Dodds Science Publishing, 1518 Walnut Street, Suite 900, Philadelphia, PA 19102-3406, USA

First published 2020 by Burleigh Dodds Science Publishing Limited

Library of Congress Control Number: 2019951731

British Library Cataloguing in Publication Data
A catalogue record for this book is available from the British Library

ISBN 978-1-78676-268-9 (Print)
ISBN 978-1-78676-271-9 (PDF)
ISBN 978-1-78676-270-2 (ePub)
ISSN 2059-6936 (print)
ISSN 2059-6944 (online)

DOI 10.19103/AS.2019.0049

Typeset by Deanta Global Publishing Services, Dublin, Ireland

Contents

Series list

Title	Series number
Achieving sustainable cultivation of maize - Vol 1 From improved varieties to local applications Edited by: Dr Dave Watson, CGIAR Maize Research Program Manager, CIMMYT, Mexico	001
Achieving sustainable cultivation of maize - Vol 2 Cultivation techniques, pest and disease control Edited by: Dr Dave Watson, CGIAR Maize Research Program Manager, CIMMYT, Mexico	002
Achieving sustainable cultivation of rice - Vol 1 Breeding for higher yield and quality Edited by: Prof. Takuji Sasaki, Tokyo University of Agriculture, Japan	003
Achieving sustainable cultivation of rice - Vol 2 Cultivation, pest and disease management Edited by: Prof. Takuji Sasaki, Tokyo University of Agriculture, Japan	004
Achieving sustainable cultivation of wheat - Vol 1 Breeding, quality traits, pests and diseases Edited by: Prof. Peter Langridge, The University of Adelaide, Australia	005
Achieving sustainable cultivation of wheat - Vol 2 Cultivation techniques Edited by: Prof. Peter Langridge, The University of Adelaide, Australia	006
Achieving sustainable cultivation of tomatoes Edited by: Dr Autar Mattoo, USDA-ARS, USA & Prof. Avtar Handa, Purdue University, USA	007
Achieving sustainable production of milk - Vol 1 Milk composition, genetics and breeding Edited by: Dr Nico van Belzen, International Dairy Federation (IDF), Belgium	008
Achieving sustainable production of milk - Vol 2 Safety, quality and sustainability Edited by: Dr Nico van Belzen, International Dairy Federation (IDF), Belgium	009
Achieving sustainable production of milk - Vol 3 Dairy herd management and welfare Edited by: Prof. John Webster, University of Bristol, UK	010
Ensuring safety and quality in the production of beef - Vol 1 Safety Edited by: Prof. Gary Acuff, Texas A&M University, USA & Prof. James Dickson, Iowa State University, USA	011
Ensuring safety and quality in the production of beef - Vol 2 Quality Edited by: Prof. Michael Dikeman, Kansas State University, USA	012
Achieving sustainable production of poultry meat - Vol 1 Safety, quality and sustainability Edited by: Prof. Steven C. Ricke, University of Arkansas, USA	013
Achieving sustainable production of poultry meat - Vol 2 Breeding and nutrition Edited by: Prof. Todd Applegate, University of Georgia, USA	014

Dedication

This book is dedicated to the global Conservation Agriculture movement and all the stakeholders engaged in this movement who have successfully generated and sustained an international momentum to transform the unsustainable and out-of-date conventional tillage agriculture into the alternate paradigm of Conservation Agriculture, representing an ecosystem approach to regenerative sustainable agriculture. The book is particularly dedicated to all the Conservation Agriculture pioneer and champion farmers and their families, researchers and extension agents as well as all the champions in the national and international public, private and civil sectors who have dedicated themselves to mobilizing policy and institutional support for making Conservation Agriculture a global mainstream reality for sustainable development in the 21st century.

Acknowledgements

This book builds on the vast amount of innovative and pioneering development-oriented work done over the past several decades in the public, private and civil sectors by those individuals and organizations who make up the global Conservation Agriculture (CA) Community of Practice. These CA stakeholders include farmers, their families and organizations, service providers along the value chain, scientific and empirical researchers, educators and trainers at all levels in schools, colleges, universities and extension services, technical assistance and development organizations and their experts at the national, regional and international levels, manufacturers of inputs, policy analysts and policy-makers, politicians and all the innovators and champions. I express my very special and sincere gratitude to all the stakeholders of this CA Community who are making sustainable agriculture real in all continents through the successful adoption and spread of the new paradigm of CA and the institutional capacity to sustain the transformation.

I would like to recognize the enormous assistance and cooperation extended to me in the preparation of this book from a large numbers of colleagues and organizations. First of all I would like to extend my sincere thanks to the Publisher Burleigh Dodds Scientific Publishing for inviting me to edit this book, and specifically to Francis Dodds who was actively engaged throughout in overseeing the project, with the able assistance from Amanda Renwick and Rennie Alphonsa at Deanta who oversaw production of the book.

Most importantly, I would like to express my deep gratitude to every lead author and all the co-authors for their valuable contributions and for spending their precious time and resources to bring together the relevant and latest global knowledge and information on CA science and systems, practice and benefits, and adoption and spread.

I acknowledge with many thanks the external reviewers of the different chapters of the book for their help to ensure that the chapters comprise the best quality material. They include: Theodor Friedrich, Rolf Derpsch, Sjoerd Duiker, Jeff Mitchell, Ivo Mello, Rafael Fuentes, Paulo Sérgio Pavinato, Denizart Bolonhezi, Telmo Amado, Enamul Haque, Frank Place, D. Pedro Freitas, Robelio Marchão, Julio César Salton, Luis Carlos Balbino, Brian Sims, Jacky Desbiolles, Willem Hoogmoed, Andrés Sylvestre Begnis, Sebastián Senesi, Gottlieb Basch, Hendrick Smith, Don Reicosky, Timothy Reeves, Yashpal Singh Saharawat, Vicenzo Tabaglio, Gualbert Gbehounou, Kevin Gallagher, Johnnie Van den Berg, Mário Carvalho, Jean-Pierre Sarthou, Emilio Gonzalez, Daryl Domitruk, David Coates, Tom Goddard, Julian Dumanski, Michele Pisante, Francis Shaxson, Ademir de Oliveira Ferreira and Zete Stumpf.

Finally, I express my very special thanks to Dr. Qu Dongyu, the Director General of the Food and Agriculture Organization of the United Nations, for his inspiring and illuminating Foreword. Dr. Dongyu's contribution is all the more important because the promotion CA in China is a national priority that has led to a very rapid growth in the area under CA and in the institutional capacity to sustain it.

Foreword

These volumes are a timely celebration of the most progressive change in farming practices that has been set in motion during the past 60 years and which is gathering momentum around the world at an extraordinary pace.

For thousands of years, soil inversion, whether by hoe or plough, has been almost universally applied by farmers and seen as essential for the successful growing of annual crops. Tilling the soil buries weeds, loosens the soil to let roots penetrate easily and allows rainfall to sink in and become available to the crop. By burying crop residues, it may also interrupt the life cycles of crop diseases and pests.

However, as the frequency and depth of tillage has increased, the negative effects have become more obvious. Every time soil is dug or ploughed, its structure is broken up and it becomes increasingly at risk to water and wind erosion. The speed with which soil organic matter content falls is accelerated, causing the surface to become susceptible to crusting, thereby reducing rainfall infiltration and increasing run-off while also restricting moisture retention capacity in the plant rooting zone below the surface. These processes tend to make crops and soils more vulnerable to drought and can ultimately lead to farmland becoming so degraded that it is abandoned.

As we understand more about the processes of climate change, we are becoming increasingly aware of the extent to which frequent tillage also contributes to global warming. As the organic matter content of soil falls this reduces the capacity of farmed land to serve as a carbon sink. Moreover, soil inversion, whether manual or mechanical, is very heavy in its energy requirements, with tractor use for ploughing accounting for a large share of the fossil fuel consumption in food production.

We are also belatedly learning that frequent soil inversion, especially when associated with heavy applications of pesticides, reduces soil biotic activity and undermines soil health with a corresponding fall in productivity.

The 'Dust Bowl' in the United States in the 1930's awoke farmers and scientists to the damaging effects of excessive tillage but it was not until the 1960s that American farmers began to adopt various no-till systems to reduce wind and water erosion. These set a precedent for the progressive emergence in the following decades of Conservation Agriculture (CA) which consists of a combination of continuous no tillage with year-round biomass soil cover and crop associations and/or rotations involving cover crops – often legumes.

The USA continues to be the country with the largest area under CA (43.5 million ha in 2015/16 or 35.1% of its arable farming area). Its farmers' associations are very active in pioneering new technologies, especially those related to the better use of cover crops in CA systems to increase Nitrogen availability and to cut herbicide dependence

One could claim that CA was ahead of its time in that it contains the main elements of what is now termed 'sustainable agriculture'. Indeed, there is much to be learnt from the CA story of the past 60 years which is relevant to inducing the global shift that must urgently be made from the currently unsustainable food production and consumption systems to ones that are truly sustainable.

Five of such lessons from the CA experience are:

Firstly, CA shows that fundamental changes in farming technologies – in this case, getting rid of the plough - can spread very rapidly throughout the world, in both developed and developing countries in which it is being taken up by large and small-scale farmers alike. The area of arable land under CA has grown from about 2 million ha in the USA in the early 1970s to 180 million ha (12.5% of global cropland) in 78 countries in 2015-16.

Secondly, the growth in CA uptake and the adaptation of CA methods to different ecological and societal conditions has been driven largely by practitioners, especially innovative farmers and machinery and hand-tool manufacturers. Approaches to CA are constantly evolving and new developments are being openly shared between all those involved. The more that farmers find that CA can boost their incomes, the faster will be the pace of change.

Thirdly, formal research has been important in developing new approaches to CA but most studies have been focussed on identifying the impact of the shift to CA on crop performance; the physical and biological conditions of soils; global warming, and farm incomes. Such validation studies have help boost the case for policy support for CA.

Fourthly, in most countries, however, CA has so far spread between farmers without explicitly supportive government policies. It seems certain that the rate of diffusion can be accelerated by targeted incentives (for instance, subsidies on appropriate machinery and equipment, payments to farmers for soil carbon accumulation and enhancement of water resources) and improvements in agricultural extension.

Lastly, although international agencies such as the FAO have only invested quite small resources in CA, they have played a valuable catalytic role, mainly by promoting exchanges of experiences between practitioners, countries and regions and nurturing the emergence of regional promotional institutions. FAO has placed CA at the core of its vision for sustainable food and agriculture which calls for 'a world in which food is nutritious, safe and accessible for everyone, where natural resources are managed sustainably, and where rural dwellers have decent livelihoods and contribute actively to economic development'. The FAO has also sponsored the foundation of a CA Community of Practice (COP) which shares new developments between its many members.

Although my own country, China, carried out trials on zero tillage from the 1980s, it was only after the turn of the century that the promotion of CA was adopted as a national priority. As a result, there has been a very rapid growth in the area under CA from just a few ha in 2002 to 9 million ha in 2015. The results have been good in terms of yield increases, especially for maize, and of reductions in erosion and river sedimentation. The government, therefore, intends to take additional measures to boost farmer uptake of CA including the recent creation of the China Institute for Conservation Tillage, technical support for the manufacture of improved machinery, subsidies for the provision of eco-systems services by farmers, and improvements in agricultural extension. It also intends to enhance the benefits of CA by linking its promotion to other aspects of its sustainable development agenda including Integrated Pest Management (IPM).

This edifying book is a compendium of much of what we have learnt across the world about CA. Its authors have all been deeply and enthusiastically involved in one way or another in the remarkable evolution and spread of Conservation Agriculture in recent decades. It will serve as an immensely valuable source of reference – and inspiration - for all those who are committed to putting the world's food systems on a truly sustainable footing.

It is with great pleasure, therefore, that I commend this book to you,

Qu Dongyu,
Director General, UN Food and Agriculture Organization (FAO),
Rome

Preface

Conservation Agriculture (CA) and its community of practice have made tremendous strides in the last two decades, particularly since the modern version of CA as we know it today was defined by the FAO at its first Regional Workshop on CA in Harare, Zimbabwe in 1998. The term CA was globalized through the 1st World Congress on Conservation Agriculture organized by the ECAF in partnership with the FAO in 2001 in Madrid, Spain. In 1999, the global spread of CA stood at about 45 M ha of cropland with some 10 countries in which CA was being practiced and promoted. In 2008/09, the global spread of CA was about 106 M ha of cropland across some 36 countries, corresponding to an annual rate of expansion of about 6.1 M ha. In 2015/16, the global spread of CA was some 180 M ha of cropland, split about equally between the Global South and the Global North, involving 78 countries. This corresponds to an annual rate of increase of some 10.5 M ha, a spectacular rate of transformation, led mainly by farmers and their associations with support from national and international champions and enthusiasts, many on a voluntary basis, and some national and international institutions. Although the spread of CA in the 1990s and 2000s was led by countries in North and South America and Australia, the spread of CA since 2008/09 across Europe (including Russia and Ukraine), Asia and Africa has been accelerating, making the transformation from conventional tillage agriculture to CA a truly global phenomenon.

The Green Revolution of the 1960s and 1970s was based on production intensification of wheat and rice, and later maize, relying on the unsustainable intensive tillage-based agriculture and expensive production inputs which brought short-term benefits to some types of farmers only in a handful of countries. The Green Revolution agriculture lost its effectiveness and appeal in the 1990s due to the unacceptably high negative economic, environmental and social impact as well as due to the loss of control by farmers of their own affairs related to production management and capital investment. The alternate CA revolution on the other hand has involved all types of smallholder and large-scale farmers, men and women, and rich and poor farming households, in all major land-based agro-ecologies in all continents. It has brought to the rural communities and society at large a wide range of productivity, economic, environmental and social benefits which cannot be harnessed with the conventional Green Revolution agriculture.

The expansion in area of CA across the world and the increase in number of smallholders adopting CA has benefitted from the growing support of the international research and development community including the FAO, the IFAD, the World Bank, the EU, CIRAD, and CGIAR as well as from many national and local level research and extension systems including NGOs, farmer associations and private sector service providers. Overall, the spread of CA since the very beginning has largely been led by farmers. Initially this was in response to the need to minimize soil erosion and land degradation, but soon this became a strategy to build and maintain soil health and productivity, reduce the high cost of production and diminishing returns, harness ecosystem services for society and nature, address climate change, and support pro-poor sustainable agricultural development strategies.

The CA Community of Practice is made up of many stakeholders and champions including farmers and their families, extension workers, development experts, researchers and academics, heads of institutions, policy analysts and decision-makers, as well as national and regional CA associations. Collectively, they have provided the pioneers and champions and support that keeps generating and sustaining the momentum to what has now become a global phenomenon, transforming conventional tillage-based agriculture into CA as the basis for sustainable agriculture and land use intensification. At the same time, they have generated enormous amounts of new knowledge, formal and experiential, as well as scientific and empirical evidence, regarding CA science and systems, CA practice and benefits and CA adoption and spread. All this knowledge and evidence constitutes the new understanding about regenerative and sustainable agriculture and represents the desire by the CA Community of Practice to move away from the degrading paradigm of tillage-based Green Revolution agriculture to the alternative paradigm of CA.

In light of the above, I was easily persuaded by the publisher Burleigh Dodds Science Publishing to take up the challenge of editing this book 'Advances in Conservation Agriculture' to bring together the latest state-of-the-art global knowledge and development-oriented information about CA science and systems, practice and benefits and adoption and spread. This book is purposely not designed to be a theoretical debate about what scientists and academics with no practical experience of real farming or of CA think about CA and how CA is performing internationally. Global scientific and empirical evidence speaks volumes about the productivity, economic, environmental and social attractiveness of CA to farmers in all continents. Globally, the rate of adoption of CA is accelerating but much remains to be explained in terms of the superior performance of CA, and also much remains to be innovated in the coming decades to maximize the wide range of benefits offered by CA to farmers, society and the natural world.

The past and current research on CA reviewed in this book is aimed at being of value to all CA stakeholders, including students, especially in the practical context of addressing global challenges related to sustainable development with effective solutions. This is particularly true in the adoption of strategies dealing with: sustainable production intensification, climate smart agriculture, regenerative agriculture, agroecology, and restoration of degraded lands including biodiversity and land-mediated ecosystem services. Thus, this book is a contribution to making sustainable agriculture development real globally. This would not have been possible without the extraordinary help of colleagues and field experts from the global CA Community of Practice who have been involved over many years in championing and bringing about the ongoing and accelerating global CA revolution. More than 120 authors based in more than 30 countries have made this book possible, building their respective chapters on the vast amount of global evidence and knowledge that is now available regarding the superior global performance of the alternate ecologically sustainable paradigm of CA.

In the long run, the expanding knowledge and education system globally must become fully engaged in generating the future human and institutional capacity and appropriate new mindset to underpin and sustain the process of mainstreaming the CA paradigm as a core component of the much-needed sustainable global food and agriculture system. Mainstreaming of any alternate paradigm in any field requires that all the relevant public, private and civil sector institutions and policies must align themselves behind it and ensure the necessary strategic and practical support to sustain its evolution. It is hoped that the information contained in this book will contribute to more rapid mainstreaming of CA globally by inspiring and generating increasing number of stakeholders and champions, particularly youngsters with practical training of CA, to become engaged in the promotion of CA as a foundation for sustainable agricultural development.

Chapter 1

Practice and benefits of Conservation Agriculture systems

Amir Kassam, University of Reading, UK; and Laila Kassam, Animal Think Tank, UK

1 Introduction

Conservation Agriculture (CA) was defined in Chapter 1 (Kassam and Kassam, 2019) and Chapter 2 (Kassam et al., 2019) of Volume 1 as an ecosystem approach to regenerative sustainable agriculture and land management based on the context-specific, locally adapted and practical application of three interlinked principles of (i) continuous, no or minimum mechanical soil disturbance (no-till seeding/planting and weeding and minimum soil disturbance with all other farm operations including harvesting); (ii) permanent maintenance of soil mulch cover (crop biomass, stubble and cover crops); and (iii) diversification of cropping system (economically, environmentally and socially adapted rotations and/or sequences and/or associations involving annuals and perennials, including legumes and cover crops), along with other complementary good agricultural production and land management practices (Hobbs et al., 2008; FAO, 2008, 2011; Friedrich et al., 2009; Kassam et al., 2009, 2013, 2018) (www.fao.org/conservation-agriculture). CA systems are present in all continents, involving rainfed and irrigated systems including annual cropland systems,

http://dx.doi.org/10.19103/AS.2019.0049.01

perennial systems, orchards and plantation systems, agroforestry systems, crop-livestock systems, pasture and rangeland systems, organic production systems and rice-based systems. Conservation Tillage, Reduced Tillage and Minimum Tillage are not CA, and nor is no-till on its own.

Conservation tillage has many variants and is often oversold for its conservation benefits; in some cases, unacceptable soil degradation continues under practices called 'conservation tillage' (Derpsch et al., 2014; Reicosky, 2015). Given the significance of no-till/direct seeding in minimizing environmental impacts, this chapter treats no-till as a separate activity with minimum soil disturbance for seed placement and uses the term conservation tillage more narrowly to denote any other type of reduced tillage practice other than no-till, after Eagle et al. (2012). Continuous no-till is preferred for optimum multiple environmental benefits (Eagle et al., 2012).

Volume 1 has focussed on CA science and systems, comprising of chapters that elaborate different CA systems in different agro-ecologies and regions, how they have developed globally and how they are managed and supported. This volume focuses more on the main practices that are core and complementary constituents of CA systems and the benefits that are harnessed by farmers and for society as a result. Given that all CA systems are integrated land-use management operations comprised of multiple components working synergistically, they do not lend themselves easily to reductive experimentation over 1 or 2 years in small plots on research stations. Often, reported research work has failed to understand that CA systems cannot be established or transformed from conventional tillage systems overnight or instantly like switching on an electric light bulb. CA is a whole new paradigm of agriculture and whatever topics and fields of interest and stakeholder involvement apply to conventional agricultural systems also apply to CA systems. Further, CA systems and their practices pay special attention *inter alia* to (i) maintaining a healthy ecological foundation to underpin sustainability, (ii) promoting biodiversity to enhance system resilience, (iii) establishing large root systems actively engaged with soil microorganisms and mesofauna, (iv) creating a healthy and regenerative soil environment for efficient soil, nutrient, water productivity, (v) harnessing ecosystem services and (vi) building adaptability to climate change. Consequently, a large range of productivity, economic, environmental and social benefits are offered to the farmer, society and nature. Given the large variations in biophysical and socio-economic conditions in farming systems, and the historical and prevailing space-time variations amongst CA systems, researchers have had a difficult challenge in establishing reliable CA research programmes and explaining the nature of their results, despite the many advantages and benefits that have been claimed by CA farmers globally.

This chapter first describes, in Section 2, CA as a basis for sustainable soil, land and natural resource management and production intensification.

Section 3 elaborates the generic process of transformation from conventional systems into CA systems. Section 4 provides an overview of some of the main practices and benefits that are possible in CA systems and their potential contribution to the Sustainable Development Goals (SDGs) (UN, 2015). Of the 17 SDGs, the following are more directly relevant to this chapter: SDG 1 (no poverty), 2 (zero hunger), 6 (water), 7 (energy), 8 (economic growth and employment), 12 (sustainable consumption and production), 13 (climate action), 14 (marine resources) and 15 (terrestrial ecosystems), although it is realized that all SDGs are interconnected. Section 5 describes some of the global evidence of benefits from CA, and Section 6 deals with constraints and enabling conditions. The final section draws some conclusions and policy implications.

2 Conservation Agriculture as a basis for sustainable soil, land and natural resource management practice and production intensification

The farming practices required to implement the above-mentioned transformation will differ according to local conditions and needs. They will however have the following required characteristics, based on optimizing conditions in the root zone in the soil and at the ground and above the ground surface as being essential to (a) biotic activity; (b) provision of water and crops; (c) assurance of self-sustainability of soil structure and porosity; (d) protection against weeds, insect pests and pathogens; and resilience to cope with extreme events especially excess water, and drought and heat stress.

These include capacities for achieving maximum rain infiltration/minimum run-off and optimum water storage; minimum compaction; reduced diurnal temperature ranges in upper soil layers; regular supply of C-rich organic matter to surface; minimal loss of soil organic matter (SOM) by oxidation; maintained N levels in the soil; optimized P availability; conditions that favour integrated management of weeds, insects and pathogens; and system resilience that can cope with extreme biotic and abiotic stresses. As described in Volume 1, Chapter 1 (Kassam and Kassam, 2019) and Chapter 2 (Kassam et al., 2019), and in Hobbs et al. (2008), FAO (2008, 2011), Friedrich et al. (2009) and Kassam et al. (2009, 2013, 2018), such capacities are best achieved by applying the following three interlinked principles of CA as a base or a foundation for sustainable soil and system management, along with other complementary good agricultural practices, namely (www.fao.org/conservation-agriculture):

1 *Continuous, no or minimum mechanical soil disturbance:* implemented by the practice of no-till seeding or broadcasting of crop seeds, and direct placing of planting material into untilled soil; no-till weeding; causing

minimum soil disturbance from any cultural operation, harvest operation or farm traffic. Sowing seed or planting crops directly into untilled soil and no-till weeding: reduces run-off and soil erosion; minimizes the loss of SOM through oxidation; reduces disruptive mechanical cutting and smearing of pressure faces; promotes soil microbiological processes; protects and builds soil structure and connected pores; avoids impairing movement of gasses and water through the soil; and promotes overall soil health.

2 *Maintaining a permanent mulch cover on the soil surface*: implemented by retaining crop biomass, root stocks and stubbles and biomass from cover crops and other sources of biomass from ex-situ sources. Use of crop residues (including stubbles) and cover crops reduces run-off and soil erosion; protects the soil surface; reduces evaporation; moderates surface soil temperatures; conserves water and nutrients; supplies organic matter and carbon to the soil system; promotes soil microbiological activity to enhance and maintain soil health, including structure and aggregate stability (resulting from glomalin production by mycorrhiza); and contributes to integrated weed, insect pest and pathogen management and to integrated nutrient and water management.

3 *Diversification of species in the cropping system:* implemented by adopting a cropping system with crops in rotations, and/or sequences and/or associations involving annuals and/or perennial crops, including a balanced mix of legume and non-legume crops and cover crops. This crop diversification contributes to diversity in rooting morphology and root compositions; enhances the number and biodiversity of the soil organisms important in organic matter decomposition and nutrient cycling; increases biological diversity and carbon storage in the soil; enhances microbiological activity; enhances crop nutrition and crop protection through the suppression of pathogens, diseases, insect pests and weeds; and builds up SOM. Crops can include annuals, short-term perennials, trees, shrubs, nitrogen-fixing legumes and pastures, as appropriate.

CA practices related to the above principles add to sustainability of production and agro-ecological systems, including soil systems, and generate a range of field-level ecosystem services important for crop productivity and for the society and nature (Friedrich et al., 2009; Kassam et al., 2013). Table 1 outlines the kinds of contribution that are made by individual and collective core practices of CA towards achieving a range of soil-mediated objectives (Friedrich et al., 2009). These core principles also improve the soil hydrological, biological, physical and chemical conditions related to soil

Table 1 Effects of CA production system components fully applied together on sustainability and soil-mediated productivity-related ecosystem services

System component ► To achieve ▼ Initially: seek, identify and alleviate any subsurface soil-compaction or pans before planting	Mulch cover (crop biomass, cover crops, stubbles)	No tillage (no or minimal soil disturbance)	Legumes (as crops for fixing nitrogen and supplying plant nutrients)	Crop rotation/association (for several beneficial purposes)
Simulate optimum 'forest-floor' conditions	√	√		
Reduce evaporative loss of moisture from soil surface	√			
Reduce evaporative loss from soil upper soil layers	√	√		
Minimize oxidation of soil organic matter, CO_2 loss		√		
Minimize compactive impacts by intense rainfall, passage of feet, machinery	√	√		
Minimize temperature fluctuations at soil surface	√			
Provide regular supply of organic matter as substrate for soil organisms' activity	√			
Increase, maintain nitrogen levels in root zone	√	√	√	√
Increase CEC of root zone	√	√	√	√
Maximize rain infiltration, minimize run-off	√	√		

(Continued)

Table 1 *(Continued)*

System component ► To achieve ▼ Initially: seek, identify and alleviate any subsurface soil-compaction or pans before planting	Mulch cover (crop biomass, cover crops, stubbles)	No tillage (no or minimal soil disturbance)	Legumes (as crops for fixing nitrogen and supplying plant nutrients)	Crop rotation/association (for several beneficial purposes)
Minimize soil loss in run-off and wind	√	√		
Permit and maintain natural layering of soil horizons by actions of soil biota	√	√		
Minimize weeds	√	√		√
Increase rate of biomass production	√	√	√	√
Speed soil-porosity's recuperation by soil biota	√	√	√	√
Reduce labour input		√		
Reduce fuel-energy input		√	√	√
Recycle nutrients	√	√	√	√
Reduce pest pressure of pathogens				√
Re-build damaged soil conditions and dynamics	√	√	√	√
Pollination services	√	√	√	√

Source: based on Friedrich and Kassam (2009) and Kassam et al. (2013).

Table 2 How CA improves soil conditions

	Key features of Conservation Agriculture ⇒			
Components of soils' productive capacity	No-till ⇓	Mulch ⇓	Rotations/associations ⇓	Legumes ⇓
Hydrological	1	4		
Physical	2	5	7	10
Biological	3	6	8	11
Chemical			9	12

Key: 1 = Water percolation; 2 = Varied soil porosity; 3 = Favours biological soil-layering; 4 = Buffers impacts of rainfall, wide diurnal ranges of surface temperature; 5 = Prevents soil-crusting; 6 = Source of energy and nutrients; 7 = Augments root channels - distribution and depth; 8 = Favours biodiversity in soil; 9 = Beneficial root exudates; 10 = Favours development of optimum soil architecture (solids × spaces); 11 = Nitrogen + C-rich organic matter; 12 = Nitrogen.
Source: adapted from Kassam et al. (2013).

productive capacity (Table 2) and result in beneficial outcomes for production, ecosystem services and socio-economic conditions (Table 3) (Kassam et al., 2013). However, to achieve sustainable *intensification,* these CA practices need to be complemented by additional best production and management practices (FAO, 2011; Lal, 2018), including:

- use of well-adapted, high-yielding cultivars and good-quality seeds;
- enhanced crop nutrition based on healthy soils;
- integrated management of insect pests, diseases and weeds;
- efficient water management;
- careful management of machines and field traffic to avoid soil compaction.

Sustainable crop production intensification is the combination of all these improved practices applied in a timely and efficient manner, underpinned by the interlinked core practices of CA. For this, the ensuring of soil stability, and the favouring of self-recuperation of appropriate soil structural conditions, are essential (see Tables 1–3). Thus, sustainable soil management depends on how and what crops are grown. However, for sustainable production *intensification* to occur, the core or foundation CA practices must integrate with other complementary practices that allow the intensification of output and the optimization of the production inputs. Such sustainable production systems, and the associated sustainable crop, soil, nutrient, water, pest and energy management practices, are knowledge and management intensive and relatively complex to learn and implement. They are dynamic systems, offering farmers many possible combinations of practices to choose from and adapt, according to their local production conditions and constraints (Kassam et al., 2009, 2017b; Godfray et al., 2010; FAO, 2011; Pretty et al., 2011).

Table 3 Some resulting beneficial outcomes with CA

For agricultural production e.g. ⇓	For ecosystem services e.g. ⇓	For socio-economic conditions e.g. ⇓
Greater security of output under varying weather conditions	Diminished water pollution by agrochemicals, eroded soil; reduced costs of water treatment	Greater efficiencies of use of labour, financial resources
Greater efficiency of rainwater use, leading to more-stable yields	Less frequency, depth, duration of flooding after unit storms of equal severity	Better health and nutrition
No/minimal soil erosion; smaller losses of applied energy, fertilizers, seeds and so forth	Longer duration of streamflow; recharge of groundwaters	Reduced frequency of flooding and severity of damages to roads, bridges and so forth
Improved soil health provides better biological controls of weeds, pests	Reduced loss of soil organic matter by tillage-induced oxidation to CO_2	More time for diverse activities on-farm (technical)
Re-circulation of carbon, micro- and macro-nutrients	Maintenance/improvement of soil-carbon content	More time for diverse activities off-farm (social)
Lesser effects of climatic drought events	Lesser damage to normal multiple functioning of soil in wider ecosystem	
Etc.	Etc.	Etc.

Source: adapted from Kassam et al. (2013).

The development of sustainable crop production intensification requires building on the core principles and practices outlined above as the production base and finding ways to support and self-empower producers to implement them all, through participatory approaches and stakeholder engagement. In addition, sustainable crop production intensification must be supported by coherent policies, institutional support and innovative approaches to overcome any barriers to adoption. Monitoring and evaluating the progress of change in production system practices and their outcomes at the farm and landscape levels is critical.

One of the main criteria for ecologically sustainable production systems such as CA is the maintenance of an environment in the root zone to optimize conditions for soil biota including healthy root function to the maximum possible depth (Mokany et al., 2006; Kell, 2011). Roots are thus able to function effectively and without restrictions to capture plant nutrients and water as well as interact with a range of soil microorganisms beneficial for soil health and crop performance (Pretty, 2002; Uphoff et al., 2006). In such systems, with the above attributes there are many similarities to resilient 'forest floor' conditions (Kassam et al., 2009). Maintenance or improvement of SOM content and soil structure and associated porosity are critical indicators for sustainable production and other ecosystem services.

A key factor for maintaining soil structure and organic matter is to limit mechanical soil disturbance in the process of crop management. For this reason, no-tillage production methods – as practiced, for example, in CA – have in many parts of the world been shown to improve soil conditions, reduce degradation and enhance productivity (Gebregziabher et al., 2006; Lal et al., 2007a). However, as a stand-alone practice, the elimination of tillage would not necessarily lead to a functioning sustainable production system. This requires a set of complementary practices to enable a functioning soil system as well as the whole agro-ecosystem to deliver a range of ecosystem services.

The contribution of practices that implement the technical principles of CA – including mulch cover, no-tillage, legume crops and crop rotations – to important ecosystem services is shown in Tables 1–3. However, for any agricultural system to be sustainable in the long term, the rate of soil erosion and degradation (loss of organic matter) must never exceed the rate of soil formation (though the steeper the slope, the greater the danger that this could happen). In most of the agro-ecosystems, this is not possible if the soil is mechanically disturbed (Montgomery, 2007). For this reason, the avoidance of mechanical soil disturbance is a starting point for moving towards sustainable production. Once it has been brought into good physical condition, no further tilling of the soil is therefore a necessary condition for sustainability, but not a sufficient condition. For sustainable crop production intensification, including ecosystem services, other complementary techniques are required as mentioned already, of which the practices related to the above three CA principles constitute the bare minimum for ecological sustainability (FAO, 2011; Kassam et al., 2013).

Considering the above, it is clear that sustainable soil management depends on both what and how crops are grown, as well as on additional aspects of soil and landscape management, which includes the horizontal integration of other production sectors such as forestry. The special role of deep-rooted legumes such as pigeon pea (*Cajanus cajan*), lablab (*Dolichos lablab*) and mucuna (*Stizolobium cinereum*) in building soil structure and biopores for drainage and aeration, in contributing biologically fixed nitrogen to improved nitrogen stocks in soils, and in generating both biomass and edible products is a case in point (Hargrove, 1991; Kell, 2011). Similarly, species diversification as the third principle of CA is related to integrated management of insect pests, pathogens and weeds, and the effectiveness of control of pests, pathogens and weeds depends on both what and how crops are grown (Liebig et al., 2014; Tsiafouli et al., 2015; Lange et al., 2015). Species diversification involving crops of different durations and complementarity is also related to the use and management of resources of different crops in space and time to maximize and optimize the production during the growing season every year to its fullest potential in an increasingly variable and unpredictable climate. Furthermore, to establish diversity of soil biological activity, it is necessary to include in the cropping system a diversity of crops instead of mono-cropping or reduced crop diversity.

CA is now (in 2015/16) adopted on about 180 million ha of arable land worldwide, which corresponds to some 12.5% of the total cropland (Kassam et al., 2019). Around 50% of this area is in the Global South. During the past decade, land under CA has been expanding at an average rate of more than 10 million ha per year. Thus, it is likely that the current area of CA globally is more than 200 million ha. The highest adoption levels, exceeding 50% of the cropland, are found in the southern part of South America, in the Canadian prairies and western Australia. Fast adoption rates are now being seen in Central Asia and China, alongside increasing policy support and early large-scale adoption taking place across Africa, particularly in Zambia, Zimbabwe, South Africa, Tanzania, Kenya, Morocco and Tunisia. Europe now has some few pockets of adoption, particularly in Finland, Spain, France, Italy, UK and Switzerland (Kassam et al., 2009; Derpsch and Friedrich, 2009; Friedrich et al., 2013).

3 Transforming conventional systems into Conservation Agriculture systems

In general, many researchers and farmers find that the best performance from CA is derived when the system quality is good and that all the core CA practices and complementary production system practices are well integrated to maximize synergies (Goddard et al., 2006; Jat et al., 2014; Kassam et al., 2013, 2017b; Farooq and Siddique, 2014).

The transformational change from the often-degraded conventional tillage agriculture conditions to good-quality responsive CA conditions is a time- and biology-related multi-year evolutionary process of ecological regeneration. The transformation from conventional system to CA cannot be done overnight in a single event. The application of the interlinked CA principles into concrete production practices depend on local situation and require the formulation of locally adapted practices based on the local biophysical, economic, social and management situation. Thus, the CA adoption process involves a system approach to managing change at the cropping system level. At the farm level, the adoption process can be implemented not on the whole farm at once but portion by portion to allow learning by doing and experiential knowledge to be generated. Such an approach also allows room for innovation, adjustments and acceptable risk-taking. The adoption process involving the implementation of a planned cropping system in a particular part of the farm can be visualized schematically as being comprised of broadly four reference phases (based on Sá, 2004; Derpsch, 2007), namely:

Initial phase (0–5 years): low SOM; low crop biomass soil cover; rebuilding of soil aggregates; re-establishing soil life (microorganisms and mesofauna); N may (mineral and/or biological) needs to be added to the system. There is a brake on degradation processes, which increase regeneration of soil health processes and improve soil moisture conditions. There is also reduced labour and energy requirement for farm operations.

Transition phase (5–10 years): increase in structural soil density (not compaction) is observed; the amount of crop biomass soil cover as well as soil carbon content and phosphorus content start to increase. Soil mulch cover provides physical protection to soil and suppresses weeds, and supplies biological substrates for soil microorganisms and mesofauna to establish a food web in the soil and connect with food chain being established above the ground, as agrobiodiversity below, at and above ground level is established, offering habitats to natural enemies of pests and pathogens. Further improvements observed in soil health and functions and in soil moisture and nutrient conditions, and there is improvement in productivity, economic and environmental performance of the system. Ecosystem services are in the process of being rehabilitated.

Consolidation phase (10–20 years): higher amounts of crop residues as well as higher carbon contents are achieved, and a higher cation exchange capacity and water retention capacity are observed. Improvements in soil health and functions continue and improved water, nutrient and carbon cycling are established. Ecosystem services are harnessed more fully at the field and landscape levels.

Maintenance phase (>20 years of continuous no-till with mulch cover in diversified cropping): ideal conditions with maximum benefits for the soil, landscape, crops and cropping systems are achieved and less production inputs of fertilizer, herbicides and other pesticides and water are needed.

The above description of the four phases is a simple conceptual illustration of what is expected to happen over a period of more than 20 years, with each phase offering benefits and changes that improve productivity (and hence profit), efficiency, resilience and ecosystem services. The four phases imply that it is necessary to keep in mind a long-term perspective when managing the transformation of conventional tillage agriculture to no-till CA. There are many factors that can affect the rate of change during the transformation process, including agroclimatic conditions, level of soil degradation, cropping and farming system, availability of production inputs, farm mechanization, farmer knowledge base and support systems.

The multi-year duration and nature of the transformation process must be understood and respected by farmers, service providers, supporting experts and decision-makers. The transformation is not like a switch to be pressed to achieve an instant change. It is during this initial and transition phases that the damage (physical, biological, chemical and hydrological) to the soil from many years of conventional tillage agriculture begins to be repaired and the soil physiological-biological-chemical-hydrological functions and soil-mediated ecosystem services are re-established. A good example is the return of soil life (microbiomes and mesofauna) as CA is established and the consequent repair and functioning of the water cycle, including water infiltration into the soil and water percolation to deeper soil layers and to groundwater, all of which are damaged under tillage agriculture. Alongside this improvement is the development of soil structure and aggregate stability, and an increase in water retention, all of which are destroyed under tillage agriculture. At the end of the initial phase after some 5 years, much of the loss in productivity potential may be restored, if the soil was not severely degraded and eroded, and often there can be an increase in yields of 100% or more with same level of production inputs, if the starting baseline yields are low and declining. By the end of the transition phase after around 10 years, many changes are expected to occur; there is an enhancement of the agro-ecological potential and yields, and factor productivity can increase further. During the next phase of consolidation, many of the soil functions and ecosystem services begin to perform more efficiently, and greater efficiency and resilience in the production system and farm system is observed (Kassam et al., 2009, 2013; Andersson and D'Souza, 2014; Cotton and Acosta-Martínez, 2018; Thierfelder et al., 2018).

Thus, it can take 10 years or more to reach a new agro-ecological 'equilibrium' but benefits and advantages begin to accrue almost immediately after the process of transformation is initiated. Beyond 20 years, soil and land productivity potentials, soil and landscape ecosystem functions and services, and below and above ground biodiversity and food webs would continue to offer all the benefits.

The above described transformation process and the possibility of different parts of the farm area undergoing transformation with different starting times and cropping systems means that at any given point in time, there can be different fields across the farm that are in different transformation phases.

4 Benefits of Conservation Agriculture systems and their potential contribution to the Sustainable Development Goals (SDGs)

CA not only offers entirely appropriate solutions to the challenges described above, it also has the potential to slow or reverse productivity losses and environmental damages. Transforming a tillage-based production system to a CA-based system is a time-related biological process. When implemented correctly, CA offers a range of benefits that correspond to the mobilization of greater crop and land potentials, and actual crop and land performance. The intensity and range of benefits generally increase over time as new and healthier soil productivity equilibrium, including a fuller range of soil functions and soil-mediated ecosystem services, is established.

4.1 Large-scale landscape-level benefits from sustainable production management

Benefits from sustainable production systems are scale independent. They do occur at the field-point scale, but benefits accrue to landscapes, farms, communities and regions. The four major sets of benefits from sustainable soil management and production systems are as follows (Kassam et al., 2009, 2013):

1 higher stable production output, productivity and profitability;
2 adaptation to climate change and reduced vulnerability;
3 enhanced ecosystem functioning and services;
4 reduced greenhouse gas emissions and 'carbon footprint' of agriculture.

All these are of direct benefit to producers and society. The relevant socio-economic indicators include farm profit, factor productivity (efficiency), amount of pesticides applied per unit of output, yield per unit area, farmers practicing sustainable intensified systems and stability of production. The relevant ecosystem service indicators include clean water provisioning from catchment areas under an intensive agriculture area, reduced erosion – both wind and water (Mello and van Raij, 2006; Laurent et al., 2011), increased biodiversity/wildlife within agricultural landscapes and increase in carbon sequestration and reduction in carbon footprint and GHG emissions of methane and nitrous oxide (Baig and Gamache, 2011; Kassam et al., 2009, 2013; FAO, 2012).

It is important to identify key indicators that would detect changes in the desired direction at the field, farm and physiographic landscape level within whose boundary the farm is located and whose management has an impact on the aggregate behaviour of the landscape. CA-based ecosystem services operate in different parts of the world and include the agricultural carbon offset scheme in Alberta, Canada; the hydrological services from Paraná III Basin in Brazil; the control of soil erosion in Andalusia, Spain; the controlling of water erosion and dust storms, and combating desertification and drought in the loess plateau of the Yellow River basin in China; and reducing susceptibility/increasing resilience to land degradation in western Australia. Controlling land degradation, particularly soil erosion, caused by tillage, exposed soils and depletions of SOM, has been a main objective of most of such initiatives (Kassam et al., 2013). Such landscape schemes are possible only when the landscape has a contiguous network of sustainable soil management that mediates such large-scale environmental and economic benefits to the producers and rural as well as urban society. With sustainable soil management practices being applied over large areas, it is then possible to overlay landscape-level development programmes to harness large-scale ecosystem services such as carbon sequestration as in Alberta, Canada, or water-related services as in Paraná Basin, Brazil, or erosion control in Andalusia, Spain.

4.2 Key benefits of Conservation Agriculture systems and their potential contribution to specific Sustainable Development Goals (SDGs)

Present and future societal needs related to food and agriculture, many of which are reflected in the SDGs, require greater crop and land potentials to be mobilized to produce more food and other agricultural products. However, if this is to be done in a truly sustainable way, we must certainly ask how economically, environmentally and socially appropriate and sustainable is the current production paradigm of tillage agriculture for meeting the SDGs related to food and agriculture, the management of natural resources and the environment? Equally, we must also consider how the SDGs can be met ethically, giving adequate attention to concerns of wastage of food and non-food commodities in the output value chain, types of food produced, diets and human health, issues of equity, food and land justice and human and animal rights. However, these questions, important as they may be, are outside the scope of this book.

The following section outlines the key benefits of CA and shows how they might contribute to the SDGs.

4.2.1 Increased productivity and profit

CA increases yields, factor productivity, farm production and profit depending on the level of initial degradation and yield, and the agro-ecological potential of the location (Lindwall and Sonntag, 2010; ECAF, 2011; Owenya et al., 2011; Soane et al., 2012; FAO, 2012, 2013; Brisson et al., 2010; Kassam et al., 2013, 2017a,b; Jat et al., 2014; Farooq and Siddique, 2014; Thierfelder et al., 2016; Li et al., 2016).

These benefits directly contribute to reducing income poverty; improving food security and reduce malnutrition; reducing vulnerability to extreme events; and economic growth and employment (SDGs 1, 2, 8 and 12).

4.2.2 Improved soil health and reduced use of fertilizer

Nutrient cycling is one of the main advantages of CA systems that involve cover crops and crop diversity. The mineral fertilizer nutrient efficiency is low in tillage agriculture. The genetic potential of modern varieties of grain crops is exploited at around 30% only. In such situations, a large amount of inputs is demanded to increase yields, resulting in high production cost and high environmental impact. It is therefore important to increase the agriculture efficiency in general and CA can be part of the solution since it restores SOM and soil health. CA can decrease fertilizer use by 50% or more if already applying higher rates and improve nutrient productivity with increased SOM level. In cases where mineral fertilizers are not available, integrated nutrient management can provide the required nutrition from local sources (Carvalho et al., 2012; FAO, 2011, 2013, 2016; Lal and Stewart, 2013; Kassam et al., 2013, 2017a,b; Sims and Kassam, 2015; Lalani et al., 2016, 2017).

These benefits can contribute to reducing income poverty; improving food security and reducing malnutrition; efficient use of natural resources; climate change mitigation by reducing energy use in manufacturing of fertilizers; reducing N_2O emissions; reducing nutrient pollution of sea water and inland fresh water systems; and increasing soil biodiversity (SDGs 1, 2, 12, 13, 14 and 15).

4.2.3 Reduced or minimized use of pesticides and herbicides

CA can decrease pesticides and herbicide use by 20–50% if already applying higher rates and increase output per unit of pesticide or herbicide. In cases where pesticides and herbicides are not used, or available, integrated weed and pest management can achieve adequate pest and weed control with less labour requirements (Lindwall and Sonntag, 2010; Owenya et al., 2011; Sims and Kassam, 2015; Lalani et al., 2016, 2017; Kassam et al., 2017a,b; Sims et al., 2018).

These benefits can contribute to improving income poverty and food security; sustainable production; climate change mitigation; and reducing pollution of sea water and inland fresh water systems (SDGs 1, 2, 12, 13, 14 and 15).

4.2.4 Reduced machinery, energy and labour use and costs

CA can reduce machinery, energy and labour costs, and time requirement by 70%. In manual production systems, there can be a 50% reduction in family labour requirement as there is much less labour required for seedbed preparation and weeding (Lindwall and Sonntag, 2010; Freixial and Carvalho,

2010; Owenya et al., 2011; Kassam et al., 2013; Sims and Kassam, 2015) (Table 4).

These benefits can contribute to reducing income poverty and food security; sustainable production, climate change mitigation and promoting economic growth and employment through greater participation from smallholders, youth, service providers and innovative manufacturing industries (SDGs 1, 2, 7, 8, 12 and 13).

4.2.5 Decreased soil erosion and land degradation

CA decreases soil erosion and water run-off (Derpsch, 2003), increases water infiltration (Fig. 1) and retention (Fig. 2). It can reduce water requirement by

Table 4 Summary of annual expenses for maintenance and repair of tractors and of tillage/drilling implements, for fuel and labour for a farm near Évora, South Portugal

	Conventional tillage (year 2000)	No-till (year 2003)	Reduction (%)
Maintenance and repair of tractors	10,450.47 €	1,507.15 €	85.6
Maintenance and repair of tillage/drilling implements	8,158.41 €	1,840.40 €	77.5
Fuel	17,460 €	7,110 €	60.0
Labour	25,000 €	15,000 €	40.0
Total	61,068.88 €	25,457.15 €	58.3

Farm power - four tractors with 384 HP under tillage and two tractors with 143 HP under no-till.
Source: adapted from Freixial and Carvalho (2010).

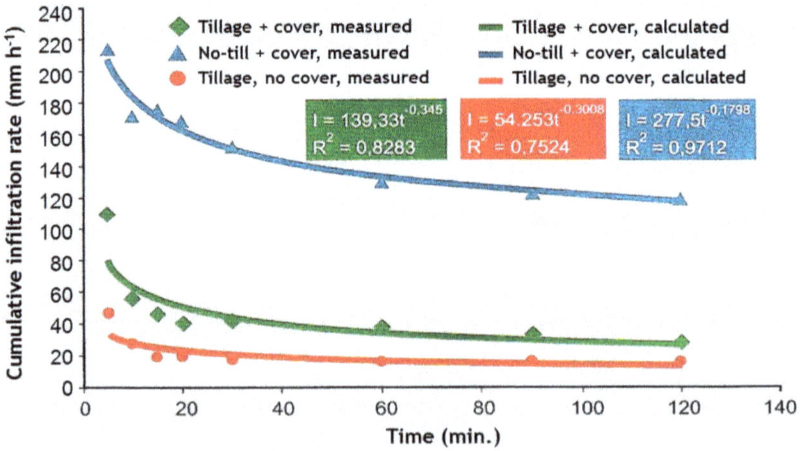

Figure 1 Gains in rainfall infiltration rates with CA. Source: adapted from Landers (2007).

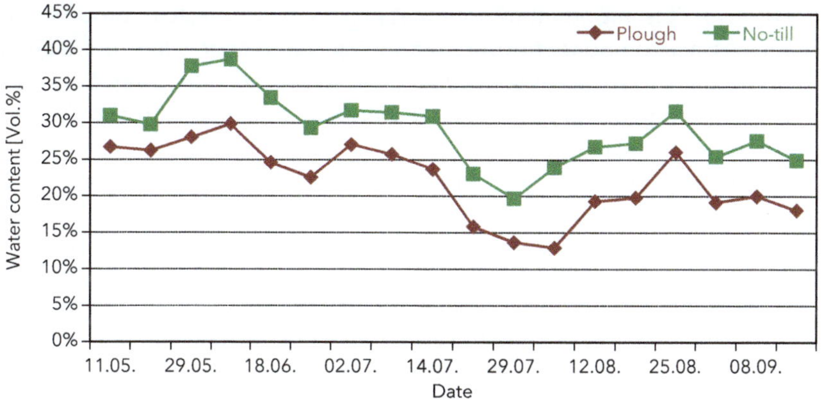

Figure 2 Soil water content in 0–30 cm soil depth; average of 1998 and 1999 under maize. Source: adapted from Derpsch (2003).

up to 40% and increase water productivity in rainfed and irrigated conditions (Lindwall and Sonntag, 2010; FAO, 2011, 2016; Basch et al., 2012; Kassam et al., 2013; Lal and Stewart, 2013; Jat et al., 2015; Reicosky, 2015; Nkonya et al., 2016; Vlek et al., 2017).

Controlling soil erosion and water run-off will improve soil health, nutrient and water retention and productivity; reduce land and ecosystem degradation; and enhance biodiversity. Reduced water requirement and increased water productivity will decrease water pollution, improve water quality and quantity and water-use efficiency (WUE).

All these benefits can contribute to reducing income poverty and food security; sustainable production and climate change adaptability; enhancing marine and fresh water resources; reducing ecosystem degradation; improving biodiversity; and regenerating ecosystems (SDGs 1, 2, 6, 12, 13, 14 and 15).

4.2.6 Increased biomass for livestock

Over time CA increases the biomass (along with greater yields) available for livestock as soil health improves, thus decreasing the initial 'conflict situation' and opening up the possibility of increased livestock carrying capacity and stocking rates (Landers, 2007; FAO, 2009, 2012b, 2013; Owenya et al., 2011; Lalani et al., 2018).

These benefits can contribute to decreasing income poverty and improving food and nutrition security; sustainable production; reducing land degradation; regenerating ecosystems; and promoting economic growth and employment through greater participation from smallholders, youth, service providers and innovative manufacturing industries (SDGs 1, 2, 8, 12 and 15).

However, it is important to note that promotion of livestock production is not a necessary component of sustainable agriculture, nor is it a core element of responsible and sustainable production and consumption, given the impact on climate change, deforestation, land grabbing, human health and the well-being and rights of animals.

Arguments for livestock contributing to sustainable production rest on recycling of nutrients leading to better plant nutrition. However, this can be achieved effectively in CA systems without livestock because CA promotes:

1 the enhancement of soil health through biomass retention, incorporation of biomass and exudates into the soil by its root system and by mesofauna such as earthworms, which also add nitrogen from the atmosphere through the nitrogen-fixing bacteria living in its gut; and
2 the transformation of biomass into humus by microorganisms. Humus comprises organic forms of plant nutrients that are retained and made available to plants. In addition, soil micro- and macro-aggregate stability is achieved through the action of mycorrhizae through the production of glomalin.

Thus, CA is capable of nutrient and carbon cycling without any need of livestock. The environmental impacts of animal agriculture, including the high land and water requirements, have been widely reported, as have the human health impact of meat-based diets (Willett et al., 2019). Grain-based livestock feed formulations compete directly with food security based predominantly on plant-based diets. Thus, the role of livestock in meeting the SDGs will remain a controversial issue given that millions of people live healthy lives on plant-based diets, and millions of hectares of agricultural lands are managed sustainably without any livestock.

4.2.7 Greater adaptability to climate change

CA contributes to greater adaptability to climate change in terms of higher and more stable yields, lower impact of climate variability on ecosystem functions and services, reduced incidence droughts, heat stress, waterlogging and floods, less damage to infrastructure such as roads, bridges and communication, navigation, dams and reservoirs.

Thus, CA lowers the environmental cost to society due to decreased levels of water pollution, and decreased damage to infrastructure, riverbanks and water bodies due to reduced erosion and flooding (ANA, 2011; ITAIPU, 2011; Mello and van Raij, 2006; FAO, 2011, 2016; ECAF, 2011; Thierfelder et al., 2016; Kassam et al., 2013, 2017a,b; Nkonya et al., 2016; Vlek et al., 2017; González-Sánchez et al., 2017, 2018).

CA is considered to be the best core component of climate-smart agriculture because it can contribute to sustainable food security, livelihoods and agricultural economic development; climate change adaptability; and climate change mitigation. The adaptability benefits from CA contribute to soil health and yield stability; improved yields; improved water cycling and soil water balance for agriculture; strengthens local, national and international climate actions on adaptability; sustainable production and development; quality of marine and fresh water resources; and regeneration of biodiversity and of degraded ecosystems. The adaptability benefits also include the promotion of economic growth and employment through greater participation from smallholders, youth, service providers and innovative manufacturing industries (SDGs 1, 2, 6, 7, 8, 12, 13, 14 and 15).

4.2.8 Climate change mitigation

CA increases the contribution of agriculture to climate change mitigation from enhanced soil carbon sequestration, reduced greenhouse gas emissions and decreased use of fossil fuel (Haugen-Kozyra and Goddard, 2009). Additionally, CA lowers agriculture's carbon and environmental footprint due to the reduced use of manufactured inputs such as agrochemicals and machinery. Also, improvement in soil structure appears to be directly linked with lower emission of all three major greenhouse gases (CO_2, CH_4, N_2O). The macroaggregates protect soil carbon and the continuous porosity improvement increases water drainage and soil aeration, which are key processes linked with lower greenhouse gas emission in CA compared to tillage agriculture (Lal et al., 2007b; ECAF, 2011; Corsi et al., 2012; González-Sánchez et al., 2012, 2017, 2018, 2019; Kassam et al., 2009, 2013, 2017a,b).

The ability of CA to convert agricultural soils into sinks for carbon sequestration and storage is becoming increasingly known and research continues to provide evidence of how much carbon can be sequestered by CA in different agro-ecologies in different continents. All three core practices of CA contribute to carbon sequestration (Corsi et al., 2012; González-Sánchez et al., 2012, 2017, 2018, 2019; Ferreira et al., 2016, 2018), as well as to lowering of greenhouse gas emissions. Carbon sequestration is due to greater amounts of biomass and root organic exudates being returned to the soil which builds soil health and productivity, thus contributing to reducing income poverty and food security. For example, Ferreira et al. (2016, 2018) showed that in six long-term CA systems (>20 years) on the farms of pioneer CA farmers in Rio Grande do Sul State, Brazil, with high crop diversity and intensive crop rotations, the SOC recovery ranged from 85% to 116% compared with virgin lands. The lowering of greenhouse gas emissions is due to various reasons including lower use and application of inputs (such as fossil fuel and mineral nitrogen) per unit of biological output, and better soil drainage which lowers the emissions of CH_4 and N_2O, thus contributing to reducing income poverty and food security.

Equally, it has been widely shown that emissions of all major greenhouse gases (CO_2, CH_4, N_2O) are decreased in CA systems (González-Sánchez et al., 2012, 2017, 2018, 2019) (SDGs 1, 2, 6, 7, 12, 13, 14 and 15).

4.2.9 Rehabilitation of degraded lands and ecosystem services

CA contributes to the rehabilitation of degraded lands and ecosystem services from all agricultural land under use, as well as from abandoned agricultural land in which the eroded topsoil and the soil profile can be rebuilt (Kassam et al., 2013; Jat et al., 2015, 2017b).

Much of the agricultural lands worldwide is degraded and large areas have become marginal or abandoned. CA offers a method to enhance soil health and regenerate soil functions and productivity. There are examples of large-scale rehabilitation initiatives that have succeeded in restoring degraded and even abandoned agricultural lands because of the adoption of CA systems such as in Brazil, Spain, China and Australia (Kassam et al., 2013).

Thus, such initiatives contribute to reducing income poverty, improving food security, ecosystem services, sustainable production, natural resource management, reducing environmental and water pollution, improving biodiversity and degraded ecosystems. These improvements can make a major contribution to economic growth and employment including for youth. They can also encourage more smallholder farmers, land managers and service providers to invest into agriculture (SDGs 1, 2, 6, 7, 8, 12, 13, 14 and 15).

4.2.10 Enhanced ecosystem services to society

With CA there is greater opportunity for establishing large-scale, community-based, cross-sectorial ecosystem service programmes, such as the watershed services programme in the Parana Basin in Brazil and the carbon offset trading scheme in Alberta, Canada (MEA, 2005; Haugen-Kozyra and Goddard, 2009; Lindwall and Sonntag, 2010; ANA, 2011; ITAIPU, 2011; Mello and van Raij, 2006; Kassam et al., 2013, 2017a,b; Juniper, 2013; Palm et al., 2014).

The role of agriculture for society and for development has been going through a considerable change in recent decades. The conventional tillage agriculture based on the Green Revolution paradigm in the industrialized and low-income countries has led to large-scale degradation and pollution of ecosystems, especially since the Second World War. The conventional paradigm is not capable of sustainable production nor is it capable of delivering the full range of ecosystem services to the society at large. The future role of agriculture will be to be productive as well as to provide all the required ecosystem services. Only CA systems can meet such a goal. Two examples of how farmers and communities are benefiting from transforming the landscape through large-scale adoption of CA come from Brazil (Section 4.2.7) and Canada (Section 4.2.8) (more details in chapter, 13, volume 2)

Benefits derived from CA have led to sustainable agricultural land management; reduced income poverty; improved food security; enhanced natural resource management including energy, nutrient and WUE; reduced degradation and water pollution; improved climate adaptability and mitigation; and improved economic growth and employment (SDGs 1, 2, 6, 7, 8, 12, 13, 14 and 15).

5 Global evidence of benefits from Conservation Agriculture

Some academic researchers have argued that the benefits of CA have been overstated and that CA proponents have presented biased evidence regarding the superior performance of CA compared to conventional tillage agriculture (Giller et al., 2009, 2015; Gowing and Palmer, 2008; Powlson et al., 2011, 2014; Sumberg and Thompson, 2012; Pittlekow et al., 2014). These critiques have often originated from a lack of experience and understanding about what CA actually is at the practical level and what it takes to establish a CA system (Derpsch et al., 2014; Reicosky, 2015). As a result, some researchers have included mixed or undifferentiated data sets in their analyses in terms of varying lengths of time the CA systems under study have been operating for, in terms of cropping system complexity, as well as using data sets from systems that are not CA compliant or genuine CA (e.g. Powlson et al., 2011, 2014; Arslan et al., 2013; Pedzisa et al., 2015a,b; Pittlekow et al., 2014; Pannell et al., 2014). This has led to a number of studies and meta-analyses drawing erroneous conclusions about CA and its global potentials, including its potential to mitigate climate change. The definition of CA makes it clear that no-till represents no or the absolute minimum soil disturbance (Eagle et al., 2012). The following discussion is an excellent explanation why the definition of continuous minimum soil disturbance is critical in CA systems. Any use of conservation tillage, minimum tillage, vertical tillage, reduced tillage, mulch tillage, strip tillage and so forth does not fit the definition of no-till or direct seeding for no or minimum soil disturbance.

Empirical evidence from the millions of farmers who practice CA globally is unequivocal about the superior performance of CA and the large array of productivity, economic, environmental and social benefits CA offers. Equally relevant is the empirical evidence represented by the annual rate of uptake by farmers of CA globally which has been more than 10 M ha since 2008/09. This evidence indicates that challenges and constraints related to CA adoption are being overcome by farmers in different ways in different agro-ecologies and in different continents through their own organized efforts and support from science, technology and innovation. In a few cases, government institutions have lent their support to farmers who have decided to move away from conventional tillage agriculture. The previous section presented evidence regarding the ability of CA to mobilize greater crop and land potentials. This evidence is further discussed in the following section along with empirical global evidence from farms that practice CA.

5.1 Scientific and empirical evidence

The above-described benefits have now been documented on large and small farms in all major ecologies throughout the world (Goddard et al., 2006; Jat et al., 2014; Farooq and Siddique, 2014; Kassam et al., 2013, 2015, 2017b, 2018). Consequently, increasingly greater attention is being paid to support the adoption and upscaling of CA by governments, international research and development organizations, national research and development bodies, NGOs and donors. They see CA as a viable option for sustainable production intensification and climate-smart production systems to support local and national food security, poverty alleviation, especially of smallholders, improve ecosystem services, reduce the cost of production and minimize land degradation (Jat et al., 2015; Farooq and Siddique, 2014; Kassam et al., 2013, 2015, 2017b, 2018).

A regularly tilled soil, whether with a hand hoe or with a plough, eventually collapses and becomes compacted, cloddy and self-sealing. Instead of having 50–60% pore volume (air space) as in a healthy undisturbed soil with 2% or more SOM, most tilled soils with degraded soil structure and lower SOM content (often less than 1%) have much lower volume of air space (10–30% when collapsed) (Shaxson, 2006). This is due to poor soil structure and aggregate stability and no significant network of functional macro and micro biopores. Of the 50–60% pore space in a healthy soil, some 50% of the pore volume can be filled with water, of which 60–75% can be considered as being available to the crop (between field capacity and wilting point) depending on soil texture, thus serving as a major buffer against climate variability (FAO, 2005; Shaxson, 2006). Soil water retention is also favoured by SOM (Hudson, 1994; Rawls et al., 2003; Murphy, 2014), an important benefit in some drought-stressed soils (Allison, 1973). While there is some debate about the benefits of SOM acting as a 'sponge' (Minasny and McBratney, 2018), in most situations, more SOM is often better with respect to WUE.

On the other hand, a regularly tilled soil would not only hold much less water due to its low pore volume and poor aggregate stability, but it will also have a low infiltration rate, leading to much of the water being lost in run-off. For example, in southern Brazil, infiltration rates of more than 100 mm hr^{-1} were measured in untilled agricultural soils with mulch cover, whereas the conventional tilled soil had infiltration rates of around 20 mm hr^{-1} (Fig. 1, Landers,

2007). Also, due to tillage, the top soil is in a destructured or pulverised state with a surface that is self-sealing when dry or wet, thus serving as an obstruction to water flow into the soil causing much of the rainwater to run off. Greater infiltration rates and moisture holding capacities in CA soils lead to better soil moisture content throughout the growing season (Fig. 2), which can often be longer by 3–5 weeks under CA soil management (Derpsch, 2003; FAO, 2008, 2009).

Scientific studies and empirical evidence worldwide have shown that the biology of the soil and all the biological processes along with the other chemical, hydrological and physical processes depend on SOM content (Shaxson, 2006; FAO, 2011). The real secret of maintaining a healthy soil is to manage the carbon cycle properly, so that the SOM content is always as high as possible, above 2%. To do this the soil should not be disturbed mechanically to minimize the decomposition of organic matter, including root biomass, and the soil surface should be protected with a permanent layer of organic mulch cover which also serves as a substrate for soil microorganisms and mesofauna (Kassam et al., 2013).

It has also been shown often that as SOM increases, there is an improvement in crop nutrient response. For example, in Portugal, it was shown by Carvalho et al. (2012) that with a soil under conventional tillage containing 1% SOM, it took 160 kg N ha^{-1} to produce 3 t ha^{-1} of wheat grain (Fig. 3). However, when the same soil built up 2% SOM through continuous CA, 3 t ha^{-1} yield was obtained with 37 kg N ha^{-1}. A crop modelling exercise showed that at 3% SOM content, a

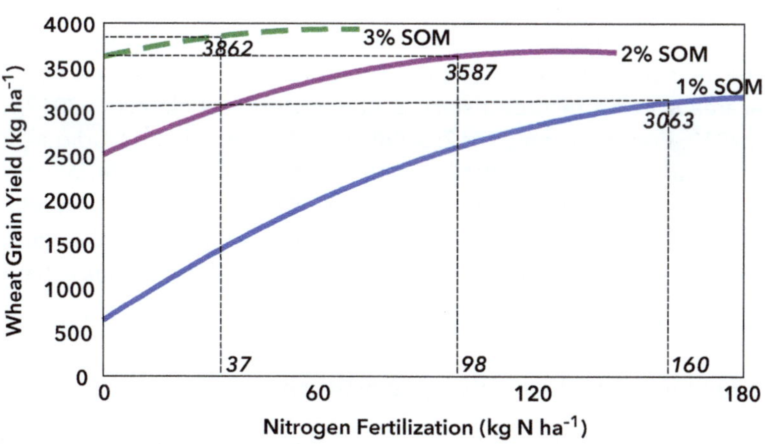

Figure 3 Yield response to nitrogen fertilization in wheat in Portugal. Values in italic represent the economically optimal N-fertilization rate and the respective yield for the different levels of soil organic matter (SOM) (0–30 cm depth). The dashed green line is the modelled yield response for a SOM level of 3%. In italic font is the most economic N-fertilization amounts and respective yields. Source: adapted from Carvalho et al. (2012).

yield of 3.5 t ha^{-1} would be achieved without any application of N, implying that the additional nitrogen was available from SOM in CA soils due to accumulation of SOM.

In addition to maintain and support natural enemies of insect pests and pathogens, a food web must be allowed to establish itself in the fields, and this can only occur naturally if there is a source of decomposing organic matter on the ground surface, including cover crops, upon which to establish a food web, above and below the ground, and to provide habitats for the natural enemies of pests including some weeds (e.g. Hajek, 2004; Khan et al., 2016; Kassam et al., 2017b).

5.2 The spread of Conservation Agriculture as evidence

The most convincing and solid evidence of the need to move away from conventional tillage agriculture to CA comes from the farmers around the world who have themselves been driving the process of adoption, uptake and innovation. This is despite the fact that relatively little attention and support from national governments and mainstream donors is being directed towards mainstreaming CA.

As noted above, in 2015/16, the global spread of CA was about 180 million ha of annual cropland (12.5% of global annual cropland), and since 2008/09, the global area under CA has expanded at an annual rate of over 10 million ha (Kassam et al., 2015, 2018). Approximately 50% of the CA area is located in low-income regions and about 50% in the industrialized regions. For historical reasons, the global distribution of CA is uneven because the initial impetus originated in North and South America. From there, it has spread to all other continents, including in Asia, Africa and Europe, where considerable expansion in CA area is being recorded since 2008/09 (Hobbs et al., 2006, 2019; Corsi et al., 2012; Erenstein et al., 2012; Jat et al., 2014; Farooq and Siddique, 2014; Kassam et al., 2009, 2013, 2017b, 2018; Bisht et al., 2016; Thierfelder et al., 2018; Reicosky and Janzen, 2019; Lal, 2018, 2018; Li et al., 2016). In addition, perennial cropping systems such as orchards and plantations in the North and the South are being transformed into CA systems in all continents (Chapter 9, Volume 1).

6 Constraints and enabling conditions

Constraints on the adoption and spread of CA practices exist in many countries and have been discussed in the CA literature (e.g. Gowing and Palmer, 2008; Friedrich and Kassam, 2009; Kassam et al., 2009, 2013, 2014a,b; Giller et al., 2009, 2015; Powlson et al., 2011, 2014; Sumberg and Thompson, 2012; Pittelkow et al., 2014; Pannell et al., 2014; Kassam and Brammer, 2016).

As discussed in Friedrich and Kassam (2009), Kassam et al. (2014a,b) and Kassam and Brammer (2016), the primary constraint to the adoption and spread of CA is the widespread belief that tillage is a necessary part of land preparation for planting. This has led to a lack of policy, research and extension support for CA. In some industrialized countries, such as in Europe, government subsidies on commodities, fuel and equipment discourage the adoption of CA, as does the widespread shift in farming from family farms which practiced diversified cropping and soil-sustaining land husbandry to commercial farms focussed solely on profit maximization (Purseglove, 2015). In many low-income countries, government subsidies on agrochemicals and fuel, and the dominance of commercial agents promoting sales of agrochemicals (often donor supported), can reduce the incentive for farmers to adopt more efficient and sustainable farming practices (Williamson, 2003).

Practical constraints to the adoption and spread of CA include a lack of appropriate machinery and equipment in the early years of CA adoption; weed competition; non-availability or high cost of herbicides in some remote interior areas of Africa; and inadequate availability of soil-protective mulches in semi-arid climates with a longer dry season where crop biomass has other competing uses such as for livestock feed (FAO, 2009; Kassam and Brammer, 2016; Lalani et al., 2017; Kassam et al., 2017b; Sims et al., 2018). However, there are many examples from different parts of the world to show that farmers, including smallholders, can increase biomass production with CA and even substantially increase livestock carrying capacity, including in semi-arid areas (Landers, 2007; Owenya et al., 2011; Lalani et al., 2018).

On severely degraded land, switching from conventional tillage systems to stable CA systems can take several years to build up SOM, soil health and system resilience (Sá, 2004). Large areas of tropical soils are severely degraded, as are large areas of soils in industrialized countries. The potentially slow pace of transformation of farming systems needs to be recognized by scientists, development agencies and programmes, and by farmers (Kassam et al., 2013; Kassam and Brammer, 2016).

However, the above notwithstanding, exemplary progress has been enabled in many countries in the South and the North by the implementation of appropriate mechanization strategies for smallholders and larger-scale farmers working collaboratively together, backed by positive policy and institutional support. Additional critical ingredients to overcome constraints have included the promotion of CA pioneers and champions, both individuals and institutions; CA farmer-led organizations; CA service providers; strong and mutually beneficial relationship between CA farmers and agricultural industries; reliable and affordable markets along the value chains; and support from CA research, education and extension (Kassam et al., 2014a,b; Dumanski et al., 2014). Thus, overall, the necessary and sufficient conditions for CA adoption and spread are becoming increasingly understood by stakeholders, and farmers continue to be at the leading edge of the global transformation of conventional tillage agriculture to CA by overcoming context-specific constraints with locally adapted STI in all continents. Indeed, globally there is already some 40 years of reliable empirical and scientific knowledge on constraints and how they are being overcome by farmers, smallholders and larger-scale farmers, in all major agro-ecologies in all continents (FAO, 2011, 2016; Kassam et al., 2013, 2014a,b, 2015, 2018).

7 Conclusion and future trends

As national economies expand and diversify, more people become integrated into the economy and can access food. However, for those whose livelihoods continue to depend on agriculture to feed themselves and the rest of the world population, the challenge for agriculture is to produce the needed food and raw material for industry with minimum harm to the environment and the society, and to produce it with maximum efficiency and resilience against abiotic and biotic stresses, including those arising from climate change. There is growing empirical and scientific evidence worldwide that the future global supplies of food and agricultural raw materials can be assured sustainably at much lower environmental and economic cost by shifting from conventional tillage-based to CA-based food and agriculture systems.

The achievement of this goal will require effective national and global policy and institutional support (including research and education). However, agriculture systems, whether they are deemed ecologically sustainable or not, are responding to the demands and pressures placed upon it by the food system, industry, government and the economy. The main political and economic drivers of the prevailing food, feed and agro-industrial systems continue to push the supporting agriculture system further down the intrusive, technology-oriented non-ecological path. These structural forces need to be overcome if CA is to become the dominant agricultural paradigm.

CA offers an opportunity to establish multifunctional agriculture with much greater crop and land potentials in agricultural land-use systems to underpin the effective implementation of the food- and agriculture-related SDGs. Modern tillage agriculture has become unfit for purpose and cannot be relied upon. CA offers an alternative approach to sustainable agriculture, an agriculture that can deliver 'more with less' resources, rehabilitate degraded lands and ecosystem services and mitigate future land, water and environmental degradation. It is unlikely that national and global food security goals, embedded in the SDGs, can be achieved sustainably without CA. Nor is it possible to achieve high output and sustain an efficient and resilient agriculture without CA. Modern tillage agriculture, with its heavy reliance on high application rates of expensive purchased inputs, is suboptimal in terms of factor and total productivities and is not climate smart. Consequently, the current agriculture paradigm is unsuitable for delivering the SDGs related to pro-poor development and poverty alleviation as well as for those aimed at protecting the environment and delivering ecosystem services.

CA has proved to be a far better option in all continents and land-based agro-ecologies. In recent years, it is fashionable to discuss and argue about the need for all nations to develop 'climate-smart' agriculture. We believe that such agriculture, as part of the SDGs, cannot be achieved unless CA is placed at the core of national food and agriculture development strategies. We further believe that such CA-based strategies will also contribute effectively to the achievement of the goals of the three international conventions namely: climate change, desertification and biodiversity that are an integral part of the SDGs.

Every effort needs to be made by all stakeholders to transform tillage agriculture to CA so that maximum impact could be made through meeting SDGs. More specifically, policy action should focus on the following areas as elaborated in FAO (2011); ECAF (2011); and Kassam et al. (2014b):

- Strengthening government capacity to support the adoption and spread of CA through updating agricultural policies, reforming or ideally removing perverse commodity support programmes and instituting policy reforms, especially in Asia, parts of Latin America and the Caribbean, Africa and Europe;
- Developing an enabling policy environment for private sector participation in developing and supporting the promotion of CA;
- Advocating for initial government support to increase farmers' access to appropriate farm equipment for CA;
- Developing large-scale programmes that would offer payments to CA farmers for harnessing ecosystem services for example carbon sequestration, watershed services and control of soil erosion;
- Including CA in all new agriculture development projects as the foundation ofsustainable production intensification;
- Funding universities and research centres to undertake innovative practical research to tackle soil and agronomic challenges; and
- Revising universities' agriculture curricula to include CA as an alternative and sustainable way of farming and providing agricultural universities and schools with relevant literature and publications about CA in the local language.

8 Where to look for further information

Volume 2 of this book is entirely on CA practice and benefits. Thus, all the other chapters in this volume serve as source of further information.

A general source of information is the FAO website at: http://www.fao.org/conservation-agriculture. The website also hosts a information exchange platform for the global CA community of practice (CA-CoP) which provides new information on CA practice and benefits on a regular basis are as follows: websites of regional CA networks such as ECAF (www.ecaf.org) and ACT (www.ACT-Africa.org) and national CA networks such as AAPRESID in Argentina (www.aapresid.org.ar) and FADERDRAP in Brazil (https://febrapdp.org.br/).

Further information regarding CA practice and benefits is also available in several books and journals. Books include: Goddard et al. (2006), Baker et al. (2007), FAO (2011, 2016), Jat et al. (2015), Farooq and Siddique (2014), and Chan and Fantle-Lepczyk (2015). Nationally-oriented information on CA practice and benefits is available for several countries including Australia (Crabtree, 2010), Canada (Lindwall and Sonntag, 2010), Brazil (Junior et al., 2012; de Freitas and Lander, 2014), Argentina (Peiretti and Dumanski, 2014) and United States (Lessiter, 2018).

Lessiter (2011) provides information on the benefits of no-till farming in North America and the backgrounds of 40 pioneers who made no-till farming possible. The International Soil and Water Conservation Research published a special issue on soil conservation and Conservation Agriculture which provides good information on the adoption, practice and benefits of CA in several countries (Dumanski et al., 2014).

9 References

Allison, F. E. 1973. *Soil Organic Matter and Its Role in Crop Production*. Elsevier Scientific Publishing Company, Amsterdam.

ANA. 2011. *Programme for Water Producers (Programa Produtor de Água)*. Available at: http://www.ana.gov.br/produagua/.

Andersson, J. A. and D'Souza, S. 2014. From adoption claims to understanding farmers and contexts: a literature review of Conservation Agriculture (CA) adoption among smallholder farmers in southern Africa. *Agriculture, Ecosystems and Environment* 187(1 April), 116–32. doi:10.1016/j.agee.2013.08.008.

Arslan, A., McCarthy, N., Lipper, L., Asfaw, S. and Cattaneo, A. 2013. Adoption and intensity of adoption of conservation farming practices in Zambia. ESA Working Paper No. 13-01. Agricultural Development Economics Division, Food and Agriculture Organization, Rome.

Baig, M. N. and Gamache, P. M. 2011. *The Economic, Agronomic and Environmental Impact of No-Till on the Canadian Prairies* (vol. 2). Alberta Reduced Tillage Linkages, Canada.

Baker, C. J., Saxton, K. E., Ritchie, W. R., Chamen, W. C. T., Reicosky, D. C., Ribeiro, M. F. S., Justice, S. E. and Hobbs, P. R. (Eds). 2007. *No-Tillage Seeding in Conservation Agriculture* (2nd edn.). CABI Publishing, Wallingford, UK.

Basch, G., Kassam, A., Friedrich, T., Santos, F. L., Gubiani, P. I., Calegari, A., Reichert, J. M. and dos Santos, D. R. 2012. Sustainable soil water management systems. In: Lal, R.

and Stewart, B. A. (Eds), *Soil Water and Agronomic Productivity*. Advances in Soil Science. CRC Press, Boca Raton, FL, pp. 229–89.

Bisht, J. K., Meena, V. S., Mishra, P. K. and Pattanayak, A. (Eds). 2016. *Conservation Agriculture: An Approach to Combat Climate Change in Indian Himalaya*. Springer Singapore, 559pp.

Brisson, N., Gate, P., Gouache, D., Charmet, G., Oury, F. X. and Huard, F. 2010. Why are wheat yields stagnating in Europe? A comprehensive data analysis for France. *Field Crops Research* 119(1), 201–12. doi:10.1016/j.fcr.2010.07.012.

Carvalho, M., Basch, G., Calado, J. M. G. and Barros, J. F. C. 2012. Long term effect of tillage system and crop residue management on soil carbon content of a luvisol under rainfed Mediterranean conditions. *Agrociencia* (Special Issue), 183–7.

Chan, C. and Fantle-Lepczyk, J. 2015. *Conservation Agriculture in Subsistence Farming: Case Studies from South-Asia and Beyond*. CABI, Wallingford, UK, 278pp.

Corsi, S., Friedrich, T., Kassam, A., Pisante, M. and Sà, J. de M. 2012. Soil organic carbon accumulation and greenhouse gas emission reductions from conservation agriculture: a literature review. Integrated Crop Management (vol. 16). FAO, Rome, 89pp.

Cotton, J. and Acosta-Martínez, V. 2018. Intensive tillage converting grassland to cropland immediately reduces soil microbial community size and organic carbon. *Agricultural and Environmental Letters Abstract – Research Letters* 3(1), 180047. doi:10.2134/ael2018.09.0047

Crabtree, W. L. 2010. *Search for Sustainability with No-till Bill in Dryland Agriculture*. Crabtree Agricultural Consulting, Western Australia, 204pp. Available at: www.no-till.com.au.

de Freitas, P. L. and Lander, J. N. 2014. The transformation of agriculture in Brazil through development and adoption of zero tillage Conservation Agriculture. *International Soil and Water Conservation Research* 1(2), 35–46.

Derpsch, R. 2003. Understanding the Process of Water Infiltration. Special Publication, No-till on the Plains Inc., Kansas, USA.

Derpsch, R. 2007. No-tillage and conservation agriculture: a progress repot. In: Goddard, T., Zoebisch, M., Gan, Y., Ellis, W., Watson, A. and Sombatpanit, S. (Eds), *No-Till Farming Systems*. Special Pub. No. 3. World Association of Soil and Water Conservation (WASWAC), Bangkok, pp. 7–39.

Derpsch, R. and Friedrich, T. 2009. Development and current status of no-till adoption in the world. *Proceedings on CD, 18th Triennial Conference of the International Soil Tillage Research Organization (ISTRO)*, Izmir, Turkey, 15–19 June 2009.

Derpsch, R., Franzluebbers, A. J., Duiker, S. W., Reicosky, D. C., Koeller, K., Friedrich, T., Sturny, W. G., Sá, J. C. M. and Weiss, K. 2014. Why do we need to standardize no-tillage research? *Soil and Tillage Research* 137, 16–22. doi:10.1016/j.still.2013.10.002.

Dumanski, J., Reicosky, D. C. and Peiretti, R. 2014. Pioneers in soil conservation and Conservation Agriculture. *International Soil and Water Conservation Research (Special Issue)* 2(1).

Eagle, A., Olander, L., Henry, L. R., Haugen-Kozyra, K., Millar, N. and Robertson, G. P. 2012. *Greenhouse Gas Mitigation Potential of Agricultural Land Management in the United States: a Synthesis of the Literature* (3rd edn.). Nicholas Institute for Environmental Policy Solutions, Duke University, Durham, NC.

ECAF. 2011. *Making Sustainable Agriculture Real in CAP 2020: the Role of Conservation Agriculture*. ECAF, Brussels, Belgium.

Erenstein, O., Sayre, K., Wall, P., Hellin, J. and Dixon, J. 2012. Conservation Agriculture in maize- and wheat-based systems in the (sub)tropics: lessons from adaptation initiatives in South Asia, Mexico, and Southern Africa. *Journal of Sustainable Agriculture* 36(2), 180-206.

FAO. 2005. *The Importance of Soil Organic Matter: Key to Drought-Resistant Soils and Sustained Food and Production*. FAO Soils Bulletin 80. FAO, Rome, 95pp.

FAO. 2008. *Investing in Sustainable Crop Intensification: The Case for Improving Soil Health*. Integrated Crop Management (vol. 6). FAO, Rome, 149pp.

FAO. 2009. *Enhancing Crop-Livestock Systems in Conservation Agriculture for Sustianble Production Intensification: A Farmer Discovery Process Going to Scale in Burkina Faso*. Integrated Crop Management (vol. 17). FAO, Rome, 42pp.

FAO. 2011. *Save and Grow: A Policymaker's Guide to the Sustainable Intensification of Smallholder Crop Production*. FAO, Rome, 102pp.

FAO. 2012. *Conservation Agriculture and Sustainable Crop Intensification in Karatu District, Tanzania*. Integrated Crop Management (vol. 15). FAO, Rome.

FAO. 2013. *Forest Management and Conservation Agriculture Experiences of Smallholder Farmers in the Eastern Region of Paraguay*. Integrated Crop Management (vol. 18). FAO, Rome.

FAO. 2016. *Save and Grow in Practice: Maize, Rice and Wheat*. FAO, Rome.

Farooq, M. and Siddique, K. H. M. 2014. *Conservation Agriculture*. Springer International, Basel, Switzerland.

Ferreira, A. O., Amado, T. J. C., Rice, C. W., Diaz, D. A. R., Keller, C. and Inagaki, T. M. 2016. Can no-till grain production restore soil organic carbon to levels natural grass in a subtropical Oxisol? *Agriculture, Ecosystems and Environment* 229, 13–20. doi:10.1016/j.agee.2016.05.016.

Ferreira, A. O., Amado, T. J. C., Rice, C. W., Diaz, D. A. R., Briedis, C., Inagaki, T. M. and Gonçalves, D. R. P. 2018. Driving factors of soil carbon accumulation in oxisols in long-term no-till systems of South Brazil. *Science of the Total Environment* 622-623, 735–42.

Freixial, R. and Carvalho, M. 2010. Aspectos prácticos fundamentales en la implantación de la agricultura de conservación/siembra directa en el sur de Portugal. *Proceedings of the European Congress on Conservation Agriculture: Towards Agro-Environmental Climate and Energetic Sustainability*, Madrid, pp. 361–9.

Friedrich, T. and Kassam, A. H. 2009. Adoption of conservation agriculture technologies: constraints and opportunities. *Proceedings of the IV World Congress on Conservation Agriculture*. ICAR, New Delhi, India, 4–7 February 2009.

Friedrich, T., Kassam, A. and Shaxson, F. 2009. Conservation Agriculture. In: *Agriculture for Developing Countries, Science and Technology Options Assessment (STOA) Project*. European Technology Assessment Group, Karlsruhe, Germany.

Friedrich, T., Derpsch, R. and Kassam, A. 2013. Overview of the global spread of Conservation Agriculture. *Field Actions Science Reports* (Special Issue 6), 1–7.

Gebregziabher, S., Mouaen, A. M., Van Brussel, H., Ramon, H., Nyssen, J., Verplancke, H., Behailu, M., Deckers, J. and De Baerdemaerker, J. 2006. Animal drawn tillage: the Ethiopian ard plough, maresha: a review. *Soil and Tillage Research* 89(2), 129–43. doi:10.1016/j.still.2005.08.010.

Giller, K. E., Witter, E., Corbeels, M. and Tittonell, P. 2009. Conservation agriculture and smallholder farming in Africa: the heretics' view. *Field Crops Research* 114(1), 23–34. doi:10.1016/j.fcr.2009.06.017.

Giller, K. E., Andersson, J. A., Corbeels, M., Kirkegaard, J., Mortensen, D., Erenstein,O. and Vanlauwe, B. 2015. Beyond conservation agriculture. *Frontiers in Plant Science* 6, 870. doi:10.3389/fpls.2015.00870.

Goddard, T., Zoebisch, M., Gan, Y., Ellis, W., Watson, A. and Sombatpanit, S. (Eds). 2006. *No-Till Farming Systems*. Special Pub. No. 3. World Association of Soil and Water Conservation (WASWAC), Bangkok.

Godfray, H. C., Beddington, J. R., Crute, I. R., Haddad, L., Lawrence, D., Muir, J. F., Pretty, J., Robinson, S., Thomas, S. M. and Toulmin, C. 2010. Food security: the challenge of feeding 9 billion people. *Science* 327(5967), 812–8. doi:10.1126/science.1185383.

González-Sánchez, E. J., Ordoñez-Fernández, R., Carbonell-Bojollo, R., Veroz-Gonzalez, O. and Gil-Ribes, J. A. 2012. Meta-analysis on atmospheric carbon capture in Spain through the use of conservation agriculture. *Soil and Tillage Research* 122, 52-60. doi:10.1016/j.still.2012.03.001.

González-Sánchez, E. J., Moreno-Garcia, M., Kassam, A., Holgado-Cabrera, A., Trivino-Tradas, P., Carbonell-Bojollo, R., Pisante, M., Veroz-Gozalez, O. and Basch, G. 2017. *Conservation Agriculture: Making Climate Change Mitigation and Adaptability Real in Europe*. European Conservation Agriculture Federation (ECAF), 154pp.

González-Sánchez, E. J., Mkomwa, S., Conway, G., Kassam, A., Ordonez-Fernandez, R., Moreno-Garcia, M., Repullo-Ruiberriz de Torres, M., Gil-Ribes, J., Basch, G., Veroz-Gozalez, O., Trivino-Tradas, P., Holgado-Cabrera, A., Miranda-Fuentes, A. and Carbonell-Bojollo, R. 2018. *Making Climate Change Mitigation and Adaptability Real in Africa with Conservation Agriculture*. European Conservation Agriculture Federation (ECAF) and African Conservation Tillage Network (ACT), 143pp.

González-Sánchez, E. J., Veroz-Gonzalez, O., Conway, G., Moreno-Garcia, M., Kassam, A., Mkomwa, Saidi, Ordoñez-Fernandez, R., Triviño-Tarradas, P. and Carbonell-Bojolloe, R. 2019. Meta-analysis on carbon sequestration through Conservation Agriculture in Africa. *Soil and Tillage Research* 190, 22–30. doi:10.1016/j.still.2019.02.020.

Gowing, J. W. and Palmer, M. 2008. Sustainable agricultural development in sub-Saharan Africa: the case for a paradigm shift in land husbandry. *Soil Use and Management* 24(1), 92–9. doi:10.1111/j.1475-2743.2007.00137.x.

Hajek, A. E. 2004. *Natural Enemies: an Introduction to Biological Control*. Cambridge University Press, Cambridge, UK, 370pp.

Hargrove, W. L. 1991. Cover crops for clean water. *Proceedings of an International Conference*. West Tennessee Experiment Station, 9–11 April 1991, Jackson, TN. Soil and Water Conservation Society, Ankeny, p. 198.

Haugen-Kozyra, K. and Goddard, T. 2009. Conservation agriculture protocols for green house gas offsets in a working carbon market. Paper presented at the *IV World Congress on Conservation Agriculture*, 3–7 February 2009, New Delhi, India.

Hobbs, P., Gupta, R. and Meisner, C. 2006. Conservation Agriculture and its application in South Asia. In: Uphoff, N., Ball, A., Fernandes, E., Herren, H., Husson, O., Lailng, M., Palm, C., Pretty, J., Sanchez, P., Sanginga, N. and Thies, J. (Eds), *Biological Approaches to Sustainable Soil Systems*. Taylor & Francis, Boca Raton, London, UK, pp. 357–71.

Hobbs, P., Sayer, K. and Gupta, R. 2008. The role of Conservation Agriculture in sustainable agriculture. *Philosophical Transactions of the Royal Society B* 363, 543–55. doi:10.1098/rstb.2007.2169.

Hobbs, P., Gupta, R., Jat, R. K. and Malik, R. K. 2019. Conservation agriculture in the IndoGangetic Plains of India: Past, present and future. *Experimental Agriculture* 55, 339–57.

Hudson, B. D. 1994. Soil organic matter and available water capacity. *Journal of Soil and Water Conservation* 49(2), 189 –94.

ITAIPU. 2011. *Cultivando Agua Boa (Growing Good Water)*. Available at: https://www.itaipu.gov.br/meioambiente/cultivando-agua-boa.

Jat, R. A., Sahrawat, K. L. and Kassam, A. H. 2014. *Conservation Agriculture: Global Prospects and Challenges*. CABI Publishing, Wallingford, UK, 393pp.

Junior, R. C., de Araujo, A. G. and Llanillo, R. F. 2012. No-till agriculture in southern Brazil: factors that facilitated the evolution of the system and the development of the mechanization of conservation farming. FAO and IAPAR.

Juniper, T. 2013. *What Has Nature Ever Done for Us? How Money Really Does Grow on Trees*. Profile Books Book Company, UK, 336pp.

Kassam, A. and Brammer, H. 2016. Environmental implications of three modern agricultural practices: Conservation Agriculture, the System of Rice Intensification and Precision Agriculture. *International Journal of Environmental Studies* 73, 702–18. doi:10.1080/00207233.2016.1185329.

Kassam, A. H. and Kassam, L. 2019. The need for Conservation Agriculture. In: Kassam, A. (Ed.), *Advances in Conservation Agriculture. Volume 1: Systems and Science*. Burleigh Dodds Science Publishing, Cambridge, UK. ISBN: 978-1-78676-264-1; www.bdspublishing.com.

Kassam, A. H., Friedrich, T., Shaxson, F. and Pretty, J. N. 2009. The spread of Conservation Agriculture: justification, sustainability and uptake. *International Journal of Agricultural Sustainability* 7(4), 292–320. doi:10.3763/ijas.2009.0477.

Kassam, A., Basch, G., Friedrich, T., Shaxson, F., Goddard, T., Amado, T., Crabtree, B., Hongwen, L., Mello, I., Pisante, M. and Mkomwa, S. 2013. Sustainable soil management is more than what and how crops are grown. In: Lal, R. and Stewart, R. A. (Eds), *Principles of Soil Management in Agro-ecosystems*. Advances in Soil Science. CRC Press, Boca Raton, FL.

Kassam, A. H., Derpsch, R. and Friedrich, T. 2014a. Global achievements in soil and water conservation: the case of Conservation Agriculture. *International Soil and Water Conservation Research* 2(1), 5–13. doi:10.1016/S2095-6339(15)30009-5.

Kassam, A., Friedrich, T., Shaxson, F., Bartz, H., Mello, I., Kienzle, J. and Pretty, J. 2014b. The spread of conservation agriculture: policy and institutional support for adoption and uptake. *Field Actions Science Report* 7, 1–12.

Kassam, A., Friedrich, T., Derpsch, R. and Kienzle, J. 2015. Overview of the worldwide spread of conservation agriculture. *Field Actions Science Reports* 8. Available at: http://factsreports.revues.org/3966.

Kassam, A., Basch, G., Friedrich, T., Gonzalez, E., Trivino, P. and Mkomwa, S. 2017a. Mobilizing greater crop and land potentials sustainably. *Hungarian Geographical Bulletin* 66, 3–11. doi:10.15201/hungeobull.66.1.1.

Kassam, A., Mkomwa, S. and Friedrich, T. 2017b. *Conservation Agriculture for Africa: Building Resilient Farming Systems in a Changing Climate*. CABI Publishing, Wallingford, UK.

Kassam, A., Friedrich, T. and Derpsch, R. 2018. Global spread of Conservation Agriculture. *International Journal of Environmental Studies* 76(1), 29–51. doi:10.1080/00207233.2018.1494927.

Kassam, A., Derpsch, R. and Friedrich, T. 2019. Development of Conservation Agriculture systems globally. In: Kassam, A. (Ed.), *Advances in Conservation Agriculture. Volume 1: Systems and Science*. Burleigh Dodds Science Publishing, Cambridge, UK. ISBN: 978-1-78676-264-1; www.bdspublishing.com.

Kell, D. B. 2011. Breeding crop plants with deep roots: their role in sustainable carbon, nutrient and water sequestration. *Annals of Botany* 108(3), 407–18. doi:10.1093/aob/mcr175.

Khan, Z. R., Midega, C. A. O., Pittchar, J. O., Murage, A. and Pickett, J. 2016. Climate-smart push-pull: a conservation agriculture technology for food security and environmental sustainability in Africa. In: Kassam, A., Mkomwa, S. and Friedrich, T. (Eds), *Conservation Agriculture for Africa: Building Resilient Farming Systems in a Changing Climate*. CABI Publishing, Wallingford, pp. 151–66.

Lal, R. 2018. Intensification of China's agroecosystems by conservation agriculture. *International Soil and Water Conservation Research* 6(1), 1–12. doi:10.1016/j.iswcr.2017.11.001.

Lal, R. and Stewart, R. A. (Eds). 2013. *Principles of Soil Management in Agro-Ecosystems*. Advances in Soil Science. CRC Press, Boca Raton, FL.

Lal, R., Reicosky, D. C. and Hanson, J. D. 2007a. Evolution of the plow over 10,000 years and the rationale for no-till farming. *Soil and Tillage Research* 93(1), 1–12. doi:10.1016/j.still.2006.11.004.

Lal, R., Follett, R. F., Stewart, B. A. and Kimble, J. M. 2007b. Soil carbon sequestration to mitigate climate change and advance food security. *Soil Science* 172(12), 943–56. doi:10.1097/ss.0b013e31815cc498.

Lalani, B., Dorward, P., Holloway, G. and Wauters, E. 2016. Smallholder farmers' motivations for using Conservation Agriculture and the roles of yield, labour and soil fertility in decision making. *Agricultural Systems* 146, 80–90. doi:10.1016/j.agsy.2016.04.002.

Lalani, B., Dorward, P. and Holloway, G. 2017. Farm level economic analysis - is Conservation Agriculture helping the poor? *Ecological Economics* 141, 144–53. doi:10.1016/j.ecolecon.2017.05.033.

Lalani, B., Aleter, B., Kassam, S., Bapoo, A. and Kassam, A. 2018. Potential for conservation agriculture in the dry marginal zone of central Syria: a preliminary assessment. *Sustainability* 10(2), 2–19. doi:10.3390/su10020518.

Landers, J. 2007. Tropical crop-livestock system in Conservation Agriculture: the Brazilian experience. Integrated Crop Management (vol. 5). FAO, Rome.

Lange, M., Eisenhauer, N., Sierra, C. A., Bessler, H., Engels, C., Griffiths, R. I., Mellado-Va´zquez, P. G., Malik, A. A., Roy, J., Scheu, S., Steinbeiss, S., Thomson, B. C., Trumbore, S. E. and Gleixner, G. 2015. Plant diversity increases soil microbial activity and soil carbon storage. *Nature Communications* 6, 6707. doi:10.1038/ncomms7707.

Laurent, F., Leturcq, G., Mello, I., Corbonnois, J. and Verdum, R. 2011. La diffusion du semis direct au Brésil, diversité des pratiques et logiques territoriales: l'exemple de la región d'Itaipu au Paraná. *Confins* 12(12). doi:10.4000/confins.7143. Available at: http://confins.revues.org/7143

Lessiter, F. 2011. 40 legends of the past. *40th Anniversary Issue of No-till Farmer*, November 2011.

Lessiter, F. 2018. *From Maverick to Mainstream: A History of No-Till Farming*. No-Till Farmer, 416p.

Li, H., He, J., Bharucha, Z. P., Lal, R. and Pretty, J. 2016. Improving China's food and environmental security with conservation agriculture. *International Journal of Agricultural Sustainability* 14(4), 377–91. doi:10.1080/14735903.2016.1170330.

Liebig, M. A., Archer, D. W. and Tanaka, D. L. 2014. Crop diversity effects on near-surface soil condition under dryland agriculture. *Applied and Environmental Soil Science* 2014, Article ID 703460. doi:10.1155/2014/703460.

Lindwall, C. W. and Sonntag, B. (Eds). 2010. *Landscape Transformed: the History of Conservation Tillage and Direct Seeding. Knowledge Impact in Society*. University of Saskatchewan, Saskatoon.

MEA. 2005. *Ecosystems and Human Well-Being: Synthesis*. Island Press, Washington DC.

Mello, I. and van Raij, B. 2006. No-till for sustainable agriculture in Brazil. *Proceedings of the World Association of Soil and Water Conservation* P1, 49–57.

Minasny, B. and McBratney, A. B. 2018. Limited effect of organic matter on soil available water capacity. *European Journal of Soil Science* 69(1), 39–47. doi:10.1111/ejss.12475.

Mokany, K., Raison, R. J. and Prokushkin, A. S. 2006. Critical analysis of root: shoot ratios in terrestrial biomes. *Global Change Biology* 12(1), 84–96. doi:10.1111/j.1365-2486.2005.001043.x.

Montgomery, D. R. 2007. *Dirt: the Erosion of Civilizations*. University of California Press, Berkeley and Los Angeles.

Murphy, B. W. 2014. *Soil Organic Matter and Soil Function – Review of the Literature and Underlying Data*. Department of the Environment, Canberra, Australia.

Nkonya, E., Mirzabaev, A. and von Braun, J. (Eds). 2016. *Economics of Land Degradation and Improvement – A Global Assessment for Sustainable Development*. IFPRI and ZEF. Springer Open, 695pp (Online). Available at: http://link.springer.com/book/10.1007/978-3-319-19168-3/page/1.

Owenya, M. Z., Mariki, W. L., Kienzle, J., Friedrich, T. and Kassam, A. 2011. Conservation Agriculture (CA) in Tanzania: the case of the Mwangaza B CA farmer field school (FFS), Rhotia Village, Karatu District, Arusha. *International Journal of Agricultural Sustainability* 9(1), 145–52. doi:10.3763/ijas.2010.0557.

Palm, C., Blanco-Canqui, H., Declerck, F., Gatere, L. and Grace, P. 2014. Conservation agriculture and ecosystems services. An overview. *Agriculture, Ecosystems and Environment* 187, 87–105. doi:10.1016/j.agee.2013.10.010.

Pannell, D. J., Llewellyn, R. S. and Corbeels, M. 2014. The farm-level economics of conservation agriculture for resource-poor farmers. *Agriculture, Ecosystems and Environment* 187, 52–64. doi:10.1016/j.agee.2013.10.014.

Pedzisa, T., Rugube, L., Winter-Nelson, A., Baylis, K. and Mazvimavi, K. 2015a. Abandonment of conservation agriculture by smallholder farmers in Zimbabwe. *Journal of Sustainable Development* 8(1), 69–82. doi:10.5539/jsd.v8n1p69.

Pedzisa, T., Rugube, L., Winter-Nelson, A., Baylis, K. and Mazvimavi, K. 2015b. The Intensity of adoption of Conservation agriculture by smallholder farmers in Zimbabwe. *Agrekon* 54(3), 1–22. doi:10.1080/03031853.2015.1084939.

Peiretti, R. and Dumanski, J. 2014. The transformation of agriculture in Argentina through soil conservation. *International Soil and Water Conservation Research* 1(2), 14–20.

Pittelkow, C., Xinqiang, L., Lindquist, B., van Groeningen, K. J., Lee, J., Lundy, M. E., van Gestel, N., Six, J., Venterea, R. T. and van Kessel, C. 2014. Productivity limits and potentials of the principles of conservation agriculture. *Nature Letter* 517, 365–8.

Powlson, D. S., Whitmore, A. P. and Goulding, K. W. T. 2011. Soil carbon sequestration to mitigate climate change: a critical re-examination to identify the true and the false. *European Journal of Soil Science* 62(1), 42–55. doi:10.1111/j.1365-2389.2010.01342.x.

Powlson, D. S., Stirling, C. M., Jat, M. L., Gerard, B. G., Palm, C. A., Sanchez, P. A. and Cassman, K. G. 2014. Limited potential of no-till agriculture for climate change mitigation. *Nature Climate Change* 4(8), 678–83. doi:10.1038/nclimate2292.

Pretty, J. 2002. *Agri-Culture: Reconnecting People, Land and Nature*. Earthscan Publications, London, UK.

Pretty, J., Toulmin, C. and Williams, S. 2011. Sustainable intensification in African agriculture. *International Journal of Agricultural Sustainability* 9(1), 5–24.

Purseglove, J. 2015. *Taming the Flood* (2nd edn.). William Collins, London.

Rawls, W. J., Pachepsky, Y. A., Ritchie, J. C., Sobecki, T. M. and Bloodworth, H. 2003. Effect of soil organic carbon on soil water retention. *Geoderma* 116(1–2), 61–76. doi:10.1016/S0016-7061(03)00094-6.

Reicosky, D. C. 2015. Conservation tillage is not Conservation Agriculture. *Journal of Soil and Water Conservation* 70(5), 103A–8A. doi:10.2489/jswc.70.5.103A.

Reicosky, D. C. and Janzen, H. H. 2019. Conservation Agriculture: maintaining land productivity and health by managing carbon flows. In: Lal, R. and Stewart, B. A. (Eds), *Soil and Climate: Advances in Soil Science*. Taylor and Francis Group, LLC, Oxford, England, pp. 131–61. Chapter 4.

Sá, J. C. M. 2004. Abnoubacao fosfatada sistema de plantio direto, pp. 201–22. In: Yamada, T., Silvia, Abdalla, R. S. (Eds), *Symposio sobre Fósforo na Agricultura Brasileira, Anais*. POTAFÓS, Piracicaba, S.P., 726pp.

Shaxson, T. F. 2006. Re-thinking the conservation of carbon, water and soil: a different perspective. *Agronomie* 26, 1–9.

Sims, B. and Kassam, A. 2015. East Anglia branch visit to Tony Reynolds at Thurlby Grange farm: no-till farming in the Lincolnshire fens. *TAA Agriculture for Development* 26(Winter), 51–2.

Sims, B., Corsi, S., Gbehounou, G., Kienzle, J., Taguchi, M. and Friedrich, T. 2018. Sustainable weed management for Conservation Agriculture: options for small farmers. *Agriculture* 8, 1–20. doi:10.3390/agriculture8080000.

Soane, B. D., Ball, B. C., Arvidsson, J., Basch, G., Roger-Estrade, F. M. J. and Roger-Estrade, J. 2012. No-till in northern, western and south-western Europe: a review of problems and opportunities for crop production and the environment. *Soil and Tillage Research* 118, 66–87. doi:10.1016/j.still.2011.10.015.

Sumberg, J. and Thompson, J. (Eds). 2012. *Contested Agronomy: Agricultural Research in a Changing World*. Routledge, London.

Thierfelder, C., Bunderson, W. T., Jere, Z. D., Mutenje, M. and Ngwira, A. 2016. Development of Conservation Agriculture (CA) systems in Malawi: lessons learned from 2005 to 2014. *Experimental Agriculture* 52(4), 579–604. doi:10.1017/S0014479715000265.

Thierfelder, C., Baudron, F., Setimela, P., Nyagumbo, I., Mupangwa, W., Mhlanga, B., Lee, N. and Gérard, B. 2018. Complementary practices supporting conservation agriculture in southern Africa. A review. *Agronomy for Sustainable Development* 38(2). doi:10.1007/s13593-018-0492-8.

Tsiafouli, M. A., Thébault, E., Sgardelis, S. P., de Ruiter, P. C., van der Putten, W. H., Birkhofer, K., Hemerik, L., de Vries, F. T., Bardgett, R. D., Brady, M. V., Bjornlund, L., Jørgensen, H. B., Christensen, S., Hertefeldt, T. D., Hotes, S., Gera Hol, W. H., Frouz, J., Liiri, M., Mortimer, S. R., Setälä, H., Tzanopoulos, J., Uteseny, K., Pižl, V., Stary, J., Wolters, V. and Hedlund, K. 2015. Intensive agriculture reduces soil biodiversity across Europe. *Global Change Biology* 21(2), 973–85. doi:10.1111/gcb.12752.

UN. 2015. *Transforming Our World: the 2030 Agenda for Sustainable Development*. United Nations, New York.

Uphoff, N., Ball, A., Fernandes, E., Herren, H., Husson, O., Lailng, M., Palm, C., Pretty, J., Sanchez, P., Sanginga, N. and Thies, J. (Eds). 2006. *Biological Approaches to Sustainable Soil Systems*. Taylor & Francis, Boca Raton, London, UK, 764pp.

Vlek, P. L. G., Khamzina, A. and Tamene, L. (Eds). 2017. *Land Degradation and the Sustainable Development Goals: Threats and Potential Remedies*. CIAT Publication No. 440. International Center for Tropical Agriculture (CIAT), Nairobi, Kenya, 67pp. Available at: http://hdl.handle.net/10568/81313

Willett, W., Rockström, J., Loken, B., Springmann, M., Lang, T., Vermeulen, S., Garnett, T., Tilman, D., DeClerck, F., Wood, A., Jonell, M., Clark, M., Gordon, L. J., Fanzo, J., Hawkes, C., Zurayk, R., Rivera, J. A., De Vries, W., Majele Sibanda, L., Afshin, A., Chaudhary, A., Herrero, M., Agustina, R., Branca, F., Lartey, A., Fan, S., Crona, B., Fox, E., Bignet, V., Troell, M., Lindahl, T., Singh, S., Cornell, S. E., Srinath Reddy, K., Narain, S., Nishtar, S. and Murray, C. J. L. 2019. Food in the Anthropocene: the EAT-Lancet Commission on healthy diets from sustainable food systems. *The Lancet* 393(10170), 447–92. doi:10.1016/S0140-6736(18)31788-4.

Williamson, S. 2003. Pesticide provision in liberalised Africa: out of control? Network Paper 126. Agricultural Research and Extension Network. ODI, London, UK.

Chapter 2

Crop and cropping systems management practices and benefits in Conservation Agriculture systems

Muhammad Farooq, Sultan Qaboos University, Oman, University of Agriculture, Pakistan, and The University of Western Australia, Australia; Ahmad Nawaz, Bahauddin Zakariya University, Pakistan; Yashpal Singh Saharawat, International Center for Agricultural Research in the Dry Areas (ICARDA), Lebanon; Timothy Reeves, The University of Melbourne, Australia; and Kadambot Siddique, The University of Western Australia, Australia

1 Introduction

Conservation Agriculture (CA) involves a set of principles that includes no or minimal soil disturbance, maintenance of soil cover, and diversified cropping to enhance soil health and minimize the costs associated with cultivation, energy use, nutrient use, and irrigation water use in conventional agriculture (Kassam et al., 2018; FAO, 2016; ICARDA, 2013; Farooq and Siddique, 2015). Enhanced soil health in CA systems is attributed to improvements in soil physical properties, including water infiltration and storage, aeration, porosity, structure, temperature, and compaction (Indoria et al., 2017), carbon contents (Govaerts et al., 2009; Abdollahi and Munkholm, 2014), and microbial biomass carbon (Nawaz et al., 2017a,b). CA can also reduce greenhouse gas (GHG) emissions, thus decreasing the global warming potential related to agriculture (Sapkota et al., 2015). But perhaps one of its greatest contributions is to the establishment of diverse farming systems, as its conservative impacts on soils

http://dx.doi.org/10.19103/AS.2019.0049.02

have facilitated more intensive cropping/forage/livestock systems with greater resilience to both bio-physical and economic risks.

Currently, CA is practiced on a range of land and soil types, from 90% sand in Australia to 80% clay in the Oxisols/Alfisols of Brazil. Interestingly, despite the high clay content in Brazilian soils, CA systems are commonly practiced due to the benefits to productivity and profitability and this is facilitated by the availability of suitable no-till seeding equipment. Moreover, some soils prone to surface sealing and crusting are also suited to CA as the mulch cover helps to prevent crust formation. In Australia and Argentina, CA systems have enabled crop production on less fertile, marginal land. Nowadays, the integration of CA into farming systems ranges from some thousands of hectares in Zimbabwe and Colombia to several million hectares in the USA, Brazil, Argentina, Australia, and Canada (Friedrich et al., 2012; Kassam et al., 2018). CA is scale-neutral and is used on both smallholder and larger farms (Kassam et al., 2018).

Most crops (including cereal, legume, oilseed, fiber, and vegetable crops including roots and tubers) can be grown successfully in CA systems, as can a range of forages (see Table 1) (Derpsch and Friedrich, 2009; Farooq and Siddique, 2015; Kassam et al., 2018). Unfortunately, in some regions, CA cropping systems are dominated by cereal mono-cropping (ICARDA, 2013) with little consideration for legume, oilseed, energy, fiber, or vegetable crops. Monoculture production systems are threatening the future sustainability of global cropping systems (Green et al., 2005; Phalan et al., 2011) including CA cropping systems. In these monoculture systems, crop diseases, herbicide-resistant weeds, and a range of pests have evolved due to the lack of break crops and persistent use of similar pesticides (Thierfelder and Wall, 2010a). Thus, crop rotations in CA systems need to be diversified to improve soil properties, energy use efficiency, carbon sequestration, and biodiversity, and effectively adapt to climate change (Kremen and Miles, 2012). Diversified and intensified cropping systems–including rotations, intercropping, cover cropping, strip cropping, and relay cropping–must be incorporated into CA systems to enhance stress resilience and to minimize the risk of developing resistant weeds, pests, and diseases (Mathews et al., 1994); to improve biodiversity and soil health; and to contribute to future food security. Diversified cropping systems have helped to reduce the incidence of nematodes (Wang et al., 2003; Govaerts et al., 2006), root rot diseases (Govaerts et al., 2007), and crop failures (Helmers et al., 2001).

Minimal soil disturbance under CA systems improves soil aggregates, microbial activities, nutrient retention, water-holding capacity, quality, and biodiversity. Also, residue retention improves soil fertility, reduces soil erosion, and increases water infiltration (Hobbs et al., 2017; Nawaz et al., 2017a,b; Reeves et al., 2016a). In this chapter, we summarize the performance of various crops in CA systems across the globe and elaborate on the management practices and benefits of including those crops in CA systems. We also highlight

Table 1 Influence of conventional and CA systems on grain yield of different crops in different regions of world grown on diverse soil types

Crop	Study duration (year)	Soil type	Grain yield (Mg ha^{-1}) CT	NT	Country	References
Alfalfa	4	Silt loam soil	7.91	7.24	USA	Sturgul et al. (1990)
Lupin	2	Chromic luvisol	1.45	1.84	Australia	Heenan et al. (2000)
Squash	1	Silty clay loam/silt loam	2.85	2.61	USA	Walters and Kindhart (2002)
Potato	1	Sandy loam	2.28	2.23	Canada	Holmstrom et al. (2006)
Tomato	1	Fine-silty, mixed, nonacid	73.87	41.56	USA	Doane et al. (2009)
Broccoli	1	Hayter loam soil	5.70	5.70	USA	Schellenberg et al. (2009)
Linseed	2	Clay loam	1.42	1.42	Greece	Bilalis and Karamanos (2010)
Oat	1	Ultic Plaxeralfs and slightly acidic	1.41	1.05	Chile	Martínez et al. (2011)
Lentil	NA	Clay loam	0.47	0.30	India	Mishra and Singh (2011)
Maize	14	Canisteo, Endoaquoll and Clarion	8.80	9.00	USA	Karlen et al. (2013)
Soybean	1	Canisteo, Endoaquoll and Clarion	1.80	1.70	USA	Karlen et al. (2013)
Rye	2	Loamy sand	3.19	2.96	Poland	Pabin et al. (2003)
Maize	11	Silt loam	9.78	9.94	China	He et al. (2011)
Maize	11	Silt loam	5.95	6.15	China	He et al. (2011)
Wheat	2	Sandy loam	3.85	3.80	Pakistan	Farooq and Nawaz (2014)
Wheat	2	Sandy loam	4.23	4.47	Pakistan	Nawaz et al. (2016)
Wheat	3	Sandy loam	4.10	4.10	Pakistan	Nawaz et al. (2017b)
Wheat	3	Sandy loam	4.10	3.90	Pakistan	Nawaz et al. (2017b)
Wheat	2	Loam	4.90	4.50	Pakistan	Nawaz et al. (2017c)

(*Continued*)

Table 1 (*Continued*)

Crop	Study duration (year)	Soil type	Grain yield (Mg ha^{-1}) CT	Grain yield (Mg ha^{-1}) NT	Country	References
Wheat	2	Loam	4.50	4.20	Pakistan	Nawaz et al. (2017c)
Wheat	2	Sandy loam	4.90	4.80	Pakistan	Nawaz et al. (2017c)
Wheat	2	Sandy loam	4.90	4.40	Pakistan	Nawaz et al. (2017c)
Wheat/oat (average)	10	Luvic Kastanozem	1.22	1.39	China	He et al., 2009
Wheat	14	Silt loam	3.04	3.61	China	Li et al. (2007)
Wheat	7	Silt loam	3.05	3.25	China	Qingjie et al. (2009)
Wheat	1	Light loam	4.40	4.80	China	Jin et al. (2009)
Maize	2	Light loam	9.25	10.05	China	Jin et al. (2009)
Wheat	1	Silt loam	3.94	4.76	China	Bai et al. (2009)
Soybean	4	Silty clay loam	2.44	2.66	China	Chen et al. (2011)
Soybean	4	Silty clay loam	2.74	3.12	China	Chen et al. (2011)
Maize	4	Silty clay loam	6.79	4.86	China	Chen et al. (2011)
Maize	4	Silty clay loam	8.10	6.82	China	Chen et al. (2011)
Maize	6	Loring (Typic Fragiudalfs)	6.91	6.94	USA	Cullum (2012)
Soybean	1	Silty clay loam	1.88	1.94	China	Liu et al. (2013)
Maize	2	Silty clay loam	5.74	4.89	China	Liu et al. (2013)
Maize	1	Silty clay loam	6.32	6.34	China	Liu et al. (2015)
Rice	2	Silty loam	3.69	4.00	India	Sutaliya and Singh (2017)

NT = no tillage with residue retention; CT = conventional tillage with no residue retention; NA = data not available.

various cropping systems that might be suitable for long-term sustainability of CA systems.

2 Crops for CA: management practices and benefits

2.1 Cereal crops

Cereal crops are generally well suited to CA systems, as observed in the cereal-based cropping systems of South Asia, North America, Europe, and Australia. Wheat is a common cereal used in CA in all of these regions.

CA systems are mainly being implemented in rice–wheat cropping systems on the Indo-Gangetic Plains (IGP), including in Bangladesh, India, Nepal, and Pakistan. Cereal-based CA systems in the IGP have enhanced farmer income, improved irrigation and water-use efficiency (WUE), moderated canopy temperatures, and reduced GHGs (White and Sapkota, 2016). In traditional rice–wheat cropping systems, wheat sowing is often delayed due to the late maturing, dominant rice varieties, and the time taken to prepare the land after rice harvest. Late-sown wheat may consequently suffer from terminal heat effects (Farooq et al., 2011; Ihsan et al., 2016). However, crops sown in a CA system may avoid heat stress due to the timely sowing of wheat after rice harvest. Moreover, crop residues left on the soil surface in CA systems help to conserve soil moisture, which may reduce yield losses associated with terminal heat effects, by creating a cooling effect (White and Sapkota, 2016). On average, CA systems for cereal-based cropping in the IGP have saved 36 L diesel ha^{-1} and reduced CO_2 emissions by 93 kg ha^{-1} (White and Sapkota, 2016), when compared to traditional cultivation systems.

In China, the adoption of a CA system increased the yields of maize by 7.5%, rice by 4.1%, and wheat by 2.9% (Zheng et al., 2014). In an 11-year experiment in China, CA produced higher yields in wheat and maize than conventional agriculture (Table 1; Jin et al., 2009).

CA is also under development in Africa. In South Africa, maize yields in a CA system increased by 12–17%, compared with conventional tillage (CT), in nine of 14 yield comparisons (Thierfelder et al., 2016). In Zimbabwe, CA enhanced maize productivity by ~235% and improved water infiltration and carbon sequestration by 64–96% and 29–97%, respectively, relative to conventional agriculture (Thierfelder et al., 2015a). In another African country, Malawi, Thierfelder et al. (2013a) studied the effect of CA systems on soil parameters and yield of maize over eight cropping seasons–maize yields increased by 24–40% compared to yields produced in conventional systems. In a further study in Malawi, Ngwira et al. (2013) found that CA in maize-based systems improved water infiltration, earthworm populations, net benefits, and overall crop productivity. While in Zambia, CA had increased water infiltration and soil water content, which subsequently increased maize yields (Thierfelder et al., 2013b).

In other African studies, a CA system with residue mulch retention produced higher yields of sorghum (*Sorghum bicolor* (L.) Moench) than the conventional system (Mupangwa et al., 2012). In another work, Rusinamhodzi et al. (2011) performed a meta-analysis of the long-term effects of CA on maize performance under rainfed conditions. Maize crops in rainfed areas performed better under CA with high input use; however, mulch cover in high rainfall areas led to lower yields due to waterlogging. In Malawi, CA systems produced 24–40% higher maize yields than the conventional system due to better water infiltration (Thierfelder et al., 2013b). Several other studies have reported better responses of maize grown in CA systems than in conventional systems in Africa (Thierfelder et al., 2013a, 2015a).

Poor stand establishment of cereals is occasionally an issue in CA systems, but this can be managed by priming seeds before sowing where required (Nawaz et al., 2016) and by strict attention to seeding depth. Weed management in CA cereal-based systems can be enhanced by integrated use of pre- and post-emergence herbicides; the incorporation of allelopathic crops and the use of allelopathic mulches (Nawaz et al., 2014; Nawaz and Farooq, 2016; Farooq et al., 2017); or by weed seed destruction techniques such as the Harrington Weed Seed Destructor (https://publications.csiro.au/rpr/pub?pid=csiro:EP175016; Lazaro et al., 2017) or the use of low-energy microwave soil heating treatments (Khan et al., 2018). In conclusion, all cereal crops can be successfully grown in CA systems across a very wide range of soil and farming conditions.

2.2 Legume crops

Long-term cereal crop monocultures in continuous cereal-based systems threaten biodiversity and increase pest- and disease-related issues. Moreover, cereal-based cropping systems are the main users of synthetic fertilizers; if poorly managed, off-site losses of these fertilizers can lead to land, water, and air pollution (Basso et al., 2019; Lassaletta et al., 2016). The incorporation of legume crops in CA systems is a pragmatic option for improving soil health and crop diversity and reducing synthetic fertilizer use (especially nitrogen), owing to their natural ability to fix atmospheric nitrogen (N). Legumes can release vast quantities of organic matter into the soil in terms of the C/N ratio, and release 5–7 times fewer GHGs per unit area than non-legume crops, or 277 kg ha^{-1} CO_2 per year (Stagnari et al., 2017).

Most of the legumes have long tap roots that enhance water infiltration and the solubilization of soil nutrients via root exudates and their recycling/uptake from deep soil layers (Stagnari et al., 2017). Many countries have incorporated legumes in their CA systems, including soybean [*Glycine max* (L.) Merr.] in Brazil, and pea (*Pisum sativum* L.), faba bean (*Vicia faba* L.), lentil (*Lens. culinaris* L.), and chickpea (*Cicer arietinum* L.) in Turkey, Australia, and North America (Table 2; Stagnari et al., 2017).

In Australia, direct seeding of crops after legumes is being practiced (Alpmann et al., 2013); this has increased soil organic carbon (SOC) stocks (Christopher and Lal, 2007) through the inclusion of forage, brown manure, and cover crop legumes in crop rotations under CA, as they return surplus C and N to the soil (Reckling et al., 2014).

Incorporating grain legumes in CA systems may also help to reduce the global warming potential of agriculture. In a study in a Mediterranean environment, Guardia et al. (2016) recorded the emissions of methane (CH_4), nitrous oxide (N_2O), and carbon dioxide (CO_2) from two crops–barley (*Hordeum vulgare* L.) and vetch (Vicia spp)–under both conventional and CA systems. The most sustainable crop tillage treatment in terms of global warming potential was non-fertilized vetch grown under CA, because of less fuel use, high carbon sequestration, and the absence of synthetic fertilizer usage. In another study on an Ultisol in a subtropical region, the soil under the legume cover crop in a CA system had more N_2O flux than a conventional system, with the flux completely eliminated by the retention of CO_2 in soil organic matter (SOM) (Bayer et al., 2016). In Zimbabwe, CA enhanced legume productivity by 173%, relative to the conventional treatment (Thierfelder et al., 2015a).

In summary, there is great potential for including legume crops in CA systems to improve soil health and mitigate GHG emissions associated with the application of synthetic fertilizers. In Australia, the major NANORP (2017) initiative (www.n2o.net.au) provided comprehensive data around these topics.

2.3 Oilseed crops

Oilseed crops can fit well into CA systems. For example, in Zambia, the cultivation of soybean for seven years in a CA system resulted in higher nodulation, SOC, and biological nitrogen fixation than the CT system. After 14 years, soil respiration, soil microbial biomass carbon (SMBC), and total porosity had also increased with a concomitant reduction in soil bulk density (Muchabi et al., 2014). In another four-year study conducted in India, the CA system produced similar seed yields of rapeseed (*Brassica campestris* L.) to a CT system, but marked improvements in soil biological, chemical, and physical properties were observed (Das et al., 2014a). In another study, CA and CT systems produced similar mustard seed yields on a Fluventic Haplustepts (Inceptisol) soil (Pradhan et al., 2016).

In India, CA improved the performance of mustard (*Brassica juncea* L. Czernj. Cosson) on a deep, clay type soil (Typical Chromusterts); (Rathore et al.,

Table 2 Influence of conventional and CA systems on grain yield and soil organic carbon concentration under diverse crop rotations on different soil types

Main crop	Crop rotation	Study duration (years)	Grain yield ($Mg\ ha^{-1}$) NT	Grain yield ($Mg\ ha^{-1}$) CT	SOC ($Mg\ ha^{-1}$) NT	SOC ($Mg\ ha^{-1}$) CT	Soil type	Country	Reference
Maize	Oat/vetch-maize + cowpea	19	4.4	-	58.1	-	Sandy clay loam (Typic Paleudult)	Brazil	Bayer et al. (2016)
Maize	Pigeon pea-maize	19	4.6	-	60.5	-	Sandy clay loam (Typic Paleudult)	Brazil	Bayer et al. (2016)
Maize	Lablab-maize	19	4.3	-	67.8	-	Sandy clay loam (Typic Paleudult)	Brazil	Bayer et al. (2016)
Maize	Oat-maize	18	2.1	2.5	46.6	41.4	Sandy clay loam (Typic Paleudult)	Brazil	Bayer et al. (2016)
Maize	Vetch-maize	18	4.5	4.4	57.5	47.2	Sandy clay loam (Typic Paleudult)	Brazil	Bayer et al. (2016)
Wheat	Fallow-wheat-vetch-barley	13	-	-	38.3	28.3	Sandy loam (Calcic Haploxeralf)	Spain	Guardia et al. (2016)
Potato	Barley-potato	2	31.1	27.8	20.7	19.2	Fine sandy loam (Orthic Podzol)	Canada	Carter and Sanderson (2001)
Barley	Barley-potato	2	4.8	4.3	-	-	Fine sandy loam (Orthic Podzol)	Canada	Carter and Sanderson (2001)
Soybean	Soybean-maize	4	2.7	2.4	-	-	Silty clay loam [Mollisols (Udolls)]	China	Chen et al. (2011)
Soybean	Soybean-maize	2	3.1	2.7	-	-	Silty clay loam [Mollisols (Udolls)]	China	Chen et al. (2011)
Maize	Maize-rapeseed	4	3.5	3.5	25.7	23.7	Clay loam (Typic Hapludalf)	India	Das et al. (2014a)
Rapeseed	Maize-rapeseed	4	0.6	0.5	-	-	Clay loam (Typic Hapludalf)	India	Das et al. (2014a)
Maize	Maize-wheat	4	2.0	1.6	-	-	Sandy loam to loam	India	Ghosh et al. (2015)
Wheat	Maize-wheat	4	3.4	2.3	-	-	Sandy loam to loam	India	Ghosh et al. (2015)
Maize	Continuous maize	4	-	-	5.4	3.4	-	Ghana	Naab et al. (2017)
Maize	Soybean-maize	4	-	-	5.4	4.3	-	Ghana	Naab et al. (2017)
Maize	Continuous maize	5	1.6	1.0	-	-	Arenosols	Zimbabwe	Thierfelder et al. (2015b)
Maize	Maize-cowpea	2	1.8	0.6	12.8	6.5	Arenosols	Zimbabwe	Thierfelder et al. (2015b)

SOC = soil organic carbon; no tillage with residue retention; CT = conventional tillage with no residue retention.

1998). On a sandy loam soil, CA improved the SOC and SMBC and reduced water use in mustard in a semiarid climate of India (Saha et al., 2010). Sarkar et al. (2007) studied the impact of CA and CT systems on water use, soil temperature, and seed yield of yellow sarson (*Brassica napus* L. var. *glauca*) in a rainfed lowland ecosystem of eastern India; the CA system produced 25% and 17% higher seed yield and WUE, respectively, than the CT system.

In northwest China, CA in soybean reduced surface runoff and improved soil water storage with a concomitant increase in soybean yield (Chen et al., 2011). In southeastern USA, cotton (*Gossypium hirsutum* L.) and peanut (*Arachis hypogaea* L.) are commonly grown in CA systems (Johnson et al., 2001). In North Cameroon, CA improved cotton productivity when compared with CT (Naudin et al., 2010). In Australia, canola is a vital component of CA systems, providing substantial benefits to profitability, disease reduction, and the control of resistant weeds (Anderson and Angus, 2011; Kirkegaard et al., 2014). In conclusion, oilseed crops can be successfully grown in CA systems with no yield penalty and with significant benefits to overall system productivity.

2.4 Energy crops

Different energy crops, including sunflower (*Helianthus annuus* L.), rapeseed (*Brassica napus* L.), biomass sorghum (*Sorghum* spp.), maize (*Zea mays* L.), kenaf (*Hibiscus cannabinus* L.), hemp (*Cannabis sativa* L.), and Ethiopian mustard (*Brassica carinata* L.), are widely cultivated across the globe (Zegada-Lizarazu and Monti, 2011). Mustard and rapeseed–two of the most important energy crops–can be successfully grown in CA systems. Energy crops could be incorporated into existing crop rotations in CA systems to improve system productivity and soil properties, as most energy crops have deep root systems that can exploit water and nutrients from deeper in the soil profile. Energy crops under CA can also contribute to greater diversification and system resilience to both biotic/abiotic stresses and to marketing strategies. Perhaps the most widely planted energy crops using CA systems are soybean for ethanol production in the USA (e.g. Malcolm et al., 2009) and rapeseed for biodiesel in Europe (https://farm-energy.extension.org/rapeseed-and-canola-for-biodiesel-production/).

There are however some potentially important conflicts between energy crop production and CA systems. If the energy crop is based on the use of all above-ground biomass–whether at the village smallholder level (for fuel) or at the industrial broadacre level–then the residues required to provide effective soil surface mulching can be compromised, with detrimental effects on productivity and sustainability.

In conclusion, energy crops may improve soil quality (Blanco-Canqui, 2010) and diversification when included in CA crop rotations; however, as residue cover is necessary for CA, growing biomass energy crops could compromise soil health benefits.

2.5 Vegetable crops

Due to the availability of no-till planters, trans-planters, strip-till cultivation equipments, advancements in the production and management of high-residue cover crop mulches, and improved integrated weed management techniques, CA has become quite effective for the sustainable production of vegetable crops. The potential advantages of using CA for vegetables include water/soil conservation, improved soil properties, reduced water/labor use, and increased nutrient recycling. However, potential disadvantages include reduced soil temperature, which may ultimately affect crop maturation, increased chemical input for weed control, carryover of pests in crop residues, and increased incidence of some diseases (Hoyt et al., 1994; Morse, 1999).

There are a number of initiatives underway to combine CA with other technologies such as drip irrigation with the resultant benefits including– reduced labor needs; enhanced soil health; reduced pesticide usage; and improved WUE (*https://horticulture.ucdavis.edu/information/growing-vegetables-conservation-agriculture*).

In the US, tomato (*Lycopersicon esculentum* Mill.) production in CA systems has been successfully adopted in North Carolina. With cooler soil temperatures from the residue cover, summer and fall-harvested tomatoes can produce similar yields in CA systems as plow/disc tillage. However, crops under CA in late fall or early spring (e.g., *Brassica* L.) may experience low soil temperatures that delay plant growth. However, appropriate selection (type and amount) of crop residue cover can enhance yield. Thus, planting short-season vegetable crops under CA systems in the warmest season will reduce the risk of delayed plant growth and, thus, harvest (Hoyt, 1999).

Numerous studies have reported the success of CA for vegetable production. Subterranean clover (*Trifolium subterraneum* L.) and hairy vetch mulch with CA enhanced tomato (*Lycopersicum esculentum* Mill.) yields and reduced the use of synthetic fertilizers and herbicides (Abdul-Baki and Teasdale, 1993). In another study, using the mulch of a cover crop improved tomato production in a CA-based furrow-irrigated system (Herrero et al., 2001). In Virginia, Morse (2000) reported successful production of broccoli (*Brassica oleracea* group *italica*) in a CA system. A CA system produced higher leaf areas and yields than a CT system for two snap bean (*Phaseolus vulgaris* L.) cultivars (Abdul-Baki and Teasdale, 1997). Hoyt et al. (1996) reported similar yields for conventional and CA systems in cabbage (*Brassica oleracea* L. var. *capitata*). Morse (1993) floated the idea that CA systems, especially strip-tillage systems, could improve cabbage production but this would be dependent on rain-irrigation patterns, planting dates, and site selection. In another study, the yield of transplanted broccoli was improved in a CA-based system (Infante and Morse, 1996).

Root and tuber vegetables are very sensitive to poor soil aeration, soil compaction, and inadequate soil drainage and thus can respond favorably to CT. However, large variation exists between crops. For example, yam (*Dioscorea* spp.), sweet potato (*Ipomoea batatas* [L.] Lam), and potato (*Solanum tuberosum* L.) need more intensive cultivation than taro (*Colocasia esculenta* L. Schott) and cassava (*Manihot esculenta* Crantz). In a three-year study on potato farming, a CA system maintained crop productivity and improved soil quality (Carter and Sanderson, 2001). In conclusion, a range of vegetable crops can be successfully grown in CA systems.

3 Cropping systems for CA: management practices and benefits

3.1 Mono-cropping

Mono-cropping is the agricultural practice of growing a single crop year after year on the same land, in the absence of rotation with other crops or the growing of multiple crops on the same land. Many herbicide-resistant weeds have evolved where monoculture production systems are widely practiced, owing to single herbicide usage for weed control (Gressel and Segel, 1990; Thierfelder and Wall, 2010a). The continued use of similar herbicides, or herbicides with the same site of action may cause a shift in weed species, making weed flora more complex and difficult to control (Tingle and Chandler, 2004). For example, in Canada, many weed biotypes have evolved resistance to triazine in maize crops (Stephenson et al., 1990); crop rotation was the dominant agronomic practice used to solve this problem. Indeed, the inclusion of diverse crops in rotations improves herbicide diversity, changes herbicide application rates, and requires alternate weed control techniques to reduce the development of herbicide-resistant weeds (Mathews et al., 1994).

Monoculture also depends almost entirely on off-farm inputs as opposed to ecosystem services from crop rotations or system diversity (Kremen and Miles, 2012). In monoculture, plants compete for space and resources, while those in intercropping or crop rotations can reduce competition through variations in spatial and temporal planting e.g. peas and barley (Kremen and Miles, 2012). Monoculture due to reduced yields may also need a larger unit area to produce the same amount of food as more diversified crop rotations, thereby potentially causing more habitat conversion and loss of biodiversity (Green et al., 2005; Phalan et al., 2011). Thus, mono-cropping should be avoided in CA-based systems.

3.2 Crop rotations

Crop rotations involve the sowing of different crops in a sequence within a field. According to the FAO, cover crops are major components of CA (Friedrich et al., 2012), improving soil and crop productivity. Along with the potential to increase soil nitrogen through the planting of legumes, crop rotations can improve ecosystem services; minimize weeds; break disease and pest cycles; and increase nutrient retention and biodiversity, with resultant benefits to crop yields (Kassam et al., 2009). In a crop rotation study of maize with cowpea (*Vigna unguiculata* (L.) Walp) and sorghum in a CA system, the maize-maize monocrop and maize-cowpea-maize rotation under CA, both produced higher grain yields than the CT maize-maize system (Mupangwa et al., 2012). In another work, maize yields also improved in rotation with cotton and sun hemp under CA, in all years except the first (Thierfelder et al., 2013c).

In a two-year study on crop rotation (maize-cotton), maize yields improved by 38–47% in a CA system compared with conventionally cultivated maize (Thierfelder and Wall, 2010a). Many studies have reported reductions in nematodes under crop rotation (e.g. Wang et al., 2003; Govaerts et al., 2006). Rotations with sun hemp prevented the carryover of nematodes into the following crops (Wang et al., 2003). In a CA system in Mexico, the prevalence of root rot declined in maize and wheat (Govaerts et al., 2007). Naab et al. (2017) reported benefits from crop rotation over continuous maize cropping, which were attributed to N fixation and other services associated with crop rotations. The inclusion of legumes in a maize rotation (either sole-cropped or intercropped) improved soil N; fertilizer-use efficiency; and grain yields (Snapp et al., 2010).

The inclusion of nitrogen-fixing legumes in rotations can also have positive soil fertility effects for succeeding crops (Giller, 2001). Under rainfed agriculture, crop rotations have more pronounced improvements in soil quality because the deep tap-rooted crops increase water infiltration and soil moisture due to better soil aggregation, structure, and porosity (Shaxson and Barber, 2003). Moreover, crop rotations can reduce the risk of crop failure and diversify farmer incomes (Helmers et al., 2001). Improvements in grain yield and SOC stocks due to the incorporation of legumes in different cropping systems are given in Table 2. Thus, crop rotations are a better option than continuous monocropping in both CA and conventional systems (Thierfelder et al., 2012).

3.3 Intercropping

Intercropping involves multiple cropping practices with two or more crops grown in proximity. The main goal of intercropping is to produce higher yields on a given piece of land by making use of resources or ecological processes that would otherwise not be used by a single crop. In CA, the intercropping of legume crops with cereals can improve soil quality, enhance economic yield, and improve system productivity. In Brazil, Bayer et al. (2016) intercropped pigeon pea (*Cajanus cajan* (L.) Millsp.), lablab (*Lablab purpureum* L.), and cowpea with maize (*Zea mays* L.) as well as vetch (*Vicia sativa* L.) with oats (*Avena strigosa* Schreb) in a CA system. Intercropping with lablab plus maize produced the highest SOC and soil nitrogen while pigeon pea plus maize produced the highest maize yields.

Intercropping legumes in a maize-based CA system improved net returns in the mid-hills of Nepal (Paudel et al., 2014). Intercropping wheat and chickpea produced higher mean grain yields than the mono-crops, increased total productivity and land use efficiency, and reduced the prevalence of weeds and pests (Banik et al., 2006). In Malawi, a CA system with maize–legume (pigeon pea) intercropping produced 3.5–4.3 Mg ha^{-1} of vegetative biomass, compared with 2.4–3.5 Mg ha^{-1} in a conventional system, and greater economic benefits (Ngwira et al., 2012). In another study in Malawi, Ngwira et al. (2013) reported higher earthworm populations and gross income in a maize–legume CA-based intercropping system than a conventional system. In Zimbabwe, pigeon pea intercropping with sorghum and cotton produced vast quantities of nitrogen-rich biomass, most of which remained in the field for the onset of the succeeding cropping season (Baudron et al., 2012).

Legume intercropping in CA-based cereal systems may have additional short- and long-term benefits, including (1) higher planting densities could result in the formation of a closed canopy to control weeds (Olsen et al., 2005); (2) mixing of N-rich residues with N-poor cereal residues (e.g. sorghum) could reduce the C:N ratio of the combined mulch, thereby avoiding potential problems of temporary nitrogen immobilization by microorganisms (Palm et al., 2001); and (3) the additional organic input could increase soil C content in the long term (Six et al., 2002; Corbeels et al., 2006; Stewart et al., 2007). Thus, intercropping legumes and cereals might be a viable option for improving productivity in CA systems.

3.4 Strip cropping

Strip cropping is the growing of crops in a systematic arrangement of strips on the contour to reduce erosion and can be successfully incorporated into CA systems. Crops are arranged so that a strip of grass or cover crop is alternated with a strip of row-crop and is used when the slope is too steep or there is no alternative method to prevent soil erosion. Strip cropping reduces runoff and thus improves soil fertility. Strip cropping uses several good farming practices including, contour cultivation, tillage where appropriate, cover cropping, stubble mulching, and crop rotations (http://www.tarahaat.com/water_stripcrop.aspx).

Adopting a modified strip-till system increased maize yield by 12–15% and enhanced net income by \$200–\$250 ha^{-1} (Gronau, 2017). Islam et al. (2015) reported that CA strip-row farming in a maize–soybean rotation increased the plant population and biomass, improved root architecture, and produced 75% more grain yield than conventional agriculture. The inclusion of onion in strip-cropping systems may inhibit herbivorous insects due to volatile compounds (Uvah and Coaker, 1984). In northwest China, strip cropping on slopes is common for moisture conservation and better soil management (Zheng et al., 2014). Strip cropping with smothering-type crops (e.g. peanut) helps to control soil erosion in CA systems (Knowler, 2005). Moreover, strip cropping using legume and allelopathic crops with cereals may improve soil fertility and weed management.

3.5 Relay cropping

Relay cropping is a specialized version of double cropping, where the second crop is planted into the first crop before harvest, rather than waiting until after harvest. Relay cropping makes efficient use of available on-farm resources and removes conflicts with sowing time and fertilizer application, and improves soil health (Tanveer et al., 2017) by avoiding preparatory tillage as practiced in conventional agriculture systems. In the IGP, relay cropping of wheat in standing cotton is a common practice. In the Punjab region of Pakistan, relay cropping of wheat in cotton improved volumetric water contents and wheat profitability (Sajjad et al., 2018). In China, the Yangtze River Basin is famous for the sequential production of rapeseed and rice. To avoid the late sowing of rapeseed, relay cropping of rapeseed in rice has become common practice, with the added benefits of improved land use efficiency and reduced winter fallow (Anwar et al., 2017). Yamane et al. (2016) found that relay cropping soybean within soybean increased soybean yields more than the use of double cropping. In Latin America, relay cropping of Brachiaria grass with maize makes optimal use of land resources and reduces land degradation (Reeves et al., 2016a).

However, studies on relay cropping in agro-ecosystems in limited; further research is needed to enable researchers and farmers to validate system performance, based on multi-year and multi-location assessments, and the impact on natural resources e.g., positive balance of SOC, nutrient cycles, and xenobiotic dynamics (Kong et al., 2016).

4 Case studies

In this section, a number of case studies involving the successful use of CA cropping systems are presented as evidence of the substantial benefits already accruing from these combinations.

4.1 Rice-wheat and rice-maize-based systems in South Asia

The most important CA cropping system in Asia is the rice-wheat system of the IGP (Gupta et al., 2002; Bhushan et al., 2007; Gupta and Seth, 2007; Jat et al., 2014). Here, wheat production after rice harvest has been revolutionized using CA. The area of wheat grown in CA systems has increased in the past two decades (Erenstein and Laxmi, 2008). Fundamental to this progress have been improvements in crop yield, soil health, and net profits, and reductions in GHG emissions and water, labor, and energy resources (Gupta et al., 2002; Ladha et al., 2009; Nawaz et al., 2017a,b; Tiwari et al., 2018). Importantly, CA has stimulated the establishment of other crops in these systems, of which maize is the most significant (Das et al., 2018; Parihar et al., 2017, 2018a).

Sapkota et al. (2015) reviewed the impact of CA on cereal-based cropping systems of the IGP in terms of climate change adaptation, economic profitability, and GHG mitigation. They reported that CA-based cropping systems reduced production costs by 23% and produced equal or even higher yields than conventional systems, especially in rice-wheat systems. CA in cereal-based cropping systems on the IGP also reduced canopy temperature by 1–4°C, thus decreasing the negative impacts of terminal heat stress in wheat. The CA-based systems also increased irrigation water productivity by 66–100% and reduced GHGs (i.e. CO_2, CH_4, N_2O) by 10–15% compared to conventional systems. Ghosh et al. (2015) studied the impact of CA on soil conservation, productivity enhancement, and profitability of a maize-wheat cropping system in the Indian sub-Himalayas; CA increased wheat yield by ~47% and reduced mean runoff coefficients and soil loss by 45% and ~54%, respectively, compared with conventional agriculture. By maize harvest, CA had conserved 108% more soil moisture (up to 90 cm) for the following wheat crop and 85% higher net benefits than conventional agriculture (Ghosh et al., 2015).

In a four-year study on a maize-rapeseed (*Brassica campestris* L.) cropping system in eastern Himalaya, India, conventional and CA systems produced similar rapeseed yields but CA substantially improved soil physical properties, SMBC, and SOC concentration (Das et al., 2014a).

In India, Das et al. (2018) also conducted a three-year study on a sandy clay loam soil in a maize-wheat cropping system; the CA systems improved water and crop productivities, SOM accumulation, and net returns. In a seven-year study in northwestern India, a CA-based maize-wheat-mungbean rotation produced the highest biomass productivity, energy use efficiency, and bio-energetic based adult equivalent yields (Parihar et al., 2018a). In another study in India, a CA-based wheat-maize-mung bean system improved WUE, biomass yield, grain yield, incident radiation conversion energy, and system productivity due to site-specific nutrient management (Parihar et al., 2017). In northwestern India, a CA-based maize cropping system performed best for cob length, grains per cob row, maize yield, SOC, energy productivity, and surplus irrigation water (Parihar et al., 2018b). After seven years of CA in a rice-wheat system in India, SOC had increased by 4.7 t C ha^{-1} compared with a decrease of 0.9 t C ha^{-1} in a conventional rice-wheat system (Sapkota et al., 2017). In Pakistan, Nawaz et al. (2017b,c) reported improved soil health, SOC, grain yield, and system productivity of a CA-based rice-wheat system. In a three-year study in the western IGP of India, an irrigated CA-based pigeon pea-wheat cropping system produced similar or higher crop yields and higher WUE than a conventional system (Das et al., 2016). In a sandy loam soil in northwestern India, Parihar et al. (2016) observed an increase in water stable aggregates, SOC, SMBC, soil enzymatic activities and a decline in soil bulk density and soil penetration resistance after seven years of CA maize-based cropping system diversified with legumes. Laik et al. (2014) also found that a rice-wheat CA system with best management practices and substitution of wheat for maize or potato in winter improved overall system productivity and farmer income in the eastern IGP of India relative to conventional farmer practices. In the western IGP of India, a CA-based system improved crop yields, water productivity, and economic profitability (Das et al., 2014b). In another study on maize-wheat rotation in northwestern IGP of India, Jat et al. (2014) found that CA-based systems improved crop yields, soil physical properties (e.g. water infiltration, soil aggregation, mean weight diameter), water productivity, and net returns relative to the conventional system. In a CA-based system on a sandy loam soil in the IGP, Kumar et al. (2013) reported increases of 81%, 17%, and 13% in operational field capacity, specific energy, and energy usage efficiency, respectively, relative to the conventional system.

4.2 Cereal-based systems in Europe

In Europe, the adoption of CA varies according to the ecological region. While Europe is behind some other regions in the adoption of CA, the indicators for future progress are encouraging. Interest in CA is growing rapidly due to increasing environmental awareness, including soil protection, and the need to reduce production costs. The European Conservation Agriculture Federation (ECAF) plays an important role in the adoption and dissemination of CA practices and in discussions involving CA and the EU Common Agricultural Policy (CAP) reform. Yield performance and stability, operating costs, environmental policies, CAP programs, and climate change will likely be the major driving forces defining the direction and extension of CA in Europe (Kertész and Madarász, 2014).

In Norway and Germany, the adoption of CA has been subsidized to mitigate soil erosion. In other European countries, the adoption process seems mainly driven by farmers and, the major driving force has been cost reductions in machinery, fuel, and labor. Soil and water conservation concerns do not appear to be the main drivers in European farmers' decisions to shift to CA (Lahmar, 2010).

Various studies have reported the success of CA in Europe. For example, Ghaley et al. (2018) assessed the benefits of CA in cereal-based (wheat, barley, oats) cropping systems in terms of soil function–improved C sequestration, primary productivity, nutrient cycling, water regulation, and purification–with the provision of the best habitat for intrinsic and functional biodiversity. In a study in Madrid, Spain on a cereal (barley)-legume (vetch)-fallow crop rotation under CA and CT systems, Guardia et al. (2016) reported a significant tillage × crop interaction for cumulative N_2O emissions; the incorporation of vetch resulted in higher N_2O emissions than barley in a conventional system, and the fluxes were similar in CA for both crops. The crop or tillage type did not affect CH_4 or CO_2 emissions. Barley had significantly higher yield-scaled emissions of N_2O than vetch. The non-fertilized CA vetch treatment was the most sustainable treatment in terms of global warming potential owing to high carbon sequestration, absence of synthetic fertilizer use, and less fuel consumption in the rainfed Mediterranean environment (Guardia et al., 2016). The success of CA-based systems in the north and south of Europe, Spain, Hungary, and Ukraine has been reported by several scientists in terms of increased yields and improved soil properties (De Tourdonnet et al., 2007; Kertész et al., 2011; Kertész and Madarász, 2014).

4.3 Maize-based systems in South America

South America is one of the main hubs of CA in the world. The success of CA in the maize-based production system is exemplified by the widespread adoption of direct-seeded, mulch-based cropping systems in Latin America. Areas permanently established under this system have increased in recent decades, reaching more than 50% of the total cropped area in Brazil, Paraguay, and Argentina (Scopel et al., 2004; Kassam et al., 2014). In Brazil, the introduction of maize under a CA-based system in rotation with soybeans helped drive the widespread adoption of CA. In the country's tropical savanna region, maize is grown between rows of trees for the first two or three years after the trees have been planted and the area is then planted with forages intercropped with maize. Once the pasture is established, it is grazed by livestock, until the trees are ready for harvest (Kluthcouski et al., 2000; Pacheco et al., 2013). This diversification mitigates the impact of both climate and market variability on farm income. It also reduces the clearing of forest for agriculture, protects biodiversity, prevents soil erosion, and improves soil structure and fertility (Vilrla et al., 2003; FAO, 2007).

Numerous long-term experiments have been conducted on CA in Brazil, which have subsequently penetrated widely into the cropping systems of South

America, including in Argentina and Paraguay. In a long-term (18–19 years) CA experiment in southern Brazil, pigeon pea and lablab were intercropped with maize in the first summer, followed by hairy vetch and oats in winter, and cowpea and maize in the following summer (Bayer et al., 2016). The performance of oat and hairy vetch as a cover crop in maize was also examined in conventional and CA systems. The authors found that N_2O emissions from the CA system were offset by CO_2 retention in SOM. The legume cover crop in the CA system also enhanced SOC, soil N, and maize grain yield more than the grass cover crop. They concluded that the CA system combined with legume cover crops is a pragmatic approach for sustainable, low-C footprint food production in subtropical regions (Bayer et al., 2016).

In another study in Brazil, Carvalho et al. (2014) examined the effect of five cover crops on the chemical fertility of a Typic Acrustox soil under conventional and CA maize systems grown in rotation with cover crops (Brazilian jack bean (*Canavalia brasiliensis* M. Benth), pigeon pea, velvet bean (*Mucuna pruriens* L.), Indian hemp (*Crotalaria juncea* L.), radish (*Raphanus sativus* L.), and pearl millet (*Pennisetum glaucum* [L.] R. Br)). The study identified significant differences in N and C stocks among all cover crops grown in succession to maize in both tillage systems after five years. Velvet bean and Brazilian jack bean enhanced soil C and N stocks, with CA accumulating more SOM than the CT system in the 0–5 cm surface layer.

Current work on CA in Latin America is focused on more integrated systems including crops/livestock/forestry; scaling up of CA systems for more perennial crops; and for vegetables; and sugar cane.

4.4 Maize-based systems in Africa

In Africa, the adoption of CA has improved the livelihoods of poor people by reducing production costs and improving soil properties. For example, maize systems under CA in sub-Saharan Africa retained more soil moisture during seasonal dry spells and were more productive than systems based on CT using plows, harrows, and hoes (Thierfelder and Mupangwa, 2014). The development of 'CA with trees' has also helped to advance the diffusion of CA in crop-livestock systems in sub-Saharan Africa (Garrity, 2011).

Various studies have reported the success of CA in Africa. For example, Naab et al. (2017) studied the impact of conventional and CA systems on the productivity, soil quality, and profitability of three different cropping systems (continuous maize, soybean–maize annual rotation, and soybean–maize intercropping) in northwestern Ghana. After four years, adoption of the full principles of CA enhanced SOC and total soil N more than a conventional system. The yields in conventional and CA systems were similar for all crops in the first three seasons. A partial budget analysis showed that the cost of producing soybean was 20–29% lower under CA than the conventional system. They concluded that adoption of the full principles of CA presents a win-win scenario due to improved crop yields, increased economic returns, and trends of increasing soil fertility. In another multi-location study, Thierfelder et al. (2015a) studied the impact of various CA-based maize-grain legume (cowpeas, soybeans, or groundnuts (*Arachis hypogaea* L.)) cropping systems in four countries (Malawi, Zimbabwe, Mozambique, and Zambia) in South Africa by combining the data from long-term on-farm trials of CIMMYT. In some cases, cowpeas and pigeon pea were also intercropped with maize. In most cases (80%), the CA-based systems produced higher yields than conventional agriculture; for example, the CA-based dibble stick planted sole maize and maize-legume intercrop systems produced yield advantages of 1152 kg ha^{-1} and 1172 kg ha^{-1}, respectively. The CA systems based on animal traction–animal traction direct-seeding and ripline seeded system–produced smaller yield benefits (458–761 kg ha^{-1}). The yield benefits increased with increasing years of practicing CA, thus highlighting the need to master critical management steps–timely planting, weeding, fertilizer application, and crop harvest residue management. Improvements in clay and silt contents were positively related to maize grain yield in CA systems and were more robust to variability in seasonal rainfall than the conventional plots. The authors suggested that the niche for CA in southern Africa is larger than expected; however, in rainfall zones below 600 mm, it will be challenging to produce enough maize biomass for effective soil cover (Thierfelder et al., 2015a).

Thierfelder et al. (2013a) examined the effects of CA-based maize production systems in Malawi on maize yield and soil properties over eight cropping seasons. The adoption of CA improved crop yields by improving water infiltration, especially during dry spells. In southern and eastern provinces of Zambia, the adoption of CA in maize-based systems improved maize yields by 78–90% with substantial improvements in soil carbon and water infiltration (Thierfelder et al., 2013b). Thierfelder et al. (2015b) conducted a long-term experiment (2004–2013) in South Zimbabwe to evaluate the performance of CA systems under low soil fertility and erratic rainfall scenarios. They found that the animal traction-based CA systems improved maize and legume productivity by 235% and 173%, respectively, compared with CT plots. The CA system increased soil carbon and soil water infiltration by 29–97% and 64–96%, respectively, more so over time.

Thierfelder and Wall (2010b) quantified the impact of CA systems in Zambia and Zimbabwe to mitigate the future effects of climate change. The CA systems resulted in higher available soil moisture, which would enable crops to overcome seasonal dry spells, mitigate the effects of drought, reduce the risk of crop failure, and secure livelihoods in the region. In another study, Nyamangara et al. (2014) investigated the effect of basin-based CA on selected soil quality parameters on smallholder farms in Zimbabwe; the CA treatments improved SOC contents and stocks and aggregate stability, reduced bulk density, and increased pore volume and moisture retention.

Despite these promising signs, a number of studies in Sub-Saharan Africa (SSA) have provided a cautionary note to expectations and further adoption and utilization by smallholder farmers and the following is taken from Reeves et al. (2016b). Strategic research on smallholder farming systems in Malawi, Zambia, Mozambique, and Zimbabwe showed that CA systems have a greater productivity and profitability as compared to traditional tillage-based agriculture systems (Thierfelder et al., 2015b, 2016; Mupangwa et al., 2016). However, the greatest benefits were only achieved if all principles of CA were applied. The results further showed that on-farm yield benefits do not occur overnight and need on average 2–5 cropping seasons until they become measurable and significant. Research results clearly showed that CA responded better in seasons of low and erratic rainfalls, thus showing its climate-smart agriculture potential. CA systems had immediate benefits in infiltration and soil moisture conservation (Thierfelder and Wall, 2009), allowing farmers to plant earlier with greater resilience to in-seasonal dry spells and terminal heat stress (Cairns et al., 2013). Established CA systems yield 30-50% higher in normal seasons, which can increase to over 100% in dryer years. Yield benefits increase when CA systems are supported by other climate-smart agriculture interventions such as drought-tolerant maize, rotations with green manure/cover crops, grain legumes, and or agro-forestry species. Profitability gains of 40–100% have been reported under the conditions of smallholder farmers (Thierfelder et al., 2015a, 2016). A systematic review of yield benefits by Wall et al. (2013) showed that in only 10% of cases, a negative yield response was realized, while in all other cases it was the same or more positive in comparison to CT systems. Depending on the farming systems, there are significant socio-economic benefits such as reduced traction and labor requirements for land preparation and for weeding if herbicides are used. This results in cost savings for manual labor, animal draft, and fuel (Ngwira et al., 2012; Johansen et al., 2012; Mazvimavi et al., 2008).

Soil quality improvements (such as increased carbon sequestration, improved nutrient use efficiency, and greater water-holding capacity) only emerge in the longer term (Cheesman et al., 2016)–which is due to the low-input agriculture systems, with a lack of sufficient groundcover if crop-livestock competition is high, the retention of residues of low quality, the long dry season, and greater decomposition rates under higher temperatures (Powlson et al., 2016). CA reduces soil erosion and runoff, which is a key benefit in reducing soil degradation (Thierfelder and Wall, 2009; Munyati, 1997). The effects of CA on greenhouse gases is more unclear (Kimaro et al.,

2016). It can be summarized that CA adapts very well to the effects of climate variability and change and increases productivity and income while reducing soil degradation. There is little evidence and confirmation about its mitigation potential under the conditions of sub-Saharan Africa, which clearly highlights a research gap.

Supported by national policies and investment frameworks in several African countries (e.g. Zambia, Malawi, Zimbabwe, and Mozambique) as well as Climate-Smart Agriculture initiatives led by NEPAD (Ethiopia, Malawi, and Zambia) and the comprehensive agriculture development program (CAADEP), CA has gained substantial support in the last decade. This led to a push in mainstreaming CA to make it the 'conventional agriculture system'. This institutionalization happened in various forms and degrees but led to a more harmonized approach in the promotion and research in development. Current evidence from southern Africa shows that there is an increasing trend in adoption of CA in Malawi (more than 80 000 farmers in 2013), and adoption of specific components of CA systems in Zimbabwe (by 371 000 farmers) and Zambia (by 250 000 farmers) (Wall et al., 2013), resulting in around one million hectares of CA managed land.

It is also important to note that the restraints to the further adoption and utilization of CA in SSA also include financial and social barriers, as well as technological barriers, all of which must be addressed if CA is to underpin sustainable intensification of agriculture in that continent (https://publications.csiro.au/rpr/pub?pid=csiro:EP176869).

4.5 Cereal-oilseed-legume-based systems in Australia

In Australia, the combined efforts of grower organizations, government agencies, farmers, and agribusiness have led to the wide-scale adoption of CA practices in different regions of the country over the last few decades (Llewellyn et al., 2012). Crop diversification, wide-scale modifications in machinery manufacturing, precision agriculture, improved crop tolerance against stubble-borne diseases and pests, and development of broad-spectrum herbicides, and other innovations have increased the uptake of CA practices in Australia (Bellotti and Rochecouste, 2014). Their report stated that '...around 80% to 90% of Australia's 23.5 million hectares of winter crops are now grown using Conservation Agriculture principles'.

There are many examples of why CA has been so popular in Australian cropping systems and these include the logistical advantages gained when sowing large crop areas (sometimes tens of thousands of hectares on one farm) in a timely manner and also the benefits to soil health and soil protection and WUE. These factors and others have been described and put into context by Kirkegaard (2017), in his excellent Farrer Memorial Oration.

Other more specific examples include a study on a Vertisol soil in Queensland, where wheat productivity increased under both CA- and CT-based systems with two levels of N (0 and 75 kg ha^{-1}) over five seasons (Dalal et al., 2013); but the CA system with N application produced the highest wheat yields. In addition, N utilization increased in the CA-based wheat system more than CT wheat, as indicated by the high values of N use efficiency (Dalal et al., 2013). In another work, Chan et al. (2003) reviewed the impact of CA systems on carbon sequestration and changes in soil quality on light-textured soils in Australia. They found significantly higher SOC in the top 2.5–10 cm soil layer under CA systems than conventional systems in wetter areas (>500 mm) as well as better soil structure and soil physical properties (e.g. macro-porosity, aggregate stability, and higher infiltration rate).

The oral history of the introduction, development, and adoption of the direct drilling technologies that underpin CA in Australia has now been recorded and archived by the National Library of Australia and is available at: https://catalogue.nla.gov.au/Record/7131702. It provides a fascinating account–through interviews with key actors from industry, research, and farming–of the essential factors in the private-public-farmer partnerships that underpinned this agricultural success story.

4.6 Corn/maize-based systems in North America

The USA accounts for 38% and 35% of global maize and soybean production, producing a respective 320 and 84 million tons of these crops annually (Grassini et al., 2015). Partially due to the introduction of genetically modified, herbicide-tolerant crops in these systems, the adoption of CA there has increased rapidly. Grassini et al. (2015) have reviewed the factors influencing these high-yield maize-soybean cropping systems and have shown that continuous genetic improvement; agronomic improvement including CA; and the synergistic interactions of improved genetics and agronomy have all been key factors in the drive for higher yields and resource-use efficiency for these crops in the 'Corn Belt' region of the USA.

CA is also used for some wheat-based systems, and Williams et al. (2009) compared soil erosion and runoff in winter wheat under inversion tillage and CA systems in northeastern Oregon, USA. The inversion tillage system comprised a summer fallow–winter wheat rotation and the CA system had a chemical fallow–winter wheat or chickpea–winter wheat rotation. The highest yields occurred in the two CA-based systems in the third year of experimentation. They also found negligible runoff in the CA-based plots relative to the inversion tillage plots. They concluded that adoption of CA in semiarid North American states (Oregon and Washington) might benefit water and soil conservation.

In eight states of the USA, averaged over 20 long-term (11 ± 1 years) studies, CA systems produced 30.2 Mg ha^{-1} SOC compared with 25.6 Mg ha^{-1} in conventional systems (Franzluebbers, 2010). In a 14-year study on the Virginia coastal plain, Spargo et al. (2008) reported a decline in soil bulk density in the 0–2.5 cm layer and an increase in the carbon stratification ratio (0–2.5 cm; 7.5–15 cm) with increasing duration of continuous CA as a result of SOM accumulation in the soil surface. In another study, Franzluebbers et al. (1998) found that a CA system on a Fluventic Ustochrept soil in south-central Texas had 10–30% more SOC to 0.2 m than the conventional system.

In Canada, the long-term and wide adoption of CA, mostly in the Western provinces, has had visible environmental benefits, including the absence of dust storms; greater biodiversity; and enhanced productivity (Baig and Gamache, 2009; Lindwall and Sonntag, 2010).

5 Conclusions and future trends

CA is being implemented in diverse climatic conditions on various soil types across the world (Kassam et al., 2018). Based on this review, it is concluded that CA can improve crop productivity/yield, support sustainable cultivation, and improve smallholder security. Various crops including broadacre crops (cereals, legumes, oilseeds, fiber, and energy crops) and horticultural (vegetables) crops can be successfully grown in CA systems. However, it is vital that legume crops be included in cereal-based CA systems as they improve soil properties and reduce the need for external synthetic N fertilizers. The incorporation of herbicide resistance strategies, including integrated pest management; weed seed destruction; cover crops; and allelopathic crops (e.g. sorghum and sunflower) in crop rotations will be essential for controlling weeds, the biggest threat to the long-term sustainability of CA systems. Use of CA for vegetable crops will also be important for sustaining the productivity of these systems globally.

It is important to avoid monocultures in CA systems as they are likely to reduce biodiversity and exacerbate losses from pests and diseases. Crop diversification in CA-based cropping systems will be essential and will include intercropping, cover cropping, and relay cropping methodologies. Intercropping with crops that produce volatile compounds (e.g. onion and garlic) may help to avoid some pests. Strip cropping could also be useful for reducing runoff losses in CA systems.

Future research must focus on the site-specific, long-term impacts of cropping systems that include legumes, oilseeds, forages, livestock, shrubs, and trees and their incorporation into CA-based cereal cropping systems under diverse soil types. Weed and nutrient dynamics and the impacts of CA on soil C sequestration should be included in these future studies.

The feasibility of different intercrops and strip/relay cropping must also be investigated.

6 Where to look for further information

Bhan, S. and Behera, U. K. 2014. Conservation agriculture in India - problems, prospects and policy issues. *Int. Soil Water Conser. Res.* 2(4), 1–12. doi:10.1016/S2095-6339(15)30053-8.

Busari, M. A., Kukal, S. S., Kaur, A., Bhatt, R. and Dulazi, A. A. 2015. Conservation tillage impacts on soil, crop and the environment. *Int. Soil Water Conser. Res.* 3(2), 119–29. doi:10.1016/j.iswcr.2015.05.002.

Chan, C. and Fantle-Lepczyk, J. 2015. *Conservation Agriculture in Subsistence Farming: Case Studies from South Asia and Beyond.* CABI, Wallingford, UK.

Jat, R. A., Sahrawat, K. L. and Kassam, A. H. 2013. *Conservation Agriculture: Global Prospects and Challenges.* CABI, Wallingford, UK.

Kassam, A., Mkomwa, S. and Friedrich, T. 2017. *Conservation Agriculture for Africa: Building Resilient Farming Systems in a Changing Climate*. CABI, Wallingford, UK.

Nawaz, A., Farooq, M., Nadeem, F., Siddique, K. H. M. and Lal, R. 2019. Rice-wheat cropping systems in South Asia: issues, options and opportunities. *Crop Past. Sci.* 70(5), 395–427. doi:10.1071/CP18383.

http://www.fao.org/conservation-agriculture/en/.

7 References

Abdollahi, L. and Munkholm, L. J. 2014. Tillage system and cover crop effects on soil quality: I. Chemical, mechanical, and biological properties. *Soil Sci. Soc. Amer. J.* 78(1), 262–70. doi:10.2136/sssaj2013.07.0301.

Abdul-Baki, A. A. and Teasdale, J. R. 1993. A no-tillage tomato production system using hairy vetch and subterranean clover mulches. Available at: http://agris.fao.org/agris-search/search.do?recordID=US9424078 (accessed on April 21, 2018).

Abdul-Baki, A. A. and Teasdale, J. R. 1997. Snap bean production in conventional tillage and in no-till hairy vetch mulch. *Hort. Sci.* 32(7), 1191–93. doi:10.21273/HORTSCI.32.7.1191.

Alpmann, D., Braun, J., and Schäfer, B. C. 2013. Analysis of a survey among successful grain legumes cultivators in conventional farming. First results from the research project Iguana. In: D.L.G. Winter Conference (Ed.), *In Focus: Domestic Grain Legumes from Cultivation to Use*. Available at: www4.fh-swf.de/de/home/forschung/forschungshighlights/leguan/leguan 3.php (accessed April 26, 2018).

Anderson, W. K. and Angus, J. F. 2011. Agronomy and cropping practices in semi-arid conditions in Australia. In: Bonjean, A. P., Angus, W. J., and van Ginkel, M. (Eds.), *The World Wheat Book* (vol. 2). Lavoisier, Paris, France, pp. 563–605.

Anwar, S., Lei, H., Kuai, J., Khan, S., Fahad, S., and Zhou, G. 2017. Soaking seeds of winter rapeseed with Quizalofop-P-Ethyl alters plant growth and improves yield in a rice-rapeseed cropping system. *Field Crop. Res.* 208, 11–17. doi:10.1016/j.fcr.2017.04.001.

Bai, Y., He, J., Li, H., Wang, Q., Chen, H., Kuhn, N. J., Hikel, H., Chen, F., and Gong, Y. 2009. Soil structure and crop performance after 10 years of controlled traffic and traditional tillage cropping in the dryland Loess Plateau in China. *Soil Sci.* 174(2), 113–19. doi:10.1097/SS.0b013e3181981ddc.

Baig, M. N. and Gamache, P. M. 2009. *The Economic, Agronomic and Environmental Impact of No-Till on the Canadian Prairies*. Alberta Reduced Tillage Linkages, Alberta, Canada.

Banik, P., Midya, A., Sarkar, B. K., and Ghose, S. S. 2006. Wheat and chickpea intercropping systems in an additive series experiment: advantages and weed smothering. *Eur. J. Agron.* 24(4), 325–32. doi:10.1016/j.eja.2005.10.010.

Basso, B., Shuai, G., Zhang, J., and Robertson, G. P. 2019. Yield stability analysis reveals sources of large-scale nitrogen loss from the US Midwest. *Sci. Rep.* 9(1), 5774. doi:10.1038/s41598-019-42271-1.

Baudron, F., Tittonell, P., Corbeels, M., Letourmy, P., and Giller, K. E. 2012. Comparative performance of conservation agriculture and current smallholder farming practices in semi-arid Zimbabwe. *Field Crop. Res.* 132, 117–28. doi:10.1016/j.fcr.2011.09.008.

Bayer, C., Gomes, J., Zanatta, J. A., Vieira, F. C. B., and Dieckow, J. 2016. Mitigating greenhouse gas emissions from a subtropical Ultisol by using long-term no-tillage in combination with legume cover crops. *Soil Till. Res.* 161, 86–94. doi:10.1016/j.still.2016.03.011.

Bellotti, B. and Rochecouste, J. F. 2014. The development of conservation agriculture in Australia–farmers as innovators. *Int. Soil Water Conser Res.* 2(1), 21–34. doi:10.1016/S2095-6339(15)30011-3.

Bhushan, L., Ladha, J. K., Gupta, R. K., Singh, S., Tirol-Padre, A., Saharawat, Y. S., Gathala, M., and Pathak, H. 2007. Saving of water and labor in a rice–wheat system with no-tillage and direct seeding technologies. *Agron. J.* 99(5), 1288–96. doi:10.2134/agronj2006.0227.

Bilalis, D. J. and Karamanos, A. J. 2010. Organic maize growth and mycorrhizal root colonization response to tillage and organic fertilization. *J. Sustain. Agric.* 34(8), 836–49. doi:10.1080/10440046.2010.519197.

Blanco-Canqui, H. 2010. Energy crops and their implications on soil and environment. *Agron. J.* 102(2), 403–19. doi:10.2134/agronj2009.0333.

Cairns, J. E., Hellin, J., Sonder, K., Araus, J. L., MacRobert, J. F., Thierfelder, C. and Prasanna, B. M. 2013. Adapting maize production to climate change in sub-Saharan Africa. *Food Sec.* 5(3), 345–60. doi:10.1007/s12571-013-0256-x.

Carter, M. R. and Sanderson, J. B. 2001. Influence of conservation tillage and rotation length on potato productivity, tuber disease and soil quality parameters on a fine sandy loam in eastern Canada. *Soil Till. Res.* 63(1–2), 1–13. doi:10.1016/S0167-1987(01)00224-0.

Carvalho, A. Md, Marchão, R. L., Souza, K. W., and Bustamante, M. MdC. 2014. Soil fertility status, carbon and nitrogen stocks under cover crops and tillage regimes. *Rev. Cienc. Agron.* 45(5spe), 914–21. doi:10.1590/S1806-66902014000500007.

Chan, K. Y., Heenan, D. P., and So, H. B. 2003. Sequestration of carbon and changes in soil quality under conservation tillage on light-textured soils in Australia: a review. *Aust. J. Exp. Agric.* 43(4), 325–34. doi:10.1071/EA02077.

Cheesman, S., Thierfelder, C., Eash, N. S., Kassie, G. T. and Frossard, E. 2016. Soil carbon stocks in conservation agriculture systems of Southern Africa. *Soil Till. Res.* 156, 99–109. doi:10.1016/j.still.2015.09.018.

Chen, Y., Liu, S., Li, H., Li, X. F., Song, C. Y., Cruse, R. M., and Zhang, X. Y. 2011. Effects of conservation tillage on corn and soybean yield in the humid continental climate region of Northeast China. *Soil Till. Res.* 115–116, 56–61. doi:10.1016/j.still.2011.06.007.

Christopher, S. F. and Lal, R. 2007. Nitrogen management affects carbon sequestration in North American cropland soils. *Crit. Rev. Plant Sci.* 26(1), 45–64. doi:10.1080/07352680601174830.

Corbeels, M., Scopel, E., Cardoso, A., Bernoux, M., Douzet, J. M., and Siqueira Neto, M. 2006. Soil carbon storage potential of direct seeding mulch-based cropping systems in the Cerrados of Brazil. *Glob. Change Biol.* 12, 1773–87. doi: 10.1111/j.1365-2486.2006.01233.x.

Cullum, R. F. 2012. Influence of tillage on maize yield in soil with shallow fragipan. *Soil Till. Res.* 119, 1–6. doi:10.1016/j.still.2011.11.011.

Dalal, R. C., Strong, W. M., Cooper, J. E., and King, A. J. 2013. Relationship between water use and nitrogen use efficiency discerned by C-13 discrimination and N-15 isotope ratio in bread wheat grown under no-till. *Soil Till. Res.* 128, 110–18.

Das, A., Ghosh, P. K., Lal, R., Saha, R., and Ngachan, S. 2014a. Soil quality effect of conservation practices in maize-rapeseed cropping system in Eastern Himalaya. *Land Degrad. Dev.* 28(6), 1862–74. doi:10.1002/ldr.2325.

Das, T. K., Bhattacharyya, R., Sudhishri, S., Sharma, A. R., Saharawat, Y. S., Bandyopadhyay, K. K., Sepat, S., Bana, R. S., Aggarwal, P., Sharma, R. K., Bhatia, A., Singh, G., Datta, S. P., Kar, A., Singh, B., Singh, P., Pathak, H., Vyas, A. K., and Jat, M. L. 2014b. Conservation Agriculture in an irrigated cotton–wheat system of the western Indo-Gangetic Plains: crop and water productivity and economic profitability. *Field Crop. Res.* 158, 24–33. doi:10.1016/j.fcr.2013.12.017.

Das, T. K., Bandyopadhyay, K. K., Bhattacharyya, R., Sudhishri, S., Sharma, A. R., Behera, U. K., Saharawat, Y. S., Sahoo, P. K., Pathak, H., Vyas, A. K., Bhar, L. M., Gupta, H. S., Gupta, R. K., and Jat, M. L. 2016. Effects of Conservation Agriculture on crop productivity and water-use efficiency under an irrigated pigeonpea–wheat cropping system in the western Indo-Gangetic Plains. *J. Agric. Sci.* 154(8), 1327–42. doi:10.1017/S0021859615001264.

Das, T. K., Saharawat, Y. S., Bhattacharyya, R., Sudhishri, S., Bandyopadhyay, K. K., Sharma, A. R., and Jat, M. L. 2018. Conservation Agriculture effects on crop and water productivity, profitability and soil organic carbon accumulation under a maize-wheat cropping system in the North-western Indo-Gangetic Plains. *Field Crop. Res.* 215, 222–31. doi:10.1016/j.fcr.2017.10.021.

Derpsch, R. and Friedrich, T. 2009. *Global Overview of Conservation Agriculture Adoption*. Proceedings, Lead Paper, 4th World Congress on Conservation Agriculture, February 4–7, 2009, New Delhi, India, pp. 429–38.

De Tourdonnet, S., Nozières, A., Barz, P., Chenu, C., Düring, R. A., Frielinghaus, M., Kölli, R., Kubat, J., Magid, J., Medvedev, V., Michels, A., Müller, L., Netland, J., Nielsen, N. E., Nieves Mortensen, C., Picard, D., Quillet, J. C., Saulas, P., Tessier, D., Thinggaard, K., and Vandeputte, E. 2007. Comprehensive inventory and assessment of existing knowledge on sustainable agriculture in the European platform of KASSA. In: Lahmar, R., Arrue, J. L., Denardin, J. E., Gupta, R. K., Ribeiro, M. F. F., and de Tourdonnet, S. (Eds.), *Knowledge Assessment and Sharing on Sustainable Agriculture*. CIRAD, Montpellier, 54pp.

Doane, T. A., Horwath, W. R., Mitchell, J. P., Jackson, J., Miyao, G., and Brittan, K. 2009. Nitrogen supply from fertilizer and legume cover crop in the transition to no-tillage for irrigated row crops. *Nutr. Cycl. Agroecosys* 85(3), 253–62. doi:10.1007/s10705-009-9264-9.

Erenstein, O. and Laxmi, V. 2008. Zero tillage impacts in India's rice-wheat systems: a review. *Soil Till. Res.* 100(1–2), 1–14. doi:10.1016/j.still.2008.05.001.

FAO. 2007. *Tropical Crop-Livestock Systems in Conservation Agriculture: the Brazilian Experience*. Integrated Crop Management 5. FAO, Rome, p. 92.

FAO. 2016. *Save and Grow in Practice: Maize, Rice and Wheat: A Guide to Sustainable Cereal Production*. FAO, Rome.

Farooq, M. and Nawaz, A. 2014. Weed dynamics and productivity of wheat in conventional and conservation rice-based cropping systems. *Soil Till. Res.* 141, 1–9. doi:10.1016/j.still.2014.03.012.

Farooq, M. and Siddique, K. H. 2015. *Conservation Agriculture*. Springer International Publishing, Switzerland.

Farooq, M., Bramley, H., Palta, J. A., and Siddique, K. H. M. 2011. Heat stress in wheat during reproductive and grain-filling phases. *Crit. Rev. Plant Sci.* 30(6), 491–507. doi:10.1080/07352689.2011.615687.

Farooq, M., Nawaz, A., Ahmad, E., Nadeem, F., Hussain, M., and Siddique, K. H. M. 2017. Using sorghum to suppress weeds in dry seeded aerobic and puddled transplanted rice. *Field Crop. Res.* 214, 211–18. doi:10.1016/j.fcr.2017.09.017.

Franzluebbers, A. J. 2010. Achieving soil organic carbon sequestration with conservation agricultural systems in the southeastern United States. *Soil Sci. Soc. Amer. J.* 74(2), 347–57. doi:10.2136/sssaj2009.0079.

Franzluebbers, A. J., Hons, F. M., and Zuberer, D. A. 1998. In situ and potential CO_2 evolution from a Fluventic Ustochrept in southcentral Texas as affected by tillage and cropping intensity. *Soil Till. Res.* 47(3–4), 303–08. doi:10.1016/S0167-1987(98)00118-4.

Friedrich, T., Derpsch, R., and Kassam, A. H. 2012. Overview of the global spread of conservation agriculture. *Facts Rep.* 6, 1–7.

Garrity, D. P. 2011. Making conservation agriculture ever green. Keynote presentation *5th World Congress on Conservation Agriculture and 3rd Farming Systems Design Conference (WCCA5 and FSD3)*, September 26–29, 2011, Brisbane Australia.

Ghaley, B. B., Rusu, T., Sandén, T., Spiegel, H., Menta, C., Visioli, G., O'Sullivan, L., Gattin, I. T., Delgado, A., Liebig, M. A., Vrebos, D., Szegi, T., Michéli, E., Cacovean, H., and Henriksen, C. 2018. Assessment of benefits of conservation Agriculture on soil functions in arable production systems in Europe. *Sustainability* 10(3), 794. doi:10.3390/su10030794.

Ghosh, B. N., Dogra, P., Sharma, N. K., Bhattacharyya, R., and Mishra, P. K. 2015. Conservation Agriculture impact for soil conservation in maize-wheat cropping system in the Indian sub-Himalayas. *Int. Soil Water Cons. Res.* 3, 112–18.

Giller, K. E. 2001. *Nitrogen Fixation in Tropical Cropping Systems* (2nd edn.). CABI Publishing, New York.

Govaerts, B., Mezzalama, M., Sayre, K. D., Crossa, J., Nicol, J. M., and Deckers, J. 2006. Long-term consequences of tillage, residue management, and crop rotation on maize/wheat root rot and nematode populations in subtropical highlands. *Appl. Soil Ecol.* 32(3), 305–15. doi:10.1016/j.apsoil.2005.07.010.

Govaerts, B., Fuentes, M., Mezzalama, M., Nicol, J. M., Deckers, J., Etchevers-Barra, J. D., Figueroa-Sandoval, B., and Sayre, K. D. 2007. Infiltration, soil moisture, root rot and nematode populations after 12 years of different tillage, residue and crop rotation managements. *Soil Till. Res.* 94(1), 209–19. doi:10.1016/j.still.2006.07.013.

Govaerts, B., Verhulst, N., Castellanos-Navarrete, A., Sayre, K. D., Dixon, J., and Dendooven, L. 2009. Conservation agriculture and soil carbon sequestration: between myth and farmer reality. *Crit. Rev. Plant Sci.* 28(3), 97–122. doi:10.1080/07352680902776358.

Grassini, P., Specht, J. E., Tollenaar, M., Ciampitti, I. and Cassman, K. G. 2015. High-yield maize-soybean cropping systems in the US Corn Belt. In: Sadras, V. and Calderini, D. (Eds), *Crop Physiology- Applications for Genetic Improvement and Agronomy.* Academic Press. doi:10.1016/B978-0-12-417104-6.00002-9.

Green, R. E., Cornell, S. J., Scharlemann, J. P., and Balmford, A. 2005. Farming and the fate of wild nature. *Science* 307(5709), 550–55. doi:10.1126/science.1106049.

Gressel, J. and Segel, L. A. 1990. Modeling the effectiveness of herbicide rotations and mixtures as strategies to delay or preclude resistance. *Weed Technol.* 4(1), 186–98. doi:10.1017/S0890037X00025215.

Gronau, I. 2017. Strip-cropping experiments yielding economic, environmental advantages. Available at: https://www.striptillfarmer.com/articles/2529-strip-cropping-experiments-yielding-economic-environmental-advantages?v=preview (accessed on April 21, 2018).

Guardia, G., Tellez-Rio, A., García-Marco, S., Martin-Lammerding, D., Tenorio, J. L., Ibáñez, M. Á., and Vallejo, A. 2016. Effect of tillage and crop (cereal versus legume) on greenhouse gas emissions and global warming potential in a non-irrigated Mediterranean field. *Agric. Ecosyst. Environ.* 221, 187–97. doi:10.1016/j.agee.2016.01.047.

Gupta, R. K., and Seth, A. 2007. A review of resource conserving technologies for sustainable management of the rice-wheat cropping systems of the Indo-Gangetic Plains (IGP). *Crop Prot.* 26(3), 436–47. doi:10.1016/j.cropro.2006.04.030.

Gupta, R. K., Naresh, R. K., Hobbs, P. R., and Ladha, J. K. 2002. Adopting conservation agriculture in the rice-wheat system of the Indo-Gangetic Plains: new opportunities for saving water. *Proceedings of the International Workshop on Water Wise Rice Production*, April 8–11, 2002, Los Banos, Philippines, pp. 207–22.

He, J., Kuhn, N. J., Zhang, X. M., Zhang, X. R., and Li, H. W. 2009. Effects of 10 years of conservation tillage on soil properties and productivity in the farming-pastoral ecotone of Inner Mongolia, China. *Soil Use Manag.* 25(2), 201–09. doi:10.1111/j.1475-2743.2009.00210.x.

He, J., Li, H., Rasaily, R. G., Wang, Q., Cai, G., Su, Y., Qiao, X., and Liu, L. 2011. Soil properties and crop yields after 11 years of no tillage farming in wheat-maize cropping system in North China Plain. *Soil Till. Res.* 113(1), 48–54. doi:10.1016/j.still.2011.01.005.

Heenan, D. P., Taylor, A. C., Chan, K. Y., McGhie, W. J., Collins, D., and Lill, W. J. 2000. The impact of long-term rotation, tillage and stubble management on lupin (*Lupinus angustifolius*) productivity. *Field Crop. Res.* 67(1), 11–23. doi:10.1016/S0378-4290(00)00078-2.

Helmers, G. A., Yamoah, C. F., and Varvel, G. E. 2001. Separating the impacts of crop diversification and rotations on risk. *Agron. J.* 93(6), 1337–40. doi:10.2134/agronj2001.1337.

Herrero, E. V., Mitchell, J. P., Lanini, W. T., Temple, S. R., Miyao, E. M., Morse, R. D., and Campiglia, E. 2001. Use of cover crop mulches in a no-till furrow-irrigated processing tomato production system. *HortTechnology* 11, 43–48. doi:10.21273/HORTTECH.11.1.43.

Hobbs, P., Gupta, R. A. J., Jat, R. K., and Malik, R. K. 2017. Conservation agriculture in the Indogangetic plains of India: past, present and future. *Exp. Agric.* 55(2), 339–57. doi:10.1017/S0014479717000424.

Holmstrom, D., Arsenault, W., Ivany, J., Sanderson, J. B., and Campbell, A. J. 2006. Effect of pre-plant tillage systems for potatoes in Prince Edward Island, Canada, on soil properties, weed control and potato yield. *J. Soil Water Conserv.* 61, 370–80.

Hoyt, G. D. 1999. Tillage and cover residue effects on vegetable yields. *HortTechnology* 9, 351–58.

Hoyt, G. D., Monks, D. W., and Monaco, T. J. 1994. Conservation tillage for vegetable production. *HortTechnology* 4, 129–35. doi:10.21273/HORTTECH.4.2.129.

Hoyt, G. D., Bonanno, A. R., and Parker, G. C. 1996. Influence of herbicides and tillage on weed control, yield, and quality of cabbage (*Brassica oleracea* L. var. capitata). *Weed Technol.* 10(1), 50–54. doi:10.1017/S0890037X00045693.

ICARDA. 2013. *The Proven Benefits of Conservation Agriculture*. Available at: https://www.icarda.org/features/provenbenefitsconservationagriculture#sthash.aualUhpB.dpbs (accessed on April 21, 2018).

Ihsan, M. Z., El-Nakhlawy, F. S., Ismail, S. M., Fahad, S., and Daur, I. 2016. Wheat phenological development and growth studies as affected by drought and late season high temperature stress under arid environment. *Front. Plant Sci.* 7, 795. doi:10.3389/fpls.2016.00795.

Indoria, A. K., Rao, S., Sharma, K. L., and Sammi Reddy, K. 2017. Conservation agriculture–a panacea to improve soil physical health. *Curr. Sci.* 112(1), 52–61. doi:10.18520/cs/v112/i01/52-61.

Infante, M. L. and Morse, R. D. 1996. Integration of no tillage and over seeded legume living mulches for transplanted broccoli production. *Hort. Sci.* 31, 376–80.

Islam, R., Glenney, D. C., and Lazarovits, G. 2015. No-till strip row farming using yearly maize-soybean rotation increases yield of maize by 75%. *Agron. Sustain. Dev.* 35(2), 837–46. doi:10.1007/s13593-015-0289-y.

Jat, R. K., Sapkota, T. B., Singh, R. G., Jat, M. L., Kumar, M., and Gupta, R. K. 2014. Seven years of conservation agriculture in a rice–wheat rotation of Eastern Gangetic Plains of South Asia: yield trends and economic profitability. *Field Crop. Res.* 164, 199–210. doi:10.1016/j.fcr.2014.04.015.

Jin, H., Qingjie, W., Hongwen, L., Lijin, L., and Huanwen, G. 2009. Effect of alternative tillage and residue cover on yield and water use efficiency in annual double cropping system in North China Plain. *Soil Till. Res.* 104(1), 198–205. doi:10.1016/j.still.2008.08.015.

Johansen, C., Haque, M. E., Bell, R. W., Thierfelder, C. and Esdaile, R. J. 2012. Conservation agriculture for small holder rainfed farming: opportunities and constraints of new mechanized seeding systems. *Field Crops Res.* 132, 18–32. doi:10.1016/j.fcr.2011.11.026.

Johnson, W. C., Brenneman, T. B., Baker, S. H., Johnson, A. W., Sumner, D. R., and Mullinix, B. G. 2001. Tillage and pest management considerations in a peanut–cotton rotation in the southeastern Coastal Plain. *Agron. J.* 93(3), 570–76. doi:10.2134/agronj2001.933570x.

Karlen, D. L., Kovar, J. L., Cambardella, C. A., and Colvin, T. S. 2013. Thirty-year tillage effects on crop yield and soil fertility indicators. *Soil Till. Res.* 130, 24–41. doi:10.1016/j.still.2013.02.003.

Kassam, A., Friedrich, T., Shaxson, F., and Pretty, J. 2009. The spread of Conservation Agriculture: justification, sustainability and uptake. *Int. J. Agric. Sustain.* 7(4), 292-320. doi:10.3763/ijas.2009.0477.

Kassam, A., Friedrich, T., Derpsch, R., and Kienzle, J. 2014. Worldwide adoption of conservation agriculture. Paper presented at the *6th World Congress on Conservation Agriculture*, June 22–25, 2014, Winnipeg, Canada.

Kassam, A., Friedrich, T., and Derpsch, R. 2018. Global spread of Conservation Agriculture. *Int. J. Environ. Stud.* 76, 49–51. doi:10.1080/00207233.2018.1494927.

Kertész, Á. and Madarász, B. 2014. Conservation agriculture in Europe. *Int. Soil Water Conserv. Res.* 2(1), 91-96.

Kertész, Á., Madarász, B., Bádonyi, K., Csepinszky, B., and Mika, J. 2011. Conservation tillage for rational water management and soil conservation. *Hungarian Geograph. Bull.* 60(2), 117–33.

Khan, M. J., Brodie, G. I., Gupta, D., and Foletta, S. 2018. Microwave soil treatment improves weed management in Australia dryland wheat. *Trans. ASABE* 61(2), 671-80. doi:10.13031/trans.12504.

Kimaro, A. A., Mpanda, M., Rioux, J., Aynekulu, E., Shaba, S., Thiong'o, M., Mutuo, P., Abwanda, S., Shepherd, K., Neufeldt, H. and Rosenstock, T. S. 2016. Is conservation agriculture 'climate-smart' for maize farmers in the highlands of Tanzania? *Nutr. Cycl. Agroecosyst.* 105(3), 217–28. doi:10.1007/s10705-015-9711-8.

Kirkegaard, J. A. 2017. Farming Systems Agronomy: it is rocket science. Available at: https://www.dpi.nsw.gov.au/__data/assets/pdf_file/0012/796665/2017-Dr-John-Kirkegaard-Farming-Systems-Agronomy-it-is-rocket-science.pdf (accessed on 19 June 2018).

Kirkegaard, J. A., Conyers, M. K., Hunt, J. R., Kirkby, C. A., Watt, M., and Rebetzke, G. J. 2014. Sense and nonsense in conservation agriculture: principles, pragmatism and productivity in Australian mixed farming systems. *Agric. Ecosyst. Environ.* 187, 133-45. doi:10.1016/j.agee.2013.08.011.

Kluthcouski, J., Cobucci, T., Aidar, H., Yokoyama, L. P., Oliveira, I. P., de Costa, J. L., da S. Silva, J. G., da Vilela, L., Barcellos, A. de O. and Magnobosco, C. de U. 2000. *Sistema Santa Fé - Tecnologia Emprapa: Integração lavoura-pecuária peolo consórcio de culturas anuais com forragerira, em áreas de lavoura, nos sistemas direto e convencional*. Circular Técnica 38. Embrapa Arroz e Feijão, Embrapa Arroz e Feijão, Santo Antônio de Goiás, Brazil, 28p.

Knowler, D. 2005. Short cut techniques to incorporate environmental considerations into project appraisal: an exploration using case studies. *J. Environ. Plan. Manag.* 48(5), 747–70. doi:10.1080/09640560500183033.

Kong, R., Sar, V., Leng, V., Trang, S., Boulakia, S., Tivet, F., and Seguy, L. 2016. Conservation agriculture for climate-resilient rainfed uplands in the Western regions of Cambodia: challenges, opportunities, and lessons from a 10-year R and D program. In: Sajise Percy, E., Cadiz Maria Celeste, H. and Bantayan Rosario, B. (Eds.), *Learning and Coping with Change: Case Stories of Climate Change Adaptation in Southeast Asia*. SEARCA, Los Baños, pp. 55–81.

Kremen, C. and Miles, A. 2012. Ecosystem services in biologically diversified versus conventional farming systems: benefits, externalities, and trade-offs. *Ecol. Soc.* 17(4), 40. doi:10.5751/ES-05035-170440.

Kumar, V., Saharawat, Y. S., Gathala, M. K., Jat, A. S., Singh, S. K., Chaudhary, N., and Jat, M. L. 2013. Effect of different tillage and seeding methods on energy use efficiency and productivity of wheat in the Indo-Gangetic Plains. *Field Crop. Res.* 142, 1-8. doi:10.1016/j.fcr.2012.11.013.

Ladha, J. K., Kumar, V., Alam, M. M., Sharma, S., Gathala, M., Chandna, P., Saharawat, Y. S., and Balasubramanian, V. 2009. Integrating crop and resource management technologies for enhanced productivity, profitability, and sustainability of the rice-wheat system in South Asia. In: Ladha, J. K. (Ed.), *Integrated Crop and Resource Management in the Rice-Wheat System of South Asia*. International Rice Research Institute, Los Baños, Philippines, pp. 69-108.

Lahmar, R. 2010. Adoption of conservation agriculture in Europe: lessons of the KASSA project. *Land Use Pol.* 27(1), 4-10. doi:10.1016/j.landusepol.2008.02.001.

Laik, R., Sharma, S., Idris, M., Singh, A. K., Singh, S. S., Bhatt, B. P., Saharawat, Y., Humphreys, E., and Ladha, J. K. 2014. Integration of conservation agriculture with best management practices for improving system performance of the rice-wheat rotation in the Eastern Indo-Gangetic Plains of India. *Agric. Ecosyst. Environ.* 195, 68-82. doi:10.1016/j.agee.2014.06.001.

Lassaletta, L., Billen, G., Garnier, J., Bouwman, L., Velazquez, E., Mueller, N. D., and Gerber, J. S. 2016. Nitrogen use in the global food system: past trends and future trajectories of agronomic performance, pollution, trade, and dietary demand. *Environ. Res. Lett.* 11(9). doi:10.1088/1748-9326/11/9/095007.

Lazaro, L., Norsworthy, J., Walsh, M., and Bagavathiannan, M. 2017. Efficacy of the Harrington weed destructor on weeds of soybean and rice production systems in the Southern United States. *Crop Sci.* 57, 1-7. doi:10.2135/cropsci2017.03.0210.

Li, H., Gao, H., Wu, H., Li, W., Wang, X., and He, J. 2007. Effects of 15 years of conservation tillage on soil structure and productivity of wheat cultivation in northern China. *Soil Res.* 45(5), 344-50. doi:10.1071/SR07003.

Lindwall, C. W. and Sonntag, B. 2010. *Landscape Transformed: the History of Conservation Tillage and Direct Seeding. Knowledge Impact in Society*. University of Saskatchewan, Saskatoon.

Liu, S., Zhang, X. Y., Yang, J., and Drury, C. F. 2013. Effect of conservation and conventional tillage on soil water storage, water use efficiency and productivity of corn and soybean in Northeast China. *Acta Agric. Scand. B* 63(5), 383-94. doi:10.1080/09064710.2012.762803.

Liu, S., Zhang, X. Y., Kravchenko, Y., and Iqbal, M. A. 2015. Maize (*Zea mays* L.) yield and soil properties as affected by no tillage in the black soils of China. *Acta Agric. Scand. B* 65, 554-65.

Llewellyn, R. S., D'Emden, F. H., and Kuehne, G. 2012. Extensive use of no-tillage in grain growing regions of Australia. *Field Crop. Res.* 132, 204-12. doi:10.1016/j.fcr.2012.03.013.

Malcolm, S. A., Aillery, M. P. and Weinberg, M. 2009. Ethanol and a changing agricultural landscape. Economic Research Report Number 86. United States Department of Agriculture, USA.

Martínez, G. I., Ovalle, C., Del Pozo, A., Uribe, H., Valderrama, V. N., Prat, C., Sandoval, M., Fernández, F. and Zagal, E. 2011. Influence of conservation tillage and soil water content on crop yield in dryland compacted alfisol of central Chile. *Chile J. Agric. Res.* 71(4), 615-22. doi:10.4067/S0718-58392011000400018.

Mathews, J. M., Powles, S. B., and Holtum, J. A. M. 1994. *Management of Herbicide Resistant Weed Populations. Herbicide Resistance Plants*. Lewis Publishers, Boca Raton, FL, pp. 317-35.

Mazvimavi, K., Twomlow, S., Belder, P. and Hove, L. 2008. An assessment of the sustainable uptake of conservation farming in Zimbabwe. Global Theme on Agroecosystems Report No. 39. International Crops Research Institute for the Semi-Arid Tropics, Bulawayo, Zimbabwe.

Mishra, J. S. and Singh, V. P. 2011. Effect of tillage and weed control on weed dynamics, crop productivity and energy-use efficiency in rice (*Oryza sativa*)-based cropping systems in Vertisols. *Indian J. Agric. Sci.* 81, 129–33.

Morse, R. D. 1993. Components of sustainable production systems for vegetables-conserving soil moisture. *HortTechnology* 3, 211–14. doi:10.21273/HORTTECH.3.2.211.

Morse, R. D. 1999. No-till vegetable production–its time is now. *HortTechnology* 9, 373–79. doi:10.21273/HORTTECH.9.3.373.

Morse, R. D. 2000. High-residue, no-till systems for production of organic broccoli. In: *Proceedings of the Southern Conservation Tillage Conference for Sustainable Agriculture*, pp. 48–50.

Muchabi, J., Lungu, O. I., and Mweetwa, A. M. 2014. Conservation agriculture in Zambia: effects on selected soil properties and biological nitrogen fixation in soya beans (*Glycine max* (L.) Merr). *Sustain. Agric. Res.* 3(3), 28. doi:10.5539/sar.v3n3p28.

Munyati, M. 1997. Conservation tillage for sustainable crop production systems: results and experiences from on-station and on-farm research (1988–1996). *Z. Sci. News* 31, 27–33.

Mupangwa, W., Twomlow, S., and Walker, S. 2012. Reduced tillage, mulching and rotational effects on maize (*Zea mays* L.), cowpea (*Vigna unguiculata* (Walp) L.) and sorghum (*Sorghum bicolor* L.(Moench)) yields under semi-arid conditions. *Field Crop. Res.* 132, 139–48. doi:10.1016/j.fcr.2012.02.020.

Mupangwa, W., Mutenje, M., Thierfelder, C. and Nyagumbo, I. 2016. Are conservation agriculture (CA) systems productive and profitable options for smallholder farmers in different agro-ecoregions of Zimbabwe? *Renew. Agric. Food Syst.* 32(1), 87–103. doi:10.1017/S1742170516000041.

Naab, J. B., Mahama, G. Y., Yahaya, I., and Prasad, P. V. V. 2017. Conservation agriculture improves soil quality, crop yield, and incomes of smallholder farmers in north western Ghana. *Front. Plant Sci.* 8, 996. doi:10.3389/fpls.2017.00996.

Naudin, K., Gozé, E., Balarabe, O., Giller, K. E., and Scopel, E. 2010. Impact of no tillage and mulching practices on cotton production in North Cameroon: a multi-locational on-farm assessment. *Soil Till. Res.* 108(1–2), 68–76. doi:10.1016/j.still.2010.03.002.

Nawaz, A. and Farooq, M. 2016. Weed management in resource conservation production systems in Pakistan. *Crop Prot.* 85, 89–103. doi:10.1016/j.cropro.2016.04.002.

Nawaz, A., Farooq, M., Cheema, S. A., and Cheema, Z. A. 2014. Role of allelopathy in weed management. In: Chauhan, B. and Mahajan, G. (Eds.), *Recent Advances in Weed Management*. Springer Science+Business Media, New York, pp. 39–61.

Nawaz, A., Farooq, M., Ahmad, R., Basra, S. M. A., and Lal, R. 2016. Seed priming improves stand establishment and productivity of no-till wheat grown after direct seeded aerobic and transplanted flooded rice. *Eur. J. Agron.* 76, 130–37. doi:10.1016/j.eja.2016.02.012.

Nawaz, A., Lal, R., Shrestha, R. K., and Farooq, M. 2017a. Mulching affects soil properties and greenhouse gases emissions under long term no-till and plough till systems in Alfisol of central Ohio. *Land Degrad. Develop.* 28(2), 673–81. doi:10.1002/ldr.2553.

Nawaz, A., Farooq, M., Lal, R., Rehman, A., Hussain, T., and Nadeem, A. 2017b. Influence of sesbania brown manuring and rice residue mulch on soil health, weeds and system productivity of conservation rice-wheat systems. *Land Degrad. Develop.* 28(3), 1078–90. doi:10.1002/ldr.2578.

Nawaz, A., Farooq, M., Lal, R., Rehman, A., and Rehman, H. 2017c. Comparison of conventional and conservation rice-wheat systems in Punjab, Pakistan. *Soil Till. Res.* 169, 35–43. doi:10.1016/j.still.2017.01.012.

Ngwira, A. R., Aune, J. B., and Mkwinda, S. 2012. On-farm evaluation of yield and economic benefit of short term maize legume intercropping systems under conservation agriculture in Malawi. *Field Crops Res.* 132, 149–57. doi:10.1016/j.fcr.2011.12.014.

Ngwira, A. R., Thierfelder, C., and Lambert, D. M. 2013. Conservation agriculture systems for Malawian smallholder farmers: long-term effects on crop productivity, profitability and soil quality. *Renew. Agric. Food Syst.* 28(4), 350–63. doi:10.1017/S1742170512000257.

Nyamangara, J., Marondedze, A., Masvaya, E. N., Mawodza, T., Nyawasha, R., Nyengerai, K., Tirivavi, R., Nyamugafata, P., and Wuta, M. 2014. Influence of basin-based conservation agriculture on selected soil quality parameters under smallholder farming in Zimbabwe. *Soil Use Manag.* 30(4), 550–59. doi:10.1111/sum.12149.

Olsen, J., Kristensen, L., Weiner, J., and Griepentrog, H. W. 2005. Increased density and spatial uniformity increase weed suppression by spring wheat. *Weed Res.* 45(4), 316–21. doi:10.1111/j.1365-3180.2005.00456.x.

Pabin, J., Lipiec, J., Wlodek, S., and Biskupski, A. 2003. Effect of different tillage systems and straw management on some physical properties of soil and on the yield of winter rye in monoculture. *Inter. Agrophys.* 17, 175–82.

Pacheco, A. R., de Queiroz Chaves, R., and Nicoli, L. 2013. Integration of crops, livestock, and forestry: a system of production for the Brazilian Cerrados. In: Hershey, C. H. and Neate, P. (Eds.), *Eco-Efficiency: From Vision to Reality (Issues in Tropical Agriculture Series).* Centro Internacional de Agricultura Tropical (CIAT), Cali, CO, pp. 51–60.

Palm, C. A., Gachengo, C. N., Delve, R. J., Cadish, G., and Giller, K. E. 2001. Organic inputs for soil fertility management in tropical agroecosystems: application of an organic resource database. *Agric. Ecosyst. Environ.* 83(1–2), 27–42. doi:10.1016/S0167-8809(00)00267-X.

Parihar, C. M., Yadav, M. R., Jat, S. L., Singh, A. K., Kumar, B., Pradhan, S., Chakraborty, D., Jat, M. L., Jat, R. K., Saharawat, Y. S., and Yadav, O. P. 2016. Long term effect of conservation agriculture in maize rotations on total organic carbon, physical and biological properties of a sandy loam soil in north-western Indo-Gangetic Plains. *Soil Till. Res.* 161, 116–28. doi:10.1016/j.still.2016.04.001.

Parihar, C. M., Jat, S. L., Singh, A. K., Ghosh, A., Rathore, N. S., Kumar, B., Pradhan, S., Majumdar, K., Satyanarayana, T., Jat, M. L., Saharawat, Y. S., Kuri, B. R., and Saveipune, D. 2017. Effects of precision conservation agriculture in a maize-wheat-mungbean rotation on crop yield, water-use and radiation conversion under a semiarid agro-ecosystem. *Agric. Water Manag.* 192, 306–19. doi:10.1016/j.agwat.2017.07.021.

Parihar, C. M., Jat, S. L., Singh, A. K., Kumar, B., Rathore, N. S., Jat, M. L., Saharawat, Y. S., and Kuri, B. R. 2018a. Energy auditing of long-term conservation agriculture based irrigated intensive maize systems in semi-arid tropics of India. *Energy* 142, 289–302. doi:10.1016/j.energy.2017.10.015.

Parihar, C. M., Yadav, M. R., Jat, S. L., Singh, A. K., Kumar, B., Pooniya, V., Pradhan, S., Verma, R. K., Jat, M. L., Jat, R. K., and Saharawat, Y. S. 2018b. Long term conservation agriculture and intensified cropping systems: effect on growth, yield, water and energy-use efficiency of maize in North-Western India. *Pedosphere* 28(6), 952–63. doi:https://doi.org/10.1016/S1002-0160(17)60468-5.

Paudel, B., Radovich, T. J. K., Chan-Halbrendt, C., Crow, S., Tamang, B. B., Halbrendt, J., and Thapa, K. 2014. Effect of conservation agriculture on maize-based farming system in the mid-hills of Nepal. *Procedia Eng.* 78, 327–36. doi:10.1016/j.proeng.2014.07.074.

Phalan, B., Onial, M., Balmford, A., and Green, R. E. 2011. Reconciling food production and biodiversity conservation: land sharing and land sparing compared. *Science* 333(6047), 1289–91. doi:10.1126/science.1208742.

Powlson, D. S., Stirling, C. M., Thierfelder, C. and White, R. P. and Jat, M. L. 2016. Does conservation agriculture deliver climate change mitigation through soil carbon sequestration in tropical agro-ecosystems? *Agric. Ecosyst. Environ.* 220, 164–74. doi:10.1016/j.agee.2016.01.005.

Pradhan, A., Idol, T., and Roul, P. K. 2016. Conservation agriculture practices in rainfed uplands of India improve maize-based system productivity and profitability. *Front. Plant Sci.* 7, 1008. doi:10.3389/fpls.2016.01008.

Qingjie, W., Hao, C., Hongwen, L., Wenying, L., Xiaoyan, W., McHugh, A. D., Jin, H., and Huanwen, G. 2009. Controlled traffic farming with no tillage for improved fallow water storage and crop yield on the Chinese Loess Plateau. *Soil Till. Res.* 104(1), 192–97. doi:10.1016/j.still.2008.10.012.

Rathore, A. L., Pal, A. R., and Sahu, K. K. 1998. Tillage and mulching effects on water use, root growth and yield of rainfed mustard and chickpea grown after lowland rice. *J. Sci. Food Agric.* 78(2), 149–61. doi:10.1002/(SICI)1097-0010(199810)78:2<149::AID-JSFA94>3.0.CO;2-U.

Reckling, M., Schläfke, N., Hecker, J. M., Bachinger, J., Zander, P., and Bergkvist, G. 2014. Generation and evaluation of legume-supported crop rotations in five case study regions across Europe. Available at: www.legumefutures.de (accessed on April 21, 2018).

Reeves, T. G., Albertengo, J., Amado, T., Bartz, M., Basch, G., Boulakia, S., Calegari, A., Day, S., Derpsch, R., Desbiollies, J., Duiker, S., Fuentes, R., Goddard, T., Gonzalez, E., Li, H., Mello, J., Muminjanov, H., Sa, J. C. M., and Thierfielder, C. 2016a. *Science and Research for the Development of Conservation Agriculture – a Global Perspective*. FAO, Cuba, October 2016.

Reeves, T. G., Thomas, G., and Ramsay, G. 2016b. *Save and Grow in Practice, Maize, Rice, Wheat: A Guide to Sustainable Cereal Production*. Food and Agriculture Organization of the United Nations, Rome.

Rusinamhodzi, L., Corbeels, M., Van Wijk, M. T., Rufino, M. C., Nyamangara, J., and Giller, K. E. 2011. A meta-analysis of long-term effects of conservation agriculture on maize grain yield under rain-fed conditions. *Agron. Sustain. Dev.* 31(4), 657–73. doi:10.1007/s13593-011-0040-2.

Saha, S., Chakraborty, D., Sharma, A. R., Tomar, R. K., Bhadraray, S., Sen, U., Behera, U. K., Purakayastha, T. J., Garg, R. N., and Kalra, N. 2010. Effect of tillage and residue management on soil physical properties and crop productivity in maize (*Zea mays*)-Indian mustard (*Brassica juncea*) system. *Indian J. Agric. Sci.* 80, 679–85.

Sajjad, A., Anjum, S. A., Ahmad, R., and Waraich, E. A. 2018. Relay cropping of wheat (*Triticum aestivum* L.) in cotton (*Gossypium hirsutum* L.) improves the profitability of cotton-wheat cropping system in Punjab, Pakistan. *Environ. Sci. Pollut. Res.* 25(1), 782–89. doi:10.1007/s11356-017-0422-3.

Sapkota, T. B., Jat, M. L., Aryal, J. P., Jat, R. K., and Khatri-Chhetri, A. 2015. Climate change adaptation, greenhouse gas mitigation and economic profitability of Conservation Agriculture: some examples from cereal systems of Indo-Gangetic Plains. *J. Integr. Agric.* 14(8), 1524–33. doi:10.1016/S2095-3119(15)61093-0.

Sapkota, T. B., Jat, R. K., Singh, R. G., Jat, M. L., Stirling, C. M., Jat, M. K., Bijarniya, D., Kumar, M., Saharawat, Y. S., Saharawat, Y. S., and Gupta, R. K. 2017. Soil organic carbon changes after seven years of conservation agriculture in a rice–wheat system of the eastern Indo-Gangetic Plains. *Soil Use Manag.* 33(1), 81–89. doi:10.1111/sum.12331.

Sarkar, S., Paramanick, M., and Goswami, S. B. 2007. Soil temperature, water use and yield of yellow sarson (*Brassica napus* L. var. glauca) in relation to tillage intensity and mulch management under rainfed lowland ecosystem in eastern India. *Soil Till. Res.* 93(1), 94–101. doi:10.1016/j.still.2006.03.015.

Schellenberg, D. L., Morse, R. D., and Welbaum, G. E. 2009. Organic broccoli production on transition soils: comparing cover crops, tillage and sidedress N. *Renew. Agric. Food Syst.* 24(2), 85–91. doi:10.1017/S1742170508002470.

Scopel, E., Triomphe, B., dos Santos Ribeiro, M. de F., Séguy, L., Denardin, J. E. and Kochhann, R. A. 2004. Direct seeding mulch-based cropping systems (DMC) in Latin America. In: Fischer, R. A. (Ed.), *New Directions for a Diverse Planet. Proceedings of the 4th International Crop Science Congress*, Brisbane, Australia.

Shaxson, T. F. and Barber, R. G. 2003. *Optimizing Soil Moisture for Plant Production: The Significance of Soil Porosity*. FAO Soils Bulletin, Rome, Italy, 79p.

Six, J., Conant, R. T., Paul, E. A., and Paustian, K. 2002. Stabilization mechanisms of soil organic matter: implications for C saturation of soils. *Plant Soil* 241(2), 155–76. doi:10.1023/A:1016125726789.

Snapp, S. S., Blackie, M. J., Gilbert, R. A., Bezner-Kerr, R., and Kanyama-Phiri, G. Y. 2010. Biodiversity can support a greener revolution in Africa. *Proc. Natl. Acad. Sci. U. S. A.* 107(48), 20840–05. doi:10.1073/pnas.1007199107.

Spargo, J. T., Alley, M. M., Follett, R. F., and Wallace, J. V. 2008. Soil carbon sequestration with continuous no-till management of grain cropping systems in the Virginia coastal plain. *Soil Till. Res.* 100(1–2), 133–40. doi:10.1016/j.still.2008.05.010.

Stagnari, F., Maggio, A., Galieni, A., and Pisante, M. 2017. Multiple benefits of legumes for agriculture sustainability: an overview. *Chem. Biol. Technol. Agric.* 4(1), 2. doi:10.1186/s40538-016-0085-1.

Stephenson, G. R., Dykstra, M. D., McLaren, R. D., and Hamill, A. S. 1990. Agronomic practices influencing triazine-resistant weed distribution in Ontario. *Weed Technol.* 4(1), 199–207. doi:10.1017/S0890037X00025227.

Stewart, C. E., Paustian, K., Conant, R. T., Plante, A. F., and Six, J. 2007. Soil carbon saturation: concept, evidence and evaluation. *Biogeochemistry* 86(1), 19–31. doi:10.1007/s10533-007-9140-0.

Sturgul, S. J., Daniel, T. C., and Mueller, D. H. 1990. Tillage and canopy cover effects on interrill erosion from first-year alfalfa. *Soil Sci. Soc. Am. J.* 54(6), 1733–39. doi:10.2136/sssaj1990.03615995005400060037x.

Sutaliya, J. M. and Singh, U. P. 2017. Effect of different crop establishment methods, tillage and residue on yield, yield attributes and economics of rice in RW system of northern plains of IGP. *Int. J. Curr. Microbiol. Appl. Sci* 6(7), 4174–83. doi:10.20546/ijcmas.2017.607.432.

Tanveer, M., Anjum, S. A., Hussain, S., Cerdà, A., and Ashraf, U. 2017. Relay cropping as a sustainable approach: problems and opportunities for sustainable crop production. *Environ. Sci. Pollut. Res.* 24(8), 6973–88. doi:10.1007/s11356-017-8371-4.

Thierfelder, C. and Mupangwa, W. 2014. Identifying new sustainable intensification pathways for smallholder farmers in Southern Africa. Paper Presented at the *World Congress of Conservation Agriculture (WCCA6)*, June 22–25, 2014, Winnipeg, Canada.

Thierfelder, C. and Wall, P. C. 2009. Effects of conservation agriculture techniques on infiltration and soil water content in Zambia and Zimbabwe. *Soil Till. Res.* 105(2), 217–27. doi:10.1016/j.still.2009.07.007.

Thierfelder, C. and Wall, P. C. 2010a. Rotation in conservation agriculture systems of Zambia: effects on soil quality and water relations. *Exp. Agric.* 46(3), 309–25. doi:10.1017/S001447971000030X.

Thierfelder, C. and Wall, P. C. 2010b. Investigating conservation agriculture (CA) systems in Zambia and Zimbabwe to mitigate future effects of climate change. *J. Crop Improv.* 24(2), 113–21. doi:10.1080/15427520903558484.

Thierfelder, C., Cheesman, S., and Rusinamhodzi, L. 2012. A comparative analysis of conservation agriculture systems: benefits and challenges of rotations and intercropping in Zimbabwe. *Field Crop. Res.* 137, 237–50. doi:10.1016/j.fcr.2012.08.017.

Thierfelder, C., Chisui, J. L., Gama, M., Cheesman, S., Jere, Z. D., Bunderson, W. T., Eash, N. S., and Rusinamhodzi, L. 2013a. Maize-based conservation agriculture systems in Malawi: long-term trends in productivity. *Field Crop. Res.* 142, 47–57. doi:10.1016/j.fcr.2012.11.010.

Thierfelder, C., Mwila, M., and Rusinamhodzi, L. 2013b. Conservation agriculture in eastern and southern provinces of Zambia: long-term effects on soil quality and maize productivity. *Soil Till. Res.* 126, 246–58. doi:10.1016/j.still.2012.09.002.

Thierfelder, C., Cheesman, S., and Rusinamhodzi, L. 2013c. Benefits and challenges of crop rotations in maize-based conservation agriculture (CA) cropping systems of southern Africa. *Int. J. Agric. Sustain.* 11(2), 108–24. doi:10.1080/14735903.2012.703894.

Thierfelder, C., Matemba-Mutasa, R., and Rusinamhodzi, L. 2015a. Yield response of maize (*Zea mays* L.) to conservation agriculture cropping system in Southern Africa. *Soil Till. Res.* 146, 230–42. doi:10.1016/j.still.2014.10.015.

Thierfelder, C., Mutenje, M., Mujeyi, A., and Mupangwa, W. 2015b. Where is the limit? Lessons learned from long-term conservation agriculture research in Zimuto Communal Area, Zimbabwe. *Food Sec.* 7(1), 15–31. doi:10.1007/s12571-014-0404-y.

Thierfelder, C., Matemba-Mutasa, R., Bunderson, W. T., Mutenje, M., Nyagumbo, I., and Mupangwa, W. 2016. Evaluating manual conservation agriculture systems in southern Africa. *Agric. Ecosyst. Environ.* 222, 112–24. doi:10.1016/j.agee.2016.02.009.

Tingle, C. H. and Chandler, J. M. 2004. The effect of herbicides and crop rotation on weed control in glyphosate-resistant crops. *Weed Technol.* 18(4), 940–46. doi:10.1614/WT-02-126.

Tiwari, A., Saharawat, Y. S., McDonald, A. J., Jat, M. L., and Kumar, B. 2018. Conservation Agriculture benefits in South Asia: a management of tillage, residue retention and crop rotation. Research & reviews. *J. Agric. Sci. Technol.* 4(2), 1-13.

Uvah, I. I. I. and Coaker, T. H. 1984. Effect of mixed cropping on some insect pests of carrots and onions. *Ent. Exper. Appl.* 36(2), 159-67. doi:10.1111/j.1570-7458.1984.tb03422.x.

Vilrla, L., Macedo, M. C. M., Júnior, G. B. M., and Kluthcouski, J. 2003. Crop livestock integration benefits. In: Kluthcouski, J., Stone, L. F., and Aidar, H. (Eds.), *Integraçã lavoura-pecuária*. Embrapa Arroz e Feijão, Santo Antônio de Goiás, Goiás, Brazil.

Wall, P. C., Thierfelder, C., Ngwira, A., Govaerts, B., Nyagumbo, I. and Baudron, F. 2013. Conservation agriculture in Eastern and Southern Africa. In: Jat, R. A., Sahrawat, K. L. and Kassam, A. H. (Eds), *Conservation Agriculture: Global Prospects and Challenges*. CABI Publishing, Wallingford, UK, 424p.

Walters, S. A. and Kindhart, J. D. 2002. Reduced tillage practices for summer squash production in southern Illinois. *HortTechnology* 12, 114-17. doi:10.21273/HORTTECH.12.1.114.

Wang, K. H., McSorley, R., and Gallaher, R. N. 2003. Effect of *Crotalaria juncea* amendment on nematode communities in soil with different agricultural histories. *J. Nematol.* 35(3), 294-301.

White, J. and Sapkota, T. 2016. Cereal systems in South Asia show diverse benefits of Conservation Agriculture. Available at: https://ccafs.cgiar.org/blog/cereal-systems-south-asia-show-diverse-benefits-conservation-agriculture (accessed on April 21, 2018).

Williams, J. D., Gollany, H. T., Siemens, M. C., Wuest, S. B., and Long, D. S. 2009. Comparison of runoff, soil erosion, and winter wheat yields from no-till and inversion tillage production systems in northeastern Oregon. *J. Soil Water Conserv.* 64(1), 43-52. doi:10.2489/jswc.64.1.43.

Yamane, K., Ikoma, A., and Iijima, M. 2016. Performance of double cropping and relay intercropping for black soybean production in small-scale farms. *Plant Prod. Sci.* 19(4), 449-57. doi:10.1080/1343943X.2016.1164574.

Zegada-Lizarazu, W., and Monti, A. 2011. Energy crops in rotation: a review. *Biomass Bioenergy* 35(1), 12-25. doi:10.1016/j.biombioe.2010.08.001.

Zheng, C., Jiang, Y., Chen, C., Sun, Y., Feng, J., Deng, A., Song, Z., and Zhang, W. 2014. The impacts of Conservation Agriculture on crop yield in China depend on specific practices, crops and cropping regions. *Crop J.* 2(5), 289-96. doi:10.1016/j.cj.2014.06.006.

Chapter 3

Soil management practices and benefits in Conservation Agriculture systems

Michele Pisante, University of Teramo, Italy; Angelica Galieni, Council for Agricultural Research and Economics and Research Centre for Vegetable and Ornamental Crops, Italy; Gottlieb Basch, University of Évora, Portugal; Theodor Friedrich, Food and Agriculture Organization of the United Nations (FAO), Italy; and Fabio Stagnari, University of Teramo, Italy

1 Introduction

Conservation Agriculture (CA) is an ecosystem approach to sustainable agriculture and land management based on the practical application of locally adapted three interlinked principles: (1) continuous no or minimum mechanical soil disturbance (no-till seeding/planting and weeding, and minimum soil disturbance with all other farm operations including harvesting); (2) permanent maintenance of mulch cover on the soil surface (crop biomass, root stocks, stubble, cover crops (CCs) and other *ex situ* biomass); and (3) diversification of cropping system (environmentally and socially adapted rotations and/or sequences and/or associations involving annuals and perennials, including legume and non-legume crops), along with other complementary good agricultural production and land management practices (i.e. use of quality seeds, integrated pest, nutrient, weed and water management) (www.fao.org/ag/ca; Kassam et al., 2019). It is important to point out that the CA approach is practiced only if all three principles are simultaneously and thoroughly applied.

The soil conservation concept was introduced for the first time in the 1930s as a result of the devastation of wide areas in North America by dust

http://dx.doi.org/10.19103/AS.2019.0049.03

bowls, and led to the development of conservation tillage[1] practices aimed at wind and water erosion regulation (Giller et al., 2015). However, the CA approach was introduced in the early 1970s, but only after 20 years it started to be taken up significantly (Kassam et al., 2019). Since then, the adoption of CA has been rapid, particularly in North America and South America, thanks to a combination of a number of factors. Among these were the release of effective herbicides (glyphosate), the availability of no-till planters to direct seed into a mulch of crop residues and the government policy incentives to promote no-till in the United States (Fuglie and Kascak, 2001; Claassen, 2012). During the 1990s, CA also spread to Europe, Asia, Africa and Australia to become a global agricultural movement; nowadays there are only a few countries where CA is not practiced. CA soil management approaches were first practiced on large-scale, mechanized farms, but more recently adoption on small farms has been increasing; in addition, there is still a major misconception in that herbicides replace tillage for weed management in CA (Friedrich, 2005).

The general interest in CA principles appears to be linked to several agronomic and economic aspects, such as soil water availability, soil physical, chemical and biological properties as well as the delivery of several ecosystem services, including, among others, climate regulation through carbon sequestration and greenhouse gas (GHG) emissions' reduction, soil conservation and biodiversity preservation and/or improvement. Moreover, crop yield levels recorded under established CA systems are similar to those obtained under 'traditional agriculture' systems. CA practices even may lead to increased crop and biomass production over time and, along with reduced nitrogen (N) fertilizer inputs, it offers farmers the opportunity to reduce costs and conserve energy in crop production.

The chapter first highlights the three CA principles as well as their agronomic implications (Section 2) and then deepens ecological and economic benefits, related to the adoption of CA practices in agroecosystems (Sections 2 and 3).

2 The principles of CA

2.1 No or minimum mechanical soil disturbance

No or minimum mechanical soil disturbance refers to tillage regimes and its exact definition is usually context specific (Nichols et al., 2015). Among

[1] It is important to specify that conservation tillage (CT) is not CA (Reicosky, 2015). CA principles emanate from CT, which includes no-tillage (NT), reduced tillage (RT) and groundcovers (GC) in perennial crops (CTIC, 1994). In particular, NT represents the CA agronomic practice for annual crops while GC is the most widely used CA agronomic practice for perennial crops, whereby the soil surface between rows of trees remains protected against erosion (Gonzalez-Sanchez et al., 2015). In general, the retention of 30% surface cover by residues characterizes the lower limit of classification for CT, jointed with conservation of time, fuel, earthworms, soil, water, soil structure and nutrients (Baker et al., 2002).

such tillage techniques, zero tillage (ZT), also called 'no-tillage' or 'no-till', is a technique that allows for the sowing of the crop without previous mechanical disturbance of the soil, except the opening of the slit for seed placement. Since the soil remains practically undisturbed, the residues left from the previous crops remain almost entirely on the soil surface, producing a layer of a variable thickness depending on climate regimes, soil type and amount and quality of residues. Along with chemical herbicide residues, crop rotations and mechanical non-soil engaging tools are preferred weed management options.

Application of ZT is not based only on a simple change of equipment, but a different method of soil as well as farm management, which requires information, practical experience and good skills. Indeed, while ploughing needs practical skills from the technical point of view, ZT implies a more forward-looking planning for the management of crop residues, weeds, pests, fertilizer application and so on. Since the factors most likely to influence the successful uptake of no-till concern crop, soil, climatic and environmental conditions, it is mainly the agronomic aspects that drive the adoption of no-till by farmers. Furthermore, a rigorous farm planning, already starting from the crops involved in the rotation is required.

Generally, all soils are suitable to the adoption of ZT, provided they are well levelled and not compacted; besides, highest improvements in terms of structure are obtained in clayey soils, especially in those with clay tending to self-structure (vertisols). On coarse textured, frequently weakly structured soils, such as Denmark, soil compaction represents a severe limitation to the success of no-till (Munkholm et al., 2003). On the other hand, sandy and sandy loam soils, especially if low in organic matter, may lack the ability to acquire a stable structure under no-till (Ehlers and Claupein, 1994); in such cases long-term structure can only be obtained through ZT application with, as consequence, the increase of organic matter and the biological formation of soil aggregates. However, the importance of soil type and permeability for successful no-till is emphasized in wet regions (Alakukku et al., 2003).

When ZT is applied within the wider context of CA systems, particular requirements are needed:

- a proper residue management (at harvesting, harvesters equipped with straw chopper or spreader and chaff spreader) and the sowing of CCs;
- adoption of appropriate crop rotations and cultivar selection;
- change in weed and fertilizer management strategies, especially during the first 2–3 years of adoption, which are particularly problematic;
- adoption of controlled-traffic techniques, in those soils particularly prone to soil compaction at deeper layers, and on large farms with heavy machinery.

Amount and type of crop residues strongly influence no-tillage techniques. Cereals, the most cultivated crops, produce amounts of straw strongly depending on climate conditions: it averages about 10–12 t/ha in Northern Europe (Christian and Ball, 1994), 6 t/ha in Central Europe (Ehlers and Claupein, 1994) and lower values under Mediterranean areas due to water shortage. Such variability poses different challenges requiring different approaches to efficiently apply no-till (Galieni et al., 2016).

In Northern Europe, characterized by moist cool climates, the thicker layers of crop residues determine lower solar radiation to the soil surface as well as lower water evaporation might delay seeding (Anken et al., 2004; Mikkola et al., 2005) and crop germination, due to a marked reduction of topsoil temperatures (Riley et al., 1994; Børresen, 1999), all amplified if the residues are still attached to the crop roots and standing (Mikkola et al., 2005). In southern Mediterranean environments, crop yields are lower, consequently also the amounts of surface residues under no-till are generally reduced and often not enough to reduce significantly soil water evaporation. Therefore, while for the former region reducing straw length is crucial (chopper application) to make sowing easier and to allow higher penetration of solar radiation to soil surface, for the latter region higher straw length is required to increase soil cover and consequently to reduce water evaporation as well as the speed of residue decomposition.

Regarding crops to be grown under no-till system in the Northern Europe regions, although spring cereals are the dominant crops (Alakukku et al., 2009), the results with winter-sown crops are more suitable (Mikkola et al., 2005). Indeed, the amount and form of residues, the height of cut and so on, affect the drilling operation and reduce access of solar radiation and the evaporation of water from the surface; these effects may cause serious delay in the drilling of spring-sown crops as the soil surface may be appreciably wetter and colder than when soils are ploughed prior to winter. In central and southern regions of Europe, because of favourable weather conditions, some spring-sown crops can be grown under no-till, such as maize, sunflowers and fava beans (Vogeler et al., 2009). The adoption of rotations in no-till has particular relevance in terms of weed population and management (Jordan et al., 1997). Weed populations under no-till change dramatically due to seed retention on soil surface with changes in their dormancy cycles (Christian and Ball, 1994). While in the northern regions perennial grass weeds are likely to be favoured by a no-till regime, becoming a serious problem on coarse textured soils; in southern areas the long, dry summers restrict perennial weed growth and most of the weed species are annual (Carvalho and Basch, 1994). Generally, where a thick layer of crop residues can be achieved, no-till and particularly CA reduce germination of weed seeds as well as the growth of weed seedlings, enabling a reduction of herbicide applications (Bräutigam and Tebrügge, 1997). A surface mulch of crop residues can also reduce the temperature at or just below the soil surface, which may adversely affect weed seed germination (Morris et al., 2010). In any case, appropriate rotations and the use of CCs are considered to be essential components in reducing weed problems and the dependence on herbicide usage under no-till systems (Derpsch and Friedrich, 2009).

It also should be considered that under no-till, thanks to the building-up of a permanent layer of crop residues as well as to higher soil moisture conditions, the survival and distribution of pathogen inoculum and the possible diffusion of leaf spot and root diseases from one season to the following could be enhanced (Mikkola et al., 2005; Knowler and Bradshaw, 2007). This can especially become a problem when the same crop is grown consecutively. Nevertheless, the application of appropriate rotations in CA systems and the increasing concentration of organic matter near the surface seem to have positive effects on suppressing some soil-borne fungal crop diseases (Ehlers and Claupein, 1994). Similar results were observed for take-all (*Gaeumannomyces graminis*) and Jordan et al. (1997) reported that the incidence of barley yellow dwarf virus on winter barley sown after no-till is only 2.7% compared to 57.3% after ploughing, thanks to the greater number of beneficial polyphagous predators (insects and spiders).

2.2 Maintenance of permanent soil cover

Soil cover can be achieved by live CCs, terminated CCs or mulches from crop residues after harvesting, which accumulate on soil surface. It appears that the inclusion of CCs into complex rotations, as well as the management of crop residues and CCs are key elements of CA systems.

Although the use of CCs has ancient roots, in recent years it has assumed a re-emerging importance due to its ecosystem services, and the renewed interest within the scientific community is demonstrated by several research activities (Blanco-Canqui et al., 2015). CCs, also named inter-crops or catch crops, are crops that replace a bare fallow period (Dabney et al., 2001) being sown in a window of time and space normally not occupied by other principal crops; in contrast to the introduction of arable-ley rotations or periodic green fallows, CCs do not reduce yields and organic manure can be applied in their cultivation. Often, the term 'green manure cover crops' is used since, besides providing ecosystem services, such crops enhance the productive capacity of soils or provide, directly, nutrients to the following crop. Following the definition of the Soil Science Society of America Glossary of Terms, CCs are classified as a 'close-growing crop that provides soil protection, seeding protection, and soil improvement between periods of normal crop production, or between trees in orchards and vines in vineyards. When plowed under and incorporated into the soil, CCs may be referred to as 'green manure crops' (SSSA, 2008).

The main difference with the traditional green manure crops is that green manure CCs in CA systems are not incorporated into the soil through mechanical tillage operations but through the activity of soil biota (Florentin et al., 2011; Van Capelle et al., 2012). CCs, indeed, are not usually grown for harvesting but only for providing important ecosystem services at field, farm and landscape level (Thorup-Kristensen et al., 2012), that is reduction of soil erosion and nutrient losses, improvement of soil quality and fertility, control of weeds and/or pests, better water quality, enhancement of biodiversity and wildlife (Derpsch et al., 2014; Schipanski et al., 2014). While some of the environmental benefits harnessed with the application of CCs are further discussed in Section 3, some more specific information is summarized in Fig. 1. In general, these benefits depend on a number of variables, which include the species of CCs as well as their management (sowing and termination techniques), soil type and climatic conditions. However, CCs are not generally adopted – uptake is only around 2% on average in large US farms (Bryant et al., 2013) – due to the costs of establishment and management, labour requirements, risk associated with timing of planting of main crops, their compatibility with residue management and irrigation practices. Mulching based on CCs, in any case, implies the need of seeding and consequent costs for labour, seed and management. Effective mulching through CCs needs the selection of species characterized

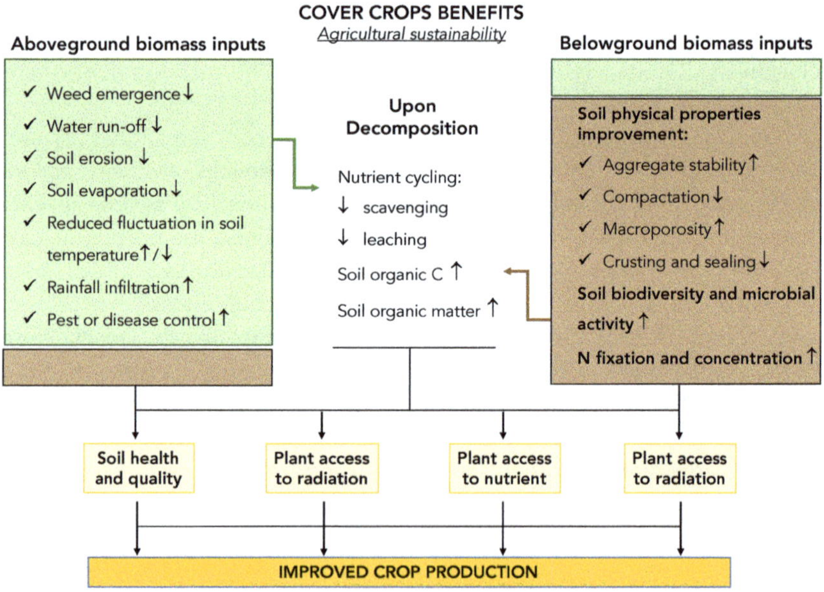

Figure 1 Potential interrelated benefits and ecosystem services of cover crops (CCs). Source: modified from Blanco-Canqui et al. (2015) and Ranaivoson et al. (2017).

by resistant mulch material compatible with the natural biomass weathering process in order to enhance mulch retention levels (see the studies of Derpsch et al. (1991) and Seguy et al. (1996) on non-leguminous CCs in Brazil). Under Mediterranean environment species belonging to the *Poaceae* (i.e. ryegrass - *Lolium multiflorum* L., oat - *Avena sativa* L., rye - *Secale cereale* L.), *Brassicaceae* and *Fabaceae* (i.e. vetch - *Vicia sativa* L., hairy vetch - *Vicia villosa* Roth., various types of clover - *Trifolium meneghinianum* Clem., *Trifolium alexsandrinum* L.) are commonly used as winter CCs (Radicetti et al., 2016). Cereals and crucifers are more efficient in catching residual N from soil (Kuo and Sainju, 1998), while legumes are known to provide additional N through biological fixation process (Galieni et al., 2017; Stagnari et al., 2017). Diversified cropping CCs patterns composed of at least three species are generally suggested (Lal, 2015): a legume–cereal CCs mixture is effective in avoiding N losses combining the N scavenging of the non-legume with the N fixing ability of the legume (Brennan and Boyd, 2012). Moreover, some CCs species (i.e. *Raphanus sativus* L., *Fagopyrum esculentum* Moench. and *Avena strigosa* L.) are known for the active release of allelochemicals during their growth, showing allelopathic efficacy on the overall weed suppression (Sturm et al., 2018).

The management of CCs' aboveground biomass generally implies the termination of the CCs and the retention of residues on the soil surface as organic dead mulches (Teasdale et al., 2008). The termination can be achieved by the use of herbicides, since spraying is relatively fast, effective and cheap, while the use of under-cutter knifes, to cut the roots leading to a 95% of mortality, is not suitable for CA systems due to soil disturbance (Creamer and Dabney, 2002). A suitable tool for mechanical control of CCs under CA is the roller crimper, which allows obtaining a high percentage of mortality (94%) - compared to synthetic herbicides (100%), if the appropriate plant growth stage is reached (Ashford and Reeves, 2003). For instance, an effective termination of legume CCs is obtained after full flowering - providing significant amounts of biomass and nutrients (Robacer et al., 2016) - while for rye the suitable stage ranges from early milk to soft dough - Zadoks growth stages 70–85 (Ashford and Reeves, 2003). Besides, some important aspects should be considered: (1) whether the created flat mat lies in the direction of travel, thus main crop and CCs sowing should be conducted in the same direction, to minimize interference between the residue and planter; (2) an eventual excessive vibration of the rollers, which have detrimental effects on the operator's health (increased heart rate, headache, stomach pain, lower back pain etc.), depending on their design (Kornecki et al., 2009).

The amount and composition of residues, coming both from CCs or main crops, directly impact on soil's physical, chemical and biological properties. Thickness of mulching, as the consequence of the produced biomass, is the main factor which assures higher soil water content, since the water vapour must diffuse through the layer of mulch (Li et al., 2012). In addition, during the transition to CA systems - when yields and biomass production, from both previous crop or CCs, are still limited - additional crop residues could be provided (Stagnari et al., 2014). In some agricultural systems such as smallholder agriculture, including sub-Saharan territories, crop residues and in particular, cereal stover, provide highly valued fodder for livestock, so competing with mulching. As a result mulching materials are often in critically low supply (Wezel and Rath, 2002). The amount of available crop residues is also affected by crop species and climate conditions. Cereals such as maize (*Zea mays* L.), wheat (*Triticum aestivum* L.), sorghum (*Sorghum bicolor* L.) and rice (*Oryza sativa* L.) produce the largest amounts (Blanco-Canqui and Lal, 2009), and northern, more humid areas in Europe are known to provide more crop residues than southern, dry environments (i.e. Mediterranean areas).

The chemical composition of plant residues (e.g. their specific C/N ratio) influence differently soil chemical, biological and physical properties: that is soil temperature is modified by the colour of the residues, where dark residues result in higher mid-day temperatures compared to light coloured ones (Sharratt and Campbell, 1994). At the same time, the efficiency of residues in terms of soil erosion reduction depends on the physical characteristics of residue. Wheat residues, characterized by a higher area-to-mass ratio than maize, rye or rice residues are most effective for soil erosion control; 1.5 and 4.5 t/ha dry matter of maize, rye and rice residues reduce soil loss by 50% compared to bare soil, while only 2 t/ha dry matter of wheat residues reduce soil loss by about 90% (Woyessa and Bennie, 2004).

2.3 Diversification of cropping systems

Diversification of cropping systems is a pillar of CA and it should ideally be based on the growing of a minimum of three different crops (Kassam et al., 2019). Crops included in the rotations, inter-, mixed- or relay-cropping involve annuals and perennials, shrubs, as well as CCs or spontaneous vegetation in perennial orchard or plantation crops in a balanced mix of leguminous and non-leguminous species, since all crops can be adequately grown in CA crop systems, including roots and tubers (Derpsch and Friedrich, 2009; Kassam et al., 2012). In Western Asia and in the Middle East, the rotation including wheat, chickpea, barley and lentil results in an attractive cropping system (Piggin et al., 2015) with the aim of increasing legumes production (Siddique et al., 2012). Under Mediterranean environments, soil quality can be improved through the adoption of biennial rotations (i.e. wheat-sunflower, wheat-chickpea, wheat-faba bean) under CA (Melero et al., 2011).

CA systems including perennial crops, such as orchards (olive, grape, fruits) and plantation systems (oil palm, cocoa, rubber, tea, coffee, coconut and sugar cane), as well as agroforestry systems and crop-pasture-tree systems, are increasing in all continents and are expected to further improve the actual CA systems in the near future (Kassam et al., 2019). In orchards, vineyards (Gómez et al., 2009, 2011) and plantations (Othman et al., 2012) a permanent soil covering is obtained with the sowing of CCs or pasture, directly beneath or between rows so affecting positively soil erosion, biodiversity, soil water retention and soil aeration (Kassam et al., 2012, 2013, 2019).

In tropical Africa and Asia, smallholder mixed systems, combining crop and livestock production, contribute to the livelihood of two-thirds of the population (Herrero et al., 2010). In these systems, crop production is destined to feed animals, implying a competitive use of crop residues (Valbuena et al., 2012) and for the adoption of CA the *in situ*-produced biomass needs to support its multiple functions (Kassam et al., 2012). Forage-based rotations in CA are useful to integrate crop–livestock systems (Sanderson et al., 2013).

Crop biodiversity provides for a better agroecosystem resilience, since a wide diversity of organisms is pre-condition for ecosystems to function properly and provide the manifold ecosystem services (Lin, 2011; Kassam et al., 2013), most of them are soil-mediated: that is crop rotations in CA reduce root rot and nematode populations (Govaerts et al., 2007), as well as decline weed pressure and other pests (Lal, 2015); certain crops in a rotation, such as sunnhemp (*Crotalaria juncea*), are reported to completely stop the carry-over of nematodes (Wang et al., 2003) or *Desmodium* spp. as intercrop or under sown CCs in maize to reduce damages and severe yield losses caused by the stem borer and *Striga* spp. (Khan et al., 2006; Kassam et al., 2017). Crop diversification and rotation have also been associated with enhanced soil fertility, especially when nitrogen-fixing legumes are involved, as well as with improved soil structure, aggregation and pore continuity, giving positive effects on infiltration and soil moisture under rainfed conditions, with the inclusion of a deep rooting crop (Thierfelder and Wall, 2010). In addition, the better nutrient distribution along soil profile represents a further advantage in response to the cultivation of crops with different rooting depths; moreover, soil structure could be ameliorated by the increase in soil organic matter content and by the exudates produced from some crops included into the rotation (Thierfelder and Wall, 2010). Diversified agroecosystems show a great adaptivity to climate variability in terms of both modified precipitation and temperature patterns (Lin, 2011) and the production of enough crop residues in CA systems can be obtained only through a balanced crop rotation or association.

3 Environmental benefits and ecosystem services

CA practices are known to promote several environmental benefits, that is improvement in soil and water quality, and their linked ecosystem services like mitigation of soil erosion and GHG (Hobbs et al., 2008). A review of environmental benefits and ecosystem services mediated through sustainable soil management under CA are reviewed by Kassam et al. (2013).

Differently to water and air quality, soil quality does not only concern the degree of soil pollution but is commonly defined as 'the capacity of a soil to function within ecosystem and land-use boundaries to sustain biological productivity, maintain environmental quality, and promote plant and animal (including human) health' (Doran and Parkin, 1994, 1996). This definition highlights the multifunctionality of soils (Nortcliff, 2002) and, consequently, its quality can be defined from environmental, agro-ecosystem or natural-ecosystem perspectives (Bünemann et al., 2018). As a consequence, a group of sensitive soil attributes – statics and/or dynamics – which reflect the ability of a soil to perform its functions must be considered as principal indicators of soil quality (Schwilch et al., 2016; Bünemann et al., 2018). Indeed, modern assessments refer to 'soil health' rather than quality or fertility (FAO, 2008). The impact of conservation practices is noticeable on several soil properties, though at different magnitudes (Busari et al., 2015), which include soil organic matter (SOM) content and composition and the activities of soil biota (Palm et al., 2014). No-tillage and residue retention significantly increase SOM content in the top soil layers (0–5 cm or 0–10 cm) (Bissett et al., 2013), specifically its particulate and more labile fractions (particulate organic matter, POM – >53 mm in size), as described in several studies (Álvaro-Fuentes et al., 2008; Martín-Lammerding et al., 2013, 2015; Blanco-Moure et al., 2016), although it depends on several factors, including amount of crop residues, species which contribute to crop residues production and soil texture (Martín-Lammerding et al., 2013, 2015; Blanco-Moure et al., 2016). In particular, in fine textured soils, where clay- and silt-sized particles with high surface activity may chemically stabilize SOM and form the building blocks for aggregates, the enhancement in SOM content (in the long term) is primarily attributable to the reduction of soil mechanical disturbance (lower SOM decomposition). Conversely, in coarse textured soils, where tillage determines the greater SOM losses, due to lack of its physical protection, more emphasis should be placed on the management of C inputs (residues retention) (Chivenge et al., 2007). SOM variation influences changes in soil biota activities – providing energy and substrate – which, in turn, affect soil organic carbon (SOC) dynamics (Álvaro-Fuentes et al., 2008; Busari et al., 2015).

CA practices, indeed, guarantee higher abundance and diversity of soil biota community composition than conventional tillage systems (Gonzalez-Chavez et al., 2010). Earthworms – the major component of the soil macrofauna, which play a significant role in increasing aggregate stability, infiltration rates, macroporosity, saturated hydraulic conductivity and nutrient cycling – are generally more abundant in soils under conservation practices (Nieminen et al.,

2011; Stagnari et al., unpublished data), probably thanks to a less disturbed environment characterized by higher humidity and a constant availability of organic residues (Pelosi et al., 2017). Soil microbial biomass, principally represented by bacteria and fungi, is generally more abundant where residues retention is applied, and mechanical disturbance is avoided, as a consequence of the ameliorated microenvironments for microbial growth and reproduction (González-Chávez et al., 2010; Bissett et al., 2013; Tiemann et al., 2015). In particular, some studies indicate a dominance of the fungal communities under no-till, probably due to the slower mineralization process and higher nutrient-use efficiency in the presence of crop residues (Verhulst et al., 2010). Also crop diversification positively impact the functional metabolic profile and diversity of soil microbial community (by 3.5% on average), which increases with the magnitude of crop variation (i.e. species and genus) (D'Acunto et al., 2018; Venter et al., 2016).

CA practices are important for SOC and nutrient - that is, N, phosphorus (P) and sulphur (S) - storage in agroecosystems (Sarker et al., 2018). The effects of CA on total N content generally retrace those observed for SOC as N and C cycles are strictly linked. Several studies have indicated the role of no-tillage in increasing directly soil total N pools in the topsoil (0–5 or 0–10 cm depth) (Luo et al., 2010; Mazzoncini et al., 2016). Besides, the effects of conservation practices on SOC and soil total nitrogen (STN) in the deeper soil layers still remain inconsistent and uncertain (Luo et al., 2010). Significant increases in STN have been observed with increasing crop residue retention; anyway, some literature results report lower N availability in CA systems due to the slower residues decomposition and higher inorganic N immobilization during the decomposition process (Verhulst et al., 2010). However, the type of residues (C/N ratio, initial residue N, lignin, polyphenols and soluble C concentrations) and the N management practices contribute to the rate of C and N mineralization. Lastly, the inclusion of legumes in CA rotations provides increasing *in situ* N availability, so reducing the application of nitrogenous fertilizers (Boddey et al., 2009). Other macronutrients - P and potassium (K) - appear to be more concentrated and available in the upper layers (0–5 cm or 0–10 cm) of no-till soils, with residue retentions, when compared to conventionally managed soils, where the less mobile mineral elements are more evenly distributed along the ploughed soil layer (Palm et al., 2014). The increase in K availability is usually attributed to the retention of cereal crop residues (due to the higher K concentration in plant tissues) while P availability is related to the reduction in P-sorption, thanks to the higher SOM content in CA topsoil. The accumulation of nutrients near the soil surface could represent a critical issue in dry years when roots should develop deeper, far from where nutrients are concentrated; anyway, the moister surface of CA soil can compensate this aspect.

Similar results are found for exchangeable calcium (Ca), magnesium (Mg) and potassium (K) (Rahman and Hasan, 2008). Yet, most of these results have not been obtained in true CA conditions and do not consider the importance of mulching covers to maintain the soil surface layers moist, enabling uptake of surface near nutrients and the incorporation of nutrients like calcium by organic acids into the soil matrix.

Positive influence of the CA practices is found also for soil physical properties, especially in the upper few soil centimetres. Significant improvements under reduced or no mechanical disturbance are found for the following: (1) hydraulic conductivity, (2) aggregate stability, (3) soil porosity (Pagliai et al., 2004) and, consequently, (4) water conservation (Stagnari et al., 2014). With respect to the latter, higher water-holding capacity or moisture content in the topsoil (0–10 cm) of no-tilled agricultural soils can reach values of 25%, with an improved water use efficiency (Su et al., 2007). High moisture contents allow better resource utilization including a 20–50% reduction in irrigation water application (Gathala et al., 2013) as well as in energy input (Al-Kaisi et al., 2017). The enhancement in soil water availability is mainly related to both the increase of infiltration (Derpsch et al., 2014) and the reduction of evaporation rates (Stagnari et al., 2014).

Increase in water infiltration allows reducing surface run-off and erosion, so indirectly contributing to one of the major ecosystem services of CA. Indeed, the acceleration of soil erosion by agricultural practices represents one of the most important and perhaps the ultimate threat posed to soils (Govers et al., 2017). The amount of human-induced agricultural erosion has been estimated at 25 000–40 000 Tg/year for water erosion, ca. 5 000 Tg/year for tillage erosion and 2 000–3 000 Tg/year for wind erosion (Govers et al., 2014), a rate of global soil reservoir depletion which is 20 times higher than the soil formation rate (Govers et al., 2017). Moreover, soil erosion has important ecological effects on aquatic environments (Stagnari et al., 2016) since soil-eroded particles like clay, silt and organic matter transport significant amounts of mineral elements to surface waters: that is, N and P annual mobilization rates correspond to 23–42 and 14–26 Tg/year, respectively (Quinton et al., 2010) – that also reflect an estimated economic loss of ca. US$35 billion for N and ca. US$80 billion for P (calculated on the basis of their annual application rate and fertilizer prices at 2013, United States) (Govers et al., 2017). Soil erosion is also responsible for the transport of heavy metals, often contained in agrochemicals such as copper (Cu), zinc (Zn), lead (Pb), nickel (Ni), cadmium (Cd), chromium (Cr), fluorine (F), mercury (Hg) and S, into rivers and lakes, resulting in the pollution of drinking water resources (Stagnari et al., 2016).

Although the rate of erosion is affected by soil type, topography and weather conditions, no-till and surface residues retention reduce erosion by an order of magnitude, when compared to conventional systems, thanks to reducing the time the soil is bare and hence exposed to wind, rainfall and run-off (Prasuhn, 2012). As a consequence, CA practices contribute to the protection from both surface and ground water pollution, that is direct drilling decreased agrochemical run-off by 70–100% (Palm et al., 2014); this is also guaranteed by the higher activity of soil microorganisms, which allows faster agrochemical degradation (Busari et al., 2015), and earthworms (Capowiez et al., 2009) which can render elevated quantity of organic matter, thus binding agrochemicals and reducing release (Alletto et al., 2012). The effects of CA practices in reducing soil erosion have yet been studied at both field and water-shed/catchment scales (Meijer et al., 2013; Montgomery, 2007; Prasuhn, 2012). Several literature results show a strong reduction in soil erosion when adopting CA practices, in a range of conditions (see the review of Palm et al., 2014). In the United States, CA has demonstrated to decrease run-off to about 15–89%, also reducing suspended agrochemicals, nutrients and sediments (Lemke et al., 2011). Conservation tillage reduced water erosion rates with respect to conventional systems, with very low (<1 t/ha/year) water erosion rates on moderate slopes – comparable to those occurring under natural vegetation (Montgomery, 2007). In India, Bhatt and Khera (2006) reported that soil loss was 22% higher, on average, under conventional tillage than minimum-till. However, it is important to specify that when no-till (or minimum tillage) is practiced without soil mulch cover, the effects can be disastrous with rapid surface sealing resulting in increased run-off and accelerated soil erosion (Giller et al., 2015). CA practices also contribute to the control of wind erosion, thanks to residue retention that (1) reduces the shear stress to which soil particles are exposed and (2) helps to keep the surface soil layer moist, so increasing its shear strength (Govers et al., 2017).

Climate regulation refers to processes contributing to mitigate the emission of GHG into the atmosphere or other factors, such as albedo, that contribute to global climate warming (Palm et al., 2014). The potential of sustainable agricultural management practices towards offsetting GHG emissions is estimated at 0.3–1.17 Pg C/year (Neufeldt et al., 2015) and represents 2.7–10.4% of the global GHGs emissions (de Moraes Sá et al., 2017). It depends on the direction and magnitude of changes in soil C, nitrous oxide (N_2O) and methane (CH_4) emissions – all expressed as CO_2 equivalents[2] – and their improvement with respect to conventional agriculture (Palm et al., 2014).

Several literature data highlight the role of CA in enhancing soil C sequestration (see for example the reviews by Grace et al., 2012; Ogle et al., 2012; Corsi et al., 2012). Minimum or no mechanical soil disturbance increase soil C content, mainly in the near-surface layers (<10 cm), thereby abating CO_2 emissions or reducing losses compared to other tillage systems (Luo

[2] N_2O has a global warming potential 298 times that of CO_2 and remaining in the atmosphere for up to 114 years; CH_4 has a global warming potential 25 times that of CO_2 and a lifetime of 12 years.

et al., 2010). The greatest reduction in CO_2 emissions was observed when no-tillage was adopted in soils with high levels of SOM (Pezzuolo et al., 2017). Nevertheless, some critical aspects have recently emerged about the C sink capacity of the soils, under no-till, to offset emissions and mitigate climate change: (1) the capacity of soil to serve as C sink is finite (Powlson et al., 2016); (2) the effects of no-tillage on C sequestration are heavily affected by crop sequences or combinations (Pittelkow et al., 2015) as well as by the amount and composition of crop residues (Govaerts et al., 2009); (3) SOC stocks are pretty difficult to be determined exactly (also depending on soil texture, porosity, bulk density, climate and weather conditions), bearing some uncertainties in effectively assessing C sequestration (Adenle et al., 2015); (4) the results on C sequestration rates under no-till are highly variable (de Moraes Sá et al., 2017) and principally refer to temperate climates (Palm et al., 2014); (5) the long-term climate change mitigation is achieved with the increase of SOC in its recalcitrant forms while several studies indicate that the additional C accumulated under CA is concentrated in particulate organic fractions or other pools considered as 'labile' forms (Hok et al., 2018; Bhattacharyya et al., 2011, 2012; Tivet et al., 2013; O'Rouke et al., 2015).

Other CA practices, that is crop rotations and surface residue retention, could contribute to mitigate global warming. The effect of rotations is directly linked to the amounts of above and below ground biomass produced and retained in the system as well as to its quality (despite affecting only short-term soil C storage and dynamics; Gentile et al., 2011), which in turn is influenced by its C/N ratio. Anyway, there is no or little indication of the amount of residues needed to maintain or increase soil C; also, the single effects of crop diversification and residues retention on soil C stocks are less pronounced than the single effect of no-tillage (West and Post, 2002). It is important to note the synergetic role of all the CA components, so that further investigation could be addressed to the adoption of a 'systemic approach' for optimizing the system.

The rate of N_2O production and emission primarily depends on the mineral N source availability – the substrate for nitrification (oxidation of ammonium to nitrate, aerobic conditions) or denitrification (reduction of nitrate to N_2O and N_2, anaerobic conditions) processes on soil temperature, water content, SOC and structure – which is a function of bulk density – all influenced by tillage practices and residue inputs. To a lower extent, legume residues can lead to high N_2O-N losses (Palm et al., 2014). Besides, since CA generally contributes to reduce fertilizers applications as well as to control leaching and erosion processes, it can be stated that it has the potential to lower N_2O emissions (Kassam et al., 2009). Soil bulk density is generally higher under NT, leading to higher water-filled pore space with anaerobic conditions which could be reached early under NT (Palm et al., 2014). In summary, no-tillage, retention of residues and crop diversification clearly impact N_2O emission magnitude on atmosphere by affecting soil aeration and water content, altering the availability of NO_3^- in the soil and enhancing the labile C concentrations and increasing the general N availability in the soil. However, results regarding N_2O emissions reported in literature are sometimes inconsistent; some authors found an increase (Yao et al., 2009) others a reduction (Smith et al., 2012) or no differences (Pelster et al., 2011) in the evaluation of no-tillage versus conventional systems. A probable explication to these contradicting findings might be the strict adherence to all CA principles and an improper soil compaction and irrigation management.

No-tillage is also expected to reduce the rate of CH_4 consumption and the potential net emission of CH_4 due to increased bulk density and water-filled pore space; nevertheless, no significant effects on CH_4 oxidation rates have been detected (Smith et al., 2012; Palm et al., 2014). The biggest effect of CA on CH_4 emissions would probably be the incorporation of rice cropping areas into CA systems by changing from permanent flooding to irrigation methods, which could allow maintaining aerobic soil conditions.

4 Economic benefits

The analysis of the economic aspects of CA should separate financial benefits from crop yield responses. From a review of Europe-wide field experiments it emerged that an average yield reduction of 8.5% when ZT was considered (Van den Putte et al., 2010; Townsend et al., 2016); as an example, in Sweden, yield reduction was around 9.8% (Arvidsson et al., 2014). Studies applied to cotton production in the Southeastern United States generally suggest little difference in yield or net revenue in the early years of establishment, but higher net revenue for CA over the long term (Bryant, 1998). However, most studies are very unspecific and do not differentiate between reduced tillage, ZT and CA or do not respect correctly the CA definition. The more serious publications on yields and profit under CA systems suggest no significant differences or rather yield and profit increases under CA (Williams et al., 2018; Bashour et al., 2016; Thierfelder et al., 2015; Kassam et al., 2013).

Analysing costs, it emerges that abolition of tillage operations lowers costs for labour, fuel and machinery, especially if the machinery is used optimally (Siemans and Doster, 1992). Considering fertilization costs, in general during the transitional phase, there is a need to improve fertilizer management. Some circumstances required an additional application of specific nutrients as well as specialized equipment for suitable placement, which reflect in higher costs; however, these additional costs only occur on severely degraded soils and only during the first years of adoption (Visioli et al., 2016). Nevertheless, several field experiments highlight that in the long term, the accumulation of crop residue from different crops, has positive impacts on organic matter accumulation, moisture retention, nutrient losses and erosion (Mengel et al., 1992).

Consequently, both the biological activity of soils and the nutrient availability would improve releasing higher nutrient amount and at better timing thanks to the role of organic matter, favouring the ability of crops to uptake nutrients; it derives that the amount of fertilizers is reduced and so has the fertilization costs and they could be further lower with the use of a suitable machinery (i.e. localized application) allowing a proper fertilization timing and placement. Besides, during the first years there can be additional costs resulting from greater incidence of weeds and diseases (Melander et al., 2013). A no-tillage approach often results for a short period in an increasing germination of newly produced weed seeds, requiring higher herbicide application costs. Conversely, later on (after few years) the dormant seed bank in the soil is not further shifted to the soil surface by tillage and hence would not germinate. Consequently, as annual weeds are controlled, the overall weed issue decreases when fields are in regimes of CA and with effective weed control application (Fawcett, 1987; Lindwall and Sonntag, 2010).

Eventually, under no-till, greater amounts of fungicides could be needed, depending on the preceding crops in the rotation (Bürger et al., 2012). Indeed, when the principle of crop rotation is not applied and farmers rely on monocropping of wheat, barley or maize the crop residues accumulated on the soil surface can host higher potential inoculum of some fungi (Bockus and Shroyer, 1998; Pittelkow et al., 2015). It is the case of *Zymoseptoria tritici*: the above cited conditions increase the expression of the symptoms on wheat and potentially increase fungicide requirements (Soane et al., 2012), although it is strongly dependent on climatic regimes being favoured by wet conditions (Kema and Annone, 1991; Te Beest et al., 2009). With a proper crop rotation a consistent reduction of *Z. tritici* leaf symptoms was observed (Van den Putte et al., 2010; Page et al., 2013; Calzarano et al., 2018).

Some studies indicate that less insecticide is used, and hence lower costs are needed for CA than for conventional tillage (Bull, 1993; Lindwall and Sonntag, 2010). Comparisons on the seed cost in the context of CA are not being addressed as several field experiments apply a constant seeding rate for both systems. Some results from unpublished studies found no statistically significant differences between the seeding rates of the tillage practice used. However, as systems advance, scientific evidence shows that in many production systems seed quantities can be reduced significantly in CA systems, especially when applied together with precision agriculture techniques (Ngwira et al., 2012; Pisante et al., 2012).

For a long time, there was confusion created regarding the economics of CA, mainly because scientific papers and meta-analysis did not separate between reduced tillage, no-till and CA systems. The fact that CA is increasing worldwide at a rate of 10 million ha/year indicates that this system must be economically viable for farmers (Kassam et al., 2019). The majority of economic studies in recent years suggest that such practices are more profitable, especially over a longer period of time and some governments (United States) provide incentives to increase their adoption (Manning, 2018); unfortunately, in some countries (France, Italy) a low degree of acceptance of alternative tillage systems, like CA, is principally attributed to the inadequate technology transfer systems, the traditional mind-set of the farmers, the lack of technical preparation and access to suitable machinery and equipment (Angás et al., 2004; Cantero-Martìnez and Gabiña, 2004).

5 Future trends

Although the correct application of CA principles is capable of resulting in the improvement of several agronomic and economic aspects, as well as to the delivering of several fundamental ecosystem services, some issues remain yet to be solved and the demand for innovations persists.

On this basis, further research activities should look at the reduction of non-selective herbicides application, since often agrochemicals usage is considered, in some countries, a barrier to the diffusion of CA. Weed dynamics, seed bank and proper weed management should be particularly investigated in relation to different pedo-climatic and crop rotation conditions. Besides, there is a growing interest in several countries to study the interaction between CA and precision agriculture to understand how the tools of the latter could improve CA performance, in order to overcome emerging issues and to analyse any additive effects resulting from such approach. The quality and nutritional properties of crops grown under CA systems deserve further investigation too.

6 Conclusion

Nowadays it is well accepted and demonstrated, mostly in the pioneer countries, that CA adoption delivers several important environmental benefits; furthermore, due to the preservation of the soil fertility, crop yields are well sustained. However, such results derive from the correct and concomitant application of all three CA principles and despite the occurring of some critical aspects, for example weed management, the use of non-selective herbicides as well as the diffusion of some diseases, the newly available knowledge and technical improvements make it easier to adopt CA even in border line situations.

Besides, it emerges that, also from economic point of view, CA systems are profitable, especially over a longer period of time, provided there exists an adequate technical level of understanding by farmers and technicians. Nevertheless, in some important agricultural countries CA still remains confined to a very small surface and its scarce adoption is principally to be found in an inadequate participatory technology innovation system, mental barriers of the farmers and the lack of technical preparation.

7 Where to look for further information

- http://www.fao.org/conservation-agriculture/en/
- http://www.ecaf.org
- https://www.mdpi.com/journal/agriculture/special_issues/Sustainable_Crop_Production

8 References

Adenle, A. A., Azad, H. and Arbiol, J. 2015. Global assessment of technological innovation for climate change adaptation and mitigation in developing world. *Journal of Environmental Management* 161, 261–75. doi:10.1016/j.jenvman.2015.05.040.

Al-Kaisi, M. M., Lal, R., Olson, K. R. and Lowery, B. 2017. Fundamentals and functions of soil environment. In: Al-Kaisi, M. M. and Lowery, B. (Eds), *Soil Health and Intensification of Agroecosystems*. Academic Press, pp. 1–23. doi:10.1016/B978-0-12-805317-1.00017-8. ISBN:978-0-12-805317-1.

Alakukku, L., Weisskopf, P., Chamen, W. C. T., Tijink, F. G. J., Van der Linden, J. P., Pires, S., Sommer, C. and Spoor, G. 2003. Prevention strategies for field traffic-induced subsoil compaction: a review: Part 1. Machine/soil interactions. *Soil and Tillage Research* 73(1–2), 145–60. doi:10.1016/S0167-1987(03)00107-7.

Alakukku, L., Ristolainen, A. and Salo, T. 2009. Grain yield and nutrient balance of spring cereals in different tillage systems. In: *Proceedings of ISTRO 18th Triennial Conference Sustainable Agriculture*, 15–19 June 2009, Izmir, Turkey, CD-ROM. T6-005-1-T6-005-7.tr.

Alletto, L., Benoit, P., Bergheaud, V. and Coquet, Y. 2012. Variability of retention process of isoxaflutole and its diketonitrile metabolite in soil under conventional and conservation tillage. *Pest Management Science* 68(4), 610–7. doi:10.1002/ps.2304.

Álvaro-Fuentes, J., López, M. V., Cantero-Martínez, C. and Arrúe, J. L. 2008. Tillage effects on soil organic carbon fractions in Mediterranean dryland agroecosystems. *Soil Science Society of America Journal* 72(2), 541–7. doi:10.2136/sssaj2007.0164.

Angás, P., Cantero-Martínez, C., Karrou, M., Benbelkacem, A. and Avci, M. 2004. Crop production technologies in Mediterranean region. Mediterranean rainfed agriculture: strategies for sustainability. *Options Méditerranéennes* 60, 99–126.

Anken, T., Weisskopf, P., Zihlmann, U., Forrer, H., Jansa, J. and Perhacova, K. 2004. Long-term tillage system effects under moist cool conditions in Switzerland. *Soil and Tillage Research* 78(2), 171–83. doi:10.1016/j.still.2004.02.005.

Arvidsson, J., Etana, A. and Rydberg, T. 2014. Crop yield in Swedish experiments with shallow tillage and no-tillage 1983–2012. *European Journal of Agronomy* 52, 307–15. doi:10.1016/j.eja.2013.08.002.

Ashford, D. L. and Reeves, D. W. 2003. Use of a mechanical roller-crimper as an alternative kill method for cover crops. *American Journal of Alternative Agriculture* 18(1), 37–45. doi:10.1079/AJAA2003037.

Baker, C. J., Saxton, K. E. and Ritchie, W. R. 2002. *No-tillage Seeding: Science and Practice* (2nd edn.). CAB International, Oxford, UK.

Bashour, I., Al-Ouda, A., Kassam, A., Bachour, R., Jouni, K., Hansmann, B. and Estephan, C. 2016. An overview of Conservation Agriculture in the dry Mediterranean environments with a special focus on Syria and Lebanon. *AIMS Agriculture and Food* 1(1), 67–84, doi:10.3934/agrfood.2016.1.67.

Bhatt, R. and Khera, K. L. 2006. Effect of tillage and mode of straw mulch application on soil erosion in the submontaneous tract of Punjab, India. *Soil and Tillage Research* 88(1–2), 107–15. doi:10.1016/j.still.2005.05.004.

Bhattacharyya, R., Fullen, M. A., Booth, C. A., Kertesz, A., Toth, A., Szalai, Z., Jakab, G., Kozma, K., Jankauskas, B., Jankauskiene, G., et al. 2011. Effectiveness of biological geotextiles for soil and water conservation in different agro-environments. *Land Degradation and Development* 22(5), 495–504. doi:10.1002/ldr.1097.

Bhattacharyya, R., Tuti, M. D., Kundu, S., Bisht, J. K. and Bhatt, J. C. 2012. Conservation tillage impacts on soil aggregation and carbon pools in a sandy clay loam soil of the Indian Himalayas. *Soil Science Society of America Journal* 76(2), 617–27. doi:10.2136/sssaj2011.0320.

Bissett, A., Brown, M. V., Siciliano, S. D. and Thrall, P. H. 2013. Microbial community responses to anthropogenically induced environmental change: towards a systems approach. *Ecology Letters* 16, 128–39. doi:10.1111/ele.12109.

Blanco-Canqui, H. and Lal, R. 2009. Crop residue removal impacts on soil productivity and environmental quality. *Critical Reviews in Plant Sciences* 28(3), 139–63. doi:10.1080/07352680902776507.

Blanco-Canqui, H., Shaver, T. M., Lindquist, J. L., Shapiro, C. A., Elmore, R. W., Francis, C. A. and Hergert, G. W. 2015. Cover crops and ecosystem services: insights from studies in temperate soils. *Agronomy Journal* 107(6), 2449–74. doi:10.2134/agronj15.0086.

Blanco-Moure, N., Gracia, R., Bielsa, A. C. and López, M. V. 2016. Soil organic matter fractions as affected by tillage and soil texture under semiarid Mediterranean conditions. *Soil and Tillage Research* 155, 381–9. doi:10.1016/j.still.2015.08.011.

Bockus, W. W. and Shroyer, J. P. 1998. The impact of reduced tillage on soilborne plant pathogens. *Annual Review of Phytopathology* 36(1), 485–500. doi:10.1146/annurev.phyto.36.1.485.

Boddey, R. M., Alves, B. J. R., Soares, L. H. B., Jantalia, C. P. and Urquiaga, S. 2009. Biological nitrogen fixation and the mitigation of greenhouse gas emissions. In: Emerich, D. W. and Krishnan, H. B. (Eds), *Nitrogen Fixation in Crop Production*. Agronomy Monograph 52. Am. Soc. Agron., Crop Sci. Soc. Am., and Soil Sci. Soc Am., Madison, WI, pp. 387–413.

Børresen, T. 1999. The effect of straw management and reduced tillage on soil properties and crop yields of spring-sown cereals on two loam soils in Norway. *Soil and Tillage Research* 51(1–2), 91–102. doi:10.1016/S0167-1987(99)00030-6.

Bräutigam, V. and Tebrügge, F. 1997. Influence of long-termed no-tillage on soil borne pathogens and weeds. Experience with the applicability of no-tillage crop production in the West-European countries. In: *Proceedings of the EC-Workshop-III-Évora*, pp. 17–29.

Brennan, E. B. and Boyd, N. S. 2012. Winter cover crop seeding rate and variety affects during eight years of organic vegetables: I. Cover crop biomass production. *Agronomy Journal* 104(3), 684–98. doi:10.2134/agronj2011.0330.

Bryant, K. J. 1998. Economic results of conservation tillage in cotton. Cooperative Extension Service, University of Arkansas, US Department of Agriculture and county governments cooperating.

Bryant, L., Stockwell, R. and White, T. 2013. *Counting Cover Crops*. National Wildlife Federation, Washington DC.

Bull, L. 1993. *Residue and Tillage Systems for Field Crops*. AGES Publications 9310. Economic Research Service, United States Department of Agriculture, Washington DC.

Bünemann, E. K., Bongiorno, G., Bai, Z., Creamer, R. E., De Deyn, G., de Goede, R., Fleskens, L., Geissen, V., Kuyper, T. W., Mäder, P., et al. 2018. Soil quality – a critical review. *Soil Biology and Biochemistry* 120, 105–25. doi:10.1016/j.soilbio.2018.01.030.

Bürger, J., de Mol, F. and Gerowitt, B. 2012. Influence of cropping system factors on pesticide use intensity – a multivariate analysis of on-farm data in North East Germany. *European Journal of Agronomy* 40, 54–63. doi:10.1016/j.eja.2012.02.008.

Busari, M. A., Kukal, S. S., Kaur, A., Bhatt, R. and Dulazi, A. A. 2015. Conservation tillage impacts on soil, crop and the environment. *International Soil and Water Conservation Research* 3(2), 119–29. doi:10.1016/j.iswcr.2015.05.002.

Calzarano, F., Stagnari, F., D'Egidio, S., Pagnani, G., Galieni, A., Di Marco, S., Metruccio, E. and Pisante, M. 2018. Durum wheat quality, yield and sanitary status under Conservation Agriculture. *Agriculture* 8(9), 140. doi:10.3390/agriculture8090140.

Cantero-Martínez, C. and Gabiña, D. 2004. Evaluation of agricultural practices to improve efficiency and environment conservation in Mediterranean arid and semi-arid production systems. MEDRATE project. Mediterranean rainfed agriculture: strategies for sustainability. *Options Méditerranéennes* 60, 21–34.

Capowiez, Y., Cadoux, S., Bouchant, P., Ruy, S., Roger-Estrade, J., Richard, G. and Boizard, H. 2009. The effect of tillage type and cropping system on earthworm communities, macroporosity and water infiltration. *Soil and Tillage Research* 105(2), 209–16. doi:10.1016/j.still.2009.09.002.

Carvalho, M. and Basch, G. 1994. Experiences with direct drilling in Portugal. In: *Proceedings of the EC-Workshop I, Experience with the Applicability of No-tillage Crop Production in the West-European Countries, Wissenschaflticher Fachverlag*, Giessen, 27–28 June 1994, pp. 105–10.

Chivenge, P. P., Murwira, H. K., Giller, K. E., Mapfumo, P. and Six, J. 2007. Long-term impact of reduced tillage and residue management on soil carbon stabilization: Implications for conservation agriculture on contrasting soils. *Soil and Tillage Research* 94(2), 328–37.

Christian, D. G. and Ball, B. C. 1994. Reduced cultivation and direct drilling for cereals in Great Britain. In: Carter, M. R. (Ed.), *Conservation Tillage in Temperate Agroecosystems*. Lewis Publishers and CRC Press, Boca Raton, FL, pp. 117–39.

Claassen, R. 2012. The future of environmental compliance incentives in US agriculture: The role of commodity, conservation, and crop insurance programs. EIB-94. U.S. Department of Agriculture, Economic Research Service, March 2012.

Corsi, S., Friedrich, T., Kassam, A. Pisante, M. and Sà, J. M. 2012. *Soil Organic Carbon Accumulation and Green House Gas Emission Reductions from Conservation Agriculture: A Literature Review*. Integrated Crop Management (vol. 16). FAO, Rome.

Creamer, N. G. and Dabney, S. M. 2002. Killing cover crops mechanically: review of recent literature and assessment of new research results. *American Journal of Alternative Agriculture* 17(1), 32–40.

CTIC. 1994. Definition of conservation tillage. In: *Conservation Impact* (vol. 12). Conservation Technology Information Centre, West Lafayette, IN. Available at: https://www.ctic.org/.

D'Acunto, L., Andrade, J. F., Poggio, S. L. and Semmartin, M. 2018. Diversifying crop rotation increased metabolic soil diversity and activity of the microbial community. *Agriculture, Ecosystems and Environment* 257, 159–64. doi:10.1016/j.agee.2018.02.011.

Dabney, S. M., Delgado, J. A. and Reeves, D. W. 2001. Using winter cover crops to improve soil and water quality. *Communications in Soil Science and Plant Analysis* 32(7–8), 1221–50. doi:10.1081/CSS-100104110.

de Moraes Sá, J. C., Lal, R., Cerri, C. C., Lorenz, K., Hungria, M. and de Faccio Carvalho, P. C. 2017. Low-carbon agriculture in South America to mitigate global climate change and advance food security. *Environment International* 98, 102–12. doi:10.1016/j.envint.2016.10.020.

Derpsch, R. and Friedrich, T. 2009. Development and current status of no-till adoption in the world. In: *Proceedings on CD. 18th Triennial Conference of the International Soil Tillage Research Organization (ISTRO)*, 15–19 June 2009, Izmir, Turkey, 13pp.

Derpsch, R., Roth, C. H., Sidiras, N., Kopke, U., Krause, R. and Blanken, J. 1991. *Controle da erosão no Paraná, Brasil: Sistemas de cobertura do solo, plantio direto e preparo conservacionista do solo* (vol. 245). GTZ, Eschborn, p. 76.

Derpsch, R., Franzluebbers, A. J., Duiker, S. W., Reicosky, D. C., Koeller, K., Friedrich, T., Sturny, W. G., Sá, J. C. M. and Weiss, K. 2014. Why do we need to standardize no-tillage research? *Soil and Tillage Research* 137, 16–22. doi:10.1016/j.still.2013.10.002.

Doran, J. W. and Parkin, T. B. 1994. Defining and assessing soil quality. In: Doran, J. W., Coleman, D. C., Bezdicek, D. F. and Stewart, B. A. (Eds), *Defining Soil Quality for a Sustainable Environment*. SSSA, Madison, WI, pp. 3–21.

Doran, J. W. and Parkin, T. B. 1996. Quantitative indicators of soil quality: a minimum data set. In: Doran, J. W. and Jones, A. J. (Eds), *Methods for Assessing Soil Quality*. Special Publication 49. Soil Science Society of America, Madison, WI, pp. 25–37.

Ehlers, W. and Claupein, W. 1994. Approaches towards conservation tillage in Germany. In: Carter, M. R. (Ed.), *Conservation Tillage in Temperate Agroecosystems*. Lewis Publishers and CRC Press, Boca Raton, FL, 141–65.

FAO. 2008. *An International Technical Workshop Investing in Sustainable Crop Intensification: The Case for Improving Soil Health*, 22–24 July 2008. Integrated Crop Management (vol. 6). Food and Agriculture Organization, Rome.

Fawcett, R. S. 1987. Overview of pest management for conservation tillage systems. In: Logan, T. J., Davidson, J. M., Baker, J. L. and Overcash, M. R. (Eds), *Effects of Conservation Tillage on Ground Water Quality*. Lewis Publishers, Inc., Chelsea, MI, pp. 19–37.

Florentin, M. A., Peñalva, M., Calegari, A. and Derpsch, R. 2011. *Green Manure Cover Crops and Crop Rotation in Conservation Agriculture on Small Farms*. Integrated Crop Management (vol. 12). Food and Agriculture Organization, Plant Production and Protection Division, Rome.

Friedrich, T. 2005. Does no-till farming require more herbicides? *Outlooks on Pest Management* 16(4), 188–91. doi:10.1564/16aug12.

Fuglie, K. O. and Kascak, C. A. 2001. Adoption and diffusion of natural-resource-conserving agricultural technology. *Review of Agricultural Economics* 23(2), 386-403. doi:10.1111/1467-9353.00068.

Galieni, A., Stagnari, F., Visioli, G., Marmiroli, N., Speca, S., Angelozzi, G., D'Egidio, S. and Pisante, M. 2016. Nitrogen fertilisation of durum wheat: A case of study in Mediterranean area during transition to conservation agriculture. *Italian Journal of Agronomy* 11(1), 12–23.

Galieni, A., Stagnari, F., Speca, S., D'Egidio, S., Pagnani, G. and Pisante, M. 2017. Management of crop residues to improve quality traits of tomato (*Solanum lycopersicum* L.) fruits. *Italian Journal of Agronomy* 12(1), 59–62. doi:10.4081/ija.2017.759.

Gathala, M. K., Kumar, V., Sharma, P. C., Saharawat, Y. S., Jat, H. S., Singh, M., Kumar, A., Jat, M. L., Humphreys, E., Sharma, D. K., et al. 2013. Optimizing intensive cereal-based cropping systems addressing current and future drivers of agricultural change in the northwestern Indo-Gangetic Plains of India. *Agriculture, Ecosystems and Environment* 177, 85–97. doi:10.1016/j.agee.2013.06.002.

Gentile, R., Vanlauwe, B., Kavoo, A., Chivenge, P. and Six, J. 2011. Residue quality and N fertilizer do not influence aggregate stabilization of C and N in two tropical soils with contrasting texture. In: Bationo, A., Waswa, B., Okeyo, J. M., Maina, F. and Kihara, J. M. (Eds), *Innovations as Key to the Green Revolution in Africa*. Springer, Dordrecht, pp. 795–805.

Giller, K. E., Andersson, J. A., Corbeels, M., Kirkegaard, J., Mortensen, D., Erenstein, O. and Vanlauwe, B. 2015. Beyond conservation agriculture. *Frontiers in Plant Science* 6, 870. doi:10.3389/fpls.2015.00870.

Gómez, J. A., Álvarez, S. and Soriano, M. A. 2009. Development of a soil degradation assessment tool for organic olive groves in southern Spain. *CATENA* 79(1), 9–17. doi:10.1016/j.catena.2009.05.002.

Gómez, J. A., Llewellyn, C., Basch, G., Sutton, P. B., Dyson, J. S. and Jones, C. A. 2011. The effects of cover crops and conventional tillage on soil and runoff loss in vineyards and olive groves in several Mediterranean countries. *Soil Use and Management* 27(4), 502–14. doi:10.1111/j.1475-2743.2011.00367.x.

González-Chávez, MdC. A., Aitkenhead-Peterson, J. A., Gentry, T. J., Zuberer, D., Hons, F. and Loeppert, R. 2010. Soil microbial community, C, N, and P responses to long-term tillage and crop rotation. *Soil and Tillage Research* 106(2), 285–93. doi:10.1016/j.still.2009.11.008.

Gonzalez-Sanchez, E. J., Veroz-Gonzalez, O., Blanco-Roldan, G. L., Marquez-Garcia, F. and Carbonell-Bojollo, R. 2015. A renewed view of conservation agriculture and its evolution over the last decade in Spain. *Soil and Tillage Research* 146, 204–12. doi:10.1016/j.still.2014.10.016.

Govaerts, B., Fuentes, M., Mezzalama, M., Nicol, J. M., Deckers, J., Etchevers-Barra, J. D., Figueroa-Sandoval, B. and Sayre, K. D. 2007. Infiltration, soil moisture, root rot and nematode populations after 12 years of different tillage, residue and crop rotation managements. *Soil and Tillage Research* 94(1), 209–19. doi:10.1016/j.still.2006.07.013.

Govaerts, B., Verhulst, N., Castellanos-Navarrete, A., Sayre, K. D., Dixon, J. and Dendooven, L. 2009. Conservation Agriculture and soil carbon sequestration: between myth and farmer reality. *Critical Reviews in Plant Sciences* 28(3), 97–122. doi:10.1080/07352680902776358.

Govers, G., Van Oost, K. and Wang, Z. 2014. Scratching the critical zone: the global footprint of agricultural soil erosion. *Procedia Earth and Planetary Science* 10, 313–8. doi:10.1016/j.proeps.2014.08.023.

Govers, G., Merckx, R., van Wesemael, B. and Van Oost, K. 2017. Soil conservation in the 21st century: why we need smart agricultural intensification. *Soil* 3(1), 45–59. doi:10.5194/soil-3-45-2017.

Grace, P. R., Antle, J., Aggarwal, P. K., Ogle, S., Paustian, K. and Basso, B. 2012. Soil carbon sequestration and associated economic costs for farming systems of the Indo-Gangetic Plain: a meta-analysis. *Agriculture, Ecosystems and Environment* 146(1), 137–46. doi:10.1016/j.agee.2011.10.019.

Herrero, M., Thornton, P. K., Notenbaert, A. M., Wood, S., Msangi, S., Freeman, H. A., Bossio, D., Dixon, J., Peters, M., van de Steeg, J., et al. 2010. Smart investments in sustainable food production: revisiting mixed crop-livestock systems. *Science* 327(5967), 822–5. doi:10.1126/science.1183725.

Hobbs, P. R., Sayre, K. and Gupta, R. 2008. The role of conservation agriculture in sustainable agriculture. *Philosophical Transactions of the Royal Society of London. Series B, Biological Sciences* 363(1491), 543–55. doi:10.1098/rstb.2007.2169.

Hok, L., de Moraes Sá, J. C., Reyes, M., Boulakia, S., Tivet, F., Leng, V., Kong, R., Briedis, C., da Cruz Hartman, D., Ferreira, L. A., et al. 2018. Enzymes and C pools as indicators of C build up in short-term conservation agriculture in a savanna ecosystem in Cambodia. *Soil and Tillage Research* 177, 125–33. doi:10.1016/j.still.2017.11.015.

Jordan, D., Stecker, J. A., Cacnio-Hubbard, V. N., Li, F., Gantzer, C. J. and Brown, J. R. 1997. Earthworm activity in no-tillage and conventional tillage systems in Missouri soils: a preliminary study. *Soil Biology and Biochemistry* 29(3-4), 489–91. doi:10.1016/S0038-0717(96)00038-7.

Kassam, A., Friedrich, T., Shaxson, F. and Pretty, J. N. 2009. The spread of Conservation Agriculture: justification, sustainability and uptake. *International Journal of Agricultural Sustainability* 7(4), 292–320. doi:10.3763/ijas.2009.0477.

Kassam, A., Friedrich, T., Derpsch, R., Lahmar, R., Mrabet, R., Basch, G., González-Sánchez, E. J. and Serraj, R. 2012. Conservation agriculture in the dry Mediterranean climate. *Field Crops Research* 132, 7–17. doi:10.1016/j.fcr.2012.02.023.

Kassam, A., Basch, G., Friedrich, T., Shaxson, F., Goddard, T., Amado, T., Crabtree, B., Hongwen, L., Mello, I., Pisante, M., et al. 2013. Sustainable soil management is more than what and how crops are grown. In: Lal, R. and Stewart, R. A. (Eds), *Principles of Soil Management in Agro-ecosystems. 2013. Advances in Soil Science*. CRC Press.

Kassam, A., Mkomwa, S. and Friedrich, T. (Eds). 2017. *Conservation Agriculture for Africa: Building Resilient Farming Systems in a Changing Climate*. CABI, Walingford, UK.

Kassam, A., Friedrich, T. and Derpsch, R. 2019. Global spread of conservation agriculture. *International Journal of Environmental Studies* 76(1), 29–51. doi:10.1080/00207233.2018.1494927.

Kema, G. H. J. and Annone, J. G. 1991. *In vitro* production of pycnidia by *Septoria tritici*. *Netherlands Journal of Plant Pathology* 97(2), 65–72. doi:10.1007/BF01974270.

Khan, Z. R., Pickett, J. A., Wadhams, L. J., Hassanalia, A. and Midega, C. A. O. 2006. Combined control of *Striga hermonthica* and stemborers by maize-*Desmodium* spp. intercrops. *Crop Protection* 25(9), 989–95. doi:10.1016/j.cropro.2006.01.008.

Knowler, D. and Bradshaw, B. 2007. Farmers' adoption of conservation agriculture: a review and synthesis of recent research. *Food Policy* 32(1), 25–48. doi:10.1016/j.foodpol.2006.01.003.

Kornecki, T. S., Price, A. J., Raper, R. L. and Arriaga, F. J. 2009. New roller crimper concepts for mechanical termination of cover crops in conservation agriculture. *Renewable Agriculture and Food Systems* 24(3), 165–73. doi:10.1017/S1742170509002580.

Kuo, S. and Sainju, U. M. 1998. Nitrogen mineralization and availability of mixed leguminous and non-leguminous cover crop residues in soil. *Biology and Fertility of Soils* 26(4), 346–53. doi:10.1007/s003740050387.

Lal, R. 2015. Cover cropping and the '4 per Thousand' proposal. *Journal of Soil and Water Conservation* 70(6), 141A–. doi:10.2489/jswc.70.6.141A.

Lemke, A. M., Kirkham, K. G., Lindenbaum, T. T., Herbert, M. E., Tear, T. H., Perry, W. L. and Herkert, J. R. 2011. Evaluating agricultural best management practices in tile-drained subwatersheds of the Mackinaw River, Illinois. *Journal of Environmental Quality* 40(4), 1215–28. doi:10.2134/jeq2010.0119.

Li, R., Hou, X. Q., Jia, Z. K., Han, Q. F. and Yang, B. P. 2012. Effects of rainfall harvesting and mulching technologies on soil water, temperature, and maize yield in Loess Plateau region of China. *Soil Research* 50(2), 105–13. doi:10.1071/SR11331.

Lin, B. B. 2011. Resilience in agriculture through crop diversification: adaptive management for environmental change. *BioScience* 61(3), 183–93.

Lindwall, W. and Sonntag, B. (Eds). 2010. *Landscapes Transformed: the History of Conservation Tillage and Direct Seeding; the Knowledge Impact in Society*. University of Saskatchewan, Saskatoon, Canada.

Luo, Z., Wang, E. and Sun, O. J. 2010. Can no-tillage stimulate carbon sequestration in agricultural soils? A meta-analysis of paired experiments. *Agriculture, Ecosystems and Environment* 139(1-2), 224–31. doi:10.1016/j.agee.2010.08.006.

Manning, L. 2018. BREAKING: conservation agriculture creates a 'cascade of cost savings' for farmers - EDF report. Available at: https://agfundernews.com/(Accessed on 14 September 2018).

Martín-Lammerding, D., Tenorio, J. L., Albarrán, M. M., Zambrana, E. and Walter, I. 2013. Influence of tillage practices on soil biologically active organic matter content over a growing season under semiarid Mediterranean climate. *Spanish Journal of Agricultural Research* 11(1), 232–43. doi:10.5424/sjar/2013111-3455.

Martín-Lammerding, D., Navas, M., del Mar Albarrán, M. d. M., Tenorio, J. L. and Walter, I. 2015. Long term management systems under semiarid conditions: influence on labile organic matter, β-glucosidase activity and microbial efficiency. *Applied Soil Ecology* 96, 296–305. doi:10.1016/j.apsoil.2015.08.021.

Mazzoncini, M., Antichi, D., Di Bene, C., Risaliti, R., Petri, M. and Bonari, E. 2016. Soil carbon and nitrogen changes after 28 years of no-tillage management under Mediterranean conditions. *European Journal of Agronomy* 77, 156–65. doi:10.1016/j.eja.2016.02.011.

Meijer, A. D., Heitman, J. L., White, J. G. and Austin, R. E. 2013. Measuring erosion in long-term tillage plots using ground-based lidar. *Soil and Tillage Research* 126, 1–10.

Melander, B., Munier-Jolain, N., Charles, R., Wirth, J., Schwarz, J., van der Weide, R., Bonin, L., Jensen, P. K. and Kudsk, P. 2013. European perspectives on the adoption of nonchemical weed management in reduced-tillage systems for arable crops. *Weed Technology* 27(1), 231–40. doi:10.1614/WT-D-12-00066.1.

Melero, S., López-Bellido, R. J., López-Bellido, L., Muñoz-Romero, V., Moreno, F. and Murillo, J. M. 2011. Long-term effect of tillage, rotation and nitrogen fertiliser on soil quality in a Mediterranean vertisol. *Soil and Tillage Research* 114(2), 97–107. doi:10.1016/j.still.2011.04.007.

Mengel, D. B., Moncrief, J. R. and Schulte, E. E. 1992. Fertilizer management. Conservation tillage systems and management. Midwest Plan Serv. Pub. MWPS-45. Iowa State University, Ames, IA, pp. 83–7.

Mikkola, H. J., Alakukku, L., Känkänen, H., Jalli, H., Lindroos, M., Huusela-Veistola, E., Nuutinen, V., Lätti, M., Puustinen, M., Turtola, E., et al. 2005. Direct drilling in Finland, a review. In: *Proceedings of the 4th International Scientific and Practical Conference, Ecology and Agricultural Machinery* (vol. 2), St. Petersburg, 25–26 May 2005, pp. 141–51.

Montgomery, D. R. 2007. Soil erosion and agricultural sustainability. *Proceedings of the National Academy of Sciences of the United States of America* 104(33), 13268–72. doi:10.1073/pnas.0611508104.

Morris, N. L., Miller, P. C. H., Orson, J. H. and Froud-Williams, R. J. 2010. The adoption of non-inversion tillage systems in the United Kingdom and the agronomic impact on soil, crops and the environment – a review. *Soil and Tillage Research* 108(1–2), 1–15. doi:10.1016/j.still.2010.03.004.

Munkholm, L. J., Schjønning, P., Rasmussen, K. J. and Tanderup, K. 2003. Spatial and temporal effects of direct drilling on soil structure in the seedling environment. *Soil and Tillage Research* 71(2), 163–73. doi:10.1016/S0167-1987(03)00062-X.

Neufeldt, H., Kissinger, G. and Alcamo, J. 2015. No-till agriculture and climate change mitigation. *Nature Climate Change* 5(6), 488–9. doi:10.1038/nclimate2653.

Ngwira, A. R., Aune, J. B. and Mkwinda, S. 2012. On-farm evaluation of yield and economic benefit of short term maize legume intercropping systems under conservation agriculture in Malawi. *Field Crops Research* 132, 149–57. doi:10.1016/j.fcr.2011.12.014.

Nichols, V., Verhulst, N., Cox, R. and Govaerts, B. 2015. Weed dynamics and conservation agriculture principles: a review. *Field Crops Research* 183, 56–68. doi:10.1016/j.fcr.2015.07.012.

Nieminen, M., Ketoja, E., Mikola, J., Terhivuo, J., Sirén, T. and Nuutinen, V. 2011. Local land use effects and regional environmental limits on earthworm communities in Finnish arable landscapes. *Ecological Applications* 21(8), 3162–77. doi:10.1890/10-1801.1.

Nortcliff, S. 2002. Standardisation of soil quality attributes. *Agriculture, Ecosystems and Environment* 88(2), 161–8. doi:10.1016/S0167-8809(01)00253-5.

O'Rourke, S. M., Angers, D. A., Holden, N. M. and McBratney, A. B. 2015. Soil organic carbon across scales. *Global Change Biology* 21(10), 3561–74. doi:10.1111/gcb.12959.

Ogle, S. M., Swan, A. and Paustian, K. 2012. No-till management impacts on crop productivity, carbon input and soil carbon sequestration. *Agriculture, Ecosystems and Environment* 149, 37–49. doi:10.1016/j.agee.2011.12.010.

Othman, H., Darus, F. M. and Hashim, Z. 2012. Best management practices for oil palm cultivation on peat: Mucuna bracteata as ground cover crop. *Malaysian Palm Oil Board* 501, 1–4.

Page, K., Dang, Y. and Dalal, R. 2013. Impacts of conservation tillage on soil quality, including soil-borne crop diseases, with a focus on semi-arid grain cropping systems. *Australasian Plant Pathology* 42(3), 363–77. doi:10.1007/s13313-013-0198-y.

Pagliai, M., Vignozzi, N. and Pellegrini, S. 2004. Soil structure and the effect of management practices. *Soil and Tillage Research* 79(2), 131–43. doi:10.1016/j.still.2004.07.002.

Palm, C., Blanco-Canqui, H., DeClerck, F., Gatere, L. and Grace, P. 2014. Conservation agriculture and ecosystem services: an overview. *Agriculture, Ecosystems and Environment* 187, 87–105. doi:10.1016/j.agee.2013.10.010.

Pelosi, C., Grandeau, G. and Capowiez, Y. 2017. Temporal dynamics of earthworm-related macroporosity in tilled and non-tilled cropping systems. *Geoderma* 289, 169–77. doi:10.1016/j.geoderma.2016.12.005.

Pelster, D. E., Larouche, F., Rochette, P., Chantigny, M. H., Allaire, S. and Angers, D. A. 2011. Nitrogen fertilization but not soil tillage affects nitrous oxide emissions from a clay loam soil under a maize–soybean rotation. *Soil and Tillage Research* 115–116, 16–26. doi:10.1016/j.still.2011.06.001.

Pezzuolo, A., Dumont, B., Sartori, L., Marinello, F., Migliorati, M. D. A. And Basso, B. 2017. Evaluating the impact of soil conservation measures on soil organic carbon at the farm scale. *Computers and Electronics in Agriculture* 135, 175–82. doi:10.1016/j.compag.2017.02.004.

Piggin, C., Haddad, A., Khalil, Y., Loss, S. and Pala, M. 2015. Effects of tillage and time of sowing on bread wheat, chickpea, barley and lentil grown in rotation in rainfed systems in Syria. *Field Crops Research* 173, 57–67. doi:10.1016/j.fcr.2014.12.014.

Pisante, M., Stagnari, F. and Grant, C. A. 2012. Agricultural innovations for sustainable crop production intensification. *Italian Journal of Agronomy* 7(4), 300–11. doi:10.4081/ija.2012.e40.

Pittelkow, C. M., Linquist, B. A., Lundy, M. E., Liang, X., Van Groenigen, K. J., Lee, J., van Gestel, N., Six, J., Venterea, R. T. and Van Kessel, C. 2015. When does no-till yield more? A global meta-analysis. *Field Crops Research* 183, 156–68. doi:10.1016/j.fcr.2015.07.020.

Powlson, D. S., Stirling, C. M., Thierfelder, C., White, R. P. and Jat, M. L. 2016. Does conservation agriculture deliver climate change mitigation through soil carbon sequestration in tropical agro-ecosystems? *Agriculture, Ecosystems and Environment* 220, 164–74. doi:10.1016/j.agee.2016.01.005.

Prasuhn, V. 2012. On-farm effects of tillage and crops on soil erosion measured over 10 years in Switzerland. *Soil and Tillage Research* 120, 137–46. doi:10.1016/j.still.2012.01.002.

Quinton, J. N., Govers, G., Van Oost, K. and Bardgett, R. D. 2010. The impact of agricultural soil erosion on biogeochemical cycling. *Nature Geoscience* 3(5), 311–4. doi:10.1038/ngeo838.

Radicetti, E., Mancinelli, R., Moscetti, R. and Campiglia, E. 2016. Management of winter cover crop residues under different tillage conditions affects nitrogen utilization efficiency and yield of eggplant (*Solanum melanogena* L.) in Mediterranean environment. *Soil and Tillage Research* 155, 329–38. doi:10.1016/j.still.2015.09.004.

Rahman, S. and Hasan, M. K. 2008. Impact of environmental production conditions on productivity and efficiency: a case study of wheat farmers in Bangladesh. *Journal of Environmental Management* 88(4), 1495–504. doi:10.1016/j.jenvman.2007.07.019.

Ranaivoson, L., Naudin, K., Ripoche, A., Affholder, F., Rabeharisoa, L. and Corbeels, M. 2017. Agro-ecological functions of crop residues under conservation agriculture. A review. *Agronomy for Sustainable Development* 37(4), 26p.

Reicosky, D. C. 2015. Conservation tillage is not conservation agriculture. *Journal of Soil and Water Conservation* 70(5), 103A–8A. doi:10.2489/jswc.70.5.103A.

Riley, H., Borresen, T., Ekeberg, E. and Rydberg, T. 1994. Trends in reduced tillage research and practice in Scandinavia. In: Carter, M. R. (Ed.), *Conservation Tillage in Temperate Agroecosystems*. Lewis Publishers, CRC Press, Boca Raton, FL, 23–45.

Robačer, M., Canali, S., Kristensen, H. L., Bavec, F., Mlakar, S. G., Jakop, M. and Bavec, M. 2016. Cover crops in organic field vegetable production. *Scientia Horticulturae* 208, 104–10. doi:10.1016/j.scienta.2015.12.029.

Sanderson, M. A., Archer, D., Hendrickson, J., Kronberg, S., Liebig, M., Nichols, K., Schmer, M., Tanaka, D. and Aguilar, J. 2013. Diversification and ecosystem services for conservation agriculture: outcomes from pastures and integrated crop-livestock systems. *Renewable Agriculture and Food Systems* 28(2), 129–44. doi:10.1017/S1742170512000312.

Sarker, J. R., Singh, B. P., Cowie, A. L., Fang, Y., Collins, D., Badgery, W. and Dalal, R. C. 2018. Agricultural management practices impacted carbon and nutrient concentrations in soil aggregates, with minimal influence on aggregate stability and total carbon and nutrient stocks in contrasting soils. *Soil and Tillage Research* 178, 209–23. doi:10.1016/j.still.2017.12.019.

Schipanski, M. E., Barbercheck, M., Douglas, M. R., Finney, D. M., Haider, K., Kaye, J. P., Kemanian, A. R., Mortensen, D. A., Ryan, M. R., Tooker, J., et al. 2014. A framework for evaluating ecosystem services provided by cover crops in agroecosystems. *Agricultural Systems* 125, 12–22. doi:10.1016/j.agsy.2013.11.004.

Schwilch, G., Bernet, L., Fleskens, L., Giannakis, E., Leventon, J., Maranon, T., Mills, J., Short, C., Stolte, J., van Delden, H., et al. 2016. Operationalizing ecosystem services for the mitigation of soil threats: a proposed framework. *Ecological Indicators* 67, 586–97. doi:10.1016/j.ecolind.2016.03.016.

Seguy, L., Bouzinac, S., Trentini, A. and Cortes, N. A. 1996. L'agriculture bresilienne des fronts pionniers. *Agricola Developpement* 12, 4–61.

Sharratt, B. S. and Campbell, G. S. 1994. Radiation balance of a soil-straw surface modified by straw color. *Agronomy Journal* 86(1), 200–3. doi:10.2134/agronj1994.00021962008600010035x.

Siddique, K. H. M., Johansen, C., Turner, N. C., Jeuffroy, M. H., Hashem, A., Sakar, D., Gan, Y. and Alghamdi, S. S. 2012. Innovations in agronomy for food legumes. A review. *Agronomy for Sustainable Development* 32(1), 45–64. doi:10.1007/s13593-011-0021-5.

Siemans, J. and Doster, D. 1992. Costs and returns. In: *Conservation Tillage Systems and Management: Crop Residue Management with No-Till, Ridge-Till, Mulch-Till*. Midwest Plan Service and Iowa State University, Ames, IA, pp. 30–52.

Smith, K. A., Mosier, A. R., Crutzen, P. J. and Winiwarter, W. 2012. The role of N_2O derived from crop-based biofuels, and from agriculture in general, in Earth's climate. *Philosophical Transactions of the Royal Society of London. Series B, Biological Sciences* 367(1593), 1169–74. doi:10.1098/rstb.2011.0313.

Soane, B. D., Ball, B. C., Arvidsson, J., Basch, G., Moreno, F. and Roger-Estrade, J. 2012. No-till in northern, western and south-Western Europe: a review of problems and opportunities for crop production and the environment. *Soil and Tillage Research* 118, 66–87. doi:10.1016/j.still.2011.10.015.

SSSA. 2008. *Glossary of Soil Science Terms*. SSSA, Madison, WI. doi:10.2136/2008.glossarysoilscienceterms.

Stagnari, F., Galieni, A., Speca, S., Cafiero, G. and Pisante, M. 2014. Effects of straw mulch on growth and yield of durum wheat during transition to Conservation Agriculture in Mediterranean environment. *Field Crops Research* 167, 51–63. doi:10.1016/j.fcr.2014.07.008.

Stagnari, F., Sumira, J., Galieni, A. and Pisante, M. 2016. Sustainable agricultural practices for water quality protection. *Water Stress and Crop Plants: a Sustainable Approach* 1, 75–85.

Stagnari, F., Maggio, A., Galieni, A.. and Pisante, M. 2017. Multiple benefits of legumes for agriculture sustainability: an overview. *Chemical and Biological Technologies in Agriculture* 4(1). doi:10.1186/s40538-016-0085-1.

Sturm, D. J., Peteinatos, G. and Gerhards, R. 2018. Contribution of allelopathic effects to the overall weed suppression by different cover crops. *Weed Research* 58(5), 331-7. doi:10.1111/wre.12316.

Su, Z., Zhang, J., Wu, W., Cai, D., Lv, J., Jiang, G., Huang J., Gao J., Hartmann R. and Gabriels, D. 2007. Effects of conservation tillage practices on winter wheat water-use efficiency and crop yield on the Loess Plateau, China. *Agricultural Water Management* 87(3), 307–14. doi:10.1016/j.agwat.2006.08.005.

Te Beest, D. E., Shaw, M. W., Pietravalle, S. and Van Den Bosch, F. 2009. A predictive model for early-warning of Septoria leaf blotch on winter wheat. *European Journal of Plant Pathology* 124(3), 413–25. doi:10.1007/s10658-009-9428-0.

Teasdale, J. R., Abdul-Baki, A. A. and Park, Y. B. 2008. Sweet corn production and efficiency of nitrogen use in high cover crop residue. *Agronomy for Sustainable Development* 28(4), 559–65. doi:10.1051/agro:2008029.

Thierfelder, C. and Wall, P. C. 2010. Rotation in conservation agriculture systems of Zambia: effects on soil quality and water relations. *Experimental Agriculture* 46(3), 309–25. doi:10.1017/S001447971000030X.

Thierfelder, C., Bunderson, W. T. and Mupangwa, W. 2015. Evidence and lessons learned from long-term on-farm research on conservation agriculture systems in communities in Malawi and Zimbabwe. *Environments* 2(4), 317–37. doi:10.3390/environments2030317.

Thorup-Kristensen, K., Dresbøll, D. B. and Kristensen, H. L. 2012. Crop yield, root growth, and nutrient dynamics in a conventional and three organic cropping systems with different levels of external inputs and N re-cycling through fertility building crops. *European Journal of Agronomy* 37(1), 66–82. doi:10.1016/j.eja.2011.11.004.

Tiemann, L. K., Grandy, A. S., Atkinson, E. E., Mari-Spiotta, E. and McDaniel, M. D. 2015. Crop rotational diversity enhances belowground communities and functions in an agroecosystem. *Ecology Letters* 18(8), 761–71. doi:10.1111/ele.12453.

Tivet, F., de Moraes Sa, J. C., Lal, R., Milori, D., Briedis, C., Letourmy, P., Pinheiro, L., Borszowskei, P.R. and Hatman, D. 2013. Assessing humification and organic C compounds by laser-induced fluorescence and FTIR spectroscopies under conventional and no-till management in Brazilian Oxisols. *Geoderma* 207-208, 71–81.

Townsend, T. J., Ramsden, S. J. and Wilson, P. 2016. Analysing reduced tillage practices within a bio-economic modelling framework. *Agricultural Systems* 146, 91–102. doi:10.1016/j.agsy.2016.04.005.

Valbuena, D., Erenstein, O., Homann-Kee Tui, S., Abdoulaye, T., Claessens, L., Duncan, A. J., Gérard, B., Rufino, M. C., Teufel, N., van Rooyen, A., et al. 2012. Conservation Agriculture in mixed crop-livestock systems: scoping crop residue trade-offs in Sub-Saharan Africa and South Asia. *Field Crops Research* 132, 175–84. doi:10.1016/j.fcr.2012.02.022.

Van Capelle, C., Schrader, S. and Brunotte, J. 2012. Tillage-induced changes in the functional diversity of soil biota–a review with a focus on German data. *European Journal of Soil Biology* 50, 165–81. doi:10.1016/j.ejsobi.2012.02.005.

Van den Putte, A., Govers, G., Diels, J., Gillijns, K. and Demuzere, M. 2010. Assessing the effect of soil tillage on crop growth: a meta-regression analysis on European crop yields under conservation agriculture. *European Journal of Agronomy* 33(3), 231–41. doi:10.1016/j.eja.2010.05.008.

Venter, Z. S., Jacobs, K. and Hawkins, H. J. 2016. The impact of crop rotation on soil microbial diversity: A meta-analysis. *Pedobiologia* 59(4), 215–23. doi:10.1016/j.pedobi.2016.04.001.

Verhulst, N., Govaerts, B., Verachtert, E., Castellanos-Navarrete, A., Mezzalama, M., Wall, P., Chocobar, A., Deckers, J. and Sayre, K. D. 2010. Conservation agriculture, improving soil quality for sustainable production systems. In: Lal, R. and Stewart, B. A. (Eds), *Advances in Soil Science: Food Security and Soil Quality*. CRC Press, Boca Raton, FL, pp. 137–208.

Visioli, G., Galieni, A., Stagnari, F., Bonas, U., Speca, S., Faccini, A., Pisante, M.. and Marmiroli, N. 2016. Proteomics of durum wheat grain during transition to conservation agriculture. *PLOS ONE* 11(6), e0156007. doi:10.1371/journal.pone.0156007.

Vogeler, I., Rogasik, J., Funder, U., Panten, K. and Schnug, E. 2009. Effect of tillage systems and P-fertilization on soil physical and chemical properties, crop yield and nutrient uptake. *Soil and Tillage Research* 103(1), 137–43. doi:10.1016/j.still.2008.10.004.

Wang, K. H., McSorley, R. and Gallaher, R. N. 2003. Effect of *Crotalaria juncea* amendment on nematode communities in soil with different agricultural histories. *Journal of Nematology* 35(3), 294–301.

West, T. O. and Post, W. M. 2002. Soil organic carbon sequestration rates by tillage and crop rotation. *Soil Science Society of America Journal* 66(6), 1930–46. doi:10.2136/sssaj2002.1930.

Wezel, A. and Rath, T. 2002. Resource conservation strategies in agro-ecosystems of semi-arid West Africa. *Journal of Arid Environments* 51(3), 383–400. doi:10.1006/jare.2001.0968.

Williams, A., Jordan, N. R., Smith, R. G., Hunter, M. C., Kammerer, M., Kane, D. A., Koide, R. T. and Davis, A. S. 2018. A regionally adapted implementation of conservation agriculture delivers rapid improvements to soil properties associated with crop yield stability. *Scientific Reports* 8(1), 8467. doi:10.1038/s41598-018-26896-2.

Woyessa, Y. E. and Bennie, A. T. P. 2004. Factors affecting runoff and soil loss under simulated rainfall on a sandy Bainsvlei Amalia soil. *South African Journal of Plant and Soil* 21(4), 203–8. doi:10.1080/02571862.2004.10635050.

Yao, Z., Zheng, X., Xie, B., Mei, B., Wang, R., Butterbach-Bahl, K., Zhu, J. and Yin, R. 2009. Tillage and crop residue management significantly affects N-trace gas emissions during the non-rice season of a subtropical rice-wheat rotation. *Soil Biology and Biochemistry* 41(10), 2131–40. doi:10.1016/j.soilbio.2009.07.025.

Chapter 4

Weed management practices and benefits in Conservation Agriculture systems

Gottlieb Basch and Fernando Teixeira, University of Évora, Portugal; and Sjoerd W. Duiker, Penn State University, USA

1 Introduction

When growing crops, the eradication of plants from a field has only a temporary effect and, soon after, it is readily recolonized from the soil seedbank, vegetative structures in the soil and by seeds transported to the field by wind, run-off water, animals, soil mass movement or even agricultural practices (irrigation, harvesting, manuring, tillage etc.). We must be grateful for the tenacity that plants show to occupy most terrestrial environments, as it allowed the evolution of terrestrial animals and ultimately us. But, in agricultural fields, most plants are out of place and will compete with crops for water, nutrients, light and space, may diminish the quality of pastures, may impair the harvest, may be a repository of diseases or pest insects and so on. From the moment humans moved from nomad gathering communities and began producing their own food through agriculture, they conceived a multitude of methods to manage weeds based on the technologies and knowledge of its time. In an ever-changing world methods of weed control that seem very promising at a given moment sooner or later will be questioned by their impact on the environment or the development of new, more efficient alternatives.

http://dx.doi.org/10.19103/AS.2019.0049.04

Conservation Agriculture (CA), as defined by FAO (FAO, 2017), means farming systems that aim at sustained crop production by the compliance with three principles: no or minimum mechanical soil disturbance, permanent organic soil cover (by both crop residues and cover crops) and species diversity (through rotations, sequences and associations). These farming systems were practised on circa 180 Mha worldwide in 2015/16 and the expansion rate since 2008/09 has been calculated at 10 Mha/year (Kassam et al., 2015, 2019). The growing adoption of CA systems worldwide shows that farmers experience benefits when converting to CA. However, lack of soil disturbance, permanent soil cover and crop rotation/association under CA are likely to have an effect on weeds' dynamics, both in terms of diversity/composition and size.

In high input systems, heavy reliance on herbicides for weed control often means that integrated weed control strategies are disregarded by the farmer. Unfortunately, this is true in many no-till systems that purely focus on minimum soil disturbance but disregard the other two pillars of CA: diverse crop rotations and cover crops to revegetate the fields in fallow periods. Herbicide-resistant weeds, withdrawal of some herbicides from the market, more restrictive environmental and health regulations and policies that promote the reduction/banning of pesticide input (e.g. organic farming promotion), all converge to question if alternative weed control methods and practices under CA are feasible and whether integrated weed control strategies tailored to reduce herbicide inputs provide the economical results farmers need to maintain their livelihood.

2 Weed control under CA

The means that farmers have to control weeds are a host of physical, chemical and biological methods, using knowledge on crops, weeds, soils and climate, and agricultural management practices and their interactions. Each method of weed control has its own merits and demerits. Under CA, as a general rule, convenient weed control strategies should be successfully accomplished by the combined use of several of the above direct methods and management practices, both preventive (e.g. cover crops) and cultural methods (e.g. intercropping, sowing time).

2.1 Physical weed control methods

Physical weed control in conventional tillage systems comprises methods that physically cut, pull, bury, kill or impair the growth and development of target plants, by the manipulation of the environment or by a direct action on the target plant. Such methods include mechanically burying, cutting or pulling weeds, the use of black plastic as a soil cover, to stop light from reaching weed seedlings, transparent plastics to practise soil solarization, heat, by direct flame or hot water and so on. These methods present different costs, and rationales, and their use under CA is not always possible or desired. In recent years, the interest in physical weed control methods research has been driven by the needs of organic farmers but early on it was realized that most of these methods have to be integrated in systems that encompass preventive and cultural methods of weed control (Melander et al., 2005). In this chapter we will keep the focus on methods that are promising under CA and that can be used to reduce the need for herbicide use.

The primary and most obvious method of weed control is mechanical, and it has been used since the dawn of agriculture. Mechanical weed control varies in the type of implement and power source. The most basic method after primary land preparation is hand-weeding, which is time consuming, laborious and can represent a substantial percentage of the total production costs of a crop. Even in mechanized agricultural systems, hand-weeding is often used to control intra-row weeds, especially where herbicides are not available, authorized or recommended, or too costly (e.g. for a particular crop or in case the herbicides available are unable to control problematic weeds). In recent years, research has been focussing on mechanized methods to obviate hand-weeding, especially for intra-row weeding, such as brush weeding, finger weeding, among others, both in permanent and annual crops. The use of image analysis and artificial intelligence applied to robots, which already led to the development of commercial implements (Garford's Robocrop, Dino, Oz etc.), is also the subject of intense research and development (Bawden et al., 2017).

Under CA, the type and amount of crop residues will impact the success of most weeding implements, and pure mechanical weed control may be much harder to achieve with high amounts of residues, especially in annual crops. An important driver for research on mechanical weed control under CA has been the adoption of CA principles by organic farmers as the herbicide option does not exist or it is very rare (e.g. pelargonic acid). In Europe, especially under humid temperate climate, adoption of reduced tillage systems is perceived as fitting better to the most common farming systems than no-till, allowing manure incorporation, faster soil warming in spring and perennial weed control (Pannacci et al., 2017; Mäder and Berner, 2012). Yet, Weber et al. (2017) found perennial dicots to become more prevalent under reduced tillage when compared to conventional tillage. In these reduced tillage approaches most of the weed control strategy relies on tillage, including the full inversion of the top soil layer at shallower depths. Therefore, the term 'reduced tillage' is rather subjective, and in many cases the respective tillage methods do not comply with the criteria or principles defined for CA. Still, the interest in CA exists and its further adoption, particularly the adoption of no-till in Europe is progressing (Friedrich et al., 2014), although the rationale for the adoption varies widely from country to country. One approach to reduce weed pressure is to use cover crops in a crop rotation, without soil incorporation. The inclusion of winter cover crops in the farming systems would harness the benefits associated with this practice, such as the retention of nutrients or prevention of erosion, while simultaneously exerting weed control by smothering weeds and later by mulch. To overcome weed outbreaks in the following row crop, the use of high residue weeders, especially designed to perform mechanical weed control underneath the mulch, without dragging mulch and preserving the weed suppression ability of mulch, should be further researched.

In some parts of the world, especially among smallholder farmers in sub-Saharan Africa (SSA), hand-weeding with the help of hand hoes and even by hand pulling are the normal practice, and only a small number of farmers (5–10%) use animal-drawn implements or tractors (Sheahan and Barret, 2014). In these cases, a recurrent question has been whether CA can be adopted and if weeds can be managed efficiently without the use of herbicides? When performed on time, hand hoe or hand pulling weeding seem to be as effective as herbicide use, with no significant differences in crop yield (e.g. Nyagumbo et al., 2016). Still, and especially in semi-arid regions, a higher initial weed pressure under CA often leads to a higher weeding frequency (e.g. Nyamangara et al., 2014; Mashingaidze et al., 2012). At locations where labour shortage exists, and alternative methods to hand hoe and hand pulling weeding are not viable due to socio-economic reasons, an increased weeding frequency may result in low adoption of CA systems even where it is proven that higher yields are possible under CA and/or a more sustainable farming system is needed because of erosion, low soil fertility and so on. Research on alternative, less labour-intensive weed control methods and weed suppressing crops and crop sequences may be pivotal for the adoption of CA and improving the livelihood of these smallholder farmers.

The use of black plastic sheets (polyethylene) as soil cover is an effective way to control weeds in vegetables and other crops (Radics and Bognar, 2004). Yet, its utilization is more generalized in row crops, and standing crop residues (e.g. cereal stubble) may restrict its application. The control is accomplished by the reduction of sunlight needed to trigger weed seed germination and/or seedling growth and by a direct physical pressure on the seedlings. Covering soil with plastic sheet has also shown other positive effects in several crops such as concentrating water for deeper infiltration, reducing evaporation, promoting early growth and harvest with transparent plastic (by increasing soil temperature), reducing erosion, increasing yields and so on (Wang et al., 2009a). Some authors, at certain locations, report poorer soil aeration and higher temperature under plastic soil cover as cause of poor crop yields and/or performance (Wang et al., 2009b). Interaction between climatic conditions and the physical properties of plastics used as soil cover should be further researched for sound agronomic recommendations.

From the environmental point of view, soil and water pollution from plastics used in agricultural fields is a growing problem. In China, where the area of soil under plastic cover was estimated to be 19.8 million ha in 2011, soil pollution from plastic particles was 72–260 kg/ha in the top layer (top 30 cm), depending on the time-lapse since adoption of plastic soil cover, resulting in a negative effect on crop performance and yield (Changrong et al., 2014). Plastic remnants were left in the soil because of the difficulties faced by farmers in collecting the degraded plastic after harvest due to its thinness and fragility (Changrong et al., 2014). The use of biodegradable films can alleviate the problem as they can be incorporated in the soil or composted off-site. Surveys conducted in the United States show that lack of adoption of biodegradable films among farmers is due to insufficient knowledge, higher costs and uncertainty in relation to film degradation, although recognizing the environmental benefits that could derive from biodegradable films (Goldberger et al., 2015). The concern with higher cost of biodegradable film is shared in other parts of the world. In the case of China, comparing the costs of biodegradable films with the use of the traditional polyethylene plastic and its cost of removal and disposal at the end of the season, Changrong et al. (2014) showed that the cost increase when adopting biodegradable plastic was between 13% and 69%, depending on its thickness. More in-depth economic studies are needed to better understand the value of biodegradable film and how best to promote its use. They should take into account (a) at the farmer's level, the need of new equipment to lay biodegradable film and to incorporate or compost it at the end of the season, (b) economies of scale in a growing market of biodegradable films and (c) the environmental benefits and other externalities of changing from traditional polyethylene to biodegradable films. The development of new biodegradable material and a better characterization of the films that would allow to predict the lifetime of the products under different agro-climatic conditions is also desirable to alleviate farmers' concerns, although, under CA, the presence of crop residues will constitute a nuisance and render an extra layer of difficulty for its use.

Thermal weed control relies on heat transfer to target plants and/or seeds and vegetative propagation tissues underground. The mechanism involved in the death of plant tissues will depend on the temperatures reached and the time of exposure. The use of fire for land clearing and stubble burning to control weeds is still practised in some parts of the world. Although effective in some instances, stubble burning to reduce the soil's weed seedbank (Narwal et al., 2006) has risks and it is not the best option, from either an environmental or from a soil conservation point of view, as it reduces soil cover and increases the emissions of CO_2, CO, NOx, particulate matter and so on (Dhammapala et al.,

2006; Miura and Kanno, 1997). Yet, under some agro-ecological conditions such as the Mediterranean, fire can even promote the germination of dormant seeds, thus contributing to the reduction of the weed seedbank through fire.

The use of direct flaming for weed control under true CA is possible, depending on the amount of crop residue, its moisture and weather conditions, but in most cases it is not advisable due to the associated fire hazards. Mojžiš et al. (2015) refer to the high cost of this method, and therefore the need to improve efficiency, as key for the adoption of the method for the control of intra-row weeds (Bond and Grundy, 2001), complemented by mechanical cultivation between rows. Other methods of thermal weed control, based on hot water, saturated steam, hot foam, electrical discharge, infrared radiators and so on are not appealing, at the moment, to CA farmers because of the investment in new equipment, but can be appealing to organic farmers or to control more resilient weeds. However, thermal-based weed control methods show limited success in the case of perennial weeds.

Soil solarization is a method that makes use of transparent polyethylene sheets that cover moist soil resulting in lethal soil temperatures and atmosphere for several pests (>50°C) for a period of 2–4 weeks. It was first designed to control soil pathogens (Katan et al., 1976), but it was soon found to reduce viable weed seeds as well (Horowitz et al., 1983). One important misconception, often mentioned, is the idea that covering soil with black plastic produces soil temperatures high enough to result in a solarization effect: this is incorrect as black plastic stops light from reaching the soil and has a very low heat capacity; on the other hand, clear plastic transmits the light that heats the soil and traps the longer wavelength infrared radiation, irradiated from the soil, increasing further soil temperature (for more information on solarization and references, see Katan, 2015). Soil solarization is limited to warm climates with abundant solar radiation, where water for irrigation and a period of 1–2 months without crops are available. At certain locations, some degree of solarization can also be attained even with black plastics that can equally be left in place as post-planting cover (e.g. Abu-Gharbieh et al., n.d.). Recent research has been focussing on the use of soil solarization as a substitute for soil fumigation at less warm and less reliable 'clear sky' locations (Samtani et al., 2017) and the use and performance of biodegradable films to minimize subsequent soil pollution (Bonanomi et al., 2008).

2.2 Chemical weed control

For most crops herbicides are the main method chosen by farmers to control weeds because of their effectiveness, low cost, ease of application and high economic return. By the end of the last century and after decades of continuous success of herbicide use, general public concern regarding the use of pesticides and their potential impact on the environment and human health prompted new legislation in many parts of the world to tackle and prevent water pollution, with more restrictive thresholds for pesticide residues in the environment (e.g. the European Union Drinking Water Directive, see reference Council Directive 98/83/EC). As a direct result some herbicides were withdrawn from the market in some parts of the world, for example, atrazine in Europe (Sass and Colangelo, 2006), and new/tighter regulations were put in practice for the sustainable use of pesticides, for example, 'Community action to achieve the sustainable use of pesticides' (Council Directive 2009/128/EC). Simultaneously, integrated pest management (IPM) started to be promoted, aiming at the reduction of pesticide-input and the use of alternative, non-chemical, weed and pest control.

The herbicide glyphosate (*N*-(phosphonomethyl)glycine), first marketed in 1974, is a non-selective systemic herbicide, absorbed mainly by leaves. For its inherent characteristics it soon replaced the herbicide paraquat as pre-seeding herbicide used in conservation tillage and especially in no-till systems. Today, it is the most used pesticide in the United States, with a usage rate range of 122–132 Gg of active ingredient in 2012, which is almost double that used in 2005 with a consumption of 67–76 Gg of active ingredient (Atwood and Paisley-Jones, 2017). This enormous increase is mainly due to the introduction of *Roundup Ready* (RR) crops in 1996 (beginning with soybeans), which are genetically modified crops tolerant to that herbicide allowing its application in post-emergence. Since its introduction in soybean, RR varieties were launched also for corn, cotton, sugar beet and rapeseed (canola). In 2016, these glyphosate-tolerant crops were cultivated on around 96 Mha, being used on one-third of the area under maize produced worldwide in 2015. While initially solving the problem of resistance to other herbicides, the almost exclusive reliance on glyphosate by farmers for weed control in RR varieties and for other purposes such as pre-harvest desiccation of crops and pre-harvest control of perennial weeds soon caused the development of weed resistance to glyphosate in an increasing number of weed species. Glyphosate-resistant weeds in 2017 totalled 38 species in 37 countries (Heap and Duke, 2018). Besides the resistance of an increasing number of weed species to glyphosate, the detection of residues in water bodies, food products and even in human tissue has triggered an unprecedented campaign against this herbicide, above all in Europe.

In recent years, glyphosate has been subject to intense scrutiny based on concerns of possible environmental and health effects (Romano-Armada et al., 2017; Myers et al., 2016). Nevertheless, it is expected that glyphosate use will continue to grow in the years to come, because of the wide range of application purposes as a broad-spectrum and cheap herbicide. Still, this growth in glyphosate use cannot be attributed to the increase in CA per se because only a part of the RR crops is managed with CA. In addition, weed management in conventional crop production must look for complementary and alternative weed management practices to reduce the dependence on glyphosate and avoid a ban of this herbicide for whatever purpose due to the pressure of public opinion.

Worldwide, the development of herbicide-resistant weeds in 2018 reached a total of 255 weed species, 148 of which are dicots and 107 monocots (International Survey of Herbicide-Resistant Weeds, see online: www.weedscience.org). The development of herbicide-resistant weeds, and the fact that no herbicides with new sites of action have been released for more than 30 years (Duke, 2012; Heap and Duke, 2018), forces many farmers to rethink their entire weed management strategy.

2.3 Herbicides under CA

It is commonly known that farming systems that rely mostly on glyphosate for weed control are bound to face increasing weed problems in the medium and even short-term. This is why CA must adopt integrated weed management strategies that encompass crop rotation, cover crops, rotation of herbicides with different sites of action and so on (Neve, 2008; Price, 2011; Heap, 2014a,b). Still, CA systems offer additional opportunities to reduce herbicide inputs when compared to conventional systems, thus maximizing agronomic, economic and environmental gains.

The use of herbicides, especially under CA, requires a good knowledge of prevalent weed species and their ecology, their sensitivity to herbicides, available herbicides, particularly their mode of action and efficacy at different crop residue levels and weed development stages, and the period of weed interference on crop yield. A better knowledge of the critical time of weed interference with crop growth would allow for better planning and more informed decision-making by farmers regarding the timing of herbicide application and efficiency, and may spur strategies for more efficient weed control. One such strategy lies in taking advantage of the higher machine bearing-capacity on no-till soils allowing the application of herbicides at times where conventionally tilled soils are often off-limit. Barros et al. (2005) studied the application of a mixture of diclofop-methyl+fenoxaprop-p-ethyl+mefen epir-diethyl in post-emergence herbicide to control *Lolium rigidum* G., in a no-till winter wheat in Portugal (Mediterranean climate), and observed that an early application of the herbicide, when 90% of *Lolium* was at the beginning of tillering, gave good weed control at lower doses, and significantly higher grain yield, when compared to later herbicide application, while higher doses were needed to get good control when applied at the complete tillering of *Lolium*. Later studies by the same authors, Barros et al. (2007, 2008), with different herbicides, revealed similar results, that is, significantly higher yields with early application of the herbicide, and an important reduction of the amount of herbicide needed when applied at the first application date, when 90% of grasses were at the beginning of tillering and at 3–4 pairs of leaves for broad-leaved weeds.

Another important aspect relates to the initial emergence potential of main weed species present in the fields. Timing planting date of annual crops after weed emergence may result in a decrease of weed density and biomass by controlling the weeds prior to planting. This is the same concept of the stale or false seedbed, using late sowing to control previously emerged weeds. Obviously, this may cause a certain yield reduction proportional to the shortening of the crop cycle length, but it may be a necessary strategy to control problem weeds and to reduce the use of herbicides. Rasmussen (2004), in Denmark (temperate climate), observed that late sowing and false seedbed were both effective in the reduction of weed biomass in organic winter wheat, and that a reduction of grain yield, significantly in some instances, could occur. Under Mediterranean conditions, and in seasons with regular autumn rainfall distribution, late sowing and pre-sowing control of emerged weeds in no-till winter wheat may be enough for adequate weed control and voids the need to apply a post-emergence herbicide (Calado et al., 2010).

2.4 Biological weed control methods

Biological control for pest management consists of applying the natural interactions/mechanisms that drive interspecific relationships to control the balance of pest populations (Tilman et al., 1999; Chidawanyika et al., 2012; Cordeau et al., 2016). Biological weed control methods encompass a wide range of agronomic approaches, from selection of cultivars for a specific purpose, such as production of allelopathic chemicals (Worthington et al., 2015) or high competitiveness (Tilman et al., 1999), to bioherbicides (Courdeau et al., 2016), obtained from microorganisms or naturally occurring molecules (e.g. certain plant extracts). Because of the scope and nature of this chapter, the use of macrofauna agents for weed control will not be explored here. For further information, please consult the catalogue *Biological Control of Weeds* edited by Winston et al. (2014), a valuable and exhaustive repository of known organisms and their targets that comprises 830 agents, 368 weed species, the accounts of 2191 releases of insects or other agents, and 2080 bibliographic references.

In the context of integrated weed management, breeding and selection of crop cultivars for their competitiveness or allelopathic potential against certain weeds is definitely a priority, not only to improve control of herbicide-resistant weeds but also to meet the needs of organic farmers and for an overall reduction of herbicide use. Competitiveness and allelopathic potential are regarded as independent factors when selecting crop cultivars for their weed-suppressive ability. For different crops and agro-ecological field conditions, research should establish which traits are best for weed-suppressive ability. Worthington et al. (2015) screened the allelopathic potential of 58 lines of winter wheat well adapted to the southeastern United States against Italian ryegrass (*Lolium perenne* L. ssp. *multiflorum* [Lam.] Husnot), and held eight lines with strong and weak allelopathic activity that were further evaluated in a field experiment for weed suppression and grain yield tolerance. The authors could not establish a correlation between the differences in allelopathic potential observed in the laboratory and weed suppression and grain yield in the field, and concluded that weed suppression correlated well with competitive traits. Olofsdotter et al. (1999) tested 111 rice cultivars in a field experiment and laboratory screening, to study their allelopathic potential to suppress *Echinochloa crus-galli*, and found significant differences between cultivars in their effect on field weed suppression due to allelopathy. Despite the great interest that breeding crop cultivars for their allelopathic potential or competitiveness generates, this field of study is still in the early days and only a few rice cultivars with commercially acceptable grain yield and quality have been released as a result (Gealy et al., 2013).

Still on the subject of allelopathy, weed control under CA has much to benefit from the allelopathic potential of some plants. Tabaglio et al. (2008) studied the effect of mulches of eight rye and one triticale cultivars on the germination of four weeds (velvetleaf, common lambsquarters, redroot pigweed and common purslane). The study revealed that rye mulches significantly suppressed the germination of redroot pigweed and common purslane, 40–52% and 40–74% respectively, while triticale mulch only suppressed common purslane by 33%, which led the authors to suggest that rye mulches could reduce the need for pre-emergence herbicides. Other uses of allelopathy that can be exploited for weed control are the inclusion of crops with high allelopathic potential in a crop rotation or intercropping system. In a study comprising five different two-crop rotations, and five tillage systems, Shahzad et al. (2016) observed a significant effect of sorghum, in a wheat–sorghum rotation, in the reduction of total weed biomass and density in wheat, when compared to the other rotations, especially under no-till. Allelopathy as an integrated tool within a holistic view of the agroecosystem will be discussed further in Section 2.4.2.

For further information on allelopathy, you may consult the excellent review by Cheng and Cheng (2015).

As an example of the different approaches regarding biological weed control methods, the following paragraphs provide a small review on the efforts that have been undertaken to control *Striga* spp. (witchweed). *Striga* spp. are among the most damaging weeds in agriculture, affecting especially cereal crops, such as maize or sorghum. These parasitic plants cause a strong slowdown of the host growth even before emerging from the soil. Seed germination of the weed is triggered by strigolactones produced by the susceptible crop, followed by an elongation of the radicle that penetrates the roots of the host, forming a haustorium (Cardoso et al., 2011). *Striga* spp. are believed to infest around 40% of the African savannah region, and according to some studies (FAO) this weed causes yield losses of between 30% and 100%, worth US$1 billion, and affects livelihoods of around 100 million people (IPBO, 2017).

One approach, although not properly biological, lies in the resistance that a natural mutant maize has to the herbicide imidazolinone (IR maize) that can be crossed with locally adapted maize to create commercially interesting cultivars. The use of imidazolinone coated seeds can then be used to control *Striga* spp. In a study on experimental stations and farms during six growing seasons (1998–2001) and across four African countries (Kenya, Tanzania, Malawi and Zimbabwe), Kanampiu et al. (2003) used Imazapyr and Pyrithiobac to coat seeds of IR maize and observed a significant reduction of *Striga* spp. (*Striga hermonthica* and *Striga asiatica*) infestation, and almost no flowering (exception for the long rainy season in Malawi, where *Striga* was observed flowering).

Another approach is the identification of cultivars with low germination stimulating activity (low production of strigolactones exudates). This has been the line of research followed for sorghum, the main cereal crop in semi-arid SSA, and some varieties have been introduced to the market in Sudan and Ethiopia (Mohemed et al., 2016).

Other researchers suggest alternatives to IR maize and to resistant sorghum to control *Striga* spp. by the use of *Desmodium* spp. as intercrop, a practice that apparently is capable of controlling *Striga* spp. (Khan et al., 2006), and in line with the principle of CA of permanent ground cover. The discovery of *Desmodium* spp.'s ability to control *Striga* spp. was totally fortuitous. Research on maize-*Desmodium* spp. intercropping, where the legume was being tested as a repellent plant for stem borer control and Napier grass (*Pennisetum purpureum* Schumach) as a trap plant, revealed the unsuspected effect of controlling *Striga hermonthica* (Khan et al., 2014). Khan et al. (2006) observed a reduction of stem borer damage in maize by 74.7%, and an even higher reduction of losses through *Striga* spp. by up to 99.2%. The authors also observed that the effectiveness of maize-*Desmodium* spp. intercropping allows gains in maize yield of up to 511% higher than in maize monocrop and maize-cowpea, resulting from *Striga* spp. suppression and lower damage from stem borers. The control mechanism of *Striga hermonthica* results from root exudates of *Desmodium* spp. containing isoflavanones that stimulate *Striga hermonthica* germination while other isoflavanones and a group of C-glycosylflavones inhibit radicle growth thus causing suicidal germination (Khan et al., 2014).

2.5 Agricultural practices

The evolution of the soil seedbank, both in terms of diversity and number of seeds, determines with few exceptions the success of weed control in a given cropping system. Depletion of the soil seedbank is governed by predation, decay and germination, and the increase of the soil seedbank by seed rain. Management practices that promote the depletion mechanisms and avoid seed-setting and seed rain will tend to lower the number of seeds in the seedbank.

2.5.1 Tillage and soil seedbank dynamics

Sosnoskie et al. (2006) observed in a long-term tillage experiment an increase of species richness and total germinable seed density under no-till. Murphy et al. (2006) also detected higher weed species diversity but a decrease in the number of seeds in the soil seedbank under no-till. Other studies show or suggest a reduction of weed species diversity under CA (e.g. Muoni et al., 2014). It is long established that under no-till the lack of soil disturbance will lead to a build-up of weed seeds in the very top layer of the soil (first 5 cm), resulting in higher concentration of weed seeds in this layer when compared to tilled soils (Swanton et al., 2000), and to the preservation of underground vegetative structures of perennial weeds (Timossi, 2009). No-till will also provide the conditions for higher weed seed predation and decay, and thus, it is foreseeable that no-till will favour, under certain conditions, the depletion of certain weed populations (Baraibar et al., 2009) but increase the pressure of perennial weeds. The differentiation of the weed pool whether showing a higher or lower species richness will depend on the local flora characteristics and time lapse since the adoption of CA principles and their management.

During the first years of adoption of CA and especially no-till, a higher concentration of weed seeds at the soil surface, when compared to traditionally tilled systems, commonly results in higher weed densities (Calado et al., 2009). The germination responses of weed seeds coming out of primary dormancy to temperature changes is apparently the same for both winter and summer annual weeds (although at opposite seasonal temperature changes), whether they are on the soil surface or buried. However, most seeds that remain on the soil surface will probably germinate the first autumn/spring after dispersal, whereas those that are buried may remain viable for many years (Baskin and Baskin, 1985). This may explain higher weed densities when starting with no-till and the need for extra care to prevent seed rain in the absence of tillage, and to favour the decrease of the soil seedbank through higher germination and subsequent control in well balanced cropping systems under no-till.

Another consequence of the lack of or minimum soil disturbance and the existence of a mulch layer is increased topsoil water content. Calado et al. (2009) studied the effects of air temperature, topsoil water matric potential of undisturbed soil samples and time (sum of daily mean temperature) on weed emergence. The authors observed a significantly higher weed emergence in the soil samples that were kept at field capacity when compared to those kept at lower matric potentials, and a consistent and, in most cases, significant decrease of weed emergence as the water matric potential decreased from two-third field capacity to one-third field capacity, and from one-third field capacity to air dried soil. Another finding of this study relates to the relative contribution of seeds at different depths, 0–2.5 cm vs. 0–5 cm, to the emergence counts: although, when compared to the 0–2.5 cm layer, a significantly higher population for 0–5 cm depth was observed, the results show that the main contribution to weed emergence came from seeds within the first 2.5 cm.

No model, if based only on a partial implementation of CA, can reliably foresee the soil seedbank dynamics. Crop rotations, crop species/varieties, management practices, soils, rain and temperature patterns, the local flora and so on, all impact seedbank size and composition. Also other practices associated with CA will have an impact on weed dynamics. Still, very few studies have focussed on the interactions of these practices, and even fewer were conducted under different agro-ecological conditions.

2.5.2 Mulches and cover crops

In the case of 100% soil cover, the use of mulches is capable of diminishing the pressure from weeds producing small seeds by cutting the light they need for breaking dormancy and initial growth, and physically restraining the small seedlings to get to the light, as well as by creating conditions for weed seed predation and higher seed decay (e.g. Pupalienė et al., 2015). Some crop residue mulches, for example, rye (*Secale cereale*), have or produce allelopathic chemicals while being decomposed by microorganisms, that inhibit weed seed germination, root growth and so on, and thus may be used to further help control weeds (Bhadoria, 2011). Nevertheless, weeds are not equal regarding their sensitivity to the allelochemicals produced by crop residues, nor are the crop residues equal in their allelopathic potential (Tabaglio et al., 2008). The duration of the allelopathic effect depends on the solubility of allelochemicals, on its degradation rate in the soil and on its detoxification capacity of weeds (Schulz et al., 2013). For example, more and more evidence is emerging that some consortia of symbiontic microorganisms may collaborate with the weeds in detoxifying allelochemicals and thus explain the different sensitivity of the different weeds to these allelochemicals released by the residue (Kia et al., 2014). Further, the weed-suppressive abilities of mulches to provide a good weed control must be established and preserved, that is, 100% cover and sufficient thickness. Mirsky et al. (2013) found that 75% inhibition of annual weed emergence was only attained when cereal rye dry matter exceeded 8 Mg/ha and when it created a mulch layer thicker than 10 cm.

Under CA, cover crops are used in the fallow period between two cash crops for the purpose of maintaining soil cover and provide subsidiary, but important services like the uptake of available nutrients and prevent them from leaching, increase soil organic matter and so on. The effect of cover crops on weed control ranges from smothering weeds, by a direct competition for resources, to provide mulch for the next crop, or to the allelopathic effect of cover crops. To terminate cover crops in no-till farming systems, design of improved roller-crimper implements may allow the termination without the need for the application of herbicides. Davis (2010), at the University of Illinois Crop Science Research and Education Center (CSREC), in the United States, observed that roller-crimper termination of a rye cover crop for the production of mulch reduced the weed biomass by 56% in a subsequent soybean crop when compared to a field without cover crop and pre-seeding spray-off of weeds. A problem identified at that time was poor soybean seed placement and seedling emergence through the heavy mulch. Grain drills and seed planters must be able to cut through the high volume of residues, which means that the coulters must perform a much cleaner cut and the equipment needs a higher weight. Basch (unpublished data), in an ongoing field demonstration in Coruche, Portugal, where a winter cereal (oat) was chemically terminated and rolled to provide mulch (8 Mg/ha) for subsequent no-till maize, observed similar results, that is, almost complete weed control by the mulch but a reduced maize stand density due to a poorer seed placement.

2.5.3 Rotations

Crop rotation is one of the pillars of CA. Studies show that crop sequence influences weed species diversity and density (Regnier and Janke, 1990; Liebman and Janke, 1990; Sosnoskie et al., 2006) as a result of different pressure over different weed species, preventing some weed species from being repeatedly successful, and hence preventing a particular weed or pool of weeds from becoming noxious at a particular site. When in monoculture, the weed pool of a field will tend towards species adapted to the management and conditions created for that crop. Rotation effects on weed populations are the result of different management (e.g. timing of weedings, herbicide application timings and modes of action), varying patterns of resource competition (periods of intense competition from the crop or cover crop) and allelopathy (Nichols et al., 2015). Thus, crop rotations are an essential tool for effective weed control as they guarantee that a specific weed or group of weeds does not become predominant.

Balanced crop rotations for weed control must be designed locally, based on the evaluation of the interactions of these pressure factors (management of the crops, resource competition and allelopathy) and weed seed emergence, weed setting and growth, seed formation/ripening (or of vegetative propagation structures) and resulting weed population. Under CA and in farming systems that rely primarily on herbicides for effective weed control, lack of soil disturbance and wider crop rotations may lead to a decrease of the soil seedbank, although to a higher weed flora diversity (Murphy et al., 2006), and diminish the pressure that a particular weed or group of weeds may cause. In addition, allowing the rotation of herbicides with different sites of action helps prevent the emergence of weed resistance to certain herbicides.

Another important aspect of crop rotation under CA is the introduction of cover crops, cereals, legumes or brassicas, by their smothering effect on weeds. When considering cover crops in warm, sub-humid agro-ecological zones, high humidity and temperature cause high biomass decomposition rates, and therefore crop sequences should include crops with high C/N ratio residues (maize, sorghum etc.) in order to produce a persistent mulch (Bayer et al., 2000). Anyway, cover crops composed of mixtures of different species and families are generally preferable due to the complementarity of their agro-ecological effects, and higher biomass production (Cardinale et al., 2007).

In mixed farming, the introduction of temporary pastures or meadows is also an effective method of weed control that combines the competition for resources from the sown species, smother effect, grazing and/or mowing.

Where wider crop rotations or cover crops are not feasible for socio-economic reasons or due to other constraints, one possible option to achieve partly the effects of a crop rotation can be the deliberate 'rotation' of early and late maturing crop varieties, with adjustments of sowing/weeding/harvest dates.

2.5.4 Intercropping and relay cropping

While crop rotations allow a sequential diversification of crops, and thus varying pressure on weeds over time, intercropping allows simultaneous and diverse pressure on weeds. Intercropping constitutes a way to introduce diversification in a farming system and thus fulfilling the requirements of CA for expected benefits. Although often considered as impracticable in modern mechanized farming systems, in a strict sense intercropping is widely used with cereal-legume mixtures for forage purposes, or the association of a cereal or cereal/legume as cover in the inter-row of permanent crops (vineyards, orchards etc.). According to Fletcher et al. (2016), the interest in intercropping is increasing worldwide. Other practices such as relay cropping, for instances of a legume sown into a standing crop, or soon after harvest of a cereal, can be of particular interest to hinder the development of weeds between cropping seasons (Amossé et al., 2013) and to extend food/forage production and ground cover into the period between seasons (e.g. Bandyopadhyay et al., 2016).

The interaction between different crop species and their effect and mechanisms on weed suppression is certainly a field that deserves further research. In recent years, maize-*Desmodium* spp. intercropping is being promoted in SSA allowing the reclamation of formerly abandoned land due to *Striga* spp. infestation. Mainly on soils of poor fertility it would allow farmers to obtain a source of high quality fodder from *Desmodium* spp. and Napier grass (see end of Section 2.3). In 2014, this mixture was already adopted by around 70 000 farmers in four countries (Kenya, Uganda, Tanzania and Ethiopia) (Khan et al., 2014).

2.5.5 Other practices

Cultural practices as simple as weed seed harvest or retaining at crop harvest, and later off-field destruction may have a profound effect on weed seedbank by the reduction of weed seed input (e.g. Walsh and Powles, 2014).

Other cultural practices that contribute to weed control and that can be managed fairly easily are crop density, spatial distribution and line orientation. Where water is not an important constraint, adjusting crop density may result in better weed control. Adigun et al. (2014) observed that a cowpea (*Vigna unguiculata* L.) seeded in narrower inter-row spacing of 60 cm, compared to 75 and 90 cm, resulted in a significant reduction of weed growth and higher grain yield in Southwest Nigeria. This aspect of spatial distribution that should be used when designing a system for better weed control is based on the competitive capacity of the crop, along with its allelopathy, and the reduction of the area available for weed development, without losing the ability to manage the crop.

A better knowledge of the interaction between weed interference period and crop yield economic optimum would allow for a better planning and informed decision-making by the farmers (Adigun et al., 2014; Chauhan and Johnson, 2011; Barros et al., 2005, 2007, 2008).

3 Smallholder farmers' strategies for weed control in developing countries: sub-Saharan Africa

CA has been promoted in SSA countries to cope with challenges such as soil erosion, soil fertility degradation, uncertain rainfall patterns in rainfed agriculture and so on (see African Conservation Tillage Network, www.act-africa.org).

Although it has been shown in many studies around the world that CA merits recognition as a soil protection and soil quality improving farming system, adoption in SSA countries is slow and most authors identify increasing labour requirements to control weeds as the main reason (Mashingaidze et al., 2012; Muoni et al., 2013; Nyamangara et al., 2014).

The aim of this section is to help identify the main agronomic and socio-economic constraints to the adoption of CA in SSA by smallholder farmers (under 2 ha), focussing on the issue of weed control under CA.

3.1 Weed control under CA in SSA by smallholder farmers

CA practices consist of a wide range of management practices embedded in the respective cropping systems and their use and/or success is largely dictated by their rationales and suitability to local conditions. To meet the scope of this section we will be focussing on weed control in the main cereal crops, associated with the following CA practices: direct seeding with the help of a dibble or other stick, planting basins, mulching, cover crops, intercropping and crop rotation.

In SSA, the most representative methods of weed control in smallholder farms are mechanical, mainly with the use of hand hoes and by hand pulling to eliminate weeds, and where farm size allows, farmers make use of the plough to bury the weeds, and tine cultivators, both animal-drawn (Sheahan and Barret, 2014).

Depending on the crop, weeding with the help of hand hoes or hand pulling may take 60 labour-hours or more per ha and it may need three or four weedings for a successful crop. In Zimbabwe, Nyamangara et al. (2014) conducted a survey on hand hoe planting basins (PB), a form of reduced tillage, and conventional mouldboard plough-based tillage (CT), across contrasting agro-ecological zones, in maize (*Zea mays* L.), and found that weeding frequency was 42.1–58.9% higher under the PB system, that is, on average weedings were performed 2.7 times/season with PB against 1.7–1.9 times under CT. A higher weeding frequency under CA (planting basins and ripper tine) in order to reduce weed infestations was also observed by Mashingaidze et al. (2012). Another problem associated with hand-weeding fields under CA, especially when using hoes, is the inherent presence of crop residues or mulch on the soil surface that slows and hardens cutting the weeds. In the first year after the conversion of the management of a field to CA, harder work and higher weeding frequency may result in low adoption of CA systems even when proven that higher yields are possible.

When promoting CA in SSA, an alternative to hand-weeding under CA, which can be achieved with a relatively small capital investment by smallholder farmers, is the use of herbicides. When executed on time, weeding with hoe or by hand pulling is as effective as herbicides, with no significant differences in crop yield, suggesting that herbicides are not a prerequisite for CA success (Nyagumbo et al., 2016; Muoni et al., 2013), but it can be argued that labour saving through the use of herbicides would allow farmers to expand their cultivated area, increase their income from off-farm activities, relieve the burden of hand-weeding from women and children and so on. For human health and environmental protection, use of herbicides should not be recommended without proper training of farmers and guidance on choosing and application of the most suitable herbicide, or mixture of herbicides, and at the right timing. Under CA, the main constraints of herbicide use result from crop residue/ mulching interception of the herbicides, decreasing the efficacy of soil applied herbicides, and in some instances increasing the herbicide concentration in run-off water (Teixeira and Basch, 1998). Nevertheless, where the three principles of CA are duly applied, lack of soil disturbance, permanent organic soil cover and crop rotation concur to create the conditions for an important reduction in the amount of herbicides needed when compared to conventional agricultural systems. In Paraguay, Kliewer et al. (2000) observed that in no-tillage CA systems with proper crop rotation and the use of green manure cover crops (GMCC), herbicides' costs could be reduced by more than 62%, when compared to the same cash crop in monoculture.

Full benefits of CA for weed suppression will only be attained where the three principles of CA are applied, and a balanced cropping system put in place. It will not suffice to keep the soil undisturbed without proper soil cover and crop diversification. Understanding the mechanisms at work is essential to properly manipulate the cropping systems towards weed suppression.

Mulching is perceived by many, and often observed, for having an impact on cereal grain yield in semi-arid SSA because of the potential effect on water conservation (e.g. Mupangwa et al., 2012). Mulching also has an effect on weed suppression. Nevertheless, the use of mulches as the single method to control weeds may not be sufficient, besides potentially posing some problems. In a study of weed growth under CA in Zimbabwe, Mashingaidze et al. (2012) observed that weed suppression by maize mulch (0, 4 and 8 Mg/ ha) in sorghum was only effective early in the cropping season with no effect later in the season, and that no effect was found in cowpea. Other problems associated with mulch cover result from the competition between soil cover and livestock for crop residues, especially in semi-arid regions where biomass production is low and demand is high. In Zimbabwe, Nyamangara et al. (2014) observed that farmers practising no-till, even when advised to conserve crop residue as mulch, would preferentially feed it to livestock. Where termites are endemic, maintaining a good soil cover with crop residues may prove difficult. In the north of Mozambique, Nyagumbo et al. (2016) observed a considerable reduction in surface cover with maize residues caused by the attack of termites when compared to Malawi. Depending on the type and amount of crop residue mulch, hand hoe weed control may become more difficult. In case crop residues for mulching are taken from other uncultivated areas, they may be contaminated with new weed seeds. Also, in drier agro-ecological zones, mulches may create surface moisture conditions that allow higher weed seed germination (Calado et al., 2009; Mashingaidze et al., 2012). Thus, in semi-arid regions, the use of mulches alone for the purpose of an effective weed control may not be satisfactory and eventually gives better results in agro-ecological zones where biomass production is plentiful and alternatives to feed the livestock exist. On the other hand, greater integration of livestock in cropping systems in dry regions may open up new possibilities to use cover crops and crop residues to feed livestock and, at the same time, improve soil health, such as is shown in semi-arid regions of the United States (Tallman, 2012).

In SSA countries, smallholder farm size is typically small, with 80% smaller than 2 ha, and in countries like Malawi and Rwanda even below 1 ha (Harris and Orr, 2014). Despite the benefits of crop rotation, the adoption of effective, weed-controlling crop rotations on very small farms seems only feasible in very specific contexts. Rotations on these very small farms would mean a smaller area with maize, their staple cereal, which would increase food insecurity, especially where markets for other subsistence and cash crops are non-existent (Ngwira et al., 2012; Thierfelder et al., 2013). As already mentioned in Section 2.4.3, in situations where crop rotations are not feasible the rotation of early and late maturing crop varieties, with adjustments of sowing date, weeding dates and so on could be an option to achieve partly the effects of crop rotation. Even where farmers can access more land, concentration of cropping and intensive rotation, based on several short-cycle crops, could be a better strategy than growing crops extensively, as long as there are functioning markets (Thierfelder et al., 2013).

Depending on the length of the growing season, several crops may be grown in succession in one year. If the rainy season is long enough, it may be possible to grow a crop (often in mixture) with a long growing period in succession with a crop with a short growing period. In regions with a bi-modal precipitation pattern different crops are already grown in the long vs. the short rainy season. Also, where land access is not the main constraint, another approach could consist in the use of crop rotations with extended planted fallow periods (2–3 years), based on carefully selected cover crops that may contribute to a significant reduction of the weed soil seedbank, particularly if the cover crop provides a complete soil cover by the canopy, high biomass and litter production. Additionally, those cover crops in the rotation would increase soil fertility, produce fodder and increase livestock income for the farmer. One such cover crop is the perennial leguminous species 'tropical kudzu' (*Pueraria phaseoloides* Roxb.) that has a high potential as fodder, both as hay or silage (Hiep et al., 2008). In a long-term fallow experiment in a humid forest-savannah transition zone in Ibadan, Nigeria, after 2–3 years of tropical kudzu planted fallow, Ekeleme et al. (2003) observed a significant reduction of the soil seedbank, when compared to leucaena (*Leucaena leucocephala* Lam.), to natural bush fallow and to continuous maize–cassava intercrop. The changes induced by the cover crop in the size and composition of the soil weed seedbank may be of interest in the design of a long-term weed control strategy. *Mucuna pruriens*, grown for 5 months after application of glyphosate, was found to be an effective control method of *Imperata cylindrica*, and more than double the yield of following cassava or yam crops in Ghana (Aflakpui and Bolfrey-Arku, 2007). Where land is scarce and the rainy season is long enough, it seems easier to find a slot to accommodate short-term cover crops (with a growing period of 50–60 days). Choosing the appropriate cover crop is site specific, and different species must be tested not only for their fitness to the agro-ecological region but also for their agronomical value (e.g. Odhiambo, 2011), namely for the suppression of weeds.

Where crop rotation is not feasible, intercropping is a way to introduce diversification in a farming system. In certain agro-ecological zones of SSA, staple cereal crops are traditionally intercropped with legumes or cassava, with little to no effect on the cereal yields (Ngwira et al., 2012; Nyagumbo et al., 2016) and appreciable improvement of household food security, cash and fodder availability (e.g. by intercropping with cowpea) (Himmelstein et al., 2017). The effects of legumes cropped in the inter-row of a cereal, like maize or sorghum, on the suppression of weeds result from a living mulch effect, smothering weeds and competing with the weeds for water, nutrients, light and so on. A comprehensive study on intercropping cereal–legume in SSA and a number of bibliographic references on weed suppression can be found in the review authored by Matusso et al. (2014). Another practice that can be of particular interest to hinder the development of weeds between cropping seasons is relay cropping, that extends food production and ground cover to the period between seasons. The benefits from both intercropping and relay cropping depend on the main crop and the crop chosen as intercrop/relay crop, and they will vary across different agro-ecological zones, timing of seeding, ease of management and so on. Intercropping and relay cropping should not be proposed to farmers before proper testing of different species/varieties that compose the intercropping system and their local adequacy.

Additionally, in most cases, the spatial configuration of a crop at a given location is based on previous work done under specific agro-industry machinery standards that are often far from technologies used by smallholders. Both different spacing between the rows and crop density should be researched for their effect on yield and weed control (Adigun et al., 2014). Other cultural practices, as simple as the removal of weed flowers in the dry season, after crop harvest, may also contribute to reduce the weed seedbank.

3.2 Discussion and prospects of change in weed control methods

Usage of herbicides in SSA is very low, in most places practically non-existent, and its use is not well documented. Working on data from FAOSTAT for the period of 1994–97, Heisey and Norton (2007) calculated the use of herbicides in SSA to be 11 700 metric tonnes, which represents around 2.3% of the world's use. As these authors point out, this value includes herbicide use in the Republic of South Africa, which represents 69% of the total amount of herbicides used in Africa.

In some crops, such as rice, the use of herbicides is known even among smallholder farmers. In a survey in the Ivory Coast, where smallholder subsistence farmers are the dominant group of rice producers, Adesina et al. (1994) found that 42% were using herbicides jointly with other methods of weed control, primarily hand pulling (only 2% used herbicides exclusively). This survey also showed that the main reasons given for the use of herbicides were the ability to cultivate a larger area and the ease of subsequent weedings; the main reasons given for the non-use of herbicides was the lack of financial resources (53%). Although outdated and sectorial (limited to the reality of the Ivory Coast), this survey allows an insight into the logic at play when it comes to the use of herbicides as a method of weed control by smallholder farmers. Herbicide usage seems to be confined to those with no land constraint, to expand the area under cultivation, and to those that can generate enough cash flow to re-invest in the farm and in order to decrease labour needs. Farmers recognize the economic value of herbicides; they just cannot afford them. A more recent review found that herbicide use is becoming more common in Africa, depending on the country. Among four countries studied in 2013, 55% of surveyed households in Ghana used herbicides, 10% in Tanzania, 8% in Kenya and 5% in Zambia. Based on their results, the authors predicted an increase in use of herbicides as agriculture becomes more commercial in Africa (Grabowski and Jayne, 2016).

SSA encompasses a wide variety of agro-ecological zones and socio-economic conditions that determine the crops produced, yields and systems used. At certain locations, even without the use of herbicides, CA systems may be less labour-consuming than traditional systems, still producing similar or higher yields. In Malawi, the use of planting basins or seeding with the dibble stick into a mulch of maize stalks of 3 Mg/ha has been found to be less time- and labour-consuming than the traditional furrow and ridge system, while providing maize yields as high or even higher (Nyagumbo et al., 2016). At these locations there is a perceived economic value in CA and adoption of CA principles by farmers seems to be the natural consequence.

Where labour demands increase and other socio-economic constraints manifest, farmers' decision-making is more complex.

In 80% of the cases, smallholder farm households in SSA dispose of less than 2 ha for cropping (Harris and Orr, 2014). In addition, FAO data show that the land-to-person ratio is declining fast since the 1960s, meaning that the potential agricultural labour force is increasing faster than land under cultivation (Jayne et al., 2003). The end result is a decline in farm size to a point where it will be no longer feasible to sustain a household. In turn, depending on the level of education, this may lead to a more or less pronounced desertion by the younger generation of the agricultural sector, and eventually to an increase in the land-to-person ratio. Surveys conducted in SSA (seven countries) and India on farm households were the basis of a study conducted by Harris and Orr (2014) that showed that the mean net income from crop production per household, in 2005, ranged from US$139 in Rwanda to US$2336 in Kenya, which represented an annual income in 2005 per adult of working age of US$52–712. These values reflect a monetization that only exists with the purpose of rending a value to the produce and allowing comparisons between different realities. Another important result of these surveys was the proportion of income from off-farm sources. The off-farm income is most probably the source of capital to be invested in farm inputs in the smaller households, and ranged from 5% in Ethiopia to 40% in Kenya, and in India 67% of total income.

In this context, the adoption of CA poses a dilemma for farmers which is: a higher weeding frequency and the associated opportunity cost of labour (hired or family), or the investment in herbicides (or in any other capital intensive method of weed control). The answer is complex and will probably be a function of available capital for investment, work opportunities outside their farms, available family labour and alternative low-cost weed-control strategies, among others. Where the lack of available land, coupled with non-existent or undeveloped markets result in little crop diversification, non-availability of inputs and consequent pure subsistence agriculture, little or no funds will be available for investment. Higher expected yields and margins per se will not translate into the adoption of new technologies that require capital.

3.3 Closing thoughts

Africa is a huge continent, and differences in the livelihoods of smallholder farmers are immense across the agro-ecological zones that are present. Yet, smallholder farmers' sustainable traditional agricultural systems and/or approaches in SSA countries are not very different from what is observed in other parts of the developing world. For example, the intercropping of a cereal crop (maize) and legumes, observed in Africa is the traditional approach in the Yucatan peninsula of Mexico (Caamal-Maldonado et al., 2001), or the plantation of cassava or maize in no-till seedbeds in Africa (Enete et al., 2002) and no-till maize in the Paraná state, Brazil (Derpsch et al.,

1991). Apart from the consequences of tillage-based agriculture promoted during colonial and post-colonial times, the need for soil conservation in Africa results from an increasing pressure on soils to produce yields (due to the declining land-to-person ratio (Jayne et al., 2003)), without proper inputs in terms of management and material inputs, such as fertilizers (organic and/ or inorganic), quality seeds and so on. Other parts of the world, South and Central America, India and other Asian countries experienced the same pressure on soils to produce yields for a growing population. The 'green revolution', although a success in terms of productivity increases, not only introduced improved cultivars, but also boosted higher agricultural inputs in the form of mineral fertilizers, pesticides and machinery for soil tillage with the known consequences regarding pollution and soil degradation. In most SSA countries, poor road networks result in underdeveloped markets or poor market access, both for farmers' produce and agricultural inputs, and farmers take refuge in subsistence crops (that they can eat and use to generate cash for other primary needs).

Weed control technologies other than traditional methods (hand hoe and hand pulling) exist and have or will have its place in SSA smallholder farms, if there is a shift in the socio-economic constraints and proper training. The use of these technologies, herbicides included, will depend not on their efficacy but on their efficiency both from an energetic point-of-view and, maybe primarily, on the perceived economic return.

There is evidence that the adoption of CA does not necessarily mean an increase in the use of herbicides in the long term. Nevertheless, there is evidence of increased weed pressure under CA in certain circumstances, especially in semi-arid areas with low biomass production potential, and that weeding frequency may be higher.

In a continent where smallholder farmers' use of herbicides to control weeds is almost non-existent, and where population growth and food demand is rising, the lack of capital, the small size of farms, the difficulties to access the markets, the absence of off-farm working opportunities and so on, all this contributes to a poverty cycle that does not envisage, in the short-term, an overall growth in the use of agricultural inputs, including machinery, fertilizers and herbicides. This may condition the ready adoption of CA by smallholder farmers unless the challenges of weed control and low soil fertility are effectively and affordably addressed through integrated approaches tailored to the local needs and agro-ecological conditions.

4 Future trends

The uncertainty surrounding the future of weed control based solely on herbicides due to the rapidly increasing number of herbicide-resistant weeds, lack of new herbicides with new sites of action, more restrictive legislation on pesticides (driven by environmental and health concerns), increased scrutiny, the withdrawal of effective herbicides from the market and the uncertainty relating to others, all contribute to the view that future weed control must move towards integrated weed management approaches where herbicides will lose their exclusiveness and partially be replaced by alternative weed control methods.

Under CA, the shift from herbicides to integrated weed management systems requires the study and tailoring of crop rotations and intercropping systems for their weed-suppressive ability, through the evaluation of the evolution of the weed population under different crop(s)-management combinations and possible manipulations (e.g. different spacing, sowing dates or use of more competitive cultivars).

Another aspect is the inclusion in the rotation or intercropping components that are particularly valuable for weed control (e.g. smothering or mulch producing crops and/or their allelopathic potential) and discern the mechanisms at play (e.g. competition for resources or allelopathy). One recent example that is creating much interest among organic farmers in the Eastern United States for its capacity to effectively control weeds, is the 'cover crop-based organic rotational no-till' (Mirsky et al., 2013), where a winter cereal (e.g. rye) or legume is terminated with a roller-crimper and a cash crop (e.g. soybean or maize) is directly seeded in the mulch. However, this approach in organic farming systems frequently still relies on the use of heavy tillage before the cover crop establishment in case perennial weeds pose a problem. In 'conventional' (non-organic) CA without any soil tillage, the same approach known as 'planting green' is also used to reduce or even avoid the application of herbicides to terminate the cover crop. It is already employed in Pennsylvania in the United States, and Brazil by some 'conventional' CA farmers.

Also, research should address the effect of cover crops on weed-crop interference (e.g. leguminous cover crops on weed development in the following crop because of different patterns of N release in relation to inorganic fertilizers); the development stage of a weed that is most effective for a certain control strategy; the stage of development that a weed is more sensitive to a certain herbicide or at lower dosages; breeding for allelopathy and competitiveness; among others.

Special attention must also be given to perennial weeds where they occur under CA and strategies for non-chemical control should be developed. For example, if cover crops alone are not able to control perennial weeds satisfactorily, mowing or grazing at specific stages of the weed development (when the underground reserve structures of the weeds are more depleted of reserves) can help enhance the suppression capacity of a cover crop and result in effective control (e.g. Graglia and Melander, 2005).

5 Conclusion

Since the birth of agriculture, the control of undesired vegetation has been managed through one form or other of mechanical soil movement (tillage). In CA systems, the minimization or even absence of soil disturbance demands alternative ways to keep weed populations within acceptable limits. This chapter elaborates on the different approaches that are used to manage weed populations in agricultural production systems, including physical, chemical, biological and agricultural management practices.

At present, physical weed control methods under CA are, in most cases, reserved to farmers that embrace CA principles but cannot use herbicides (e.g. organic farmers) or where herbicides are not available for a specific crop or are not able to control specific weeds, or simply not affordable. Research on mechanical weed control, by burying, cutting or pulling weeds, has been focussing, in recent years, on the mechanized intra-row weeding in order to obviate the need of hand-weeding, both in permanent and annual crops, complemented with reduced tillage techniques for inter-row. This research trend is further enhanced by the development of image analysis and artificial intelligence applied to robots. The use of plastic sheets (polyethylene) as soil cover, is growing but not without concerns on soil and environmental pollution. Thermal methods, such as direct flaming or soil solarization, have limited use due to equipment/energy costs, adequacy and climatic or other environmental constraints.

The use of herbicides by farmers is generalized, although it is not a 'sine qua non' condition to control weeds under CA, because of the effectiveness, low cost, ease of use and high return. Despite the existing concerns in society on the agricultural use of herbicides, new and more restrictive legislation and regulations on their use and environmental exposure to pesticides, their use, at present, is of paramount importance for crop production and world food security. That said, the fact that herbicides with new sites of action have not been released to the market, and that the development of new herbicide-resistant weeds is increasing rapidly, is forcing many farmers to rethink their strategy for managing weeds. The principles underlying CA, including a higher soil bearing-capacity, especially under no-till, and integrated weed management strategies that encompass crop rotation, cover crops, rotation of herbicides with different sites of action and so on increases the opportunities to reduce herbicide usage when compared to conventional systems, thus maximizing agronomic, economic and environmental gains. On the other hand, it may require a better knowledge by the farmer of prevalent weed species and their ecology, their sensitivity to herbicides, available herbicides, particularly their mode of action and efficacy at different crop residue levels and weed development stages, and the period of weed interference on crop yield.

Driven by the need to reduce herbicides inputs in agricultural systems, biological methods of weed control have been the subject of intensive study, with promising results and, in some instances, has led to new cropping systems (e.g. cereal-*Desmodium* spp. intercropping for the control of the weed *Striga* spp.). Biological methods of weed control are based on the natural interactions/mechanisms that drive interspecific relationships, to control the balance of pest populations, and approaches range from selection of cultivars for specific purposes, such as the allelopathic potential, or high competitiveness, to bioherbicides obtained from microorganisms or plant extracts.

Different agricultural practices have different effects on the evolution of the soil seedbank, both in terms of density and diversity. Under CA, minimum soil disturbance will promote the accumulation of weed seeds near or at the soil surface and lack of tillage will promote the survival of subterranean propagation of vegetative structures of perennial weeds, explaining why higher weed densities are observed when starting with no-till and the difficulty associated with the control of perennial species. Nevertheless, CA will also promote higher weed seed decay and predation. The net effect of lack of tillage on soil weed seedbank dynamics depends on how the whole system works, stressing the need to prevent seed rain under the absence of tillage.

Permanent soil cover through mulches plays an important role on the control of weeds by reducing the light that weed seeds need to break dormancy and for the initial growth of the plantlets, by physically restraining them and by creating a propitious environment for weed seed decay and predation. This effect depends on the amount of crop residue, soil cover that it offers and thickness. Another aspect of mulches results from the presence or production of allelopathic substances while being decomposed by microorganisms (e.g. rye mulch) that can have a profound effect on weed germination and growth; although mulches of different crop residues have different allelopathic potential and weeds have different sensitivity. Cover crops (cereals, legumes and brassicas) offer soil cover between seasons, smothering weeds and acting like a living mulch; at the beginning of next season they can be used as mulch for the next crop, and weed suppression attained, depending on cover crop species, biomass produced and correct management of the cover crop, may be enough to avoid the need for herbicide usage.

Along with the lack of soil disturbance and permanent organic soil cover, the third differentiating aspect of CA is crop diversification that also contributes to weed suppression. Crop rotation and the inherent succession of different management practices at different timings will put different pressure (timing of weedings, different herbicides, varying patterns of resource competition etc.) on different weeds, preventing a particular weed or group of weeds to be repeatedly successful at a particular site, and reducing the occurrence of weed-resistance to certain herbicides. Crop rotation allows the introduction of cover crops that will further enhance the weed suppression ability of the crop rotation, by their smothering and mulch effect. Where crop rotation is not feasible, intercropping and relay cropping are ways to achieve crop diversification and higher pressure on weeds. Other ways to increase the pressure on weeds are the use of early/late maturing crop varieties, altering the timing and management practices of the crops, affecting the weeds that mimic the crop and preventing them from becoming too successful.

Other management practices, like weed seeds harvest/retaining and off-field destruction, or crop density, spatial distribution and line orientation, may be easily managed and have a profound effect on soil weed seedbank, crop competitiveness, allelopathy and surface area for weed development.

Weed control under CA in the developing countries of SSA by smallholder farmers is achieved mainly by hand-weeding. The labour increase that is often associated with hand-weeding crops under CA when compared to conventional tillage, is the main reason pointed out by many authors for the low adoption of CA in SSA, mainly in semi-arid regions, where biomass for mulching is unavailable and preferentially saved to feed the livestock. In other regions, conventional techniques may be far more labour-consuming than CA, as is the case where ridges and furrows are used as a water conservation technique, and in many such regions, no-till systems and the use of the dibble stick have proven to be as productive or even produce higher yields than conventional systems, and a higher adoption of CA can be foreseen. There is no single approach for CA in SSA, and most probably there will be situations where the three principles of CA cannot be fully implemented, and the full benefits from weed control will not be attained. Where modern agriculture is practised, the most obvious alternative to mechanical weeding is the use of herbicides, and the same is valid for SSA agriculture. For the smallholder farmer, with less than 2 ha for cropping, non-existent or poor markets for produce and inputs, non-existence of roads, of off-farm work opportunities, of rural extension services and so on narrows the choices and opportunities, and lack of financial resources to buy the inputs needed for alternative methods of weed control (e.g. herbicides and spraying equipment), and/or knowledge on how to use them may condition further the ready adoption of CA.

At present, the best strategies under CA to cope with the ubiquity and resilience of weeds, and to keep them at a level that allows a low impact on the income of the farmer and maximum sustainability of the farm, have to be developed at the cropping system level (rotations/crop sequences; intercropping). The key issues that must be addressed are keeping the seedbank at a low level (although with a diverse composition), impeding the prevalence of a single or small group of weeds well adapted to the crops and management practices. This requires crop diversification and diverse management practices.

The balance between depletion/inputs of weed seeds in the soil seedbank must be monitored. Over time, the absence of tillage will lead to a much faster depletion of the soil seedbank than traditional tillage, if seed set/rain is prevented, but on the other hand, strategies to control perennial weeds must be further developed.

Concordant with the previous paragraph, when designing crop rotations/intercropping, choosing a crop sequence or crop mix for a desired weed control effect should allow a certain degree of manipulation (e.g. early-late sowing, or varieties' competitiveness), and the inclusion of components in the rotation with marked effects on weed control (e.g. cover crops or meadows) should be favoured. The study of regionally best adapted cover crops/meadows and their management to achieve the desired impact on weeds would increase the options available for non-chemical weed control.

Another aspect, although only indirectly related to the subject, is the need to promote the diversity of diets, especially in modern societies, and to decrease the dependency on a few cereals or other staple crops. This dietary diversification would create the market and the opportunities to commercialize a range of today's 'less' important crops and amplify the options available in terms of crop rotation and the improvement of soil fertility.

This text and the information reviewed indicate that weed control under CA farming systems, in a frame of reduction of herbicide inputs, is possible. Under CA, where weed control through tillage is not an option, crop diversification and soil cover management are key for keeping the weed population at a manageable size, thus avoiding negative impacts on yields, especially in the long term. Future research should not be limited to whether crop diversification (crop rotation and/or intercropping) and soil cover can help maintaining weed population under control but should also focus on the mechanisms of weed suppression that are in action, allowing further refinement of the cropping systems. Depletion of the soil weed seedbank and perennial weeds' vegetative propagation structures are essential features in the design of crop rotations and intercropping approaches.

6 Where to look for further information

CA is relatively new, and most weed management experiments that stray from conventional approaches are recent, and results are mainly disseminated in scientific papers. As there is no 'one-size-fits-all' solution, a good starting point to obtain further information is to get acquainted with the numerous solutions that have been developed worldwide to tackle the weed problem when adopting CA. The book *Conservation Agriculture: Global Prospects and Challenges*, edited by Ram A. Jat, Kanwar L. Sahrawat and Amir H. Kassam (CAB International), in 2014, is a comprehensive textbook that covers CA adoption and challenges, including weed management, in all continents and main agricultural countries. Another interesting textbook is *Conservation Agriculture*, edited by Muhammad Farooq and Kadambot H. M. Siddique (Springer), in 2015, with a dedicated chapter on weed management in CA systems and a similar but distinct approach on CA adoption and evolution in different regions worldwide.

A more detailed discussion on herbicide use under CA when compared to tillage-based systems, and the common idea that herbicide use will increase where weeds are not controlled by tillage, can be found in Friedrich and Kassam's paper 'No-till farming and the environment: do no-till systems require more chemicals?' (2012).

Information on herbicide-resistant weeds can be found at the International Survey of Herbicide-Resistant Weeds site (http://www.weedscience.org/). A good introductory text on herbicide-resistant weeds is Ian Heap's *Herbicide Resistant Weeds* (Heap, 2014a).

Rising technologies, like the use of high-voltage electrical currents, may reduce in the near future the overall need for herbicides and allow full automation of weed control (see https://zasso.eu/en/technologies-digital-weeding/).

New management practices with potential benefits for weed management under CA, like *planting green*, have only been consistently studied in the United States for 3–4 years. A good English language source of information is PennState Extension, Pennsylvania State University (https://extension.psu.edu/). For Portuguese speaking readers, EMBRAPA (Empresa Brasileira de Pesquisa Agropecuária) has a vast collection of documents on *plantio verde* (planting green) and *plantio directo* (no-till) (https://www.embrapa.br/).

Professional and scientific organizations are also good resources if looking for information on current research and on many aspects of weed management. To name only a few: European Weed Research Society (EWRS) (http://www.ewrs.org/); Weed Science Society of America (WSSA) (http://wssa.net/); and the Council of Australasian Weed Societies (CAWS) (http://caws.org.nz/).

The best source for CA in SSA is the African Conservation Tillage Network (http://www.act-africa.org/).

7 References

Abu-Gharbieh, W., Saleh, H. and Abu-Blan, H. n.d. Use of black plastic for soil solarization and post-plant mulching. Available at: https://bit.ly/2Ugbn7v (Accessed on 25 January 2019).

Adesina, A., Johnson, D. and Heinrichs, E. 1994. Rice pests in the Ivory Coast, West Africa: farmers' perceptions and management strategies. *International Journal of Pest Management* 40(4), 293–9.

Adigun, J., Osipitan, A. O., Lagoke, S. T., Adeyemi, R. O. and Afolami, S. O. 2014. Growth and yield performance of cowpea (*Vigna unguiculata* (L.) Walp) as influenced by row-spacing and period of weed interference in South-West Nigeria. *Journal of Agricultural Science* 6(4), 188–98. doi:10.5539/jas.v6n4p188.

Aflakpui, G. and Bolfrey-Arku, G. 2007. Integrated management of *Imperata cylindrica* (Speargrass) in yam and cassava: weed pressure in crop, crop growth, and yield. *Journal of Plant Sciences* 2(1), 14–24. doi:10.3923/jps.2007.14.24.

Amossé, C., Jeuffroy, M., Celette, F. and David, C. 2013. Relay-intercropped forage legumes help to control weeds in organic grain production. *European Journal of Agronomy* 49, 158–67. doi:10.1016/j.eja.2013.04.002.

Atwood, D. and Paisley-Jones, C. 2017. *Pesticides Industry Sales and Usage 2008–2012 Market Estimates*. EPA United States Environmental Protection Agency, Washington DC.

Bandyopadhyay, P. K., Singh, K. C., Mondal, K., Nath, R., Ghosh, P. K., Kumar, N., Basu, P. S. and Singh, S. S. 2016. Effects of stubble length of rice in mitigating soil moisture stress and on yield of lentil (*Lens culinaris* Medik) in rice-lentil relay crop. *Agricultural Water Management* 173, 91–102. doi:10.1016/j.agwat.2016.05.009.

Baraibar, B., Westerman, P. R., Carrión, E. and Recasens, J. 2009. Effects of tillage and irrigation in cereal fields on weed seed removal by seed predators. *Journal of Applied Ecology* 46(2), 380–7. doi:10.1111/j.1365-2664.2009.01614.x.

Barros, J. F. C., Basch, G. and Carvalho, M. 2005. Effect of reduced doses of a post-emergence graminicide mixture to control *Lolium rigidum* G. in winter wheat under direct drilling in Mediterranean environment. *Crop Protection* 24(10), 880–7. doi:10.1016/j.cropro.2005.01.020.

Barros, J. F. C., Basch, G. and Carvalho, M. 2007. Effect of reduced doses of a post-emergence herbicide to control grass and broad-leaved weeds in no-till wheat under Mediterranean conditions. *Crop Protection* 26(10), 1538–45. doi:10.1016/j.cropro.2006.12.017.

Barros, J. F. C., Basch, G. and Carvalho, M. 2008. Effect of reduced doses of a post-emergence graminicide to control *Avena sterilis* L. and *Lolium rigidum* G. in no-till wheat under Mediterranean environment. *Crop Protection* 27(6), 1031–7. doi:10.1016/j.cropro.2007.12.006.

Basch, G. (unpublished data 2018).

Baskin, J. M. and Baskin, C. C. 1985. The annual dormancy cycle in buried weed seeds: a continuum. *BioScience* 35(8), 492–8. doi:10.2307/1309817.

Bawden, O., Kulk, J., Russell, R., McCool, C., English, A., Dayoub, F., Lehnert, C. and Perez, T. 2017. Robot for weed species plant-specific management. *Journal of Field Robotics* 34(6), 1179–99. doi:10.1002/rob.21727.

Bayer, C., Mielniczuk, J., Amado, T. J. C., Martin-Neto, L. and Fernandes, S. V. 2000. Organic matter storage in a sandy clay loam acrisol affected by tillage and cropping systems in southern Brazil. *Soil and Tillage Research* 54(1–2), 101–9. doi:10.1016/S0167-1987(00)00090-8.

Bhadoria, P. 2011. Allelopathy: a natural way towards weed management. *American Journal of Experimental Agriculture* 1(1), 7–20. doi:10.9734/AJEA/2011/002.

Bonanomi, G., Chiurazzi, M., Caporaso, S., Del Sorbo, G., Moschetti, G. and Felice, S. 2008. Soil solarization with biodegradable materials and its impact on soil microbial communities. *Soil Biology and Biochemistry* 40(8), 1989–98. doi:10.1016/j.soilbio.2008.02.009.

Bond, W. and Grundy, A. C. 2001. Non-chemical weed management in organic farming systems. *Weed Research* 41(5), 383–405. doi:10.1046/j.1365-3180.2001.00246.x.

Caamal-Maldonado, J. A., Jiménez-Osornio, J. J., Torres-Barragán, A. and Anaya, A. L. 2001. The use of allelopathic legume cover and mulch species for weed control in cropping systems. *Agronomy Journal* 93(1), 27–36. doi:10.2134/agronj2001.93127x.

Calado, J. M. G., Basch, G. and de Carvalho, M. 2009. Weed emergence as influenced by soil moisture and air temperature. *Journal of Pest Science* 82, 81–8.

Calado, J. M. G., Basch, G. and de Carvalho, M. 2010. Weed management in no-till winter wheat (*Triticum aestivum* L.). *Crop Protection* 29(1), 1-6. doi:10.1016/j.cropro.2009.09.011.

Cardinale, B. J., Wright, J. P., Cadotte, M. W., Carroll, I. T., Hector, A., Srivastava, D. S., Loreau, M. and Weis, J. J. 2007. Impacts of plant diversity on biomass production increase through time because of species complementarity. *Proceedings of the National Academy of Sciences* 104(46), 18123–8. doi:10.1073/pnas.0709069104.

Cardoso, C., Ruyter-Spira, C. and Bouwmeester, H. J. 2011. Strigolactones and root infestation by plant-parasitic *Striga*, *Orobanche* and *Phelipanche* spp. *Plant Science: an International Journal of Experimental Plant Biology* 180(3), 414–20. doi:10.1016/j.plantsci.2010.11.007.

Changrong, Y., Wenqing, H., Turner, N., Enke, L., Qin, L. and Shuang, L. 2014. Plastic-film mulch in Chinese agriculture: importance and problems. *World Agriculture* 4(2), 32–6.

Chauhan, B. S. and Johnson, D. E. 2011. Row spacing and weed control timing affect yield of aerobic rice. *Field Crops Research* 121(2), 226–31. doi:10.1016/j.fcr.2010.12.008.

Cheng, F. and Cheng, Z. 2015. Research progress on the use of plant allelopathy in agriculture and the physiological and ecological mechanisms of allelopathy. *Frontiers in Plant Science* 6, Article 1020. doi:10.3389/fpls.2015.01020.

Chidawanyika, F., Mudavanhu, P. and Nyamukondiwa, C. 2012. Biologically based methods for pest management in agriculture under changing climates: challenges and future directions. *Insects* 3(4), 1171–89. doi:10.3390/insects3041171.

Cordeau, S., Triolet, M., Wayman, S., Steinberg, C. and Guillemin, J.-P. 2016. Bioherbicides: dead in the water? A review of the existing products for integrated weed management. *Crop Protection* 87, 44–9. doi:10.1016/j.cropro.2016.04.016.

Council Directive 2009/128/EC of 21 October 2009 establishing a framework for Community action to achieve the sustainable use of pesticides. Available at: https://bit.ly/2BdtWlV (Accessed on 25 January 2019).

Council Directive 98/83/EC of 3 November 1998 on the quality of water intended for human consumption. Available at: https://bit.ly/2Ht96nW (Accessed on 25 January 2019).

Davis, A. S. 2010. Cover-crop roller–crimper contributes to weed management in no-till soybean. *Weed Science* 58(3), 300–9. doi:10.1614/WS-D-09-00040.1.

Derpsch, R., Roth, C., Sidiras, N. and Köpke, U. 1991. *Controlo da erosão no Paraná, Brasil: sistemas de cobertura do solo, plantio directo e preparo conservacionista do solo*. Deutsche Gesellschaft für Technische Zusammenarbeit (GTZ) GmbH, Eschborn.

Dhammapala, R., Claiborn, C., Corkill, J. and Gullett, B. 2006. Particulate emissions from wheat and Kentucky bluegrass stubble burning in eastern Washington and northern Idaho. *Atmospheric Environment* 40(6), 1007–15. doi:10.1016/j.atmosenv.2005.11.018.

Duke, S. O. 2012. Why have no new herbicide modes of action appeared in recent years? *Pest Management Science* 68(4), 505–12. doi:10.1002/ps.2333.

Ekeleme, F., Akobundu, I. O., Isichei, A. O. and Chikoye, D. 2003. Cover crops reduce weed seedbanks in maize-cassava systems in southwestern Nigeria. *Weed Science* 51(5), 774–80. doi:10.1614/0043-1745(2003)051[0774:CCRWSI]2.0.CO;2.

Enete, A., Nweke, F. and Tollens, E. 2002. Contributions of men and women to food crop production labour in Africa: information from COSCA. *Outlook on Agriculture* 31(4), 259–65. doi:10.5367/000000002101294155.

FAO. 2017. Conservation Agriculture. Available at: http://www.fao.org/3/a-i7480e.pdf (Accessed on 25 January 2019).

Fletcher, A. L., Kirkegaard, J. A., Peoples, M. B., Robertson, M. J., Whish, J. and Swan, A. D. 2016. Prospects to utilise intercrops and crop variety mixtures in mechanised, rain-fed, temperate cropping systems. *Crop and Pasture Science* 67(12), 1252–67. doi:10.1071/CP16211.

Friedrich, T. and Kassam, A. 2012. No-till farming and the environment: do no-till systems require more chemicals? *Outlooks on Pest Management* 23, 153–7.

Friedrich, T., Kassam, A. and Corsi, S. 2014. Conservation Agriculture in Europe. In: Jat, R., Sahrawat, K. and Kassam, A. (Eds), *Conservation Agriculture: Global Prospects and Challenges*. CAB International, Wallingford, pp. 127–79. Chapter 6.

Gealy, D., Moldenhauer, K. and Duke, S. 2013. Root distribution and potential interactions between allelopathic rice, sprangletop (*Leptochloa* spp.), and barnyardgrass (*Echinochloa crus-galli*) based on 13C isotope discrimination analysis. *Journal of Chemical Ecology* 39(2), 186–203. doi:10.1007/s10886-013-0246-7.

Goldberger, J. R., Jones, R. E., Miles, C. A., Wallace, R. W. and Inglis, D. A. 2015. Barriers and bridges to the adoption of biodegradable plastic mulches for US specialty crop production. *Renewable Agriculture and Food Systems* 30(2), 143–53. doi:10.1017/S1742170513000276.

Grabowski, P. and Jayne, T. 2016. Analyzing trends in herbicide use in sub-Saharan Africa. MSU International Development Working Paper 142. Michigan State University, East Lansing, MI.

Graglia, E. and Melander, B. 2005. Mechanical control of *Cirsium arvense* in organic farming. Paper at *European Weed Research Society Symposium*, Bari, Italy, 20–23 June. Available at: http://orgprints.org/4600/1/4600.pdf (accessed 25 January 2019).

Harris, D. and Orr, A. 2014. Is rainfed agriculture really a pathway from poverty? *Agricultural Systems* 123, 84–96. doi:10.1016/j.agsy.2013.09.005.

Heap, I. 2014a. Herbicide resistant weeds. In: Pimentel, D. and Peshin, R. (Eds), *Integrated Pest Management*. Springer Science+Business Media, Dordrecht, pp. 281–301. Chapter 12.

Heap, I. 2014b. Global perspective of herbicide-resistant weeds. *Pest Management Science* 70(9), 1306–15. doi:10.1002/ps.3696.

Heap, I. and Duke, S. O. 2018. Overview of glyphosate-resistant weeds worldwide. *Pest Management Science* 74(5), 1040–9. doi:10.1002/ps.4760.

Heisey, P. and Norton, G. W. 2007. Fertilizers and other farm chemicals. In: Evenson, R. and Pingali, P. (Eds), *Handbook of Agricultural Economics* (vol. 3). Elsevier B.V., pp. 2741–77. Chapter 53.

Hiep, N., Wiktorsson, H. and Man, N. 2008. The effect of molasses on the quality of Kudzu silage and evaluation of feed intake and digestibility of diets supplemented with Kudzu silage or Kudzu hay by heifers. *Livestock Research for Rural Development* 20(Suppl.).

Himmelstein, J., Ares, A., Gallagher, D. and Myers, J. 2017. A meta-analysis of intercropping in Africa: impacts on crop yield, farmer income, and integrated pest management effects. *International Journal of Agricultural Sustainability* 15(1), 1–10. doi:10.1080/14735903.2016.1242332.

Horowitz, M., Regev, Y. and Herzlinger, G. 1983. Solarization for weed control. *Weed Science* 31, 170–9.

IPBO. 2017. Maize in Africa. Fact Series. International Plant Biotechnology Outreach. Available at: http://www.vib.be/en/about-vib/Documents/VIB_MaizeInAfrica_EN_2017.pdf (Accessed on 25 January 2019).

Jayne, T. S., Yamano, T., Weber, M. T., Tschirley, D., Benfica, R., Chapoto, A. and Zulu, B. 2003. Smallholder income and land distribution in Africa: implications for poverty reduction strategies. *Food Policy* 28(3), 253–75. doi:10.1016/S0306-9192(03)00046-0.

Kanampiu, F. K., Kabambe, V., Massawe, C., Jasi, L., Friesen, D., Ransom, J. K. and Gressel, J. 2003. Multi-site, multi-season field tests demonstrate that herbicide seed-coating herbicide-resistance maize controls Striga spp. and increases yields in several African countries. *Crop Protection* 22(5), 697–706. doi:10.1016/S0261-2194(03)00007-3.

Kassam, A., Friedrich, T., Derpsch, R. and Kienzle, J. 2015. Overview of the worldwide spread of Conservation Agriculture. Field Actions Science Reports (Online), vol. 8. Available at: http://factsreports.revues.org/3966 (Accessed on 25 January 2019).

Kassam, A., Friedrich, T. and Derpsch, R. 2019. Global spread of Conservation Agriculture. *International Journal of Environmental Studies* 76(1), 29–51. doi:10.1080/00207233.2018.1494927.

Katan, J. 2015. Soil solarization: the idea, the research and its development. *Phytoparasitica* 43(1), 1–4. doi:10.1007/s12600-014-0419-0.

Katan, J., Greenberger, A., Alon, H. and Grinstein, A. 1976. Solar heating by polyethylene mulching for the control of diseases caused by soil-borne pathogens. *Phytopathology* 66(5), 683–8. doi:10.1094/Phyto-66-683.

Khan, Z. R., Pickett, J. A., Wadhams, L. J., Hassanalia, A. and Midega, C. A. O. 2006. Combined control of *Striga hermonthica* and stemborers by maize-*Desmodium* spp. intercrops. *Crop Protection* 25(9), 989–95. doi:10.1016/j.cropro.2006.01.008.

Khan, Z. R., Midega, C. A. O., Pittchar, J. O., Murage, A. W., Birkett, M. A., Bruce, T. J. A. and Pickett, J. A. 2014. Achieving food security for one million sub-Saharan African poor through push-pull innovation by 2020. *Philosophical Transactions of the Royal Society B: Biological Sciences* 369(1639), 20120284. doi:10.1098/rstb.2012.0284.

Kia, S. H., Schulz, M., Ayah, E., Schouten, A., Müllenborn, C., Paetz, C., Schneider, B., Hofmann, D., Disko, U., Tabaglio, V., et al. 2014. *Abutilon theophrasti*´s defense against the allelochemical benzoxazolin-2(3H)-one: support by *Actinomucor elegans*. *Journal of Chemical Ecology* 40(11–12), 1286–98. doi:10.1007/s10886-014-0529-7.

Kliewer, I., Casaccia, J., Vallejos, J. and Derpsch, R. 2000. Cost and herbicide reduction in the no-tillage system by using green manure cover crops in Paraguay. In: *Proceedings 15th ISTRO Conference, Fort Worth*, Texas, USA.

Liebman, M. and Janke, R. 1990. Sustainable weed management practices. In: Francis, C. A., Flora, C. B. and King, L. D. (Eds), *Sustainable Agriculture in Temperate Zones*. John Wiley & Sons, New York, pp. 111–43. Chapter 4.

Mäder, P. and Berner, A. 2012. Development of reduced tillage systems in organic farming in Europe. *Renewable Agriculture and Food Systems* 27(1), 7–11. doi:10.1017/S1742170511000470.

Mashingaidze, N., Madakadze, C., Twomlow, S., Nyamangara, J. and Hove, L. 2012. Crop yield and weed growth under Conservation Agriculture in semi-arid Zimbabwe. *Soil and Tillage Research* 124, 102–10. doi:10.1016/j.still.2012.05.008.

Matusso, J., Mugwe, J. and Mucheru-Muna, M. 2014. Potential role of cereal-legume intercropping systems in integrated soil fertility management in smallholder farming systems of sub-Saharan Africa. *Research Journal of Agriculture and Environmental Management* 3(3), 162–74.

Melander, B., Rasmussen, I. A. and Bàrberi, P. 2005. Integrating physical and cultural methods of weed control – examples from European research. *Weed Science* 53(3), 369–81. doi:10.1614/WS-04-136R.

Mirsky, S. B., Ryan, M. R., Teasdale, J. R., Curran, W. S., Reberg-Horton, C. S., Spargo, J. T., Wells, M. S., Keene, C. L. and Moyer, J. W. 2013. Overcoming weed management challenges in cover crop-based organic rotational no-till soybean production in the Eastern United States. *Weed Technology* 27(1), 193–203. doi:10.1614/WT-D-12-00078.1.

Miura, Y. and Kanno, T. 1997. Emissions of trace gases (CO_2, CO, CH_4, and N_2O) resulting from rice straw burning. *Soil Science and Plant Nutrition* 43(4), 849–54. doi:10.1080/00380768.1997.10414651.

Mohemed, N., Charnikhova, T., Bakker, E. J., van Ast, A., Babiker, A. G. and Bouwmeester, H. J. 2016. Evaluation of field resistance to *Striga hermonthica* (Del.) Benth. in *Sorghum bicolor* (L.) Moench. The relationship with strigolactones. *Pest Management Science* 72(11), 2082–90. doi:10.1002/ps.4426.

Mojžiš, M., Vitázek, I., Varga, F. and Lindák, S. 2015. Experimental determination of lethal doses of heat in thermal weed control. *Research in Agricultural. Engineering* 61(Special Issue), 9–12.

Muoni, T., Rusinamhodzi, L. and Thierfelder, C. 2013. Weed control in Conservation Agriculture systems of Zimbabwe: identifying economical best strategies. *Crop Protection* 53, 23–8. doi:10.1016/j.cropro.2013.06.002.

Muoni, T., Rusinamhodzi, L., Rugare, J. T., Mabasa, S., Mangosho, E., Mupangwa, W. and Thierfelder, C. 2014. Effect of herbicide application on weed flora under Conservation Agriculture in Zimbabwe. *Crop Protection* 66, 1–7. doi:10.1016/j.cropro.2014.08.008.

Mupangwa, W., Twomlow, S. and Walke,S. 2012. Reduced tillage, mulching and rotational effects on maize (*Zea mays* L.), cowpea (*Vigna unguiculata* (Walp) L.) and sorghum (*Sorghum bicolor* L. (Moench)) yields under semi-arid conditions. *Field Crops Research* 132, 139–48. doi:10.1016/j.fcr.2012.02.020.

Murphy, S. D., Clements, D. R., Belaoussoff, S., Kevan, P. G. and Swanton, C. J. 2006. Promotion of weed species diversity and reduction of weed seedbanks with conservation tillage and crop rotation. *Weed Science* 54(1), 69–77. doi:10.1614/WS-04-125R1.1.

Myers, J., Antoniou, M., Blumberg, B., Carroll, L., Colborn, T., Everett, L. G., Hansen, M., Landrigan, P. J., Lanphear, B. P., Mesnage, R., et al. 2016. Concerns over use of glyphosate-based herbicides and risks associated with exposures: a consensus statement. *Environmental Health* (published online) 15, 19. doi:10.1186/s12940-016-0117-0.

Narwal, S., Sindel, B. and Jessop, R. 2006. Tillage and residue burning affects weed populations and seed banks. *Communications in Agricultural and Applied Biological Sciences* 71(3 Pt A), 715–23.

Neve, P. 2008. Simulation modelling to understand the evolution and management of glyphosate resistance in weeds. *Pest Management Science* 64(4), 392–401. doi:10.1002/ps.1495.

Ngwira, A. R., Aune, J. B. and Mkwinda, S. 2012. On-farm evaluation of yield and economic benefit of short term maize legume intercropping systems under Conservation Agriculture in Malawi. *Field Crops Research* 132, 149–57. doi:10.1016/j.fcr.2011.12.014.

Nichols, V., Verhulst, N., Cox, R. and Govaerts, B. 2015. Weed dynamics and Conservation Agriculture principles: a review. *Field Crops Research* 183, 56–68. doi:10.1016/j.fcr.2015.07.012.

Nyagumbo, I., Mkuhlani, S., Pisa, C., Kamalongo, D., Dias, D. and Mekuria, M. 2016. Maize yield effects of Conservation Agriculture based maize-legume cropping systems in contrasting agro-ecologies of Malawi and Mozambique. *Nutrient Cycling in Agroecosystems* 105(3), 275–90. doi:10.1007/s10705-015-9733-2.

Nyamangara, J., Mashingaidze, N., Masvaya, E. N., Nyengarai, K., Kunzekweguta, M., Tirivavi, R. and Mazvimavi, K. 2014. Weed growth and labor demand under hand-hoe based reduced tillage in smallholder farmers' fields in Zimbabwe. *Agriculture, Ecosystems and Environment* 187, 146–54. doi:10.1016/j.agee.2013.10.005.

Odhiambo, J. 2011. Potential use of green manure legume cover crops in smallholder maize production systems in Limpopo province, South Africa. *African Journal of Agricultural Research* 6(1), 107–12.

Olofsdotter, M., Navarez, D., Rebulanan, M. and Streibig, J. 1999. Weed-suppressing rice cultivars – does allelopathy play a role? *Weed Research* 39(6), 441–54. doi:10.1046/j.1365-3180.1999.00159.x.

Pannacci, E., Tei, F. and Guiducci, M. 2017. Mechanical weed control in organic winter wheat. *Italian Journal of Agronomy* 11(4). doi:10.4081/ija.2017.900.

Price, A. J., Balkcom, K. S., Culpepper, S. A., Kelton, J. A., Nichols, R. L. and Schomberg, H. 2011. Glyphosate-resistant *Palmer amaranth*: A threat to conservation tillage. *Journal of Soil and Water Conservation* 66(4), 265–75. doi:10.2489/jswc.66.4.265.

Pupalienė, R., Sinkevičienė, A., Jodaugienė, D. and Bajorienė, K. 2015. Weed control by organic mulch in organic farming system. In: Pilipavicius, V. (Ed.), *Weed Biology and Control*. IntechOpen. doi:10.5772/60120.

Radics, L. and Bognár, E. 2004. Comparison of different mulching methods for weed control in organic green bean and tomato. In: *Proceedings of the XXVI IHC – Sustainability of Horticultural Systems. Acta Horticulturae* 638(638), 189–96. doi:10.17660/ActaHortic.2004.638.24.

Rasmussen, I. A. 2004. The effect of sowing date, stale seedbed, row width and mechanical weed control on weeds and yields of organic winter wheat. *Weed Research* 44(1), 12–20. doi:10.1046/j.1365-3180.2003.00367.x.

Regnier, E. and Janke, R. 1990. Evolving strategies for managing weeds. In: Edwards, C., et al. (Eds), *Sustainable Agricultural Systems*. Soil and Water Conservation Society, Ankeny, IA, pp. 174-202.

Romano-Armada, N., Amoroso, M. J. and Rajal, V. 2017. Effect of glyphosate application on soil quality and health under natural and zero tillage field conditions. *Soil and Environment* 36(2), 141–54. doi:10.25252/SE/17/51241.

Samtani, J. B., Derr, J., Conway, M. A. and Flanagan III, R. D. 2017. Evaluating soil solarization for weed control and strawberry (*Fragaria xananassa*) yield in annual plasticulture production. *Weed Technology* 31(3), 455–63. doi:10.1017/wet.2017.4.

Sass, J. B. and Colangelo, A. 2006. European union bans atrazine, while the United States negotiates continued use. *International Journal of Occupational and Environmental Health* 12(3), 260–7. doi:10.1179/oeh.2006.12.3.260.

Schulz, M., Marocco, A., Tabaglio, V., Macias, F. A. and Molinillo, J. M. G. 2013. Benzoxazinoids in rye allelopathy - from discovery to application in sustainable weed control and organic farming. *Journal of Chemical Ecology* 39(2), 154–74. doi:10.1007/s10886-013-0235-x.

Shahzad, M., Farooq, M., Jabran, K. and Hussain, M. 2016. Impact of different crop rotations and tillage systems on weed infestation and productivity of bread wheat. *Crop Protection* 89, 161–9. doi:10.1016/j.cropro.2016.07.019.

Sheahan, M. and Barrett, C. 2014. Understanding the agricultural input landscape in sub-Saharan Africa: recent plot, household, and community-level evidence. Policy Research Working Paper 7014. World Bank Group.

Sosnoskie, L. M., Herms, C. P. and Cardina, J. 2006. Weed seedbank community composition in a 35-yr-old tillage and rotation experiment. *Weed Science* 54(2), 263–73. doi:10.1614/WS-05-001R2.1.

Swanton, C. J., Shrestha, A., Knezevic, S. Z., Roy, R. C. and Ball-Coelho, B. R. 2000. Influence of tillage type on vertical weed seedbank distribution in a sandy soil. *Canadian Journal of Plant Science* 80(2), 455–7. doi:10.4141/P99-020.

Tabaglio, V., Gavazzi, C., Schulz, M. and Marocco, A. 2008. Alternative weed control using the allelopathic effect of natural benzoxazinoids from rye mulch. *Agronomy for Sustainable Development* 28(3), 397–401. doi:10.1051/agro:2008004.

Tallman, S. 2012. *No-Till Case Study, Brown's Ranch: Improving Soil Health Improves the Bottom Line*. National Sustainable Agriculture Information Service, United States Department of Agriculture, Washington DC.

Teixeira, F. and Basch, G. 1998. Effects of the traditional and no-tillage systems upon runoff and herbicide transport. In: Pereira, L. and Gowing, J. (Eds), *Water and the Environment - Innovation Issues in Irrigation and Drainage*. E & FN Spon, London, pp. 27–34.

Thierfelder, C., Cheesman, S. and Rusinamhodzi, L. 2013. Benefits and challenges of crop rotations in maize-based Conservation Agriculture (CA) cropping systems of southern Africa. *International Journal of Agricultural Sustainability* 11(2), 108–24. doi :10.1080/14735903.2012.703894.

Tilman, E. A., Tilman, D., Crawley, M. J. and Johnston, A. E. 1999. Biological weed control via nutrient competition: potassium limitation of dandelions. *Ecological Applications* 9(1), 103–11. doi:10.1890/1051-0761(1999)009[0103:BWCVNC]2.0.CO;2.

Timossi, P. C. 2009. Manejo de Rebrotes de *Digitaria insularis* no Plantio Direto de Milho. *Planta Daninha* 27(1), 175–9. doi:10.1590/S0100-83582009000100022.

Walsh, M. J. and Powles, S. B. 2014. High seed retention at maturity of annual weeds infesting crop fields highlights the potential for harvest weed seed control. *Weed Technology* 28(3), 486–93. doi:10.1614/WT-D-13-00183.1.

Wang, Y., Xie, Z., Malhi, S. S., Vera, C. L., Zhang, Y. and Wang, J. 2009a. Effects of rainfall harvesting and mulching technologies on water use efficiency and crop yield in the semi-arid Loess Plateau, China. *Agricultural Water Management* 96(3), 374–82. doi:10.1016/j.agwat.2008.09.012.

Wang, F. X., Feng, S., Hou, X., Kang, S. and Han, J. 2009b. Potato growth with and without plastic mulch in two typical regions of Northern China. *Field Crops Research* 110(2), 123–9. doi:10.1016/j.fcr.2008.07.014.

Weber, J., Kunz, C., Peteinatos, G., Zikeli, S. and Gerhards, R. 2017. Weed control using conventional tillage, reduced tillage, no-tillage, and cover crops in organic soybean. *Agriculture* 7(5), 43. doi:10.3390/agriculture7050043.

Winston, L., Schwarzländer, M., Hinz, H., Day, M., Cock, M. and Julien, M. (Eds.) 2014. *Biological Control of Weeds: A World Catalogue of Agents and Their Target Weeds* (5th edn.). USDA Forest Service, Forest Health Technology Enterprise Team, Morgantown, WV, 838pp. Available at: www.ibiocontrol.org/catalog/(Accessed on 25 January 2019).

Worthington, M., Reberg-Horton, S. C., Brown-Guedira, G., Jordan, D., Weisz, R. and Murphy, J. P. 2015. Relative contributions of allelopathy and competitive traits to the weed suppressive ability of winter wheat lines against Italian ryegrass. *Crop Science* 55(1), 57–64. doi:10.2135/cropsci2014.02.0150.

Chapter 5

Insect pest and disease management practices and benefits in Conservation Agriculture systems: a case of push-pull practice

Z. R. Khan, International Centre of Insect Physiology and Ecology (icipe), Kenya; A. W. Murage, Kenya Agricultural and Livestock Research Organization (KALRO), Kenya; and J. O. Pittchar and C. A. O. Midega, International Centre of Insect Physiology and Ecology (icipe), Kenya

1 Introduction

National and international agricultural research institutions are engaged in developing technologies and formulating dissemination packages in order to manage constraints, to enhance production, and to guarantee food security. The main crops grown in sub-Saharan Africa (SSA) are cereals which include maize, rice, sorghum, and millet supporting livelihoods of over 80% small-scale farmers. These staple cereals form the principal food and cash crops for millions of the poorest people. Their production has however been characterized by small holdings, low capitalization, and low yield per unit of land.

Low productivity is attributed to unreliable rainfall, pests, diseases, poor soil fertility, and weeds. In particular, cereals are adversely affected by stemborers (mainly *Chilo partellus* (Swinhoe) and *Busseola fusca* (Füller)) (Kfir, 2002) and parasitic Striga weeds (mainly *Striga hermonthica* (Del.) Benth. and *Striga asiatica* (L.) Kuntze) (Khan et al., 2006b; Hassanali, 2008). Furthermore, the recent outbreak of fall armyworms (FAWs) (*Spodoptera frugiperda*) (J. E. Smith) has posed a serious threat to food security, causing crop losses estimated to be more than US$6.2 billion annually in SSA, while

http://dx.doi.org/10.19103/AS.2019.0049.05

the combination of stemborers and Striga are estimated to cause losses of up to US$14 billion annually (Goergen et al., 2016; Midega et al., 2018). The FAW invasion, in particular, has adverse economic impacts at the household level as it directly increases yield losses, production costs, requires specialized knowledge to control, and confounds the ability of cropping systems to respond to production constraints all of which have an overall impact on household incomes (Midega et al., 2018).

In addition, the natural resource base for agricultural production is severely degraded (FAO, 2004; Khan et al., 2014; Kassam et al., 2017). Land degradation in tillage agriculture further constrains the production of cereals, particularly maize and sorghum, the main staple and cash crops for millions of smallholder farmers in SSA. Low and declining yields cause food and nutrition insecurity, poor incomes, rising poverty and poor health. The resource-constrained smallholder farmers living in the arid and semi-arid regions who practice mixed crop-livestock systems are particularly affected. As a result of these constraints, every year there is a critical shortage of cereals in many smallholder households leading to high prices of grain, hunger, and widespread poverty (Khan et al., 2006b). Future projections indicate that without urgent intervention, SSA will have more than 500 million food insecure people by 2020 (USDA, 2010).

Farmers have attempted to minimize the adverse effects of the above-mentioned pests through conventional control strategies such as hand weeding, direct uprooting, use of nitrogen fertilizers, and other chemical means. However, research findings have shown that these methods are insufficient, expensive, unaffordable, and unfriendly to the environment (Berner et al., 1995; Woomer et al., 2004). Developing climate-resilient, adaptable, and productive agricultural systems such as Conservation Agriculture (CA) is imperative to meet future demands for food (Pretty, 2011; Khan et al., 2014; Kassam et al., 2017). Part of resilience building needs to involve climate-smart agricultural technologies such as those practiced in CA, which also protect and enhance natural resources and ecosystem services in ways that mitigate future climate change effects (Tittonel and Giller, 2013; Kassam et al., 2017; Gonzalez-Sanchez et al., 2018).

Significant and sustainable increases in grain yields and animal production require ecologically sound methods of managing weeds and pests, including CA systems and practices, with a strong focus on a systems approach that maximizes soil quality, moisture conservation, and crop productivity. Sustainable agricultural intensification and resource-conserving technologies such as push-pull (www.push-pull.net), which produce more output per unit area of land while reducing the negative environmental impacts, and increasing agricultural system contributions to natural capital and the flow of environmental services, are necessary for agricultural development (Royal Society, 2009; Conway and Waage, 2010; Godfray et al., 2010).

In 2015/16, CA covered more than 1.5 million ha in Africa (Kassam et al., 2019). CA promotes integrated insect pest and disease control through the practical application of its three interlinked principles. Push-pull practice of insect pest control fits well into the CA cropping systems as it utilizes the CA principles in building and optimizing the effectiveness of the practice, adding further to system output, factor productivity, resilience, and profit margins. This chapter elaborates on how the CA-based push-pull practice works at the farm level and what benefits are harnessed. The chapter also addresses how disease management works in CA systems and with what benefits.

2 Push-pull technology: a sustainable innovation in Conservation Agriculture

The push-pull production and protection system was developed by the International Centre of Insect Physiology and Ecology (*icipe*) and partners as a control measure for stemborers, the striga weed, and low soil fertility. The push-pull practice harnesses resource-conserving integrated pest management (IPM) and integrated soil fertility management (ISFM) approaches, using natural processes and locally available bio-resources to increase farm productivity by controlling both biotic and abiotic constraints to smallholder agriculture (Cook et al., 2007; Hassanali et al., 2008). The practice design originated from tillage-based cereal-legume intercropping strategies practiced in Africa to maximize staple crop productivity.

As part of a CA-based approach, the push-pull practice was further developed for no tillage-based systems, providing continuous soil cover with a perennial cover crop (live mulch) and plant residue, and a diversified cereal-legume-fodder intercropping practice. The push-pull CA practice is based on companion cropping, which effectively controls stemborer and fall armyworm insect pests as well as parasitic Striga weeds, while improving soil fertility by fixing nitrogen, sequestering carbon, and conserving soil moisture. Farmers practicing push-pull have realized substantial grain yield increases with minimal use of external synthetic inputs. The push-pull's diversified cereal-legume-fodder intercropping strategy conforms to the CA principles of minimum soil disturbance, mulching through continuous soil cover by Desmodium, which is a perennial cover crop that generates biomass. The perennial legume intercrop improves above-ground and below-ground arthropod abundance, agro-biodiversity, and the food web of natural enemies of stemborers (Midega et al., 2006, 2009, 2015). This practice effectively controls the major insect pests of cereals in SSA, mainly the lepidopteran stemborers, and more recently the invasive FAW, as well as parasitic Striga weeds. Furthermore, it improves soil health and conserves soil moisture. The technology involves the use of inter- and trap-crops in a mixed cropping system (Khan et al., 2006a). These companion plants release chemicals in a stimulo-deterrent tactic that mitigates the behavior of both stemborers and beneficial insects, thus controlling their distribution and abundance as a pest-management strategy. The push-pull practice is based on an in-depth understanding of chemical ecology, agro-biodiversity, and plant-plant and insect-plant interactions (Miller and Cowles, 1990; Cook et al., 2007) and is well suited to African mixed cropping practices.

In the push-pull practice, cereals are intercropped with Desmodium (e.g. *Desmodium uncinatum* (Jacq)), and Napier grass (*Pennisetum purpureum* Schumach) is planted as a border crop around this intercrop (Khan et al., 2001, 2003; Midega et al., 2010, 2015). The Desmodium repels stemborers moths (push), while the surrounding Napier grass attracts them (pull) (Khan et al., 2001). In addition, Desmodium suppresses Striga weeds through several mechanisms, with allelopathy (chemical growth inhibition) being the most important (Tsanuo et al., 2003). Due to the adverse effects of climate change on the companion crops, the push-pull technology was adapted by selecting and incorporating drought-tolerant companion crops in a new strategy, termed 'climate-smart push-pull' where the cereal crops are intercropped with the drought-tolerant Greenleaf Desmodium, (*D. intortum* (Mill.) Urb.), with the drought-tolerant Brachiaria cv 'Mulato' (*Brachiaria* spp.) grass planted as a border crop (Khan et al., 2014; Midega et al., 2015; Cheruiyot et al., 2018) (Fig. 1).

Figure 1 Maize under conventional push-pull technology, Kisumu West, Kenya.

This chapter presents a review of information on the dissemination, adoption, and benefits of the push-pull practice at the household level. The study reviews various studies that have been conducted on the push-pull technology since its inception in order to elicit the significance of the practice.

3 Dissemination and adoption of push-pull practice

Dissemination and adoption of the push-pull practice have widely been studied and documented (e.g. Amudavi et al., 2008, 2009; Khan et al., 2008a,b; Murage et al., 2011, 2012, 2015a,b; Hailu et al., 2017). Since its inception, the practice has been disseminated and widely adopted by farmers in East Africa and beyond. The dissemination of the push-pull practice started in early 2000 in Vihiga County and has progressively diffused in other counties in western Kenya and the neighboring countries in eastern and southern Africa including Uganda, Tanzania, Ethiopia, Malawi, Zambia, Zimbabwe, Rwanda, Bururndi, Burkina Faso, and Ghana. *icipe* and partners have employed a wide range of dissemination pathways to reach as many farmers as possible. These include mass media, the print media, and the interpersonal pathways such as field days, farmer field schools (FFSs), farmer teachers/trainers, fellow farmers, and public meetings.

Dissemination pathways play a key role in determining the adoption of new innovations. It has been shown that farmer preferences for certain dissemination pathways do exist and farmers are likely to be persuaded to adopt a technology if they perceive information pathways to be credible and reliable (Rogers, 1997; Gloy et al., 2000; Tucker and Napier, 2002; Roderick et al., 2008; Murage et al., 2011). Previous studies by Murage et al. (2011) showed that adopters of the push-pull practice mostly preferred field days as the pathway through which they could effectively receive information about a new agricultural technology. Indeed, further analysis by Murage et al. (2012) showed that use of field days as the desired dissemination pathway triggered the highest impact on both the probability and intensity of push-pull adoption, followed by use of FFSs, and farmer teachers in that order. The results were corroborated by Kassie et al. (2018), who noted that the adoption of push-pull practice appeared to increase according to the number of field days a farmer attended.

Under the push-pull production system, field days were organized by the participating farmers groups usually at one of the members' plot. It is through these interactions that publicity and knowledge about the practice were increased from the on-farm experiments and gave an opportunity to solicit feedback from the participating farmers (Oswald, 2005; Amudavi et al., 2008). Furthermore, field days provide a forum for sharing information, exchanging experiences, and encouraging farmer-to-farmer dissemination (Doss, 2003; Minja et al., 2004).

Field days are used to catalyze interactive learning among participants and they have a higher tendency to elicit farmer's interest compared to other dissemination pathways (Doss, 2003). They have been predominantly used as extension events where farmers interact with the facilitators as well as with other farmers and exchange ideas and experiences (Madukwe, 2006). In some cases, hands-on training and physical participation of the farmers are encouraged. Farmers' propensity to seek new agricultural knowledge motivates them to attend field day sessions (Amudavi et al., 2008) and in view, they favorably rated them in terms of effectiveness in information dissemination. In the study by Amudavi et al. (2008), the effectiveness of using field days in disseminating the push-pull practice was highly rated by the participating farmers particularly in being able to acquire knowledge and skills related to the technology component. Further analysis by Amudavi et al. (2008) showed that the effectiveness of field days was dependent on the knowledge and the capacity of the host farmer and the facilitator. This implies that the choice of the field day site and the facilitator is critical in encouraging farmers' participation and adoption particularly in making the farmers feel at ease during the sessions.

The farmer-to-farmer extension model using the farmer teachers has also been studied and was found to be very effective in disseminating the push-pull technology. Amudavi et al. (2009) showed that a farmer teacher was able to directly influence 17 other farmers to adopt the push-pull practice, while the follower farmers influenced on average two farmers each over a period of 2–3 years. This implies that, through a multiplier effect, the farmer-to-farmer extension model was able to drastically increase awareness and adoption of the practice. Most farmer teachers acknowledged that their desire to share knowledge with other farmers motivated them to become trainers.

Other pathways such as the FFS and print and the electronic media were also found to have niches within different farmers segments with selected socio-economic characteristics implying that a 'one-size fits all' approach would clearly not be appropriate in dissemination of the push-pull practice. Recommendations have been made to initially understand the socio-economic and demographic factors within a given region prior to choosing an appropriate information transfer mechanism (Murage et al., 2011, 2012).

The wide adoption of the push-pull practice has also been triggered by various factors. Studies have shown that the main drivers of adoption of the push-pull practice are to control Striga and stemborers, increase cereal yields, provide livestock fodder, control soil erosion, and improve soil fertility (Khan et al., 2014; Midega et al., 2015). Farmers who adopted a CA-based push-pull system reported lower proportions of stemborer-damaged plants. The studies by Khan et al. (2014) and Midega et al. (2015) further showed that farmers who practiced the push-pull technology were able to establish new initiatives in dairy and poultry farming. Moreover, Khan et al. (2014) noted that the increase

Figure 2 Sorghum under climate-smart push-pull technology, Siaya, Kenya.

in grain yields and availability of the leguminous Desmodium gave farmers the impetus to keep more poultry. In addition, farmers have been able to start organic farming through preparation and use of animal manure, thus allowing nutrient cycling and reduction in the use of chemical fertilizers (Fig. 2).

4 Benefits of push-pull practice in Conservation Agriculture systems

The push-pull practice fits well with the traditional mixed farming practiced under CA and is appropriate for the resource-poor farmers as it is based on locally available plants, not expensive external inputs like fertilizers (icipe, 2018). With its increased adoption of push-pull as a CA system, a wide range of direct and indirect benefits to farmers and the economy at large have been realized and documented.

4.1 Control of pests and weeds

Prior to the invention of the push-pull practice, the conventional control of stemborers, Striga weeds, and other pests in cereal farming using chemicals and other conventional methods had serious setbacks. Apart from being expensive, the use of chemicals to control pests often failed to reach the inside of plant stems where the stemborer larvae are found (Khan et al., 2006a). Furthermore, the behavior of the adult moth often makes it difficult to kill it as it lays eggs after dusk, making the use of pesticides ineffective in addition to being harmful to the environment (Khan et al., 2001, 2003). Pest larvae that are the damaging stage of all lepidopteran borers as well as the FAW are cryptic in their feeding behavior and feed inside plant whorls and stems where they cannot be effectively reached by insecticides (Midega et al., 2018). Similarly, the use of herbicides against Striga is neither effective nor feasible among smallholders because they are not affordable and do not prevent the buildup of Striga seed density in the soil.

Use of the push-pull practice has been proven very effective as it simultaneously controls the stemborers and Striga weeds significantly (Khan et al., 2006c). Recently, Midega et al. (2018) noted that use of the push-pull practice leads to a significant reduction of infestation by FAW in maize leading to lower damage levels. This novel practice uses a combination of behavior-modifying stimuli to manipulate the distribution and abundance of insect pests by repelling them away from the main crop (push) and simultaneously attracting them (pull) using other trap crops where the insect pest becomes concentrated and therefore facilitate their control (Khan et al., 2001, 2003, 2014; Midega et al., 2015). On the other hand, Desmodium suppresses Striga weed through a combination of mechanisms ranging from increased availability of nitrogen, soil shading, and by an allelopathic root exudation, which diminishes Striga seeds through suicidal germination, thus providing a novel means of *in situ* reduction of the Striga seed bank in the soil (Khan et al., 2003; Tsanuo et al., 2003; Midega et al., 2015).

4.2 Increasing grain yields

The main reason as to why the push-pull practice was developed was to minimize stemborer damage on cereals such as maize and sorghum, especially, in the SSA where it had caused a major yield loss. With the adoption of the push-pull practice, cereal production has significantly increased with minimal input use thus making it available in the household and providing surplus for the market at affordable prices. Khan et al. (2011) noted that the push-pull system is highly used by smallholders in Africa as it addresses the major constraints to achieving higher yields and it is economical. So far, over 197 000 farmers are achieving higher yields of maize and sorghum and over 75% reporting three-fold to four-fold increase, having harvested more than five tons of maize per hectare, up from previous yields of below one ton per hectare (Khan et al., 2006b; Fischler, 2010; icipe, 2018).

A study by Khan et al. (2008b) showed that although the initial costs of establishing the push-pull practice were significantly higher compared to the maize-bean intercrop and the monocrop, these costs were significantly reduced in the subsequent years. Further analysis from the same study showed that farmers were able to recoup the initial costs within the first year of establishment following the higher grains and fodder yields that they were able to harvest. Data from all the districts where the study was done showed a significantly higher total variable cost under push-pull in the initial years that significantly dropped in the subsequent years; despite the higher costs, the gross benefits from the push-pull system were reported to be higher during the same period. Furthermore, the returns to land and labor for the push-pull practicing farmers were higher than that for the non-push-pull farmers (Khan et al., 2008b). These results are corroborated by a recent study by Chepchirchir et al. (2017, 2018), who noted that the total revenues from sale of farm produce, total variable costs, and material input costs were significantly higher with push-pull than without push-pull in eastern Uganda. Chepchirchir et al. (2017) demonstrated that the intensity of adoption of the push-pull practice determined the average maize yields obtained, which ranged from 2.6 t ha^{-1}, if a farmer had a small plot of 0.01 ha, to 3.5 t ha^{-1} for farmers with 0.4 ha of land. A similar study by Kassie et al. (2018) showed that the unconditional mean of maize yield from plots with the push-pull practice was 3.9 t ha^{-1} while that for the non-push-pull plot was 2.3 t ha^{-1}, and that the per capita consumption of maize was higher for push-pull farmers (132 kg) compared to the non-push-pull farmers (113 kg).

4.3 Increasing dairy milk production

Most farmers in SSA and any other part of Africa practice mixed farming that includes cultivation of crops and livestock keeping; thus, lack of fodder can be a major constraint to productivity (Khan et al., 2006b). Provision of livestock feeds from push-pull fields became one of the main entry points for adopting the practice by most farmers (Khan et al., 2008a). The push-pull system generates quality fodder for livestock thus stimulating increased milk production and enhanced growth rate (Khan et al., 2006b). Furthermore, farmers practicing the push-pull system noted that the maize plants were relatively tall and hence apart from the feed from the companion crops, they were also able to harvest higher volumes of crop residues that were fed to livestock. In the study by Kassie et al. (2018), the annual mean milk production for push-pull adopters was approximately 460 liters per cow, relative to 263 liters per cow for the non-adopters. Farmers who were interviewed in this study observed that the companion plants (Desmodium and Brachiaria fodders) more than doubled their cows' daily milk production. Desmodium being a legume contains high protein, dry matter, and fiber, which are good for rumen digestibility hence leading to higher milk yield and improved growth of the livestock. On the other hand, Napier grass is very important in milk production thus increasing the farmers' economic returns. Farmers acknowledged that when animals are fed with quality fodder, they were also able to produce products that are healthy to human beings for consumption (icipe, 2018). Many farmers referred to the push-pull as a 'springboard' for diversifying farming systems, especially through the incorporation of dairy operations using Napier and Desmodium as fodder (Fig. 3).

4.4 Improved market participation

Compared to non-adopters, farmers who practiced push-pull farming can participate in the market to sell their excess products such as grain, milk, fodder, and manure. Kassie et al. (2018) noted that, on average, a push-pull adopter sold approximately 406 liters of milk per year while non-adopters sold 161 liters. Furthermore, adopters of the push-pull practice sold more maize than the non-adopters. In terms of net income from maize, adopters achieved a 55% higher return in comparison with non-adopters demonstrating a higher productivity and income from the adoption of the push-pull practice, which can translate into improved household food security and reduced poverty. Previous studies by Khan et al. (2008b) showed that even though the total labor cost and total variable cost were lower in farmers' practice as compared to the push-pull fields, the total gross revenue and gross benefit of push-pull were significantly higher and that farmers were able to earn more by being able to sell their products in the

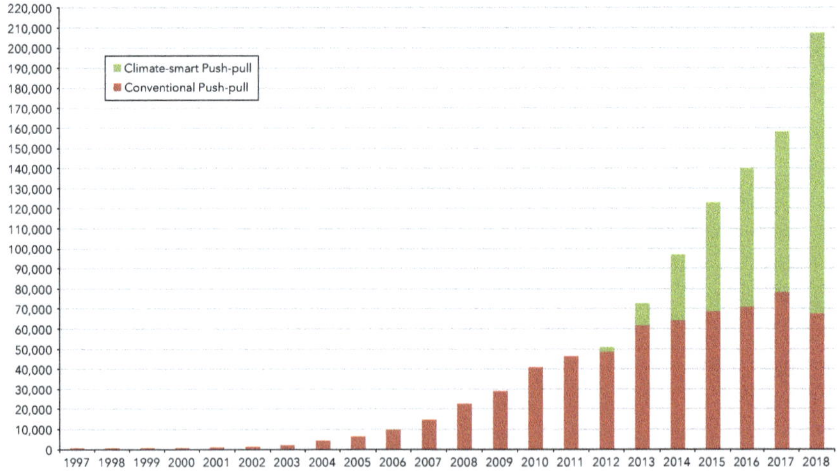

Figure 3 The trend in adoption of conventional and climate-smart push-pull technology.

market. Market participation is important to economic growth and farm level poverty reduction as farmers can benefit from welfare gains derived from trade (Barrett, 2008).

4.5 Increased household economic returns

Agriculture is classified as the highest contributor to the Gross Domestic Product (GDP) in Kenya in terms of food productivity and employment. The contribution of the push-pull strategies to household food security and economic returns cannot be over-emphasized. The intercropping of maize, grasses, and fodder legumes has enabled farmers to increase crop yields and thus improve their food security and gross benefits. Not only are farmers able to use the resultant yields for their household consumption, but also are able to sell the excess to the market. Furthermore, the increased production of quality fodder from push-pull farms enhances household economic returns through sale of excess milk and fodder (Khan et al., 2008b; Kassie et al., 2018; Chepchirchir et al., 2018). In a study by Khan et al. (2008b), it was noted that with effective control of stemborers, striga, and other biotic stresses, a significant cost-benefit return of 2.2 was reported in the push-pull technology relative to 0.8 for the maize monocrop. This study further reported a sustainable increase in maize grain yields and higher returns to labor.

Using long-term researcher-managed trial data and partial budget and marginal analysis, De Groote et al. (2010) found that the push-pull practice was more profitable than other practices used to control Striga weed and stemborer. Further economic studies by Kassie et al. (2018) observed that although adoption of push-pull increased the cost of maize production by 15.3%, the average net maize income increased by 38.6%, representing an additional US$20.34 per capita per year (approximately US$20). Overall, Kassie et al. (2018) found that the push-pull technology generates economic benefits of US$73 million, thus contributing to lifting out of poverty 75 790 people per year. This compares closely with Chepchirchir et al. (2018) who through a 20-year period of the simulation of the economic model observed that push-pull farmers in eastern Uganda would have an overall net gain of US$3.8 million with a net present value of US$1.61 million, internal rate of return of 51%, and the benefit cost ratio of 1.54.

Additionally, the companion crops have an added advantage of being useful fodder for livestock; hence, farmers can confine their livestock into zero-grazing units and therefore accumulate excess manure for use in their farms and also for sale if in excess. Further analysis by Chepchirchir et al. (2018) showed a higher gross margin per hectare of US$725 from maize and fodder in a push-pull field, a gross margin of US$405 from maize without fodder, and a gross margin of US$239 from a non-push-pull farm. This implies that the fodder in a push-pull farm on average contributed an additional US$320 to the farm income, which was much higher than a gross margin from a non-push-pull farm. These are additional farm incomes on top of extra profits arising from cost savings from reduced time, labor, and fuel input in land preparation in CA systems.

All these form a wide range of economic benefits that enhances the livelihoods of smallholder farmers. Another economic benefit of the push-pull practice is the ability to use fewer or no pesticides and other chemicals, hence making it a low-cost practice that is affordable to smallholders. Since there is less use of costly chemical inputs such as fertilizers as well as labor and fuel in push-pull CA farming, farmers can minimize their production expenses and hence increase their income without causing any negative impact to the environment. Furthermore, the push-pull practice is based on locally available plants that are not expensive, and thus fits well with traditional mixed cropping systems under CA. This cuts on the cost of training about the technology since most of the farmers are familiar with the methods, that is, the planting of Napier grass and Desmodium and its techniques.

4.6 Improving soil health

The push-pull practice in CA systems has been proven to ecologically enhance soil biodiversity, thereby improving soil health and fertility. Being a leguminous crop, Desmodium fixes the atmospheric nitrogen to improve the nitrogenous component in the soil in addition to organic matter from the fallen leaves, hence improving soil fertility (icipe, 2018). Increasing nitrogen into the soil is important in the food chain where nitrates are converted into proteins and produce ammonia (Vitousek et al., 2002). Vitousek et al. (1997) had earlier noted that microbial breakdown of ammonium and nitrates in the soil enhanced the release of nitrous oxide. Nitrates are important in tissue development and building immune systems in the plants (Liu et al., 2015). Most of the fertilizers contain high concentrations of nitrate to help in production of seeds and plant development. The use of the push-pull practice limits fertilizers and pesticides application thus reducing the harmful chemicals that go into the soil. As a result, the push-pull practice contributes to reduced levels of greenhouse gasses that originate from conventional tillage-based agricultural practices. Furthermore, the increased use of mineral fertilizers especially during the rainy season is harmful to the water bodies if soil erosion is not controlled (Moss, 2008). In CA systems, runoff and soil erosion are minimized.

Besides improving fertility, Desmodium together with the surrounding Napier grass protects the soil against erosion Khan et al. (2003). Furthermore, because of its ground cover, Desmodium provides a mulch that reduces soil temperatures. Therefore, farms under the push-pull practice become more sustainable and resilient, with improved potential to mitigate the effects of climate change. Majority of push-pull adopters integrate the technology with crop-livestock production in mixed farming systems. This close association with livestock means that farmyard manure can be added to the soil, increasing the fertility benefits already gained from nitrogen fixation by the Desmodium intercrop. Most of the farmers thus notice an improvement in their soil within a very short time of adopting the technology (Khan et al., 2011). Therefore, farmers who are using the push-pull method can improve their potential to mitigate climate change (Midega et al., 2015).

Other benefits of CA-based push-pull practice to the soil include conserving soil moisture and improving soil carbon sequestration, biomass, and soil biota (Midega et al., 2015). The push-pull practice when applied in CA systems contributes to conserving and enhancing the production resource base. The technology improves the overall agro-ecosystem resilience through the practical application of the CA principles of continuous minimum mechanical soil disturbance, continuous soil cover, addition of organic matter, prevention of loss of top soil through soil erosion, improved water conservation, and other ecosystem services (Kassam et al., 2017).

4.7 Improving human health

The improvement of human health is one of the indirect benefits of CA-based push-pull practice to the livelihoods of the community. Enhancing the production of fodder indirectly impacts on health indices, especially for children through increased protein uptake in milk and milk products. Farm households practicing push-pull have higher dietary diversity scores than non-push-pull households (Ogot et al., 2018). Moreover, the additional income realized from the sale of push-pull products and by-products enables many farm households to purchase foods that they are not able to produce for themselves. Additionally, the push-pull practice allows less or no use of mineral fertilizers and chemicals in controlling the pests and weeds and hence the products from the push-pull technology offer nutritional health to smallholder households. Mineral fertilizer contains toxic chemicals that are consumed by human beings through the vegetables and cereals and are very harmful when the chemicals flow into ground water used for drinking. Therefore, the CA-based push-pull practice contributes to reduced levels of chemicals in the environment that could cause human ill-health. The reduction in the use of costly synthetic insecticides and herbicides also potentially enhances human health (Pickett et al., 2014). The control in soil erosion also reduces the potential hazards of human displacement and a high rate of death as a result of land and mudslides caused by loose soils.

4.8 Environmental health

The adoption of the push-pull practice in CA systems has enormous positive impacts to the environment, in general, which are basically because of other benefits such as improved soil health, crop and livestock health, and household economy. Climate-resilient technologies contribute to maintaining agro-ecosystems functions and services that are usually provided by natural systems. This means integrating instead of segregating, closing water and nutrient cycles, increasing biological and genetic diversity, and regenerating instead of degrading bio-resources (Pretty, 2011).

Push-pull practice in CA systems contributes to stable and climate-resilient agro-ecosystems by providing farmers with a tool for on-farm diversification which is in line with these underlying principles. With increased economic returns from farming, there is a probability of reduction in rural urban migration since households can make more money from farming than in towns. Urbanization has been a major cause of heavy agricultural mining, deforestation, and overpopulation in the urban areas leading to negative environmental effects. By use of the push-pull practice in CA systems, rural communities can earn a better income from crop and livestock farming (Khan et al., 2006b). Besides, increase in fodder production allows farmers to intensify dairy production and hence there is reduced soil erosion since the animals do not trample and compact the soil. Furthermore, intensification of dairy farming leads to less attack by ticks, hence the low use of acaricides, which would otherwise be harmful to the environment.

4.9 Gender equality

The basic principle in the dissemination of the push-pull practice has been targeting all gender groups equally. These include men, women, youth, and people living with disabilities. Adoption of the push-pull practice has however had particular benefits for women farmers and has been demonstrated that being a female farmer accelerated the speed of uptake (Murage et al., 2011). A quick and rapid adoption of the push-pull practice by women farmers was observed, while men were seen to allocate larger portions of land to the practice (Murage et al., 2015). It has been argued that women often try out new innovations to tackle the overarching constraints such as the Striga and stemborers, which affected them more directly than men. Indeed, the study by Murage et al. (2015) showed that women rated the Striga and stemborers constraints higher than men and hence their desire to adopt the push-pull technology. Specifically, adopting the practice ensured more grain using less labor, hence reducing the time women must spend digging and weeding. As a result, women are able to save extra time to invest in other productive activities and therefore increase and diversify their income (Khan et al., 2015; Grassi et al., 2015).

Women seem to be more affected by Striga infestation as they provide the bulk of labor for manual uprooting. Subsequently, the reduction in yields, hunger, and malnutrition disproportionately affect women and children more than men (World Bank, 2008). In addition, women bear the burden of feeding their livestock with the limited sources of fodder given the constraining land sizes. The adoption of the push-pull practice in CA systems avails women with options to increase cereal yields, control Striga, and increase fodder for their livestock *in situ*. However, men also tend to take over enterprises that seem to have promising financial benefit with less labor requirement as do all CA systems. Hence, with the expected increase in cereal yield, some of which can be sold to the market and the expected income from the sale of excess milk and fodder coming, it is expected that men would take up the practice.

Households with increased incomes, particularly for women, are able to educate both boys and girls. Previous studies have shown that girls' education tended to be given less emphasis, especially in poor families (e.g. Stromquist, 2001; Hossain, 2004). In such cases, boys were taken to school while girls remained behind to help their mothers with household chores. However, this scenario has changed in most families that have adopted the push-pull practice in eastern Africa.

The dissemination of the practice has also built the skills, confidence, and networks of many women farmers, hence improving their social status in the community; they are seen as intellectuals who offer advice to upcoming farmers (icipe, 2018). Empowering women and girls has a multiplier effect in the economy and brings about growth and development in the country (OECD, 2010). Kassie et al. (2015) stated that women who are empowered are able to access financial services, develop new skills, increase income-earning opportunities, and access information about the market.

4.10 General economic welfare

The push-pull practice has led to improvements in the general welfare of the households practicing through increased incomes (Khan et al., 2008a). The practice has been rated as efficient and low cost; hence, families are now able to use the extra income to pay school fees for their children, purchase household items, and improve their overall nutrition and health (Khan et al., 2014). The general food security of push-pull adopters has improved with a proven increase in per capita consumption of grains and milk. Kassie et al. (2018) noted that the per capita milk and maize consumption was higher for push-pull adopters than the non-adopters. A similar study by Chepchirchir et al. (2017) in eastern Uganda showed that on average, the household income ranged from US$135 for a farmer with 0.01 ha of push-pull to an average of US$273 for a farmer who plants 0.4 ha of the practice implying that income varied with the intensity of production. The study further demonstrated that the per capita food consumption increased from US$15 for a farmer with 0.01 ha area to US$27 for a farmer with 0.4 ha, with a clear indication that the extent of poverty declined significantly with the intensity of adoption. Increasing food production especially for staple crops can essentially lead to food security and sustainable economic growth (AGRA, 2014). Kassie et al. (2018) and Chepchirchir (2018) observed a general welfare improvement in gains from income and poverty reduction with the adoption of the push-pull practice.

4.11 Contribution to the Sustainable Development Goals (SDGs)

The above information clearly shows that the push-pull practice in the CA system has generated a wide range of benefits either directly or indirectly to the rural households. The chapter has demonstrated that the practice has immensely contributed toward attainment of the sustainable development goals (SDGs) – http://www.push-pull.net/sdgs.shtml. For example, the practice directly contributes to SDG 1 on ending extreme poverty through generation of incomes from the sale of excess grains, milk, fodder, and manure. This has been alluded to by studies by Chepchirchir et al. (2018) and Kassie et al. (2018). We have seen that the participating households were able to produce excess grains, milk, manure, and other by-products from the technology, all of which contributed to household income which in turn led to reduction in poverty.

It is also evident that the practice has contributed to SDG 2 by ending hunger, achieving food security and improved nutrition, and promoting sustainable agriculture. Families are able to have timely access to enough quality food either through their own production or through purchase from the market. Moreover, they have nutrition security through consumption of diverse diets such as proteins from milk and from other purchased food products. With good-quality food and diverse nutrition, household members can live a quality and healthy life thus contributing to SDG 3 in ensuring health lives and promoting well-being at all ages. Consumption of quality and healthy food free of toxic chemicals helps in managing preventable diseases.

Hunger and extreme poverty have been some of the reasons children were kept out of school. With a well-fed family and good-quality health, children are able to fully participate in schooling. Notably, most of the households practicing push-pull have also indicated that they are now able to give their children proper education by being able to pay school fees promptly and buy other school requirements; their children's performance in school has improved, hence contributing to SDG 4.

Evidently, the push-pull practices in CA systems have also in a special way contributed to SDG 5 on achievement of gender equality and empowerment of women and girls. This has been made possible by using a dissemination strategy that equally targets all gender groups in the society. It has been shown that women, men, youth, and people living with disabilities have unequivocally been able to participate in the push-pull production system each sharing their positive experience and benefits from the practice. Furthermore, this review has already demonstrated how women farmers have benefited through capacity building and skills development, being able to address the day-to-day challenges at the household level such as food availability and education, which often are the responsibility of women farmers in the rural community.

The use of the climate-smart push-pull production system has contributed to SDG 13 in the action to combat climate change and its impacts. This has been achieved through the use of drought-resilient local plants and natural processes to control Striga and stemborers without introducing toxic chemicals that have a high carbon footprint and negatively impact the environment. This has in addition helped in improving soil health and conserving biodiversity, hence contributing to SDG 14 focusing on conservation and sustainable use of life under water as well as SDG 15 on protecting restoration, using sustainably the life on land. Indeed, the United Nations General Assembly recognized push-pull as one of the practices that have benefited farmers by doubling yields through IPM, soil conservation (United Nations General Assembly, 2010), and by making cereal cropping systems resilient to climate change (United Nations General Assembly, 2015).

4.12 Disease management and benefits in Conservation Agriculture systems

The push-pull CA practice exploits the chemical ecology and natural plant genetic diversity of repellent intercrops and trap plants (Khan et al., 2006b; Cook et al., 2007) and is based on an understanding of chemical ecology, agro-biodiversity, and plant-plant and insect-plant interactions (Miller and Cowles, 1990; Cook et al., 2007). The companion plants used in push-pull release chemicals in a stimulo-deterrent tactic that mediate the behavior of both stemborers and beneficial insects, thus controlling their distribution and abundance as a pest-management strategy (Khan et al., 2006c). Behavior of insect pests is mediated by herbivore-induced plant volatiles (HIPVs) emitted by these plants. This approach is exploited to reduce pest, disease, and parasitic weed – related constraints to the production of staple cereals, principally maize and sorghum in Africa (Pickett and Khan, 2016). The push-pull CA practice has been proven to reduce populations of stemborers and FAWs and subsequent damage on maize plants (Khan et al., 2014; Midega et al., 2015).

In a study conducted in western Kenya on the effect of cropping system on the incidence and severity of maize ear rots, the authors found that the push-pull CA practice had a significant effect ($p \leq 0.001$) on the incidence of ear rots by reducing stemborer infestation of maize (Owuor et al., 2018). Maize ear rots fungal infections are found in all agro-ecologies where maize is grown (Dragich and Nelson, 2014), with the most prominent genera found in maize grown in SSA being Aspergillus, Fusarium, Sternocarpella, and Penicillium, causing estimated yield losses ranging between 10% and 30% (Kapindu et al., 1999; Ajanga and Hillocks, 2000; Bigirwa et al., 2007). Stemborer insect infestation and mechanical damage of maize have been shown to predispose the maize grains to ear rots and mycotoxin attack. Stemborer infestation has been observed to have a positive and high correlation with incidence of ear rots in maize (Ajanga and Hillocks, 2000). Maize crop losses due to ear rot damage is associated with mycotoxins, the most prevalent of which are fumonisins, zearalenone, deoxynivalenol, and aflatoxins (Gxasheka et al., 2015). Mycotoxins cause serious diseases both in humans and animals (Zain, 2011) and are thus stringently regulated (Otsuki, 2001).

The significant reduction in the incidence of ear rots observed in the push-pull fields resulted from the multiple ecological benefits provided by the practice (Owuor et al., 2018), particularly stemborer control as well as improvement of soil organic matter (Midega et al., 2005). Moreover, interaction of other ecological factors such as buildup of soil organic matter (Alakonya et al., 2008), cover cropping (Tédihou et al., 2012), and intercropping (Vincelli, 1997; Flett and Ncube, 2015) have been observed to reduce the incidence of ear rots in maize. The push-pull CA practice deploys an intercropping strategy, increases soil organic matter, and Desmodium used in push-pull is a cover crop. Maize grown under the push-pull CA practice had significantly less ear rots than monocrop maize, reducing the incidence level to 7.3%. Monocropped maize fields had significantly higher ($p \leq 0.001$) incidences of all the types of ear rots than maize planted under push-pull (Owuor, et al., 2018). Ear rot incidences in monocropped maize and push-pull, respectively, were 7.31% and 3.33%, Diplodia; 4.48% and 1.30%, Gibberella; 2.09% and 0.65%, Aspergillus; 0.51% and 0.21%, Fusarium; and 0.40% and 0.11%, Penicillium (Owuor et al., 2018). The severities of maize ear rots were also significantly different ($p \leq 0.001$) between the two cropping systems but different ear rot types had different severity levels. For example, Diplodia and Gibberella ear rots were the most severe, 1.85 and 1.15 $\mu g\ kg^{-1}$ in sole maize, and 0.84 and 0.62 $\mu g\ kg^{-1}$ in push-pull, when compared to other ear rots. Aspergillus had higher incidence than Fusarium, yet it had the lower severity rating (0.25 and 0.09 $\mu g\ kg^{-1}$) than Fusarium (0.68 and 0.19 $\mu g\ kg^{-1}$) in monocropped maize and push-pull, respectively. Penicillium ear rot rated less severely, 0.05 and 0.03$\mu g\ kg^{-1}$ in monocrop maize and push-pull, respectively.

5 Future trends and conclusion

This chapter describes the benefits of the push-pull practice in CA systems that has been developed by *icipe* and partners. Since its inception almost two decades ago, the practice has continuously been improved to address the challenges facing smallholder farming in SSA. The conventional push-pull that was originally using Napier grass was challenged by climate change conditions and hence was improved by selecting drought-tolerant crops. Other chemical experiments on the practice continue to be implemented in order to enhance its benefits. The two versions of push-pull production systems have been steadily adopted over the years and the diffusion is still expanding to other regions in SSA.

The push-pull CA practice exploits the chemical ecology and natural plant genetic diversity of repellent intercrops and trap plants and is based on understanding of chemical ecology, agro-biodiversity, and plant-plant and insect-plant interactions. The companion plants used in push-pull release chemicals that mediate the behavior of both stemborers and beneficial insects, thus controlling their distribution and abundance as a pest-management strategy that reduces pest, disease, and parasitic weed - related constraints to the production of staple cereals in Africa. Stemborer insect infestation and mechanical damage of maize have been shown to predispose the maize grains to ear rots and mycotoxin attack. Mycotoxins cause serious diseases both in humans and animals. The push-pull CA practice has been proven to reduce populations of stemborers and FAWs and subsequent damage on maize plants and thus had a significant effect ($p \leq 0.001$) on the incidence of ear rots by reducing stemborer infestation of maize. The significant reduction in incidence of ear rots observed in the push-pull fields resulted from the multiple ecological benefits provided by the CA practice, particularly stemborer and FAW control as well as cover cropping, intercropping, and improvement of soil organic matter. Maize grown under the push-pull CA practice had significantly less ear rots than monocrop maize, reducing the incidence level to 7.3%. Monocrop maize fields had significantly higher ($p \leq 0.001$) incidences of all the types of ear rots than maize planted under push-pull.

While the initial proposition for the CA-based push-pull practice was to control stemborers in cereal production, the adoption of push-pull production systems led to other multiple benefits, which have brought welfare to smallholder farmers. The system has brought under control the parasitic Striga weed as well as the new threats posed by the outbreak of FAW. Further, farmers who have adopted the CA-based push-pull practice have experienced more benefits including increase in production of cereals, improved dairy enterprise, improved soil, environmental and human health, market participation, gender equality, and general household economic welfare. This has led to direct contributions to the attainment of the SDGs making the system a unique pathway to come out of poverty. With the practice addressing critical constraints to smallholder production, it is expected that its expansion will continue both at the farm level and in the region as one way of scaling up the benefits of the push-pull practice in CA systems.

There are at least 350 million smallholders in sub-Saharan Africa (SSA) whose cereal crops suffer from stemborers, fall army worm and striga, and they could immediately benefit from adoption of push-pull technology. The *icipe's* technology transfer unit is disseminating push-pull in SSA by proving training.

Efforts are underway to further adapt push-pull to more drought and rising temperature in Africa by replacing 'Mulato II' with new Brachiaria cultivars, for example, 'Piata' and 'Xaeres' as trap plants and by replacing *D. intortum* with more drought-tolerant *D. ramosissimum or D. Incanum*. The push-pull technology has been demanded in Asia also due to recent invasion of fall armyworm in the continent.

Although push-pull was originally developed for maize production systems, it has been applied successfully to sorghum, millet, upland rice and sugarcane. Research is also being carried out to adapt its application to cotton, coffee and horticultural crops and this adaptability will give it enormous potential for its reproduction elsewhere.

6 References

Ajanga, S. and Hillocks, R. J. 2000. Maize cob rot in Kenya and its association with stalk borer damage. *Crop Protection* 19(5), 297–300. doi:10.1016/S0261-2194(00)00020-X.

Alakonya, A. E., Monda, E. O. and Ajanga, S. 2008. Management of *Fusarium verticillioides* root infection court in maize using organic soil amendments. *World Applied Sciences Journal* 5(2), 161–70.

Alliance for a Green Revolution in Africa (AGRA). 2014. *Africa Agriculture Status Report: Climate Change and Smallholder Agriculture in sub-Saharan Africa*. Alliance for a Green Revolution in Africa (AGRA), Nairobi, Kenya.

Amudavi, D. M., Khan, Z. R., Wanyama, J. M., Midega, C. A. O., Pittchar, J., Hassanali, A. and Pickett, J. A. 2008. Evaluation of farmers' field days as a dissemination tool for push-pull technology in western Kenya. *Crop Protection* 28(3), 225–35. doi:10.1016/j.cropro.2008.10.008.

Amudavi, D. M., Khan, Z. R., Wanyama, J. M., Midega, C. A. O., Pittchar, J., Nyangau, I. M., Hassanali, A. and Pickett, J. A. 2009. Assessment of technical efficiency of farmer teachers in the uptake and dissemination of push-pull technology in western Kenya. *Crop Protection* 28(11), 987–96. doi:10.1016/j.cropro.2009.04.010.

Barrett, C. B. 2008. Smallholder market participation: concepts and evidence from eastern and Southern Africa. *Food Policy* 33(4), 299–317. doi:10.1016/j.foodpol.2007.10.005.

Berner, O. K., Kling, J. G. and Singh, B. B. 1995. *Striga* research and control: a perspective from Africa. *Plant Disease* 79(7), 652–60. doi:10.1094/PD-79-0652.

Bigirwa, G., Kaaya, A. N., Sseruwu, G., Adipala, E. and Okanya, S. 2007. Incidence and severity of maize ear rots and factors responsible for their occurrence in Uganda. *Journal of Applied Sciences* 7(23), 3780–5. doi:10.3923/jas.2007.3780.3785.

Chepchirchir, R. T., Macharia, I., Murage, A. W., Midega, C. A. O. and Khan, Z. R. 2017. Impact assessment of push-pull pest management on incomes, productivity and poverty among smallholder households in Eastern Uganda. *Food Security* 9(6), 1359–72. doi:10.1007/s12571-017-0730-y.

Chepchirchir, R. T., Macharia, I., Murage, A. W., Midega, C. A. O. and Khan, Z. R. 2018. Ex-post economic analysis of push-pull technology in Eastern Uganda. *Crop Protection* 112, 356–62. doi:10.1016/j.cropro.2018.07.001.

Cheruiyot, C., Midega, C. A. O., Van den Berg, J., Pickett, J. A. and Khan, Z. R. 2018. Genotypic responses of brachiaria grass (*Brachiaria* spp.) accessions to drought stress. *Journal of Agronomy* ISSN 1812-5379 doi:10.3923/ja.018.136.146.

Conway, G. R. and Waage, J. 2010. *Science and Innovation for Development*. UKCDS, London.

Cook, S. M., Khan, Z. R. and Pickett, J. A. 2007. The use of push-pull strategies in integrated pest management. *Annual Review of Entomology* 52, 375–400. doi:10.1146/annurev.ento.52.110405.091407.

De Groote, H., Vanlauwe, B., Rutto, E., Odhiambo, G. D., Kanampiu, F. and Khan, Z. R. 2010. Economic analysis of different options in integrated pest and soil fertility management in maize systems of western Kenya. *Agricultural Economics* 41(5), 471–82. doi:10.1111/j.1574-0862.2010.00459.x.

Doss, C. R. 2003. Understanding farm level technology adoption: Lessons learned from CIMMYT's Micro-surveys in Eastern Africa. CIMMYT Economics Working Paper 03-07, Mexico, DF.

Dragich, M. and Nelson, S. 2014. *Gibberella and Fusarium Ear Rots of Maize in Hawaii*. College of Tropical Agriculture and Human Resource, Office of Communication Services, Hawaii, pp. 1–8.

Fischler, M. 2010. *Impact Assessment of Push-Pull Technology Promoted by icipe and Partners in Eastern Africa*. International Centre of Insect Physiology and Ecology/icipe Science Press, Nairobi.

Flet, B. and Ncube, E. 2015. *Ear Rots of Maize: A Continuous Threat to Food Safety and Security April 2015, ARC-Grain Crops Institute*.

Food and Agriculture Organization, (FAO). 2004. *Status of Food and Agriculture*. Food and Agriculture Organization, Rome, Italy.

Gonzalez-Sanchez, E. J., Mkomwa, S., Conway, G., Kassam, A., Ordonez-Fernandez, R., Moreno-Garcia, M., Repullo-Ruiberriz, de Torres, M., Gil-Ribes, J., Basch, G., Veroz-Gozalez, O., Trivino-Tradas, P., Holgado-Cabrera, A., Miranda-Fuentes, A. and Carbonell-Bojollo, R. 2018. *Making climate change mitigation and adaptability real in Africa with conservation agriculture*. European Conservation Agriculture Federation (ECAF) and African Conservation Tillage Network (ACT), Brussels, Belgium, 143p.

Gloy, B. A., Akridge, J. T. and Whipker, L. D. 2000. Sources of information for commercial farms: usefulness of media and personal sources. *The International Food and Agribusiness Management Review* 3(2), 245–60. doi:10.1016/S1096-7508(01)00046-5.

Godfray, H. C., Beddington, J. R., Crute, I. R., Haddad, L., Lawrence, D., Muir, J. F., Pretty, J., Robinson, S., Thomas, S. M. and Toulmin, C. 2010. Food security: the challenge of feeding 9 billion people. *Science* 327(5967), 812–8. doi:10.1126/science.1185383.

Goergen, G., Kumar, P. L., Sankung, S. B., Togola, A. and Tamò, M. 2016. First report of outbreaks of the fall armyworm *Spodoptera frugiperda* (J.E. Smith) (Lepidoptera, Noctuidae), a new alien invasive pest in West and Central Africa. *PLoS ONE* 11(10), e0165632.

Grassi, F., Landberg, J. and Huyer, S. 2015. *Running Out of Time: The Reduction of Women's Work Burden in Agricultural Production*. Food and Agriculture Organization, Rome, Italy. ISBN 978-92-5-108810-4.

Gxasheka, M., Wang, J., Tyasi, T. L. and Gao, J. 2015. Scientific understanding and effects on ear rot diseases in maize production: a review. *International Journal of Soil and Crop Sciences* 3(4), 77–84.

Hailu, G., Khan, Z. R., Pittchar,J. O. and Ochatum, N. 2017. Radio and mobile ownership or access by smallholder farmers of eastern Uganda and its potential use for push-pull technology dissemination.*Int. J. Agric. Ext.* 5(2), 19–28.

Hassanali, A., Herren, H., Khan, Z. R., Pickett, J. A. and Woodcock, C. M. 2008. Integrated pest management: the push-pull approach for controlling insect pests and weeds of cereals, and its potential for other agricultural systems including animal husbandry. *Philosophical Transactions of the Royal Society of London. Series B, Biological Sciences* 363(1491), 611–21. doi:10.1098/rstb.2007.2173.

Hossain, N. 2004. *Access to Education for the Poor and Girls: Educational Achievements in Bangladesh; Reducing Poverty, Sustaining Growth; What Works, What Doesn't, and Why - A Global Exchange for Scaling Up Success Scaling Up Poverty Reduction: A Global Learning Process and Conference Shanghai*, 25–27 May 2004.

International Centre of Insect Physiology and Ecology (icipe). 2018. *Push-Pull A Novel Farming System for Ending Hunger and Poverty in sub-Saharan Africa*. International Centre of Insect Physiology and Ecology, Nairobi, Kenya.

Kapindu, S. J., Saka, V. W., Julian, A. M., Hillocks, R. and Msuku, W. A. B. 1999. The significance and management of maize cob rots in smallholder farms in Central Malawi. *Africa Crop Science Journal* 7, 531–7.

Kassam, A., Mkomwa, S. and Friedrich, T. (Eds). 2017. *Conservation Agriculture for Africa: Building Resilient Farming Systems in a Changing Climate*. CABI Publishing, Wallingford, UK.

Kassam, A., Friedrich, T. and Derpsch, R. 2019. Global spread of Conservation Agriculture. *International Journal of Environmental Studies* 76(1), 29–51. doi:10.1080/00207233.2018.1494927.

Kassie, M., Teklewold, H., Marenya, P., Jaleta, M. and Erenstein, O. 2015. Production risks and food security under alternative technology choices in Malawi: application of a multinomial endogenous switching regression. *Journal of Agricultural Economics* 66(3), 640–59. doi:10.1111/1477-9552.12099.

Kassie, M., Stage, J., Diiro, G., Muriithi, B., Muricho, G., Ledermann, S. T., Pittchar, J., Midega, C. and Khan, Z. 2018. Push-pull farming system in Kenya: implications for economic and social welfare. *Land Use Policy* 77, 186–98. doi:10.1016/j.landusepol.2018.05.041.

Kfir, R., Overholt, W. A., Khan, Z. R. and Polaszek, A. 2002. Biology and management of economically important Lepidopteran cereal stemborers in Africa. *Annual Review of Entomology* 47(1), 701–31. doi:10.1146/annurev.ento.47.091201.145254.

Khan, Z. R., Pickett, J. A., Wadhams, L. J. and Muyekho, F. 2001. Habitat management strategies for the control of cereal stem borers and *Striga* weed in maize in Kenya. *Insect Science and its Application* 21, 375–80.

Khan, Z. R., Hassanali, A., Pickett, J. A., Wadhams, L. J. and Muyekho, F. 2003. Strategies for control of cereal stemborers and striga weed in maize-based farming systems in eastern Africa involving 'push-pull' and allelopathic tactics, respectively. *African Crop Science Society* 6, 602–8.

Khan, Z. R., Hassanali, A. and Pickett, J. A. 2006a. Managing polycropping to enhance soil system productivity: a case study from Africa. In: Uphoff, N., Ball, A. S., Palm, C., Fernandes, E., Pretty, J., Herren, H., Sanchez, P., Husson, O., Sanginga, N., Laing, M. and Thies, J. (Eds), *Biological Approaches to Sustainable Soil Systems*. CRC Press, Taylor and Francis, Boca Raton, FL, pp. 575–86.

Khan, Z. R., Midega, C. A. O., Hutter, N. J., Wilkins, R. M. and Wadhams, L. J. 2006b. Assessment of the potential of Napier grass (*Pennisetum purpureum*) varieties as trap plants for management of *Chilo partellus*. *Entomologia Experimentalis et Applicata* 119(1), 15–22. doi:10.1111/j.1570-7458.2006.00393.x.

Khan, Z. R., Pickett, J. A., Wadhams, L. J., Hassanali, A. and Midega, C. A. O. 2006c. Combined control of striga and stemborers by maize-*Desmodium* spp. intercrops. *Crop Protection* 25, 989–95.

Khan, Z. R., Midega, C. A. O., Amudavi, D. M., Hassanali, A. and Pickett, J. A. 2008a. On-farm evaluation of the 'push-pull' technology for the control of stem borers and *Striga* weed on maize in western Kenya. *Field Crops Research* 106(3), 224–33. doi:10.1016/j.fcr.2007.12.002.

Khan, Z. R., Midega, C. A. O., Njuguna, E. M., Amudavi, D. M., Wanyama, J. M. and Pickett, J. A. 2008b. Economic performance of the push-pull technology for stemborer and *Striga* control in smallholder farming systems in western Kenya. *Crop Protection* 27(7), 1084–97. doi:10.1016/j.cropro.2008.01.005.

Khan, Z. R., Midega, C. A. O., Pittchar, J., Pickett, J. and Bruce, T. 2011. Push-pull technology: a conservation agriculture approach for integrated management of insect pests, weeds and soil health in Africa. UK government's Foresight Food and Farming Futures Project. *International Journal of Agricultural Sustainability* 9(1), 162–70. doi:10.3763/ijas.2010.0558.

Khan, Z. R., Midega, C. A. O., Pittchar, J. and Pickett, J. A. 2014. Push-pull: a novel IPM strategy for the green revolution in Africa. In: Peshin, R. and Pimentel, D. (Eds), *Integrated Pest Management Experiences with Implementation. Global Overview* (vol. 4). Springer Science + Business Media, Dordrecht and New York, 574pp.

Khan, Z. R., Midega, C. A. O., Pittchar, J. O., Murage, A. W., Birkett, M. A., Bruce, T. J. A. and Pickett, J. A. 2015. Achieving food security for one million sub-Saharan African poor through push-pull innovation by 2020. *Philosophical Transactions of the Royal Society B* 369(1639), 20120284. doi:10.1098/rstb.2012.0284.

Liu, Q., Chen, X., Wu, K. and Fu, X. 2015. Nitrogen signaling and use efficiency in plants: what's new? *Current Opinion in Plant Biology* 27, 192–8. doi:10.1016/j.pbi.2015.08.002.

Madukwe, M. C. 2006. Delivering agricultural extension services to farmers in developing countries: knowledge for development. Available at: http://knowledge.cta.int/en/content/view/full/3009.

Midega, C. A. O., Khan, Z. R., van den Berg, J. and Ogol, C. K. P. O. 2005. Habitat management and its impact on maize stemborer colonisation and crop damage levels in Kenya and South Africa. *African Entomology* 13, 333–40.

Midega, C. A. O., Khan, Z. R., van den Berg, J., Ogol, C. K. P. O., Pickett, J. A. and Wadhams, L. J. 2006. Maize stemborer predator activity under 'push-pull' system and *Bt*-maize: a potential component in managing *Bt* resistance. *International Journal of Pest Management* 52(1), 1–10. doi:10.1080/09670870600558650.

Midega, C. A. O., Khan, Z. R., Van den Berg, J., Ogol, C. K. P. O., Bruce, T. J. and Pickett, J. A. 2009. Non-target effects of the 'push-pull' habitat management strategy: parasitoid activity and soil fauna abundance. *Crop Protection* 28(12), 1045–51. doi:10.1016/j.cropro.2009.08.005.

Midega, C. A. O., Khan, Z. R., Amudavi, D. M., Pittchar, J. and Pickett, J. A. 2010. Integrated management of *Striga hermonthica* and cereal stemborers in finger millet (*Eleusine coracana* (L.) Gaertn.) through intercropping with *Desmodium intortum*. *International Journal of Pest Management* 56(2), 145–51. doi:10.1080/09670870903248843.

Midega, C. A. O., Bruce, T. J. A., Pickett, J. A., Pittchar, J. O., Murage, A. and Khan, Z. R. 2015. Climate-adapted companion cropping increases agricultural productivity in East Africa. *Field Crops Research* 180, 118–25. doi:10.1016/j.fcr.2015.05.022.

Midega, C. A. O., Pittchar, J. O., Pickett, J. A., Hailu, G. W. and Khan, Z. R. 2018. A climate-adapted push-pull system effectively controls fall armyworm, *Spodoptera frugiperda* (J E Smith), in maize in East Africa. *Crop Protection* 105, 10–5. doi:10.1016/j.cropro.2017.11.003.

Miller, J. R. and Cowles, R. S. 1990. Stimulo-deterrent diversion: a concept and its possible application to onion maggot control. *Journal of Chemical Ecology* 16(11), 3197–212. doi:10.1007/BF00979619.

Minja, E., Ulicky, E., Mfoi, M., Marawiti, M. and Mziray, H. 2004. Farmer group activity reports for the DFID Crop Protection Programme (CPP) Bean IPM Promotion Project in Eastern and southern Africa. Available at: http://www.ciat.cgiar.org/africa/pdf/ffd_sanjajuu_Tan_Jun04.pdf.

Moss, B. 2008. Water pollution by agriculture. *Philosophical Transactions of the Royal Society of London. Series B, Biological Sciences* 363(1491), 659–66. doi:10.1098/rstb.2007.2176.

Murage, A. W., Amudavi, D. M., Obare, G., Chianu, J., Midega, C. A. O., Pickett, J. A. and Khan, Z. R. 2011. Determining smallholder farmers' preferences for technology dissemination pathways: the case of 'push-pull' technology in the control of stemborer and striga weeds in Kenya. *International Journal of Pest Management* 57(2), 133–45. doi:10.1080/09670874.2010.539715.

Murage, A. W., Obare, G., Chianu, J., Amudavi, D. M., Midega, C. A. O., Pittchar, J. and Pickett, A. J. 2012. The effectiveness of dissemination pathways on adoption of 'push-pull' technology in western Kenya. *Quarterly Journal of International Agriculture* 51(1), 51–71.

Murage, A. W., Pittchar, J. O., Midega, C. A. O., Onyango, C. O. and Khan, Z. R. 2015a. Gender specific perceptions and adoption of the climate-smart push-pull technology in eastern Africa. *Crop Protection* 76, 83–91. doi:10.1016/j.cropro.2015.06.014.

Murage, A. W., Midega, C. A. O., Pittchar, J. O., Pickett, J. A. and Khan, Z. R. 2015b. Determinants of adoption of climate smart push-pull technology for enhanced food security through integrated pest management in eastern Africa. *Food Security* 7(3), 709–24. doi:10.1007/s12571-015-0454-9.

Ogot, N. O., Pittchar, J. O., Midega, C. A. O. and Khan, Z. R. 2018. Attributes of push-pull technology in enhancing food and nutrition security. *African Journal of Agriculture and Food Security* 6, 229–42.

Organization for Economic Co-operation and Development (OECD). 2010. *Investing in Women and Girls: The Breakthrough Strategy for Achieving All the MDGs*. Based on a speech by Jon Lomoy, Director of the OECD's Development Co-operation Directorate, at the Helsinki High-level Symposium, United Nations 2010 Development Co-operation Forum, 4 June 2010.

Oswald, A. 2005. Striga control-technologies and their dissemination. *Crop Protection* 24(4), 333–42. doi:10.1016/j.cropro.2004.09.003.

Otsuki, T. 2001. What price precaution? European harmonisation of aflatoxin regulations and African groundnut exports. *European Review of Agriculture Economics* 28(3), 263–84. doi:10.1093/erae/28.3.263.

Owuor, M. J., Midega, C. A. O., Obonyo, M. and Khan, Z. R. 2018. Impact of companion cropping on incidence and severity of maize ear rots and mycotoxins in Western Kenya. *African Journal of Agricultural Research* 13(41), 2224–31. doi:10.5897/AJAR2018.13396.

Pickett, J. A. and Khan, Z. R. 2016. Plant volatile-mediated signalling and its application in agriculture: successes and challenges. *The New Phytologist* 212(4), 856–70 doi:10.1111/nph.14274.

Pickett, J. A., Woodcock, C. M., Midega, C. A. and Khan, Z. R. 2014. Push-pull farming systems. *Current Opinion in Biotechnology* 26, 125–32. doi:10.1016/j.copbio.2013.12.006.

Pretty, J., Toulmin, C. and Williams, S. 2011. Sustainable intensification in African agriculture. *International Journal of Agricultural Sustainability* 9(1), 5–24. doi:10.3763/ijas.2010.0583.

Roderick, R. M., Knight, T. O., Jaramillo, M., Coble, K. H., Patrick, G. F. and Baquet, A. 2008. Preference for risk management information sources: implications for extension and outreach programming. *Agricultural and Resource Economics Review* 37(1), 106–16. doi:10.1017/S1068280500002185.

Rogers, M. E. 1997. The diffusion of innovations, model and outreach from National network of libraries of medicine to Native American communities. Department of Communication and Journalism, University of New Mexico, Albuquerque, New Mexico. Available at: http://nnlm.gov/pnr/eva/roger.html (accessed on 9 August 2018).

Royal Society. 2009. *Reaping the Benefits: Science and the Sustainable Intensification of Global Agriculture*. Royal Society, London.

Stromquist, N. P. 2001. What poverty does to girls' education: the intersection of class, gender and policy in Latin America. *Compare: A Journal of Comparative and International Education* 31(1), 39–56.

Tédihou, E., Olatinwo, R., Hell, K., Hau, B. and Hoogenboom, G. 2012. Effects of variety, cropping system and soil inoculation with *Aspergillus flavus* on aflatoxin levels during storage of maize. *Tropical Plant Pathology* 37(1), 25-36. doi:10.1590/S1982-56762012000100003.

Tittonell, P. and Giller, K. E. 2013. When yield gaps are poverty traps: the paradigm of ecological intensification in African smallholder agriculture. *Field Crops Research* 143, 76–90. doi:10.1016/j.fcr.2012.10.007.

Tsanuo, M. K., Hassanali, A., Hooper, A. M., Khan, Z. R., Kaberia, F., Pickett, J. A. and Wadhams, L. J. 2003. Isoflavanones from the allelopathic aqueous root exudates of *Desmodium uncinatum*. *Phytochemistry* 64(1), 265–73. doi:10.1016/S0031-9422(03)00324-8.

Tucker, M. and Napier, T. L. 2002. Preferred sources and channels of soil and water conservation information among farmers in three Midwestern US watersheds. *Agriculture, Ecosystems and Environment* 92(2–3), 297–313. doi:10.1016/S0167-8809(01)00293-6.

United Nations General Assembly. 2010. Report submitted by the Special Rapporteur on the right to Food. Human Rights Council, 16th Session, New York, A/HRC/16/49, pp. 8–9; 14–16.

United Nations General Assembly. 2015. Agricultural technology for development. Report of the Secretary-General, 70th Session – Sustainable Development, New York, A/70/298, p. 11.

United States Department of Agriculture (USDA). 2010. Food security assessment 2010–2020 report. USDA Economic Research Service, Washington DC.

Vincelli, P. 1997. Ear rot of corn caused by *Stenocarpella maydis*. Cooperative Extension Service, University of Kentucky, pp. 1–4.

Vitousek, P. M., Aber, J., Howarth, R. W., Likens, G. E., Matson, P. A., Schindler, D. W., Schlesinger, W. H. and Tilman, G. D. 1997. Human alteration of the global nitrogen cycle: causes and consequences. Issues in Ecology No. 1. Ecological Society of America, Washington DC, 17p.

Vitousek, P. M., Cassman, K., Cleveland, C., Crews, T., Field, C. B., Grimm, N. B., Howarth, R. W., Marino, R., Martinelli, L., Rastetter, E. B. and Sprent, J. I. 2002. Towards an ecological understanding of biological nitrogen fixation. *Biogeochemistry* 57(1), 1–45. doi:10.1023/A:1015798428743.

Woomer, P. L., Mukhwana, E. J., Odhiambo, G., Okello, D., Omare, M. and Sanginga, N. 2004. Striga control in western Kenya: Raising awareness, containing and reducing the infestation and developing strategies for eradication. A report on a collaborative project involving the African Agricultural Technology Foundation (AATF) and the Western Regional Alliance for Technology Evaluation (We RATE), Kenya.

World Bank. 2008. *The Growth Report: Strategies for Sustained Growth and Inclusive Development. Commission on Growth and Development*. World Bank, Washington DC.

Zain, M. E. 2011. Impact of mycotoxins on human and animals. *Journal of Saudi Chemical Society* 15(2), 129–44. doi:10.1016/j.jscs.2010.06.006.

Chapter 6

Nutrient management practices and benefits in Conservation Agriculture systems

Stephane Boulakia, Florent Tivet and Olivier Husson, Centre de coopération Internationale en Recherche Agronomique pour le Développement (CIRAD), France; and Lucien Séguy, AgroécoRiz, France

1 Introduction

Can we - and do we need to - consider *specific* nutrient management practices within CA systems? If yes, should such practices target short-term benefits, at crop and/or cover crop-crop succession scales, or long-term benefits aimed at ensuring soil with improved and accessible nutrient reserves? Alternatively, should we consider that nutrient cycling efficiency within CA systems is primarily linked to the intensity and frequency of fresh biomass inputs? This approach suggests looking at nutrient management as subservient to the carbon cycle, the Fresh Organic Matter (FOM) inputs being the main driver to create a positive Soil Organic Carbon (SOC) balance and increasing amounts of high activity Organic Matter (OM) in soils.

From this perspective, CA system design and management calls for trade-offs between production needs, linked to biomass exportation, and agroecosystem services delivery, including nutrient cycling, based on

http://dx.doi.org/10.19103/AS.2019.0049.06

maximizing nutrient recovery and enhancing biodiversity. This is, for instance, the case when CA systems integrate livestock production (Naudin, 2012), with part of the biomass being used as animal feed and recovered through manure (biomass by-pass through animal feed – manure).

We propose in this chapter to examine some issues specific to nutrient management in CA-based systems. Our approach comes from our practical experience of CA systems design, *on-farm and with farmers*, in various tropical and subtropical agroecosystems (Seguy, 2012; Husson et al., 2016a,b; Kong, 2016). It can be seen as an 'engineering' approach, where field practices depend on and determine management at the farm level and beyond. The overall cropping system (i.e. crops–cover-crops relationships/successions/rotations and management) generates the framework in which sub-components of the cropping systems can be adjusted. These sub-components include crop varieties, weed and pest control management etc. as well as organic and mineral fertilizer inputs.

In the first section, we review how integrated nutrient management in CA derives from the a 'forest model' based on OM dynamics and discuss its strengths and weaknesses in optimizing cropping systems practices. In the second section, we highlight the key processes that improve nutrient accessibility for plants under CA management. Case studies from contrasting agroecosystems in subtropical regions of Brazil and southern France illustrate how CA, associated with key OM inputs from cover crops and crops residues, impacts and modifies plant nutrient management.

2 Integrated nutrient management and the conceptual 'forest model' of CA

2.1 Current trends in CA systems design

CA developed from pre-existing no-till (NT) practices aimed at solving problems of farming sustainability. However, while NT practices are mostly driven by practical objectives (such as reducing costs, rapid crops implementation), CA firstly aimed at solving problems induced by conventional tillage (CT)-based practices such as soil degradation caused by erosion and OM mineralization (Seguy, 2006). In terms of Soil Organic Carbon (SOC) balances, CA has proven its ability to establish sustainable agricultural production in most of the world's agroecosystems and even restore environments degraded by years of CT (Tivet et al., 2013). Through the recent 4p1000 initiative (Soussana, 2019), this capacity to store carbon in soil has led many to consider CA as a key method of mitigating global climate change (CC).

While still based on ensuring a positive SOC balance in soil, CA now seeks to integrate new agronomical practices to improve resilience in adapting to CC as well as limiting farmers' dependency on external inputs, especially agrochemicals. This has led to important new trends in the implementation of the three interlinked principles of CA:

- *Minimization or avoidance of soil disturbance* is progressing with new implements such as the single disc no-till planter/seeder, or no disturbance at all - where possible - with broadcast seeding practices for both crops and cover crops;
- *Permanent soil cover* is improved through seeding while preceding crops or cover crops are still growing (*planting green*, Duiker, 2015); the time lapse with dead cover before crop sowing is reducing, in order to prolong the period of cover, while weed and cover crop control is undertaken using mechanical means to replace or complement herbicides; and
- Use of mix species cover enhances species diversity in rotations; these mixes are blending grass and legumes species with other species chosen for their adaptation to climatic conditions and specific functional traits. These mixes integrate up to around 10 species with genera from diverse families, for instance, *Cruciferae, Polygonaceae, Asteraceae, Amaranthaceae, Lamiaceae, Malvaceae, Chenopodiacea*, etc.

These trends affect cycling and dynamics of nutrients within CA systems and require targeted research to optimize timing and quantities of particular inputs.

2.2 The forest model and nutrients dynamic framework

The 'Forest model' concept (see Fig. 1) shows how the sustainability of CA systems is achieved through high and continuous inputs of fresh biomass from plants (wood, leaves, roots, exudates), animals and microbes. This fresh organic matter (FOM) is decomposed and used as an energy and carbon source. A dense and deep network of roots and mycorrhiza fungi helps to filter and retain the mineral nutrients released across the soil food web.

2.2.1 FOM inputs, mineralization processes and kinetics of nutrients release

The biological processes driving FOM degradation and the mineral products that derive from these processes are dependent on both the living components of the SOM, their activity and functional diversity, and the composition of OM inputs. Litter degradation is driven by the macrofauna that shred tissue into particles accessible to bacteria (easily degradable compounds such as sugars and proteins) and fungi (more resilient macromolecule, cellulose, hemicellulose, lignin or polyphenols). Simple models based on C/N ratio, with an emphasis on the lignin content of FOM inputs, accurately describe C and N mineralization

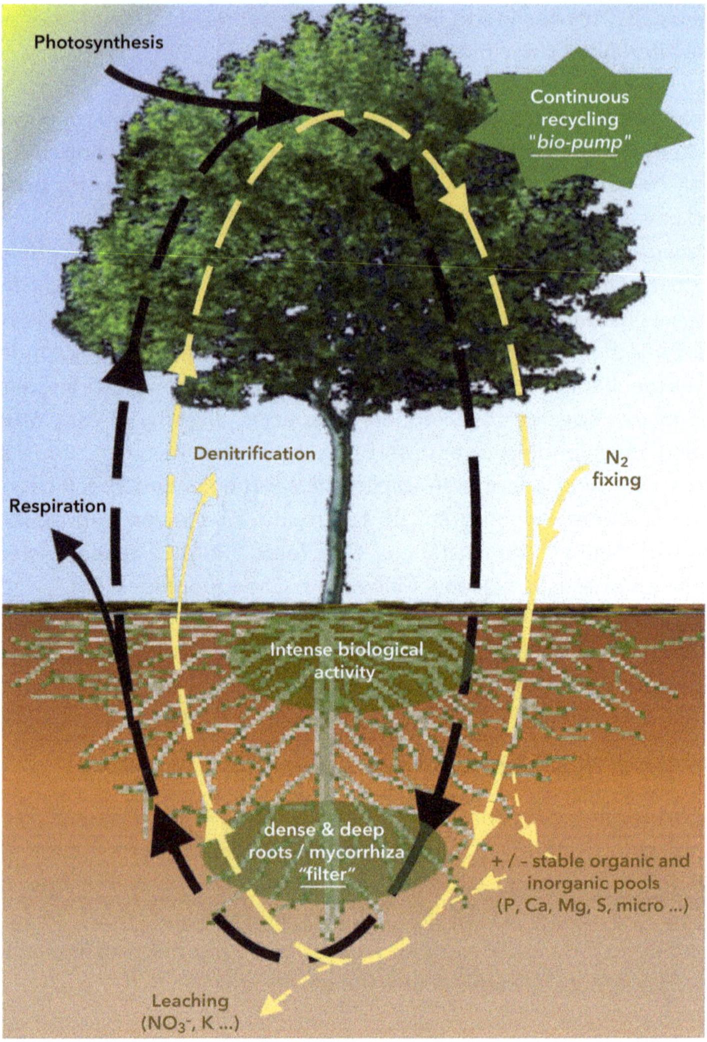

Figure 1 Conceptual CA 'Forest model' and coupling of carbon and nutrient cycling. Source: adapted from Seguy (2008).

in decomposition of crops and cover crops residues (Frankenberger and Abdelmagid, 1985; Nicolardot, 2001). Palm (2001) developed a database of organic resources focusing on N, lignin and polyphenol content for tropical agroecosystems management as a tool to support cropping system design and modelling. Biomass inputs with a low C/N ratio allow the rapid build-up of microbial biomass while those with a C/N ratio higher than 30 undergo a decomposition process limited by N availability (see Fig. 2). The composition and nature of the biomass will thus lead to different types of decomposition

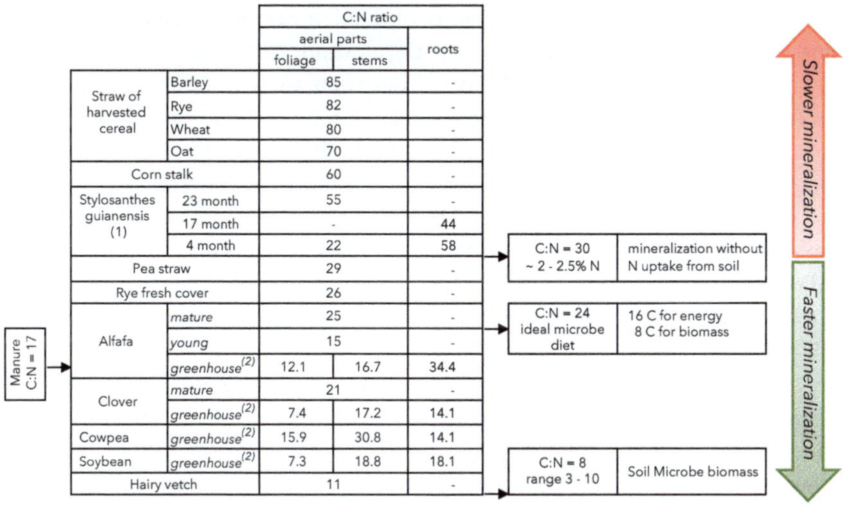

Figure 2 C:N ratio of residues and cover biomass and general influence on mineralization kinetic. Source: adapted from USDA Agronomy technical note 76; (1) Zemek et al. (2018); (2) Frankenberger and Abdelmagid (1985).

process (see Table 1) and nutrient releases, as shown in field studies using resin cores (Schomberg et al., 2006).

Numerous studies based on monitoring litterbags have determined nutrient release from residues decomposition, leading to the adjustment of models of nutrient release (Shi, 2013; Salume, 2017). Giacomini (2003), studying the P and K releases from common vetch (*Vicia sativa*), forage radish (*Raphanus sativus*) and black oat (*Avena strigosa*) in pure or mix cover, has highlighted:

- contrasted kinetics for the two elements, very fast release for K with half-life time varying from 4.2 (vetch) to 12.5 (oat) days and slower for P (from 6.5 - vetch to 51.2 days - radish);
- a slower release from the mix vetch + oat compared to pure vetch; and
- a P release kinetic in relation to the amount of soluble phosphorus in the residues, with the presence of a resistant P-pool, still un-mineralized at 180 days, comprised between 20 and 40% of the initial organic P.

Table 1 Corresponding range of C/N ratio and time (in weeks) of incubation to mineralize half of the total N initial content of legumes crop residue

	Foliage	Stems	Roots
C:N ratio	7.3-15.9	16.7-30.8	14.1-34.4
N half-life time	2.1-4.8	7.6-11.4	5.0-10.7

Source: adapted from Frankenberger and Abdelmagid (1985); study conducted on soil columns with incorporated residues; see also Fig. 2.

The amount and dynamics of macro and micronutrients released through mineralization processes are mostly dependent on initial composition of the organic inputs (Maluf, 2015). It is important to keep in mind that these studies, for obvious methodological reasons, focus mostly on above-ground biomass inputs and rarely on the 'root litter'. Khalaf (2013), in a greenhouse study with seven cover crop species commonly found in Sweden, showed that roots represent 23–57% and 15–49% of total biomass and P content, respectively. As with above-ground parts of plants (leaf, stem), decomposition and mineralization kinetics of root litter (fine vs. coarse exudate) depend on the structure and composition of OM tissues (C/N and lignin/N ratios) and edaphic (soil-related) conditions.

The release of abundant and diversified nutrients by FOM mineralization allows for the rapid growth of bacterial communities. This development can induce temporary N immobilization during the peak growth phase, which is called the priming effect (Löhnis, 1926), before further mineral N ($N\text{-}NO_3$ and $N\text{-}NH_4$) release by protozoa and nematodes predation and SOM degradation by released enzymes (Fontaine, 2003). These processes have led to studies stressing the importance of managing cover crop timing to match the phase between microbial immobilization and crop demands and the necessity of applying banded N fertilizer at crop onset (Sa, 1999). This temporary N immobilization induced by FOM inputs with low N content favors mix-species cover which provide organic inputs with differentiated C/N ratio. Such composite amendments smooth peaks of immobilization and prolong recovery phases (Giacomini, 2003), thanks to the additional and relayed effects of heterogeneous FOM inputs (Redin, 2014). However, McDaniel (2016) qualified this direct incidence of chemical composition and tissue structure of OM inputs and showed that long-term diversity in crop rotation has greater influence, most likely through OM degradability and microbial community diversity, on the soil microbial biomass, respiration and mineral N release than the chemical characteristics and diversity of the last FOM inputs.

2.2.2 Nitrogen fixing and other natural nutrients entrances

CA systems allow a 'densification' of the nitrogen fixing process within rotations through crops and cover crops combinations. Integrating legume species (built upon the expertise developed in green manure management) has always been a central objective in CA systems design (Derpsh and Calegari, 1985; Florentín et al., 2010) to increase nitrogen input through symbiotic N_2-fixing pathways. In a recent field study conducted in the middle-west region of Madagascar on low-input rice-based cropping systems, incorporating a year-long *Stylosanthes guianensis* fallow, Zemek (2018) found that, on the 225 kg N uptake ha^{-1} of the cover crop, about 55 and 45% originated from the soil and atmosphere, respectively. Coarse roots on the 0–30 cm top layer weighed 1.3 Mg ha^{-1} and contained around 16 kg N ha^{-1} while the inputs from rhizodeposition (exudation, excretion, lysis of root cap cells, etc.) and turnover from root hairs remained unknown, estimated at 8–173 kg N ha^{-1} according to the cited literature.

Mobilization of nonsymbiotic nitrogen fixing processes, based on aerobic and heterotrophic bacteria (*Azotobacter* sp., *Azospirillum* sp., *Klebsiella* sp., etc.), can be boosted through specific cover crop species introduction, potentially associated with microbial inoculation (Kelemu et al., 2011). For instance, finger millet (*Eleusine coracana*) fixes nitrogen through a rhizosphere association with *Azospirillum brasilense*. Rai (1991) highlights that its nitrogenase is active in low-pH and Al-rich conditions; Weinhard (1971) suggesting a strong dependency of nitrogenase activity levels on SOM content. Dart (1986) reviewed studies showing nitrogenase activity enhancement by straw restitution and farmyard manure application, and Husson (2013) hypothesized that symbiotic nitrogen fixation could depend on redox homeostasis in the nodules. CA, through the insertion of cover crops, allows for increasing intensity and regularity of symbiotic and nonsymbiotic N fixation processes in crop succession and rotation.

CA also creates favorable conditions for the development of beneficial symbiosis with arbuscular mycorrhizal fungi (AMF). The symbiotic relationship with AMF increases the plants' capacity to absorb water, immobile nutrients such as P and Zn, and mobile nutrients such as N and S, especially in low-concentration environments. Non-disturbance of the soil, permanent soil cover and use of mixed cover, including various species of legumes, lead to a greater rate of inoculation and colonization by diverse AMF communities under CA management (Arihara and Karazawa, 2001; Verzeaux, 2017). Brito (2014), in a two-phase greenhouse experiment, showed that wheat experiences doubled growth in undisturbed soil in comparison to disturbed soil, where different species are chosen for their contrasting mycotrophic dependencies and ability to enhance the development of extra-radical mycelium (ERM). According to research by de Varennes and Goss (2007) and Brito (2013), improved wheat growth is due to faster and earlier root colonization by AMF in the presence of intact ERM and correlates with reduced Mn absorption and augmented P and S uptake. This experiment suggests that ERM developer species influence AMF potential for Mn protection and that the ERMs developed by mycotrophic species are associated with functional diversity. Turrini (2016), using a molecular approach, characterizes a higher AMF diversity under multispecies cover compared to mono-specific cover; however, if subsequent maize crops, sown after incorporation in the soil of the cover crop, present different AMF communities, these appear independent from the species profile detected under the preceding cover crop. Understanding the drivers behind AMF diversity and abundance will require more studies to differentiate between the multiple potential influences, such as season or host preferences in relation to functional significance, itself dependent on biotic and abiotic conditions. In any case, cover crops blending multiple species with high phylogenetic diversity appear to lead to a higher diversity of indigenous AMF species in the agroecosystem, which increases the potential for efficient combinations.

CA activates biological pathways to access nutrients present under forms which cannot be assimilated (e.g. acidolysis or alkalinolysis of substratum minerals) in the view of conventional fertility management; but do these enzymatic pathways allow access to resilient nutrient pools permanently, or only temporarily activated by specific plants species and associated microbes, possibly under specific physical soil conditions? The balance and kinetics of elements immobilized and released from resilient pools challenge the notion of 'exchangeability', and may require new laboratory extraction methods, combined with classification of pools in functional classes (Turner, 2002), related to accessibility and dynamics under CA management.

2.2.3 Uncertainties about lixiviation and denitrification

Views on the incidences of CA on denitrification and loss of nitrogen by N_2 and N_2O emissions remain diverse. This is most likely due to the variety of practices and contexts that encompass the term CA. Once again, contextualization, type of past and present systems, and the number of years under CA are required to understand and better assess the impact of CA-based management on denitrification processes.

A modeling (Lugato, 2018) study based on 8000 examples across the EU compared NT scenarios (minimum soil disturbance and residue retention) and CA-based scenarios (= NT scenarios + use of legumes cover). It predicted that that legumes would boost the capacity to sequestrate SOC in the initial decades (up to around 2040) before shifting to net GHG emitters with N_2O emissions, even with a reduction in mineral N supply, offsetting the benefits of incorporating CO_2 in soil. However, in the Midwest United States, in a three-decade-long experiment, Omonode et al. (2011) found about 40% and 57% lower N_2O emissions under NT compared to the use of a moldboard plow and chisel plow respectively. In the same experiment, N_2O emissions appeared to be limited by 20% in corn-soybean rotation compared to continuous corn. Rochette (2008), reviewing 25 field studies, suggested that NT is associated with increased N_2O emissions in poorly aerated conditions and is neutral in medium or well-aerated soils. Corsi (2012), in a review of SOC sequestration and GHG emissions, stressed that nitrous oxide emissions under CA are regarded by some authors as a consequence of more anoxic soil conditions. Proper CA practices lead to globally improved physical parameters, but this trend, at the soil profile scale, is not necessarily inconsistent with the creation of an O_2 limited atmosphere at the micro- and meso-aggregate scales, where C and N substrate are available for microbial processes. However, from an engineering perspective, it should be possible to progressively improve CA system management, based on organic input chemistry, through mix cover composition, according to N-total and N-mineralized soil content. These multi-specific covers should allow experimentation with C/N stoichiometry and the nature of the C compounds (lignin, polyphenol etc.). In situations with a high denitrification potential, inclusion of nitrification-inhibiting traits (Subbarao, 2009, 2013) to regulate nitrification pathways through roots exudation or biomass compound released along the decomposition process (Subbarao, 2015; Coskun, 2017), could also be tested. Further investigation and CA-based cropping systems design could draw inspiration from wet tropical grass savanna where nitrification regulation occurs through plant assemblies and associated microbial communities (Lata, 1999, 2004; Srikanthasamy, 2018).

Looking at possible nutrient loss by lixiviation (e.g. through leaching), data collected under CA with high biomass amendments is scarce. In the UK, Goss (1993) compared N-NO_3^- losses under winter crops and found that NT limits winter drainage loss compared to moldboard plowing. The capacity to retain nutrients in the soil-microorganism-plant system is considered a key beneficial effect of CA (Hobbs, 2008) but is better suggested by indications such as nutrient stratification (Rivera-Zayas, 2018), deep cover crop root development and so on than direct measurement methods and/or deep soil solution sampling. Field studies assessing effluxes below rooting zones along crop/cover-crop sequences or across the overall rotation of the most mobile anions (NO_3^-, SO_4^{2-}) and bases (Ca^{2+}, Mg^{2+} and K^+) are lacking. For instance, is the potassium that is quickly released from mulch mineralization caught and recycled by succeeding crops and cover crops? Some increasingly used techniques, such as planting on green cover crops, and applying manure on living cover crops rather than prior crops, may improve cycling performance by preventing mineral flushing prior to crop establishment.

2.3 Timing and development of accessible nutrients pools

Time-related changes are particularly significant when considering nutrient cycling efficiency under CA. At the crop cycle scale, the kinetics and timing of nutrient release by mineralization, immobilization by microbial biomass and plant uptake have been well studied for nitrogen and, to a lesser extent, for P and other macro and microelements. At the rotation scale, Hu (2015) showed in a field study that the recovery rate in the soil-plant system of an N-based organic input (maize residue) is strongly dependent on the repetition of residue input in subsequent years. In the fourth year, the cumulative retention rates in plant biomass and soil are 61.9% and 91.9% without and with continued residue amendment respectively, highlighting the importance of avoiding disruption in FOM inputs during rotation. Over a longer period, nutrient use efficiency improves with a steady transformation of soil conditions; continuous fluxes of FOM amendments associated with plant and microbial diversity provide soil with growing pools of available nutrients. In the Brazilian Cerrados, Maltas et al. (2007), comparing SOC and total N contents of fields cultivated using CA for periods varying from two to 14 years, obtained a good correlation between C and N stocks and duration of CA practice, with an estimated rate, for a 0–30cm layer, of 1.91Mg ha^{-1} $year^{-1}$ and 103kg ha^{-1} $year^{-1}$ of C and total N respectively. Studying the evolution of N mineralization, she found an increase of 2.4kg N ha^{-1} $year^{-1}$. Schoenau and Campbell (1996) reviewed a 15-year-long study by Greer and Schoenau (1992) of a Canadian Chernozem soil. They found a mineralizable organic pool for N-OM of 3.0% and 6.0% and for S-OM of 1.7% and 5.1% under conventional and no-till managements respectively. The authors attributed these differences to the build-up of a more easily mineralizable nutrient pool under long-term NT management.

Processes regulating nutrient cycling in CA systems have begun to be identified and described. The short-term kinetics of OM input mineralization and mineral elements releases are being quantified. However, when it comes to a sequence of cover-crops/crops/cover crops over time, there is still insufficient information to establish gross balances for major nutrients. A full understanding of the soil–microorganisms–plant system under CA remain insufficiently informed across the diversity of agroecosystems and practices.

Previous reviews of key aspects of nutrient dynamics under CA management have shown the difficulty in building up operational input-output models integrating the complexity of the biophysical processes driving nutrient cycling, their evolution across diversified crop rotations, and across the range of agroecosystems in which CA has been undertaken. There are too few studies showing how nutrient sources evolve when CA is adopted, including the relative share of crop nutrition represented by release from primary FOM and secondary SOM mineralization and what this means for mineral fertilizer applications.

3 Bio-availability of elements and mineral use efficiency in CA systems

CA not only modifies nutrient cycling by coupling it to organic matter inputs and metabolic processes in trophic networks, it also induces profound changes in the soil's physical (Thierfelder, 2012), chemical and biological (Lienhard, 2013) attributes.

Improving the soil's interrelated physical properties, such as porosity, aggregate stability, suppression of plow-pan, water infiltration rate and water holding capacity, reduction of anoxic conditions, etc. leads to an extension of the soil volume colonized by the roots and associated mycorrhiza and to enhanced access to the soil solution. Combined with reduced evaporation, this contributes to improving water reserves and the mineral nutrition of the plants.

CA modifies the availability of nutrients through the alteration of chemical and biological parameters which results in the emergence of new interrelations between the soil, plant and microorganisms.

3.1 CA and soil chemical parameter evolution

3.1.1 Direct effects of fresh mulch on acidity and other soil chemical parameters

Early studies focusing on mineral release from mulch mineralization have highlighted the resulting rapid modification of the soils chemical parameters.

Franchini (2001), working on soil columns with plant extracts from various crops (soybean, wheat) and cover crop species (oil seed radish, *Raphanus sativus*, blue lupine, *Lupinus angustifolius*, black oat, *Avena strigosa*) noticed an increase in pH, improved availability of calcium, magnesium and potassium and a decrease in exchangeable aluminum. Comparing fresh and incubated plant extracts, they noted that the effects are much more limited in 15-day incubated plant extracts. Chromatographic analysis demonstrated the evolution of the chemical composition of the extract during incubation. Particularly noticeable was a reduction in the concentration of short chain fatty acids (malic, maleic, succinic, phenyl-acetic etc. acids) which quickly oxidized into long chain and aromatic compounds (also released by slower process of lignin degradation). These ultimate compounds are less reactive and have a limited capacity to form a mobile organic complex with Al and Ca. The study highlights the marked differences between plant extracts, with stronger effects in blue lupine and oil seed radish compared to black oat, soybean and wheat. Cassiolato (2002), utilizing a similar experimental approach with acid oxisol, reached convergent conclusions regarding the strong reactivity of low molecular organic acids through the carboxyl and phenolic groups, and their active role in metal complexing reactions, ion sorption and rhizosphere chemistry. They also noted higher acidity and aluminum neutralization by the radish and black oat extracts compared to white lupine (*Lupinus albus*). Viera (2008) showed similar impacts on the pH and aluminum of organic amendments applied to an acidic sandy soil. Regarding phosphorus, Pavinato (2008) studied the effect on P availability of fresh plant extracts of various species (millet, sorghum, maize, oat, radish and soybean grown for 60 days), associated or not with soluble P fertilization. All extracts, with radish and soybean being the most effective, had a positive effect on P availability by substituting inorganic complexation of soluble P with less stable binding to organic compounds. In acid oxisol, the metal affinity of the organic acids released from plant residues in solution were observed to limit the availability of Cu and Zn, and to foster the accessibility of S and B through the competition of organic anions for soils adsorption sites of sulfate and borate cations.

For historical reasons, there are far more studies of the impact of mulch use on soil chemical parameters and nutrient availability for tropical acid soil classes than on alkaline or saline soils. Nevertheless, we can assume that CA with strengthened inputs from mix cover have the capacity to buffer and regulate other types of unbalanced Cation Exchangeable Capacity (CEC). Hasbullah and Marschner (2015) highlighted that organic additives, more specifically water extractable organic compounds, alleviate the depressive effect of soil salinity on microbial activity. Regular inputs of FOM should decrease the sodium exchangeable percentage (Na_{exch}/CEC) and limit its dispersive effect on soil aggregate, as does the regular addition of poultry manure (Tejada, 2006). In the experience of our R&D group in New Caledonia (Jullien F., unpublished), CA has enabled agricultural production on the brown hypermagnesian soil (Inceptisol) that has developed at a base of ultramafic rock massifs. These soils present a high CEC, linked to the presence of smectite clay, dominated by magnesium; mineral perspective, considering their Ca/Mg ratio below 1, previously lead to consider them unsuitable for crop production. Notably, trials conducted in the 1980s showed that applications of up to 12 t ha^{-1} of CaO combined with 380kg N ha^{-1}, 220kg P_2O_5 ha^{-1} and 100kg K_2O ha^{-1} failed to enable maize productivity (Collet, 1985).

3.1.2 CA and redox (Eh) potential regulations

As with pH, CA can regulate the redox potential in soil and helps to regulate pH-Eh conditions in a favorable range for soil-plant-microorganism synergies (Husson et al., 2016a,b). Progressive physical structuration limits hydromorphic reductive conditions in the case of high rainfall levels. In our experience of rainfed lowland rice, the adoption of CA-based systems with high cover induces a transformation of the soil's physical condition, with a progressive impact from the surface of the soil to deeper layers. In the meantime, plant residues release polyphenol compounds such as tannins that, through their antioxidant properties, regulate the Eh potential that, in turn, controls the destructive chain reactions of free radicals (Rimmer and Smith, 2009; Rimmer, 2006). Keluweit (2015) described a fungi-mediated Mn-based redox reaction involved in oxidative decomposition processes. This 'Eh-pH' regulation, which occurs especially in the rhizosphere and the immediate vicinity of the mulch, conditions the chemical forms of nutrients with high redox numbers range such as N, P, S and elements in high concentration such as Fe (Husson et al., 2016a,b).

3.1.3 Soil organic matter, Cationic Exchangeable Capacity (CEC) and bases saturation rate

CA adoption allows the SOC dynamics to revert from a negative balance (losses greater than gains) in CT management to a positive one in CA management (Tivet et al., 2013; de Moraes Sá, 2015). These SOC evolutions positively impact CEC due to the negative charges and high reactivity of SOM; Gobat (1998), based on multiple sources, reminds us that the OM-associated CEC ranges are 60–280, 485–870 and up to 1400 $cmol^+ kg^{-1}$ for SOM, pure humic acids and pure fulvic acids, respectively. These OM negative charges increase with pH and SOM may account for up to 90% of the total CEC in sandy soil.

Sá (2009) highlighted the intricate evolutions of SOM, CEC, bases saturation rate and pH, comparing sites with long-term CA vs CT management systems in Brazilian subtropical conditions. CEC evolution is positively correlated to SOC, with an increase of 0.37 $cmol^+ kg^{-1}$ soil for a 1 g C kg^{-1} soil increase under NT. This SOM evolution positively affects pH, limits potential acidity and is associated with P and K increases of 9.7 and 17.9kg ha^{-1} $year^{-1}$ respectively for 0–10cm depth. In the tropical central regions of Brazil, the use of CA-based cropping systems, with live *Arachis pintoi* cover or desiccated mulch of various species, for five years restored the SOC stocks depleted by 5 years of CT-based system to levels observed under the native forest (Seguy, 2008; see Fig. 3). This SOM restoration is consistent with CEC and exchangeable potassium improvements but not with improvements in the base saturation rates. In this experiment, live *Arachis pintoi* cover appears to improve strongly exchangeable phosphorus concentration.

In another long-term study of soil evolution comparing CA and CT management in the subtropical southern regions of Brazil, Nicolodi et al. (2008) suggested that CA alters the relations between pH, base saturation rate and exchangeable aluminum established under CT. For instance, at a similar pH level, exchangeable Al tends to be lower under CA management, possibly due to the tighter bonds between OM and hydrolyzed Al ion ($AlOH^{2+}$). She noted in particular that, once SOM exceeds 3%, high CEC values are no longer positively correlated with OM content. The absence of a correlation between SOM and Al_{exch} led her to suggest that interactions between Al and OM are limited to fluctuating labile fractions, placing more emphasis on SOM quality rather than quantity in understanding these interrelations and how they are impacted by the medium- to long-term soil transformations induced by CA. The record high maize yield in the presence of elevated Al_{exch} concentration brings into question the validity, in CA conditions, of the criteria developed in CT for adjusting nutrient supplies and defining toxicity thresholds. Hok (2018) highlights, for a reddish clay oxisol in Cambodia, an early effect of CA on SOM labile pools associated with significantly greater microbial enzymatic activities.

		(cm)	OM%	CEC (meq/100g)	V%	P (ppm)
	Forest	0-10	3.4	10.3	16.5	1.2
		10-20	2.0	5.8	7.0	0.5
5 years → CT	Soybean monocropping	0-10	2.4	6.5	49.5	9.7
		10-20	1.5	5.8	14.2	0.2
5 years → AC	Rice on *Arachis pintoi*	0-10	3.2	8.6	54.5	23.0
		10-20	1.8	5.3	35.0	0.3
	Soybean on *Cynodon dactylon*	0-10	2.8	6.6	45.0	2.2
		10-20	2.3	6.1	19.0	0.9
	diversif. rotation*	0-10	3.5	8.9	48.0	10.1
		10-20	1.9	5.5	12.0	0.6

		(cm)	K (mg/dm3)	Mn (mg/dm3)	Cu (mg/dm3)	Zn (mg/dm3)
	Forest	0-10	2.3			
		10-20	1.9			
5 years → CT	Soybean monocropping	0-10	48	8.8	0.6	4.9
		10-20	24	2.9	0.1	3.0
5 years → AC	Rice on *Arachis pintoi*	0-10	210	16.3	1.2	13.3
		10-20	44	2.6	0.2	0.4
	Soybean on *Cynodon dactylon*	0-10	93	6.6	0.6	2.6
		10-20	30	2.1	0.3	0.3
	diversif. rotation*	0-10	220	14.6	0.7	6.8
		10-20	37	1.4	0.2	0.4

Figure 3 Evolution of the soil organic matter (OM%), Cationic Exchangeable Capacity (CEC), bases saturation rate (V%), exchangeable phosphorus, potassium, manganese, copper and zinc along 10-year chronosequence with 5 years of CT-based soybean mono-cropping succeeded by five years of various CA systems. Source: adapted from Seguy (2008). * Rice + Eleusine c.//Soybean + Eleusine c.//Maize + Stylosanthes g.// Stylosanthes g.//Rice.

3.2 CA-induced biological modifications influencing nutrient use efficiency

3.2.1 Developing 'soil breeding'

In the short-term, soil microbiota composition and activity is dependent on residue amendment types. In the Madagascan highlands, in a long-term field experiment on andic Dystrupstept soil, CA on crop residues of a bi-annual maize/soybean rotation led to higher soil enzymatic activity and microbial respiration than under CA systems using living mulch of *Desmodium uncinatum* (Rabary, 2008). Marschner and Bell (1994) showed the beneficial effect of symbiosis with AMF for the uptake of P, N, K, Ca, S, Zn and Cu. Absorption capacity by roots and mycorrhizae is strengthened by synergies developed in the rhizosphere with bacteria which solubilize nutrients into exchangeable forms. These processes of 'bio-mediation' are particularly essential for phosphorus, which is present mostly in unavailable forms such as primary mineral apatite, Ca-bind compounds in alkaline soils, adsorbed in hydrated oxides of Al, Fe and Mn in acid soils or found in very stable organic compounds such as phytate. Enzymes able to hydrolyze those substrates, such as acid phosphatase or phytase, for instance, are mostly produced and released by bacteria present in the rhizosphere (Rodriguez and Fraga, 1999; Richardson and Simpson, 2011).

The composition of the pools that serve as a nutrient source depends both on microbial enzymatic diversity and the soil aggregation dynamics, which regulate substrate access and metabolic kinetics through connectivity and gradients of moisture, oxygen and nutrients (Dungait et al., 2012). Hence, the combined application of CA principles leads to the creation of interconnected niches colonized and functionalized by soil microbial communities that can be active or latent, depending on the environmental conditions, substrate availability and the exchanged signals. In this way, CA, with diverse, substantial and continuous biomass inputs, progressively endows the soil with plastic metabolic capacities that reflect more the long-term history of the plot than the direct influence of the preceding crops and the composition of the latest residue (see Case study 1).

3.2.2 The effect of plant varieties with microbiome genetic

Habbib (2016) explored the impact of tillage, cover crop and mineral N supply on the Nitrogen Use Efficiency (NUE) of maize in multiannual crop sequence, including winter and summer mixed cover, in northern France. She highlighted that CA improves different traits related to NUE, influencing nitrogen uptake by roots and the internal remobilization of nitrogen from the aerial biomass to the grain, likely due to better water access, which impacts grain filling and late, post-anthesis N uptake. Pursuing a similar study on winter wheat (Habbib, 2017), she obtained different results, suggesting that the positive impact on NUE is dependent on the presence of cover, impacting N remobilization more than uptake. These results (see Case study 2) illustrate how CA can reduce N fertilizer application and its associated losses. Le Gouis (2000) showed under a conventional management system that, among the modern winter wheat varieties used in the making of bread and selected for their high yield in mineral resource-rich conditions, some consistently perform well with a low N supply and offered comparable results to the old varieties grown in the pre-green revolution era. However, breeding crops adapted to ongoing CA practices, and which perform well in organic resource-rich conditions, require new types of materials with specific traits. Rooting speed and depth, synchronicity between crop nutrient uptake and release from soil pools, physiological traits related to 'poorly' accessible pools (Akhtar, 2009) and so on will lead to a new variety profiles, selected for organic and microbe-mediated resource-rich agroecosystems. Crop genotype selection should thus be based on the capacity to develop synergies with enriched soil microbiota and attention should be paid to the contribution of cv to the enrichment (diversity) and maintenance (redundancy) of the soil's functional properties (Johnston-Monje and Raizada, 2011; Peiffer and Ley, 2013; Lucero, 2014).

CA, specifically when it includes regular, diverse and important biomass production-amendment loops based on multi-specific cover, progressively builds up edaphic properties that profoundly redesign the biotic and abiotic environment of crops. Such evolution requires a departure from the 'mineral' views, inherited from the green revolution, when looking at improving nutrient use efficiency (NtUE). This paradigm, coupling nutrient addition and breed plants for high NtUE within best management practices (Baligar, 2001), becomes inoperative and potentially binding when implementing CA-based approaches. This of course does not mean that mineral supply reduction is inherent in CA design; mineral fertilizers contribute to microbial community development through direct nutrient supply or via increased food sources, thanks to enhanced biomass production. However, this may not justify considering the appropriate use of mineral fertilizers in the African context as the fourth *principle* of CA (Vanlauwe, 2014). Even if in weathered soils, fertilizers are needed to boost the production of primary biomass, it might be confusing to promote mineral supply as an inherent component of CA, as already contested by Sommer (2014). In fact, in Africa, as elsewhere, the capacity of CA to substantially reduce or eliminate mineral fertilizer applications depends on its capacity to produce and retain sufficient biomass on the soil surface and on the integration of CA-induced nutrient cycling at plot level within the larger management of organic resources at farm-, community- and territory-level and beyond. Carvalho (2012) quantified, on a Mediterranean luvisol, the impact of SOC content improvement on the efficiency of mineral nitrogen supplies. In this 11-year-long field experiment, wheat crop responses to N dressings varied from 19.1 to 104kg grain ha^{-1} per kg N applied under CT (0.58% SOC in 0–30cm soil depth) and NT with residue retention (1.74% SOC) respectively. Such results indicate that the primary socio-political action should target the creation of arrangements prone to permit FOM retention and cycling, a challenging evolution in many African contexts, but a prerequisite to strengthen the efficacy of fertilizer access support.

Case study 2, based on a farm trajectory in SW France, illustrates how, once efficient nutrient cycling at plot level is achieved, CA eases the integration of organic resources within extended circular management at farm level, and possibly beyond at farm group or community territory scale, for instance.

4 Conclusions and future trends

4.1 On general trends of nutrients management under CA

- Continuous practice of CA systems integrating diverse and high biomass inputs leads to profound transformations in soil conditions and the agroecosystem as a whole. The soil-microorganism-plant system acts as an interface between the atmosphere and water (from soil solution to phreatic table). Hence, efficient CA systems affect major nutrient cycles and can limit losses and negative externalities such as N_2O emission and NO_3^- leaching.

- The transformation of soil structure and functions is induced by complex auto-organization processes triggered and sustained by continuous fluxes of matter and energy, across the soil food web, from the *mulchospheric* sources of OM.
- NtUE is related to microbial activity and diversity while redundancy supports functional resilience.
- CA endows soils and agroecosystems with their own nutrient supply capabilities. This development of *auto-fertility margins* relies on three inter-related groups of processes that improve nutrient supplies to crops (see Fig. 4), including:
    - nutrients issued from fresh biomass decomposition and mineralization;
    - nutrients released from SOM labile pools mineralization and;
    - growing organic and inorganic pools with increased accessibility due to extended soil microbial enzymatic diversity and activity.
- These processes come into play in different phases in the agroecological transition, immediately with the first biomass applications for the FOM mineralization, and more progressively for the development and activation of SOM and nutrient pools.

4.2 Knowledge gaps and improvements of the nutrient use efficiency

- Nutrient cycling services attached to CA practices have been less studied than other ecosystem services such as SOC dynamics and CC mitigation potential. Some shadow zones remain and call for complementary studies to be conducted under efficient and 'mature' CA systems, to assess the existence and extent of nutrient losses, notably through leaching and denitrification. Studies focusing on how regularity, intensity and diversity of FOM addition are related to system's capacities of retaining nutrients in multiple organic and inorganic pools are lacking. These type of insights into process are needed for practice adjustments; for instance:
    - Are permanent living covers sufficient to prevent *any* losses by leaching under regular and significant FOM inputs?
    - Can leached elements be further retained using deep rooting cover species or integrating trees into the system?
    - Are losses enhanced - and to what extent - in the case of rupture or reduction of the biomass supply (e.g. due to erratic climate conditions)?

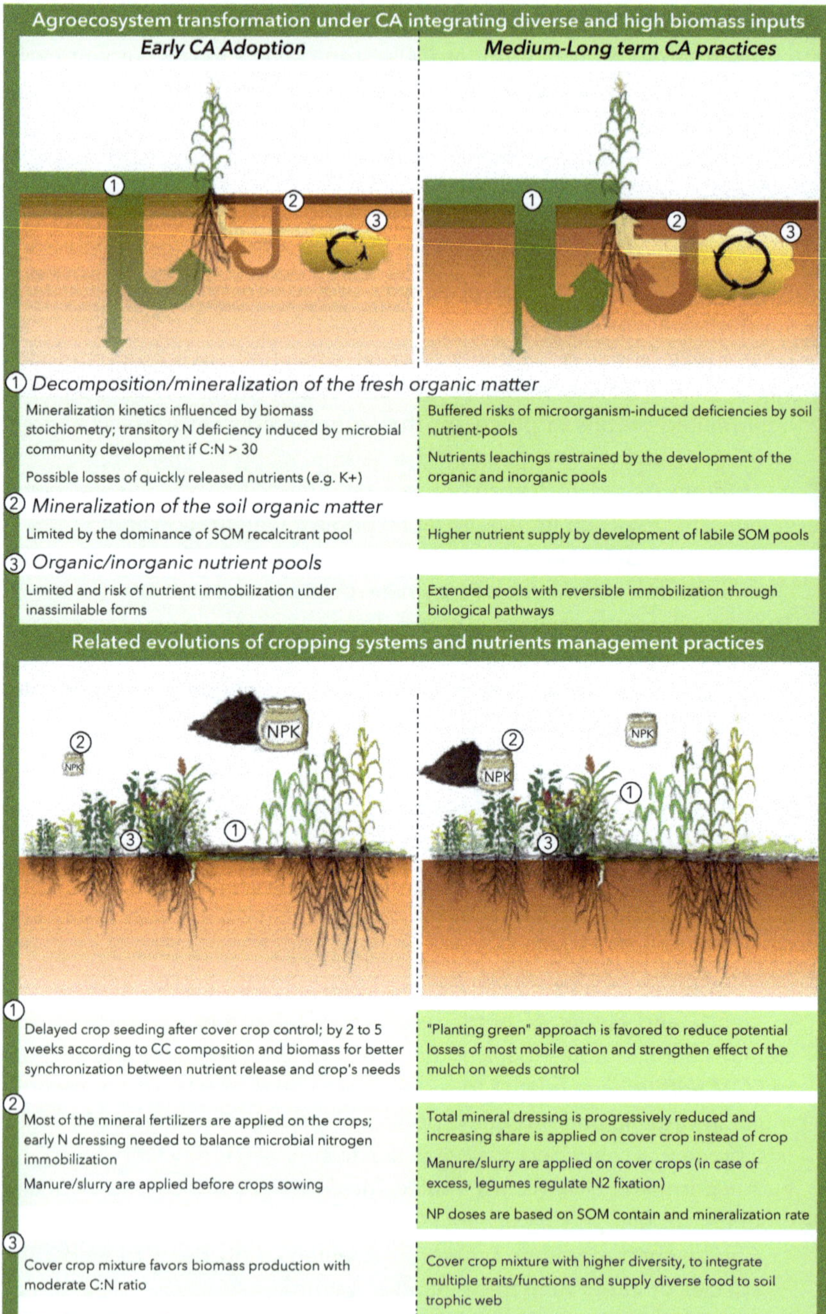

Figure 4 Three major processes leading to the progressive elaboration of *autofertility* margins under medium- to long-term CA-based management and correlated evolutions of organic and mineral nutrients inputs.

- ◦ How to sustain nutrient cycling service, across trade-off management, when part of the biomass is exported for production purposes (livestock, bio-energy etc.) and later when reintroduced in the form of an organic by-product?
- Regarding the discussed losses by denitrification, the possibility of using cover crop assemblies to influence nitrification (and denitrification) pathways, through biomass and/or exudate inputs, constitutes an apparently new field of research for CA systems with an improved CC mitigation effect. Mimicking natural wet tropical savanna functions could inspire and orientate the conception of such systems.
- Research into the method of inoculation will be useful to foster endophytic, rhizospheric and *mulchospheric* microbial communities and their roles in plant mechanisms of mineral nutrition and defense against pests and diseases.
- CA calls for ad hoc methods to evaluate soil nutrient status, taking into account the multiple and accessible pools. Extraction methods should be adjusted to better reflect direct and bio-mediated nutrient availability. Information on the assimilability of the nutrients contained in the cover crop biomass at sowing could facilitate the development of recommendations for complementary mineral dressing.
- CA also requires a specific breeding program for the creation and selection of varieties that are adapted to CA- designed agroecosystems. Works dedicated to mixed cover blending and cover crop cv selection for targeted traits are also needed to improve agronomical performance and ecosystem services delivery.

5 Case study 1: A Brazilian Fazenda initiating a transition from 'generic' CA to CA based on multifunctional mix species cover

The Fazenda 'Saint Isidore' of Polinghi-Savencia is located in Angatuba Municipe, in the south of Sao Paulo State (23° 34′ S, 48° 23′ W), and cultivates about 1 150 ha, preserving a legal land reserve of 500 ha. Soils are predominantly clayey red oxisol. Field crops cover about 800 ha including 130 ha under irrigation pivot, 120 ha of pasture (*Brachiaria brizantha*) and 210 ha of eucalyptus.

Located near the cradle region of CA, the Fazenda has practiced NT for decades. An annual succession of winter and summer crops gives structure to the cropping system. Winter crops include about 50% of black oat cover crops, 40% of wheat and 10% of rapeseed-based diversification. Oat surfaces serve as pasture to balance the low productivity of the *Brachiaria* in the cool season. Soybean and maize occupy most of the land in the summer, with the potential introduction of bean, a highly speculative option. Crop choice is determined by the market, which was favorable to maize until 2013 before it shifted back to soybean. Today, despite the economic attractiveness of soybean, the Fazenda maintains 25% of its land under maize for agronomical reasons. For the 2017/18 harvest, rainfed maize yielded 10 tha^{-1} for a zero profit margin. A recent option, inspired by 'safrinha' ('small crop') practices developed in central tropical regions, consists of sowing a secondary maize crop right after the harvest of early sown and short cycle soybean. Exposed to the hazard of night frost at the beginning of the cool season, this practice carries a high economic risk when maize is sown after mid-February.

Due to a progressive decline in maize productivity since early 2010 - from 10–11 t/ha to 9–10 t/ha - and despite high mineral fertilizer application (950kg ha^{-1}-190kg N ha^{-1}, 100kg P_2O_5 ha^{-1} and 90kg K_20 ha^{-1}), the farm management considers the cropping and production systems to be fragile and barely resilient in the face of market and climate variability. It considers that the successions have scarce biomass inputs when soybean dominates, and a lack of legumes when maize dominates. It therefore decided, in 2016, to include short-term biomass production based on mixed species between seasons. After seed multiplication, the first significant pilot was sown on four plots over 150 ha in 2017 (Fig. 5).

The NPK content of these cover crops represent an equivalent of about 650kg ha^{-1} of mineral fertilizers (using urea, mono-ammonium phosphate and chloride potassium) for an approximate value of 250 USD ha^{-1} in May 2017, compared to a cover crop management cost of 100–125 USD ha^{-1}.

These preliminary achievements have led to further evolutions. At the cropping system scale, recycled nutrients contained in the biomass allow for

Cover mix	Growth period		Fresh biomass	Dry matter	Composition						Macronutrient contents (kg/ha)			
		day	kg/ha	kg/ha	N (%)	P (%)	K (%)	Ca (%)	Mg (%)	S (%)	N	P2O5	K20	S04
Mix 1 (1)	25/1 - 12/4	77	31 340	4764	2.12	0.31	1.90	0.57	0.44	0.26	101.0	33.2	108.6	37.0
Mix 2 (2)	12/3 - 01/5	50	27 838	3819	2.51	0.54	3.62	0.72	0.54	0.56	95.9	46.4	165.9	64.0
Mix 2 (2)	13/3 - 03/5	51	29 150	3667	3.10	0.47	3.38	1.00	0.58	1.52	113.7	38.8	148.7	166.7
Mix 2 (2)	10/3 - 01/5	52	32 406	3746	2.88	0.54	3.27	0.90	0.59	0.87	107.9	45.5	147.0	97.4
min.		50	27 838	3667	2.12	0.31	1.90	0.57	0.44	0.26	95.9	33.2	108.6	37.0
max.		77	32 406	4764	3.10	0.54	3.62	1.00	0.59	1.52	113.7	46.4	165.9	166.7
mean		**58**	**30 184**	**3999**	**2.65**	**0.47**	**3.04**	**0.80**	**0.54**	**0.80**	**104.6**	**41.0**	**142.6**	**91.3**

Mix 1	
Species	kg/ha
Avena sativa	16.0
Helianthus annuus	1.3
Crotalaria ochroleuca	9.5
Raphanus sativus	5.0
Fagopyrum esculentum	22.0
TOTAL	53.8

Mix 2	
Species	kg/ha
idem Mix 1 +	53.8
Vicia sativa	7.0
Pisum sativum	7.0
TOTAL	67.8

Figure 5 Biomass production, mineral contents and specific composition of short-duration mix species cover grown prior to winter crops. (1) Broadcast sowing. (2) Direct seeding.

a reduction of basal fertilizer use (by about 30% in the first year). Following the decision to boost diversity and nitrogen symbiotic fixation, the farm management intends to replace oat with mixed species cover for a growing period of up to 160 days, with a composition that is adjusted according to the succeeding crops. When considering medium-term evolution at the production system level, successful introduction of various multifunctional mix cover options should permit improved integration with cattle production.

6 Case study 2: A dairy farm evolution in SW France restoring soil potential and animal feed autonomy as a result of diversification and intensification of the biomass inputs

Mr Christian Abadie took over the family farm located in SW France (43° 24 N, 17° 00 E) in the early 1980s. Soils are shallow *Boulbène*, a gleyic luvisol developed on pebble bed, typical of the Garonne river basin (Favrot, 1992) in a hilly topography prone to erosion. The climate is temperate with an annual rainfall of 850mm and an average temperature of about 13.5°C.

In 2002, observing advanced soil degradation marked by the loss of SOM, Mr Abadie decided to convert all 109 ha of the dairy farm to CA. Supported and guided by the Agrod'Oc association (Ms Marie-José Blazian, Mr Sylvain Hypolite) and a CA expert (Mr Lucien Seguy), the cropping system evolved from a conventional and poorly diversified rotation system based on irrigated maize to a quasi-permanent cropping system, integrating mix covers with maize, soybean and winter cereal as the main crops (Fig. 6). Maximizing photosynthesis (347 days per year with living plant cover) to improve biomass recovery and the quest for protein autonomy associated with livestock production structure the evolution of the system.

Summer cover crops sown after the cereal harvest consist of a mix of sorghum, sunflower and radish (with other tropical species being tested).

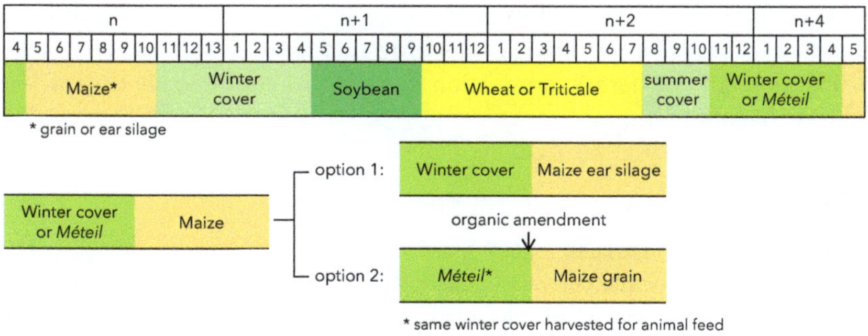

Figure 6 Current CA systems integrated with livestock production.

Winter cover composition varies according to the succeeding crops, featuring mainly poaceae (oat and triticale) before soybean, and legumes (Broad bean - *Vicia faba*, common vetch - *Vicia sativa* + oat, radish etc.) before maize. This legume-based winter cover is entirely restored to soil when the succeeding maize is devoted to ear silage production or is harvested as protein rich silage (*méteil*) before sowing maize for grain production. Solid organic (manure) is applied to the soil after the *méteil* harvest to 'balance' the biomass loss.

After 15 years of continuous evolution based on permanent trials, cropping and farming systems reached high levels of agroenvironmental and economic performance. In 2014, a favorable climatic year, maize yield reached 14 tha^{-1} (compared to 10–11 tha^{-1} under the conventional system); an 0 N control plot produced almost 9 tha-1, and mineral N application was reduced from ~22 in CT to 14kg Nt^{-1} grain in CA. Irrigation decreased by about 30%, which together with the yield increase led to water productivity multiplying by a factor of almost two; there is now no application of insecticide or fungicide. In 2016, the 0 N control plots yielded 11tha^{-1} and mineral N applications passed below 12kg Nt^{-1} grain.

At farm level, SOM contents increased from 1.5% to 2.6% on a 0–20cm layer with an assessed SOC increase of +1.6 $Mgha^{-1}$ $year^{-1}$. Mineral fertilizer represents less than 50% of the nitrogen input, the remainder coming from organic amendments (22%) and legumes (30%). The farm has a positive protein balance and produces 98% of its animal feed.

Further evolutions consist of the continuation of trials focusing on CA improvement (cover crop species selection and mix elaboration, crop variety selection, herbicide reduction to total substitution by mechanical and organic control) and the introduction of agroforestry by planting tree hedges (90 trees ha^{-1}) on a 4 ha plot. Most of the species (*Populus* sp, *Fraxinus excelsior, Morus nigra, Acer campestris*) are fast growing so that wood chips from regular branch pruning may be used as litter in stables (animal comfort, reduction of straw importation, increase and C/N adjustment of the organic amendment from livestock production). This evolution will lead to greater biodiversity, strengthened soil C input, improved energy autonomy, as well as contributing to upscaling of the farm's production system to integrated landscape and territory management.

7 References

Akhtar, M. S., Oki, Y. and Adachi, T. 2009. Mobilization and acquisition of sparingly soluble *P*-Sources by *Brassica* cultivars under *P*-starved environment II. Rhizospheric pH changes, redesigned root architecture and pi-uptake kinetics. *Journal of Integrative Plant Biology* 51(11), 1024–39. doi:10.1111/j.1744-7909.2009.00873.x.

Arihara, J. and Karasawa, T. 2001. Phosphorus nutrition in cropping systems through arbuscular mycorrhizal management. In: Ae, N., Arihara, J., Okada, K. and Srinivasan, A. (Eds), *Plant Nutrient Acquisition*, pp. 319–37.

Baligar, V. C., Fageria, N. K. and He, Z. L. 2001. Nutrient use efficiency in plants. *Communications in Soil Science and Plant Analysis* 32(7–8), 921–50. doi:10.1081/CSS-100104098.

Brito, I., Carvalho, M. and Goss, M. J. 2013. Soil and weed management for enhancing arbuscular mycorrhiza colonization of wheat. *Soil Use and Management* 29(4), 540-6. doi:10.1111/sum.12069.

Brito, I., Carvalho, M., Alho, L. and Goss, M. J. 2014. Managing arbuscular mycorrhizal fungi for bioprotection: Mn toxicity. *Soil Biology and Biochemistry* 68, 78–84. doi:10.1016/j.soilbio.2013.09.018.

Carvalho, M., Basch, G., Calado, J. M. G. and Barros, J. F. C. 2012. Long term effect of tillage system and crop residue management on soil carbon content of a Luvisol under rainfed Mediterranean conditions. *Agrociencia (Special Issue)*, 183–7.

Cassiolato, M. E., Miyazawa, M., Meda, A. R. and Pavan, M. A. 2002. A laboratory method to estimate the efficiency of plant extract to neutralize soil acidity. *Brazilian Archives of Biology and Technology* 45(2), 183–7. doi:10.1590/S1516-89132002000200009.

Collet, L., Boucaron, C., Gourdon, F., Bonzon, B., André, P. and Huelvan, Y. 1985. *Effets des amendements calciques sur sol sodique et sur un vertisol hypermagnésien*. Orstom and Dider, Noumea, 37p.

Corsi, S., Friedrich, T., Kassam, A., Pisante, M. and Sà, J. C. M. 2012. *Soil Organic Carbon Accumulation and Greenhouse Gas Emission Reductions from Conservation Agriculture: A Literature Review*. Integrated Crop Management (vol. 16). FAO, Rome, 103p.

Coskun, D., Britto, D. T., Shi, W. and Kronzucker, H. J. 2017. How plant root exudates shape the nitrogen cycle. *Trends in Plant Science* 22(8), 661–73. doi:10.1016/j.tplants.2017.05.004.

Dart, P. J. 1986. Nitrogen fixation associated with non-legumes in agriculture. *Plant and Soil* 90(1–3), 303–34. doi:10.1007/BF02277405.

de Moraes Sá, J. C., Séguy, L., Tivet, F., Lal, R., Bouzinac, S., Borszowskei, P. R., Briedis, C., dos Santos, J. B., da Cruz Hartman, D., Bertoloni, C. G., Rosa, J. and Friedrich, T. 2015. Carbon depletion by plowing and its restoration by no-till cropping systems in oxisols of subtropical and tropical agro-ecoregions in Brazil. *Land Degradation and Development* 26(6), 531–43. doi:10.1002/ldr.2218.

Derpsch, R. and Calegari, A. 1985. Guia de plantas para adubação verde de inverno. Documentos IAPAR, 9. IAPAR, Londrina, 96p.

de Varennes, A. and Goss, M. J. 2007. The tripartite symbiosis between legumes, rhizobia and indigenous mycorrhizal fungi is more efficient in undisturbed soil. *Soil Biology and Biochemistry* 39(10), 2603–7. doi:10.1016/j.soilbio.2007.05.007.

Duiker, S. W. 2015. 9 reasons to No-Till Plant into green cover crops. *No-Till Farmer* (online). Available at: https://www.no-tillfarmer.com/articles/4664-reasons-to-no-till-plant-into-green-cover-crops.

Dungait, J. A. J., Hopkins, D. W., Gregory, A. S. and Whitmore, A. P. 2012. Soil organic matter is governed by accessibility not recalcitrance. *Global Change Biology* 18(6), 1781–96. doi:10.1111/j.1365-2486.2012.02665.x.

Favrot, J. C., Bouzigues, R., Tessier, D. and Valles, V. 1992. Contrasting structures in the subsoil of the boulbènes of the Garonne Basin, France. *Geoderma* 53(1–2), 125–37. doi:10.1016/0016-7061(92)90026-4.

Florentín, M. A., Peñalva, M., Calegari, A. and Derpsch, R. 2010. *Green Manure/Cover Crops and Crops Rotation in Conservation Agriculture on Small Farms*. Integrated Crop Management (vol. 12). FAO, Rome, 109p.

Fontaine, S., Mariotti, A. and Abbadie, L. 2003. The priming effect of organic matter: a question of microbial competition? *Soil Biology and Biochemistry* 35(6), 837–43. doi:10.1016/S0038-0717(03)00123-8.

Franchini, J. C., Gonzalez-Vila, F. J., Cabrera, F., Miyazawa, M. and Pavan, M. A. 2001. Rapid transformations of plant water-soluble organic compounds in relation to cation mobilization in an acid Oxisol. *Plant and Soil* 231(1), 55–63. doi:10.1023/A:1010338917775.

Frankenberger Jr., W. T. and Abdelmagid, H. M. 1985. Kinetic parameters of nitrogen mineralization rates of leguminous crops incorporated into soil. *Plant and Soil* 87(2), 257–71. doi:10.1007/BF02181865.

Giacomini, S. J., Aita, C., Hübner, A. P., Lunkes, A., Guidini, E. and do Amaral, E. B. 2003. Liberação de fósforo e potássio durante a decomposição de resíduos culturais em plantio direto. *Pesquisa Agropecuária Brasileira* 38(9), 1097–104.

Gobat, J.-M., Aragno, M. and Matthey, W. 1998. *Le sol vivant*. Presses polytechniques et universitaires romandes, 519pp.

Goss, M. J., Howse, K. R., Lane, P. W., Christian, D. G. and Harris, G. L. 1993. Losses of nitrate-nitrogen in water draining from under autumn-sown crops established by direct drilling or mouldboard ploughing. *Journal of Soil Science* 44(1), 35–48. doi:10.1111/j.1365-2389.1993.tb00432.x.

Greer, K. and Schoenau, J. 1992. Soil organic matter and nutrient turnover in thin Black Oxbow soils after intensive conservation management. In: Anon. (Ed.), *Proceedings of the Soils and Crops Workshop*. University of Saskatchewan, Saskatoon, Canada.

Habbib, H., Verzeaux, J., Nivelle, E., Roger, D., Lacoux, J., Catterou, M., Hirel, B., Dubois, F. and Tétu, T. 2016. Conversion to no-till improves maize nitrogen use efficiency in a continuous cover cropping system. *PLoS ONE* 11(10), e0164234. doi:10.1371/journal.pone.0164234.

Habbib, H., Hirel, B., Verzeaux, J., Roger, D., Lacoux, J., Lea, P., Dubois, F. and Tétu, T. 2017. Investigating the combined effect of tillage, nitrogen fertilization and cover crops on nitrogen use efficiency in winter wheat. *Agronomy* 7(66), 15p.doi:10.3390/agronomy7040066.

Hasbullah, H. and Marschner, P. 2015. Residue properties influence the impact of salinity on soil respiration. *Biology and Fertility of Soils* 51(1), 99–111. doi:10.1007/s00374-014-0955-2.

Hobbs, P. R., Sayre, K. and Gupta, R. 2008. The role of conservation agriculture in sustainable agriculture. *Philosophical Transactions of the Royal Society of London. Series B, Biological Sciences* 363(1491), 543–55. doi:10.1098/rstb.2007.2169.

Hok, L., Sá, J. C. M., Reyes, M., Boulakia, S., Tivet, F., Leng, V., Kong, R., Briedis, C., da Cruz Hartman, D., Ferreira, L. A., Inagaki, T. M., Gonçalves, D. R. P. and Bressand, P. T. 2018. Enzymes and C pools as indicators of C build up in short-term conservation agriculture in a savanna ecosystem in Cambodia. *Soil and Tillage Research* 177, 125–33. doi:10.1016/j.still.2017.11.015.

Hu, G., Liu, X., He, H., Zhang, W., Xie, H., Wu, Y., Cui, J., Sun, C. and Zhang, X. 2015. Multi-seasonal nitrogen recoveries from crop residue in soil and crop in a temperate agro-ecosystem. *PLoS ONE* 10(7), e0133437. doi:10.1371/journal.pone.0133437.

Husson, O. 2013. Redox potential (Eh) and pH as drivers of soil/plant/microorganism systems: a transdisciplinary overview pointing to integrative opportunities for agronomy. *Plant and Soil* 362(1-2), 389–417. doi:10.1007/s11104-012-1429-7.

Husson, O., Husson, B., Brunet, A., Babre, D., Alary, K., Sarthou, J. P., Charpentier, H., Durand, M., Benada, J. and Henry, M. 2016a. Practical improvements in soil redox potential (Eh) measurement for characterisation of soil properties. Application for comparison of conventional and conservation agriculture cropping systems. *Analytica Chimica Acta* 906, 98–109. doi:10.1016/j.aca.2015.11.052.

Husson, O., Tran Quoc, H., Boulakia, S., Chabanne, A., Tivet, F., Bouzinac, S., Lienhard, P., Michellon, R., Chabierski, S., Boyer, J., Enjalric, F., Rakotondramanana, Moussa, N., Jullien, F., Balarabe, O., Rattanatray, B., Castella, J. C., Charpentier, H. and Séguy, L. 2016b. Co-designing innovative cropping systems that match biophysical and socio-economic diversity: the DATE approach to conservation agriculture in Madagascar, Lao PDR and Cambodia. *Renewable Agriculture and Food Systems* 31(5), 452–70. doi:10.1017/S174217051500037X.

Johnston-Monje, D. and Raizada, M. N. 2011. Conservation and diversity of seed associated endophytes in Zea across boundaries of evolution, ethnography and ecology. *PLoS ONE* 6(6), e20396. doi:10.1371/journal.pone.0020396.

Keiluweit, M., Nico, P., Harmon, M. E., Mao, J., Pett-Ridge, J. and Kleber, M. 2015. Long-term litter decomposition controlled by manganese redox cycling. *Proceedings of the National Academy of Sciences of the United States of America* 112(38), E5253-60. doi:10.1073/pnas.1508945112.

Kelemu, S., Fory, P., Zuleta, C., Ricaurte, J., Rao, I. and Lascano, C. 2011. Detecting bacterial endophytes in tropical grasses of the *Brachiaria* genus and determining their role in improving plant growth. *African Journal of Biotechnology* 10(6), 965–76.

Khalaf, R. A. 2013. Potential uptake of phosphorus by catch crops in greenhouse conditions and release after freezing-thawing. Master's thesis. Environmental Science, SLU Uppsala, 48p. Available at: http://stud.epsilon.slu.se.

Kong, R., Sar, V., Leng, V., Trang, S., Boulakia, S., Tivet, F. and Séguy, L. 2016. Conservation agriculture for climate-resilient rainfed uplands in the Western regions of Cambodia: challenges, opportunities, and lessons from a 10-year R and D program. In: Sajise, P. E., Cadiz, M. C. H. And Bantayan, R. B. (Eds), *Learning and Coping with Change: Case Stories of Climate Change Adaptation in Southeast Asia*. SEARCA, Los Baños, pp. 55–81. Available at: http://www.searca.org/index.php/news-and-events/searca-news/2708-new-publication-available.

Lata, J. C., Durand, J., Lensi, R. and Abbadie, L. 1999. Stable coexistence of contrasted nitrification statuses in a wet tropical savanna ecosystem. *Functional Ecology* 13(6), 762–8. doi:10.1046/j.1365-2435.1999.00380.x.

Lata, J. C., Degrange, V., Raynaud, X., Maron, P. A., Lensi, R. and Abbadie, L. 2004. Grass populations control nitrification in savanna soils. *Functional Ecology* 18(4), 605–11. doi:10.1111/j.0269-8463.2004.00880.x.

Le Gouis, J., Beghin, D., Heumez, E. and Pluchard, P. 2000. Genetic differences for nitrogen uptake and nitrogen utilisation efficiencies in winter wheat. *European Journal of Agronomy* 12(3–4), 163–73. doi:10.1016/S1161-0301(00)00045-9.

Lienhard, P., Terrat, S., Mathieu, O., Levêque, J., Chemidlin Prévost-Bouré, N., Nowak, V., Régnier, T., Faivre, C., Sayphoummie, S., Panyasiri, K., Tivet, F., Ranjard, L. and Maron, P.-A. 2013. Soil microbial diversity and C turnover modified by tillage and cropping in Laos tropical grassland. *Environmental Chemistry Letters* 11(4), 391-8. doi:10.1007/s10311-013-0420-8.

Löhnis, F. 1926. Nitrogen availability of green manures. *Soil Science* 22(4), 253-90. doi:10.1097/00010694-192610000-00001.

Lucero, M. E., DeBolt, S., Unc, A., Ruiz-Font, A., Reyes, L. V., McCulley, R. L., Alderman, S. C., Dinkins, R. D., Barrow, J. R. and Samac, D. A. 2014. Using microbial community interactions within plant microbiomes to advance an evergreen agricultural revolution. In: Oelbermann, M. (Ed.), *Sustainable Agroecosystems in Climate Change Mitigation*. Wageningen Academic Publishers, pp. 183-202.

Lugato, E., Leip, A. and Jones, A. 2018. Mitigation potential of soil carbon management overestimated by neglecting N_2O emissions. *Nature Climate Change* 8(3), 219-23. doi:10.1038/s41558-018-0087-z.

Maltas, A., Corbeels, M., Scopel, E., Oliver, R., Douzet, J.-M., da Silva, F. A. M. and Wery, J. 2007. Long-term effects of continuous direct seeding mulch-based cropping systems on soil nitrogen supply in the Cerrado region of Brazil. *Plant and Soil* 298(1-2), 161-73. doi:10.1007/s11104-007-9350-1.

Maluf, H. J. G. M., Soares, E. M. B., da Silva, I. R., Neves, J. C. L. and Silva, L. O. S. 2015. Decomposição de Resíduos de Culturas e Mineralização de Nutrientes em Solo com Diferentes Texturas. *Revista Brasileira de Ciência Do Solo* 39, 1681-9.

Marschner, H. and Dell, B. 1994. Nutrient uptake in mycorrhizal symbiosis. *Plant and Soil* 159(1), 89-102. doi:10.1007/BF00000098.

McDaniel, M. D., Grandy, A. S., Tiemann, L. K. and Weintraub, M. N. 2016. Eleven years of crop diversification alters decomposition dynamics of litter mixtures incubated with soil. *Ecosphere* 7(8), e01426. doi:10.1002/ecs2.1426.

Naudin, K., Scopel, E., Andriamandroso, A. L. H., Rakotosolofo, M., Andriamarosoa Ratsimbazafy, N. R. S., Rakotozandriny, J. N., Salgado, P. and Giller, K. E. 2012. Trade-offs between biomass use and soil cover. the case of rice-based cropping systems in the Lake Alaotra region of Madagascar. *Experimental Agriculture* 48(2), 194-209. doi:10.1017/S001447971100113X.

Nicolardot, B., Recous, S. and Mary, B. 2001. Simulation of C and N mineralisation during crop residue decomposition: a simple dynamic model based on the C:N ratio of the residues. *Plant and Soil* 228(1), 83-103. doi:10.1023/A:1004813801728.

Nicolodi, M., Gianello, C., Anghinoni, I., Marré, J. and Mielniczuk, J. 2008. Insuficiëncia do conceito mineralista para expresar a fertilidade do solo percebida pelas plantas cultivadas no sistema plantio direto. *Revista Brasileira de Ciência Do Solo* 32, 2735-44.

Omonode, R. A., Smith, D. R., Gal, A. and Vyn, T. J. 2011. Soil nitrous oxide emissions in corn following three decades of tillage and rotation treatments. *Soil Science Society of America Journal* 75(1), 152-63. doi:10.2136/sssaj2009.0147.

Palm, C. A., Gachengo, C. N., Delve, R. J., Cadisch, G. and Giller, K. E. 2001. Organic inputs for soil fertility management in tropical agroecosystems: application of an organic resource database. *Agriculture, Ecosystems and Environment* 83(1-2), 27-42. doi:10.1016/S0167-8809(00)00267-X.

Pavinato, P. S., Merlin, A. and Rosolem, C. A. 2008. Organic compounds from plant extracts and their effect on soil phosphorus availability. *Pesquisa Agropecuária Brasileira* 43(10), 1379-88. doi:10.1590/S0100-204X2008001000017.

Peiffer, J. A. and Ley, R. E. 2013. Exploring the maize rhizosphere microbiome in the field: a glimpse into a highly complex system. *Communicative and Integrative Biology* 6(5), e25177. doi:10.4161/cib.25177.

Rabary, B., Sall, S., Letourmy, P., Husson, O., Ralambofetra, E., Moussa, N. and Chotte, J.-L. 2008. Effects of living mulches or residue amendments on soil microbial properties in direct seeded cropping systems of Madagascar. *Applied Soil Ecology* 39(2), 236-43. doi:10.1016/j.apsoil.2007.12.012.

Rai, R. 1991. Isolation, characterization and associative *N*-fixation of acid-tolerant *Azospirillum brasilense* strains associated with *Eleusine coracana* in low pH-*Al*-rich acid soil. In: Wright, R. J., Baligar, V. C. and Murrmann, R. P. (Eds), *Plant-Soil Interactions at Low pH. Developments in Plant and Soil Sciences* (vol. 9), pp. 663–71.

Redin, M., Recous, S., Aita, C., Dietrich, G., Skolaude, A. C., Ludke, W. H., Schmatz, R. and Giacomini, S. J. 2014. How the chemical composition and heterogeneity of crop residue mixtures decomposing at the soil surface affects C and N mineralization. *Soil Biology and Biochemistry* 78, 65–75. doi:10.1016/j.soilbio.2014.07.014.

Richardson, A. E. and Simpson, R. J. 2011. Soil microorganisms mediating phosphorus availability update on microbial phosphorus. *Plant Physiology* 156(3), 989–96. doi:10.1104/pp.111.175448.

Rimmer, D. L. 2006. Free radicals, antioxidants, and soil organic matter recalcitrance. *European Journal of Soil Science* 57(2), 91–4. doi:10.1111/j.1365-2389.2005.00735.x.

Rimmer, D. L. and Smith, A. M. 2009. Antioxidants in soil organic matter and in associated plant materials. *European Journal of Soil Science* 60(2), 170–5. doi:10.1111/j.1365-2389.2008.01099.x.

Rivera-Zayas, J., Hansel, F. D. and Ruiz-Diaz, D. 2018. Soil nutrients stratification affected by tillage and nitrogen source in a corn long-term experiment. In: *Great Plains Soil Fertility Conference* (vol. 17), Denver, CO, 143–9.

Rochette, P. 2008. No-till only increases N_2O emissions in poorly-aerated soils. *Soil and Tillage Research* 101(1–2), 97–100. doi:10.1016/j.still.2008.07.011.

Rodríguez, H. and Fraga, R. 1999. Phosphate solubilizing bacteria and their role in plant growth promotion. *Biotechnology Advances* 17(4–5), 319–39. doi:10.1016/S0734-9750(99)00014-2.

Sá, J. CdM. 1999. Manejo da fertilidade do solo no sistema plantio direto. In: Siqueira, J. O., Moreira, F. M. S., Lopes, A. S., Guilherme, L. R. G., Faquin, V., Furtini Neto, A. E. and Carvalho, J. G. (Eds), *Soil Fertility, Soil Biology, and Plant Nutrition Interrelationships*. Soc. Bras. Ci. Solo. UFLA, Lavras, MG, Brazil, pp. 267–320.

Sá, J. CdM., Cerri, C. C., Lal, R., Dick, W. A., Piccolo, M. C. and Feigl, B. E. 2009. Soil organic carbon and fertility interactions affected by a tillage chronosequence in a Brazilian Oxisol. *Soil and Tillage Research* 104(1), 56–64. doi:10.1016/j.still.2008.11.007.

Salume, J. A. 2017. Decomposição e mineralização de nutrientes de resíduos de plantas de cobertura, espontâneas e implantadas, em solo de pomares de pereira (*Pyrus communis* L.). Brunetto G. (orientador) and Comin J. J. and Giacomini S. J (co-Orientadors). Florianópolis, SC, 71pp. Available at: https://repositorio.ufsc.br/xmlui/bitstream/handle/123456789/179784/348916.pdf?sequence=1&isAllowed=y (accessed on 15 April 2018).

Schoenau, J. J. and Campbell, C. A. 1996. Impact of crop residues on nutrient availability in conservation tillage systems. *Canadian Journal of Plant Science* 76(4), 621–6. doi:10.4141/cjps96-111.

Schomberg, H. H., Endale, D. M., Calegari, A., Peixoto, R., Miyazawa, M. and Cabrera, M. L. 2006. Influence of cover crops on potential nitrogen availability to succeeding crops in a Southern Piedmont soil. *Biology and Fertility of Soils* 42(4), 299–307. doi:10.1007/s00374-005-0027-8.

Séguy, L., Bouzinac, S., Trentini, A. and Cortez, N. A. 1996. L'agriculture brésilienne des fronts pionniers. *Agriculture et Développement* 12, 2–61.

Séguy, L., Bouzinac, S. and Husson, O. 2006. Direct-seeded tropical soil systems with permanent soil cover: learning from Brazilian experience. In: Uphoff, N., Ball, A. S., Fernandes, E., Herren, H., Husson, O., Laing, M., Palm, C., Pretty, J., Sanchez, P., Sanginga, N. and Thies, J. (Eds), *Biological Approach to Sustainable Soil Systems*. CRC Press, Taylor and Francis.

Séguy, L., Bouzinac, S., et al. 2008. La symphonie inachevée du semis direct dans le Brésil central: le système dominant dit de 'semi-direct' – Limites et dégâts, éco-solutions et perspectives: la nature au service de l'agriculture durable. CIRAD, Montpellier, 214p. Online access on CIRAD open-library Part 1 – http://open-library.cirad.fr/files/2/39__Symphonie_inachevee_10_Mo_A.pdf; Part 2 – http://open-library.cirad.fr/files/2/40__Symphonie_inachevee_10_Mo_B.pdf; Part 3 – http://open-library.cirad.fr/files/2/41__Symphonie_inachevee_10_Mo_C.pdf.

Séguy, L., Husson, O., Charpentier, H., Bouzinac, S., Michellon, R., Chabanne, A., Boulakia, S., Tivet, F., Naudin, K., Enjalric, F., Chabierski, S., Rakotodralambo, P., Ramaroson, I. and Rakotondramanana 2012. Principles, functioning and management of ecosystems cultivated under Direct seeding Mulch-based Cropping systems (DMC). CIRAD – GSDM – TAFA, 76p. Online access on CIRAD open-library http://open-library.cirad.fr/files/2/18__Manuel_SCV_v_eng_v_2012_04_25_finale_web_mini.pdf.

Shi, J. 2013. Decomposition and nutrient release of different cover crops in organic farm systems. Dissertations & Theses in Natural Resources, 75. Available at: http://digitalcommons.unl.edu/natresdiss/75.

Sommer, R., Thierfelder, C., Tittonell, P., Hove, L., Mureithi, J. and Mkomwa, S. 2014. Fertilizer use should not be a fourth principle to define conservation agriculture: response to the opinion paper of Vanlauwe et al. (2014) 'A fourth principle is required to define conservation agriculture in sub-Saharan Africa: The appropriate use of fertilizer to enhance crop productivity'. *Field Crops Research* 169, 145–8. doi:10.1016/j.fcr.2014.05.012.

Soussana, J.-F., Lutfalla, S., Ehrhardt, F., Rosenstock, T., Lamanna, C., Havlík, P., Richards, M., Wollenberg, E., Chotte, J.-L., Torquebiau, E., Ciais, P., Smith, P. and Lal, R. 2019. Matching policy and science: rationale for the '4 per 1000 – soils for food security and climate' initiative. *Soil and Tillage Research* 188, 3–15. doi:10.1016/j.still.2017.12.002.

Srikanthasamy, T., Leloup, J., N'Dri, A. B., Barot, S., Gervaix, J., Koné, A. W., Koffi, K. F., Le Roux, X., Raynaud, X. and Lata, J.-C. 2018. Contrasting effects of grasses and trees on microbial *N*-cycling in an African humid savanna. *Soil Biology and Biochemistry* 117, 153–63. doi:10.1016/j.soilbio.2017.11.016.

Subbarao, G. V., Nakahara, K., Hurtado, M. P., Ono, H., Moreta, D. E., Salcedo, A. F., Yoshihashi, A. T., Ishikawa, T., Ishitani, M., Ohnishi-Kameyama, M., Yoshida, M., Rondon, M., Rao, I. M., Lascano, C. E., Berry, W. L. and Ito, O. 2009. Evidence for biological nitrification inhibition in Brachiaria pastures. *Proceedings of the National Academy of Sciences of the United States of America* 106(41), 17302–7. doi:10.1073/pnas.0903694106.

Subbarao, G. V., Nakahara, K., Ishikawa, T., Ono, H., Yoshida, M., Yoshihashi, T., Zhu, Y., Zakir, H. A. K. M., Deshpande, S. P., Hash, C. T. and Sahrawat, K. L. 2013. Biological nitrification inhibition (BNI) activity in sorghum and its characterization. *Plant and Soil* 366(1–2), 243–59. doi:10.1007/s11104-012-1419-9.

Subbarao, G. V., Yoshihashi, T., Worthington, M., Nakahara, K., Ando, Y., Sahrawat, K. L., Rao, I. M., Lata, J. C., Kishii, M. and Braune, H. J. 2015. Suppression of soil nitrification by plants. *Plant Science: an International Journal of Experimental Plant Biology* 233, 155–64. doi:10.1016/j.plantsci.2015.01.012.

Tejada, M., Garcia, C., Gonzalez, J. L. and Hernandez, M. T. 2006. Use of organic amendment as a strategy for saline soil remediation: influence on the physical, chemical and biological properties of soil. *Soil Biology and Biochemistry* 38(6), 1413–21. doi:10.1016/j.soilbio.2005.10.017.

Thierfelder, C. and Wall, P. C. 2012. Effects of conservation agriculture on soil quality and productivity in contrasting agro-ecological environments of Zimbabwe. *Soil Use and Management* 28(2), 209–20. doi:10.1111/j.1475-2743.2012.00406.x.

Tivet, F., de Moraes Sá, J. C. M., Lal, R., Briedis, C., Borszowskei, P. R., dos Santos, J. B., Farias, A., Eurich, G., Hartman, DdC., Nadolny Junior, M., Bouzinac, S. and Séguy, L. 2013. Aggregate C depletion by plowing and its restoration by diverse biomass-C inputs under no-till in sub-tropical and tropical regions of Brazil. *Soil and Tillage Research* 126, 203–18. doi:10.1016/j.still.2012.09.004.

Turner, B. L., McKelvie, I. D. and Haygarth, P. M. 2002. Characterisation of water-extractable soil organic phosphorus by phosphatase hydrolysis. *Soil Biology and Biochemistry* 34(1), 27–35. doi:10.1016/S0038-0717(01)00144-4.

Turrini, A., Sbrana, C., Avio, L., Njeru, E. M., Bocci, G., Bàrberi, P. and Giovannetti, M. 2016. Changes in the composition of native root arbuscular mycorrhizal fungal communities during a short-term cover crop-maize succession. *Biology and Fertility of Soils* 52(5), 643–53. doi:10.1007/s00374-016-1106-8.

Vanlauwe, B., Wendt, J., Giller, K. E., Corbeels, M., Gerard, B. and Nolte, C. 2014. A fourth principle is required to define Conservation Agriculture in sub-Saharan Africa: the appropriate use of fertilizer to enhance crop productivity. *Field Crops Research* 155, 10–3. doi:10.1016/j.fcr.2013.10.002.

Verzeaux, J., Hirel, B., Dubois, F., Lea, P. J. and Tétu, T. 2017. Agricultural practices to improve nitrogen use efficiency through the use of arbuscular mycorrhizae: basic and agronomic aspects. *Plant Science: an International Journal of Experimental Plant Biology* 264, 48–56. doi:10.1016/j.plantsci.2017.08.004.

Vieira, F. C. B., He, Z. L., Bayer, C., Stoffella, P. J. and Baligar, V. C. 2008. Organic amendment effects on the transformation and fractionation of aluminum in acidic sandy soil. *Communications in Soil Science and Plant Analysis* 39(17–18), 2678–94. doi:10.1080/00103620802358813.

Weinhard, P., Balandreau, J., Rinaudo, G. and Dommergues, Y. 1971. Fixation non symbiotique de l'azote dans la rhizosphère de quelques non-légumineuses tropicales. *Revue d'Ecologie, et de Biologie du Sol* VIII(3), 367–73.

Zemek, O., Frossard, E., Scopel, E. and Oberson, A. 2018. The contribution of *Stylosanthes guianensis* to the nitrogen cycle in a low input legume-rice rotation under conservation agriculture. *Plant and Soil* 425(1–2), 553–76. doi:10.1007/s11104-018-3602-0.

Chapter 7

Carbon management practices and benefits in Conservation Agriculture systems: carbon sequestration rates

João Carlos de Moraes Sá, State University of Ponta Grossa, Brazil; Florent Tivet, Centre de coopération Internationale en Recherche Agronomique pour le Développement (CIRAD), France; Rattan Lal, The Ohio State University, USA; Ademir de Oliveira Ferreira, Federal Rural University of Pernambuco, Brazil; Clever Briedis, Brazilian Agricultural Research Corporation, Agricultural Instrumentation Center, Brazil; Thiago Massao Inagaki, Technical University of Munich, Germany; and Daniel Potma Gonçalves and Jucimare Romaniw, State University of Ponta Grossa, Brazil

1 Introduction

Soils can be a sink or source of atmospheric CO_2 depending on climate (i.e. temperature and precipitation), soil mineralogy, net primary production (NPP), land use and management. Land-use change, particularly conversion of native vegetation (NV) to agricultural land, exacerbates CO_2 emissions through deforestation, biomass burning, and soil organic carbon (SOC) depletion by conventional plow-based tillage (CT). The primary driving forces for the decline of SOC under CT are the disruption of soil aggregates, and marked changes in soil environment (temperature, moisture, and oxygen) which affect the microbial activity which affects SOC processes. In addition, the bare soil surface under CT is exposed to frequent wet-dry cycles, enhancing the turnover rate of aggregates and promoting soil erosion (Lal, 1976; Beare et al., 1994). In contrast, adoption of Conservation Agriculture (CA) with no soil disturbance,

http://dx.doi.org/10.19103/AS.2019.0049.08

and where inputs of plant residues (aboveground biomass and roots) and associated turnover of soil biological activity are enhanced, can increase the SOC pool and its stability by physical protection within stable aggregates (Six et al., 2000; Tivet et al., 2013b; De Oliveira Ferreira et al., 2018a). Recognizing the role of soils as a potential sink for atmospheric CO_2 (Lal, 2004, 2006, 2018), there is a growing interest in adopting NT to store carbon (C) in soils for off-setting anthropogenic emissions of CO_2, but also to restore SOC pool lost by the conversion of NV and to improve and sustain productivity of agroecosystems (Sá et al., 2013, 2015).

Currently, Brazil has 36.8 million ha under CA (Sá and De Oliveira Ferreira, 2018), corresponding to around 75% of total grain production. South Brazil (Paraná and Rio Grande do Sul state) was a pioneer in CA adoption in the early 1970s but the system was only adopted on a large scale in the 1990s. In 1990 the area expansion under CA was 0.9 million ha (M ha), whose annual rate of 0.081 M ha yr^{-1}, and in 2000 had significant expansion, expanding to 14.0 million ha (1.64 M ha yr^{-1}). The surface doubled in 2010 to 30.3 M ha (1.51 M ha yr^{-1}) and in 2016–17 reached the current values (Sá and De Oliveira Ferreira, 2018, Fig. 1).

The Cerrado (together with the South is the focus of the present study) is of great importance in the national scenario and presents more than 12 million ha cultivated with annual crops, of which more than 7 million ha are cultivated under NT (APDC, 2007). These regions of Brazil (South and Cerrado) experienced severe soil degradation and SOC depletion because of intensive tillage, and high rate of SOC mineralization (Sá et al., 2001; Bayer et al., 2004;

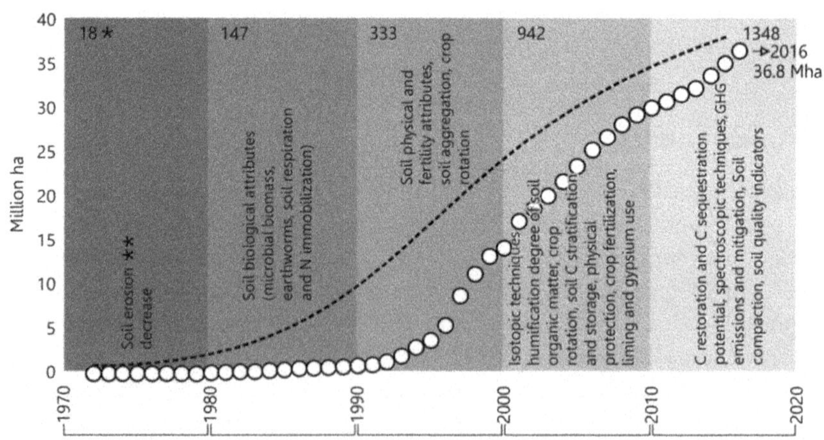

Figure 1 Timeline by decades, published articles (*), most relevant events in each decade (**), and CA expansion rate (M ha yr^{-1}) by decade (***). Source: Sá and De Oliveira Ferreira (2018).

Dieckow et al., 2009). The principal challenge under hot and humid Cerrado climate with a rapid turnover of soil organic matter (SOM) and nutrients was to develop high biomass-C inputs managed under NT with complex crop rotation (Tivet et al., 2013a; Sá et al., 2015).

Several reports have demonstrated C sequestration in soils managed by CA in conjunction with complex crop rotation (Sá et al., 2001, 2015; Six et al., 2002b; Sisti et al., 2004; Dieckow et al., 2005; Bayer et al., 2006b; Cerri et al., 2007; Sá and Lal 2009; Boddey et al., 2010; De Oliveira Ferreira et al., 2012). Important factors in increasing CO_2 mitigation and the SOC stock are the amount, quality and frequency of the crop residues added to soil under a wide range of climate-driven decomposition rates (Sá et al., 2015; Tivet et al., 2013b), soil mineralogy and profile characteristics (Paustian et al., 1997; Six et al., 2002a; Bayer et al., 2006b; Ogle et al., 2012; Virto et al., 2012; De Oliveira Ferreira et al., 2018b).

2 The potential for carbon (C) sequestration by Conservation Agriculture (CA) in subtropical and tropical agroecosystems: a case study

2.1 Context

The data on SOC sequestration rates for tropical (−0.03 to 1.98 Mg ha^{-1} yr^{-1}; (Bayer et al., 2006b; Corbeels et al., 2006; Cerri et al., 2007; Neto et al., 2010; Sá et al., 2015) and subtropical (−0.07 to 1.4 Mg ha^{-1} yr^{-1}; (Bayer et al., 2000, 2006a; Sá et al., 2001, 2015; Amado et al., 2006; Cerri et al., 2007; Boddey et al., 2010; De Oliveira Ferreira et al., 2012) regions of Brazil vary widely because of differences in the amount of biomass-C input and the agronomic potential of agroecosystems (Bayer et al., 2006b; Corbeels et al., 2006; Cerri et al., 2007). Batlle-Bayer et al. (2010) reported that changes in SOC stock, when converting from CT to CA, ranged from 0.13 to 1.91 Mg ha^{-1} yr^{-1} for 0–30 cm depth, and 1.98 Mg ha^{-1} yr^{-1} for 0–40 cm depth (Sá et al., 2014). A large variability in SOC sequestration rate may be attributed to a high diversity of cropping systems, amount and frequency of biomass-C input, and soil properties (Batlle-Bayer et al., 2010). For example, the rate of SOC sequestration is negligible in those CA with low biomass-C input and only one crop (summer) per year (Bayer et al., 2006b; Jantalia et al., 2007; Carvalho et al., 2009). In comparison, in subtropical and tropical ecosystems in Brazil, the rate is high for those CA which involve a high biomass-C input along with a summer crop followed by one or a combination of several winter crops (Séguy et al., 2006).

Furthermore, the rate of SOC sequestration by NT can be higher also in the subsoil layers (Chabbi et al., 2009; Boddey et al., 2010; Rumpel and Kogel-Knabner, 2011; Lal, 2018). For a double cropping system (legume–cereal cropping sequence) under an extensive farm-land network, Neto et al. (2010) measured the SOC sequestration rate of 1.98 Mg ha^{-1} yr^{-1} for 0–30 cm depth. In the long-term NT experiment associated with high biomass-C input, Sá et al. (2015) reported a rate of 1.98 Mg ha^{-1} yr^{-1} for 0–40 cm depth. Cropping systems with a high biomass input to maintain a permanent soil cover mimic the environment under undisturbed ecosystems (e.g. forest, savanna, prairies). These cropping systems support a continuous flow of mass and energy, which release organic compounds, accentuate soil biodiversity, and enhance SOM (Séguy et al., 2006; Six et al., 2006). CA with continuous input of large amounts of biomass to the soil surface creates a positive C budget, enhances the C sequestration, and restoration SOC (De Oliveira Ferreira et al., 2018b), increasing soil quality and agronomic productivity (Sá et al., 2015).

In this case study the specific objectives were to (i) assess the impact of the continuous plow-based CT on SOC stock for subtropical and tropical agroecosystems vis-à-vis native vegetation (NV) as baseline; (ii) compare SOC balance among plow-based CT, CA cropping systems, and NV; and (iii) evaluate the redistribution of SOC stock in the soil profile in relation to soil resilience and impact on the agronomic productivity.

2.2 Experimental procedures

2.2.1 Site description, land uses management, and experimental design

Field experiments were conducted at two research sites: (i) near the town of Ponta Grossa, in Paraná State, Southern Brazil, at the experimental station of IAPAR (Instituto Agronômico do Paraná), and hereafter called the PG site, and (ii) near the city of Lucas do Rio Verde, in Mato Grosso State, center western Brazil, at the experimental station of Fundação Rio Verde, and hereafter called the LRV site (Table 1).

2.2.1.1 Ponta Grossa site (PG)

In 1967, some of the NV areas were converted to pastureland by plowing to 20-cm depth, and then disking twice to break the clods. In 1978, after 11 years of pasture, this area was converted to cropland and cultivated for three years: rice (*Oryza sativa* L.) for two and soybean (*Glycine max,* L.Merril) for one. The present experiment was initiated in 1981 and the experimental design consisted of three tillage treatments involving conventional plow tillage (CT), minimum tillage (MT) and NT laid out as whole plots. The CT consisted of plowing with a 70-cm disk after summer harvest and one after winter harvest to 20-cm depth followed by two 60-cm harrowing. The MT comprised of one chisel plowing to

Table 1 Sites description: Location, soil type, parent material, climate, land use and management, duration of experiment, sampling depth, chemical properties and clay content

Description	Ponta Grossa, PR (PG site)				Lucas do Rio Verde, MT (LRV site)		
Location-coordinates	25°09′S-50°09′W				13°00′S-55°58′S		
Altitude	865 m				380 m		
Soil type (FAO)	Red Latosol, Oxisol				Red Yellow Latosol, Oxisol		
Soil Type (Soil Taxonomy)	Rhodic Hapludox				Typic Haplustox		
Parent material	Shale				Shale and Sandstone		
Climate - type	Mesothermic, Summer and Winter wet, cold winter (Cfa)				Humid tropic, Summer hot and very wet, winter hot and dry (Aw)		
Mean annual temperature (MAT)	18.5°C				25.2°C		
Mean annual rainfall (MAR)	1545 mm				1950 mm		
Land use[a]	NV, CT, MT and NT				NV, CT and NT1 to NT6		
Years of experiment	29				8		
Sampling depth (cm)	0–5, 5–10, 10–20, 20–40, 40–60, 60–80, 80–100				0–5, 5–10, 10–20, 20–40, 40–60, 60–80, 80–100		
Soil properties	**NV**	**CT**	**MT**	**NT**	**NV**	**CT**	**NT[b]**
Chemical (0-40 cm)							
pH ($CaCl_2$, 0.01M)	4.01	4.54	4.64	4.47	3.95	4.80	4.57
Al^{3+} ($cmol_c\ dm^{-3}$)	2.12	0.48	0.35	0.50	1.00	0.17	0.31
H + Al ($cmol_c\ dm^{-3}$)	14.34	9.77	8.59	9.47	8.26	5.45	6.15
Ca^{2+} ($cmol_c\ dm^{-3}$)	2.50	3.40	3.35	3.06	0.33	1.62	1.83
Mg^{2+} ($cmol_c\ dm^{-3}$)	0.90	1.27	1.51	1.31	0.26	0.63	0.68
K^+ ($cmol_c\ dm^{-3}$)	0.16	0.25	0.19	0.19	0.09	0.19	0.19
$P_{available}$ ($mg\ dm^{-3}$)	4.16	4.46	3.29	4.56	0.47	18.96	8.10
Clay content (g kg⁻¹)							
0–20 cm	593	681	637	619	393	381	398
20–40 cm	673	713	690	667	437	425	444
40–100 cm	717	731	719	692	529	455	490

[a] NV: native vegetation of forest (PG) and Cerrado (LRV), CT: conventional plow-based tillage, MT: minimum tillage, NT: no-tillage.
[b] Mean values of the six no-till systems were used for the quantification of the chemical properties and clay content.
Source: adapted from Sá et al. (2015).

25-cm depth and one 60-cm narrow disking. In contrast to CT and MT, there was no pre-sowing soil disturbance in NT. Each plot was 100 × 100 m for the NT, and 50 × 140 m for the CT and MT treatments with six subplots (33 × 50 m for NT, and 25 × 46 m for MT and CT). A 3-years cropping sequence was used for all tillage treatments over the 10-years period, involving two crops per year (see Table 2) with soybean (*Glycine max,* L. Merril) grown in six and maize in four summers alternating with oats (*Avena strigosa* Schreb), wheat (*Triticum aestivum* L.), and vetch (*Vicia sativa* L.) in the winter.

An area under NV and adjacent to the experimental plots was selected as a baseline to assess the management-induced changes in SOC stock. Six subplots were demarcated for soil sampling on an area of 100 × 140 m under NV. The mineral composition of the soil is similar among plots and comprises of 21.7%

Table 2 Site, tillage systems, crop sequence, cumulative and annual biomass-C input in the 29 years experiment period at the PG site and 8 years experiment period at the LRV site

			Biomass-C input[d] (Mg ha^{-1})	
Site	Tillage systems[a]	Crop sequence[b,c]	Cumulative	Annual
PG	CT	S/O - M/O - S/W - S/O+V - M/O - S/O - M/O - S/O - M/R - S/V	86.1	3.07
	MT	S/O - M/O - S/W - S/O+V - M/O - S/O - M/O - S/O - M/R - S/V	83.7	2.99
	NT	S/O - M/O - S/W - S/O+V - M/O - S/O - M/O - S/O - M/R - S/V	116.1	4.15
LRV	CT	S/M - S/M - S/Ct - S/Ct - S/Ct - S/ Ct-S/Ct-S/Ct	32.1	4.01
	NT_1	S/M+Brz - Rc - M - S/M+Brz - S/ M+Brz - S/M+Brz - M - S/Cs	60.8	7.60
	NT_2	S/Fm - Rc - M - S/Fm - S/Fm+Pp - S/Fm+Pp - M - S/M+Cs	58.0	7.25
	NT_3	S/Fm+Pp - Rc - M - S/Fm+Pp - S/ Fm+Cs - S/Fm+Cs - M - S/G+Cs	54.7	6.84
	NT_4	S/Fm+Cs - Rc - M - S/Fm+Cs - S/ G+Brz - S/G+Brz - M - S/Sg+Cs	58.7	7.34
	NT_5	S/Sg+Brz - Rc - M - S/Sg+Brz - S/Sg+Brz - S/Sg+Brz - M -S/ Sg+Brz	67.0	8.38
	NT_6	S/Mt - Rc - M - S/Mt - S/M+Brz - S/M+Brz - M - S/M	59.3	7.41

[a] Tillage systems: CT = conventional tillage, MT = minimum tillage, NT = no-tillage.
[b] Crop sequence for the last 10 years at the PG site: S = soybean (*Glycine max* (L.) Merr.), O = black oats (*Avena strigosa* Schreb), M = maize (*Zea mays* L.), W = wheat (*Triticum aestivum* L.), V = vetch (*Vicia sativa*).
[c] Crop sequence and tillage systems for the 8 years experiment at the LRV site: Ct = cotton (*Gossypium hirsutum* L.), Brz = *Brachiaria ruziziensis* cv. ruzi, Rc = rice (*Oriza sativa*); Cs = *Crotalaria spectabilis* Roth, Fm = finger millet (*Eleusine coracana* (L.) Gaertn), Pp = pigeon pea (*Cajanus cajan* (L.) Millsp), G = sunflower (*Helianthus annuus* L.), Sg = *Sorghum bicolor* (L.) Moench, Mt = millet (*Pennisetum typhoides* Burm.).
[d] C input by the crop residues: cumulative = sum of C input along the 29 years of experiment at PG site and 8 years at LRV site; Annual C input = Cumulative C input/29 and 8 years, for PG site and LRV site, respectively.
Source: adapted from Sá et al. (2015).

SiO_2, 36.4% Al_2O_3, 13.1% Fe_2O_3, and 2.1% TiO_2. The average concentration of Fe, Si, and Al oxides did not vary among treatments or depth (Goncalves et al., 2008), and soil of all plots has a similar textural composition (Table 1).

2.2.1.2 Lucas do Rio Verde site (LRV)

The NV was converted to cropland in 1986 by clearing, burning, and plowing. Immediately after conversion of NV between 1988 and 2000, soil acidity was managed by incorporating 4.5 Mg ha^{-1} of dolomite lime to 20-cm depth. The soil was cultivated to rice for one year and soybean as monoculture for 10 years under CT involving a disk harrow to 20-cm depth followed by a fallow during the winter. A uniform area of 10.9 ha was selected in 2001 to initiate the present experiment.

The experiment was designed to compare the standard tillage management of the region represented by one crop per year in the summer by alternating soybean with cotton (*Gossypium hirsutum* L.) under plow-based CT, compared with cropping systems involving different biomass-C input under NT. Tillage treatments were assigned as whole plots (see Table 2) and cropping sequences consisted of: (i) soybean alternating with cotton under plow-based CT designated as CT-S/Ct [C-input = 4.01 Mg ha^{-1} yr^{-1}]; (ii) NT1 - soybean in the summer followed by maize (*Zea mays* L.) + *Brachiaria ruziziensis* as second crop [C-input = 7.60 Mg ha^{-1} yr^{-1}]; NT2 - soybean in the summer followed by finger millet (*Eleusine coracana*) or finger millet + pigeon pea (*Cajanus cajan*) as second crop [C-input = 7.25 Mg ha^{-1} yr^{-1}]; NT3 - soybean in the summer followed by finger millet + pigeon pea or finger millet + *Crotalaria spectabilis* as second crop [C-input = 6.84 Mg ha^{-1} yr^{-1}]; NT4 - soybean in the summer followed by finger millet + *C. spectabilis* or sunflower (*Helianthus annuus*) + *B. ruziziensis* as second crop [C-input = 7.34 Mg ha^{-1} yr^{-1}]; NT5-soybean in the summer followed by sorghum (*Sorghum bicolor*) + *B. ruziziensis* as second crop [C-input = 8.38 Mg ha^{-1} yr^{-1}]; and NT6 - soybean in the summer followed by millet (*Pennisetum glaucum*) or maize + *B. ruziziensis* as second crop [C-input = 7.41 Mg ha^{-1} yr^{-1}].

2.2.2 Total biomass (above and belowground) and carbon (C) input

The annual inputs of C from crop residues (roots and shoots) for each tillage treatment and summer crop, was calculated on the basis of the harvest index (HI) and root and shoot ratio (Sá et al., 2001) during the experimental period (Table 2). Aboveground biomass of the second crop (oat, vetch, *B. ruziziensis*, finger millet, pigeon pea, *C. spectabilis*) was measured during the first year of the experiment and the belowground biomass was estimated from the previous studies (Sá et al., 2001).

2.2.3 Soil sampling

Soil samples at the end of the 29-yr (September 2009) for the PG and 8-yr (October 2009) of experimentation for the LRV site were collected from seven depths: 0–5, 5–10, 10–20, 20–40, 40–60, 60–80, and 80–100 cm. Bulk soil samples for each subplot were obtained for 0–5, 5–10, and 10–20 cm depths by digging 15 × 15 cm trenches.

Soil samples for 20–40, 40–60, 60–80 and 80–100 cm depths were obtained with an auger (4.5 cm diameter) from the same trench. Soil samples were obtained from five points for each subplot, and composited.

Soil bulk density (ρ_b) was measured by the core method (Blake and Hartge 1986) using steel cylinders of 5 × 5 cm. Cores from 10–20 to 80–100 cm layers were taken in the middle of each layer. Bulk samples were oven-dried at 40°C, gently ground, sieved through a 2-mm sieve, and mixed.

3 Key results

3.1 The Ponta Grossa (PG) site

Concentrations of SOC and N decreased significantly with the increase in soil depth, and exhibited different patterns of distributions in the surface layer among land uses. These concentrations were highly stratified with depth under NV (Table 3), and were 80.7 g C kg^{-1} and 5.2 g N kg^{-1} in the 0–5 cm layer, decreasing sharply to 30.6 g C kg^{-1} and 2.06 g C kg^{-1} in 10–20 cm depth and 16.8 g C kg^{-1} and 0.88 g N kg^{-1} in 80–100 depth, respectively. However, there were no significant differences in SOC and N concentrations among tillage treatments and the NV for 10–100 cm depth. Further, a decreasing trend in SOC concentration was observed under MT and NV in comparison with that under NT or CT. Among tillage treatments, greater concentrations of SOC and N, respectively, were recorded in soil under NT with 27.5 g kg^{-1} and 20.9 g kg^{-1} more SOC and 2.86 g kg^{-1} and 1.96 g kg^{-1} more N than those under CT and MT in the 0–5 cm depth. These data represent a decrease of 44% and 33% in SOC, and 60% and 41% in N, respectively, in CT and MT compared with CA. Expectedly, SOC concentration under NV, CT, MT, and NT, was positively correlated with N concentration ($SOC = 11.4N + 10.\ 5$, $R^2 = 0.86$, $n = 188$, $P < 0.0001$) across all soil depths, and indicate C:N ratio of 11.4.

Relatively rapid changes in TOC concentration with intensive CA at the LRV site indicate that there is a considerable potential to reverse the process of soil degradation and SOC depletion by historic CT, and to enhance the rate of SOC restoration.

Measurements of ρ_b and concentrations of SOC under NV in an adjacent undisturbed soil to 1-m depth (Table 3) indicated that the SOC stock in

Table 3 Soil organic carbon (SOC) and total nitrogen (N) concentrations (g kg^{-1}) managed under CT, MT and NT, at the PG site, and under CT and NT cropping systems (NT1 to NT6) at the LRV site, and under the neighboring native vegetation (NV) for both sites

Sites	Land use	Soil depth (cm) 0–5			5–10			10–20			20–40			40–60			60–80			80–100		
		SOC (g kg^{-1})																				
PG	NV	80.7	4.9a	a[b]	43.5	4.9	ns	30.6	2.0	ns	28.9	2.2	ns	22.6	2.0	ns	18.5	1.0	b	16.8	1.7	ns
	CT	35.0	0.6	c	35.0	1.0		33.8	0.6		28.5	1.1		24.2	1.5		21.8	1.2	a	20.2	1.2	
	MT	41.7	1.7	c	40.6	1.1		30.3	1.0		26.4	1.0		21.8	0.7		19.8	0.7	ab	18.5	0.4	
	NT	62.5	1.4	b	44.3	1.0		32.1	0.4		28.5	0.5		24.1	0.4		21.8	0.6	a	20.3	0.7	
LRV	NV	38.3	3.9	a	21.3	1.1	ns	16.1	1.0	ns	12.2	0.6	a	9.5	0.5	ns	8.3	0.6	ns	7.1	0.2	ns
	CT	18.3	0.4	b	17.5	0.5		13.4	0.6		9.5	0.1	c	7.8	0.1		7.1	0.2		7.1	0.4	
	NT1	25.2	2.5	b	20.8	1.4		18.2	1.5		11.0	0.5	abc	8.2	0.3		7.5	0.4		7.0	0.2	
	NT2	21.5	1.6	b	18.8	1.1		16.7	0.6		11.4	0.6	abc	8.6	0.6		7.1	0.0		6.8	0.2	
	NT3	22.4	0.8	b	17.8	1.0		15.3	1.0		10.4	0.2	bc	8.3	0.4		7.0	0.2		7.0	0.1	
	NT4	19.7	0.9	b	18.3	0.8		16.2	0.8		12.1	1.4	ab	8.9	0.6		7.6	0.5		6.8	0.2	
	NT5	23.6	2.3	b	20.8	2.1		18.4	2.8		12.3	0.7	ab	8.7	0.3		7.1	0.2		6.6	0.3	
	NT6	24.0	1.9	b	18.8	0.7		16.7	0.3		11.7	0.2	ab	9.1	0.3		7.5	0.3		6.9	0.4	

(Continued)

Table 3 *(Continued)*

Sites	Land use	Soil depth (cm) 0–5			5–10			10–20			20–40			40–60			60–80			80–100		
		TN (g kg⁻¹)																				
PG	NV	5.20	0.29	a	2.81	0.26	a	2.06	0.08	ns	1.70	0.17	ns	1.13	0.22	ns	1.00	0.18	ns	0.88	0.19	ns
	CT	1.93	0.07	b	1.84	0.14	b	1.68	0.07		1.30	0.15		1.07	0.16		0.87	0.12		0.82	0.09	
	MT	2.83	0.43	b	2.64	0.43	a	1.68	0.31		1.44	0.25		1.23	0.21		1.25	0.20		1.16	0.18	
	NT	4.79	0.19	a	2.99	0.12	a	1.66	0.07		1.51	0.10		1.11	0.09		0.98	0.12		0.95	0.12	
LRV	NV	1.61	0.17	a	1.06	0.09	ns	0.76	0.06	ns	0.59	0.06	ns	0.49	0.09	ns	0.37	0.05	ns	0.34	0.03	bc
	CT	1.03	0.05	b	0.86	0.07		0.62	0.07		0.39	0.03		0.37	0.03		0.38	0.02		0.35	0.02	ab
	NT1	1.40	0.21	ab	1.04	0.19		0.76	0.14		0.47	0.03		0.38	0.02		0.31	0.01		0.31	0.02	bc
	NT2	1.04	0.21	b	0.95	0.14		0.73	0.06		0.41	0.03		0.37	0.02		0.31	0.02		0.27	0.02	c
	NT3	1.28	0.05	ab	0.77	0.05		0.65	0.05		0.43	0.05		0.41	0.02		0.36	0.02		0.42	0.01	a
	NT4	0.93	0.05	b	0.76	0.03		0.68	0.03		0.54	0.05		0.45	0.04		0.36	0.03		0.31	0.05	bc
	NT5	1.19	0.17	ab	0.88	0.02		0.78	0.03		0.52	0.01		0.46	0.04		0.36	0.02		0.30	0.01	bc
	NT6	1.33	0.17	ab	0.92	0.12		0.83	0.11		0.53	0.06		0.48	0.04		0.37	0.05		0.33	0.02	bc

[a] Superscript numbers refers the standard error.

[b] Lower case letters (columns) refers the comparison among tillage treatments within depth.

NT[1] = soybean (Glycine max (L.) Merr.)/maize (Zea mays L.) + Brachiaria ruziziensis cv. ruzi - rice (Oryza sativa L.) - maize - soybean/maize + Brachiaria ruziziensis - soybean/maize + Brachiaria - soybean/maize + Brachiaria - maize - soybean/Crotalaria spectabilis Roth; NT[2] = soybean/finger millet (Eleusine coracana (L.) Gaertn) - rice (Oryza sativa L.) - maize (Zea mays L.) - soybean/finger millet - soybean/finger millet + pigeon pea (Cajanus cajan (L.) Millsp) - soybean/finger millet + pigeon pea - maize - soybean/maize + Crotalaria spectabilis Roth; NT[3] = soybean/finger millet + pigeon pea - rice - maize - soybean/finger millet + pigeon pea - soybean/finger millet + Crotalaria spectabilis Roth - soybean/finger millet + Crotalaria spectabilis Roth - maize - soybean/sunflower (Helianthus annuus L.) + Crotalaria spectabilis Roth; NT[4] = soybean/finger millet + Crotalaria spectabilis Roth - rice - maize - soybean/finger millet + Crotalaria spectabilis Roth - soybean/sunflower + Brachiaria ruziziensis - soybean/sunflower + Brachiaria ruziziensis - maize - soybean/Sorghum bicolor (L.) Moench + Crotalaria spectabilis Roth; NT[5] = soybean/Sorghum bicolor (L.) Moench + Brachiaria ruziziensis - rice - maize - soybean/Sorghum bicolor (L.) Moench + Brachiaria ruziziensis - soybean/Sorghum bicolor (L.) Moench + Brachiaria ruziziensis - soybean/Sorghum bicolor (L.) Moench + Brachiaria ruziziensis - maize - soybean/Sorghum bicolor (L.) Moench + Brachiaria ruziziensis; NT[6] = soybean/millet (Pennisetum glaucum Burm.) - rice - maize - soybean/millet - soybean/maize + Brachiaria ruziziensis - soybean/maize + Brachiaria ruziziensis - maize - soybean/maize.

Source: adapted from Sá et al. (2015).

0–20 cm depth represented 36% of the total compared with only 27% under CT SOC stock (Table 4). Further, land-use changes significantly impacted on SOC and N stocks (Table 4). For example, the average SOC stock in 0–20 cm depth decreased from 92.0 Mg ha^{-1} in NV (Table 4) to 67.4 Mg ha^{-1} in CT plot ($P < 0.001$), a decline of ~ 27% over 42 years since the conversion of NV into cultivated field land use of CT for >29 years. This decline represents a loss of ~ 0.58 Mg C ha^{-1} yr^{-1} and ~ 62.0 kg N ha^{-1} yr^{-1} for the clayey Rhodic Hapludox. The soil under NT in 0–20 cm depth contained 17.0 and 14.3 Mg ha^{-1} more SOC and 2.02 and 1.2 Mg ha^{-1} more N than those under CT and MT, respectively (Fig. 2, Table 4).

These data represent an annual accumulation rate of 0.59 and 0.49 Mg C ha^{-1} for SOC and 69.7 and 41.2 kg ha^{-1} of N over 29 years under CT and MT, respectively. The rate of SOC sequestration is relatively low when compared with previous results reported by Sá et al. (2001) in an Oxisol near this experiment and with Amado et al. (2006) in a similar subtropical region. The main reason was due to the low amount of biomass-C added through residues of summer crop and winter species during the 29-yr of experiment and the near-absence of legumes as a winter cover crop. De Oliveira Ferreira et al. (2012) reported for this region the minimum amount of biomass-C input to maintain the steady-state that was 3.21 Mg ha^{-1} yr^{-1}, and in this study the biomass-C input was 4.15 Mg ha^{-1} yr^{-1}.

The percentage of biomass-C converted into SOC in this study was 14.2 and was very similar to the findings obtained by De Oliveira Ferreira et al. (2012) in this region that was ~ 12.63–14.26%. There were no significant differences among tillage treatments even with a trend of soils under CT and NT containing relatively more SOC than that under MT in the 40–100 cm depth.

3.2 The Lucas do Rio Verde (LRV) site

Concentrations of SOC and N decreased with an increase in depth and exhibited differential distribution in the soil profile among the land-use treatments (Table 3). Concentrations of SOC and N, respectively, under NV were 38.3 g C kg^{-1} and 1.6 g N kg^{-1} in 0–5 cm depth, compared with 7.1 g C kg^{-1} and 0.34 g N kg^{-1} in 80–100 cm depth. Similarly, SOC and N concentrations under CA also declined with depth, but the stratification was less pronounced than that under the Cerrado NV. The SOC concentration under cropped fields (CT and NT) was significantly lower ($P < 0.05$) than that under NV in 0–5 cm depth. Total N concentration was likewise higher under the Cerrado NV than that under CT and NTs, but no difference among land uses was observed below 40 cm depth. The soil under NT, with a predominance of cereal as second crops [maize + *B. ruziziensis*, sorghum + *B. ruziziensis*, millet] during the dry season, and higher

Table 4 Effects of land use and management practices on soil organic C (SOC) and total nitrogen (TN) stocks (Mg ha^{-1}) at the PG and the LRV sites for 0–20, 20–40, 40–60, 60–80 and 80–100 cm depths, and under the neighboring native vegetation (NV) for both sites

		Soil depth (cm)														
Sites	Land use	0–20			20–40			40–60			60–80			80–100		
		SOC stock (Mg ha^{-1})														
PG	NV	92.0	5.8a	a^b	53.7	4.2	ns	44.0	2.3	ns	35.9	2.7	ns	32.3	3.0	ns
	CT	67.4	1.5	b	52.9	0.9		48.1	3.4		42.1	2.2		39.0	2.9	
	MT	70.2	2.4	b	49.4	2.4		43.3	2.0		38.2	1.5		35.7	1.3	
	NT	84.4	1.0	a	53.2	0.9		47.8	2.0		42.0	1.7		39.1	2.0	
LRV	NV	48.0	2.6	a	27.7	1.1	a	20.9	1.4	ns	17.7	1.4	ns	16.0	0.5	ns
	CT	33.8	1.4	c	21.8	0.8	c	17.3	0.3		15.0	0.3		16.2	0.9	
	NT1	44.2	4.5	ab	25.1	1.0	abc	18.1	0.9		16.1	1.0		15.7	0.3	
	NT2	39.5	1.8	bc	25.8	1.6	abc	18.8	1.0		15.3	0.4		15.2	0.3	
	NT3	37.7	2.2	bc	23.7	0.8	bc	18.1	1.0		15.1	0.6		15.6	0.1	
	NT4	37.7	1.7	bc	27.5	2.9	ab	19.5	1.6		16.2	0.8		15.3	0.0	
	NT5	43.3	5.0	ab	28.1	1.9	ab	19.0	0.3		15.3	0.7		14.8	0.9	
	NT6	40.7	1.9	ab	26.5	0.7	ab	20.0	0.4		16.0	0.4		15.5	0.8	
		TN stock (Mg ha^{-1})														
PG	NV	6.03	0.33	a	3.22	0.41	ns	2.21	0.28	ns	1.91	0.21	ns	1.72	0.29	ns
	CT	3.49	0.14	b	2.38	0.21		2.14	0.34		1.67	0.23		1.58	0.19	
	MT	4.32	0.68	b	2.68	0.44		2.41	0.40		2.39	0.36		2.22	0.34	
	NT	5.51	0.17	a	2.81	0.16		2.18	0.17		1.87	0.24		1.85	0.27	
	P values	0.001			0.305			0.929			0.295			0.426		

LRV	NV	2.20	0.14	ns	1.33	0.13	ns	1.07	0.19	ns	0.79	0.09	ns	0.76	0.07	ns
	CT	1.72	0.11		1.00	0.06		0.91	0.11		0.87	0.05		0.91	0.04	
	NT1	2.13	0.42		1.07	0.05		0.84	0.05		0.66	0.01		0.69	0.03	
	NT2	1.85	0.20		0.94	0.07		0.82	0.05		0.66	0.02		0.60	0.04	
	NT3	1.78	0.07		0.97	0.10		0.89	0.07		0.76	0.06		0.94	0.04	
	NT4	1.62	0.07		1.24	0.12		0.99	0.07		0.77	0.04		0.69	0.09	
	NT5	1.93	0.11		1.19	0.03		1.00	0.09		0.78	0.02		0.68	0.01	
	NT6	2.07	0.16		1.21	0.12		1.06	0.11		0.80	0.10		0.74	0.04	

[a] Superscript numbers refers the standard error.
[b] lower case letters (columns) refers the comparison among tillage treatments within depth. ns = no significant.
NT[1] = soybean (Glycine max (L.) Merr.)/maize (Zea mays L.) + Brachiaria ruziziensis cv. ruzi - rice (Oryza sativa L.) - maize - soybean/maize + Brachiaria ruziziensis - soybean/maize + Brachiaria - soybean/maize + Brachiaria - maize - soybean/Crotalaria spectabilis Roth; NT2 = soybean/finger millet (Eleusine coracana (L.) Gaertn) - rice (Oryza sativa L.) - maize (Zea mays L.) - soybean/finger millet - soybean/finger millet + pigeon pea (Cajanus cajan (L.) Millsp) - soybean/finger millet + pigeon pea - maize - soybean/maize + Crotalaria spectabilis Roth; NT3 = soybean/finger millet + pigeon pea - rice - maize - soybean/finger millet + pigeon pea - soybean/finger millet + Crotalaria spectabilis Roth - soybean/finger millet + Crotalaria spectabilis Roth - maize - soybean/sunflower (Helianthus annuus L.) + Crotalaria spectabilis Roth; NT4 = soybean/finger millet + Crotalaria spectabilis Roth - rice - maize - soybean/finger millet + Crotalaria spectabilis Roth - soybean/sunflower + Brachiaria ruziziensis - soybean/sunflower + Brachiaria ruziziensis - maize - soybean/Sorghum bicolor (L.) Moench + Crotalaria spectabilis Roth; NT5 = soybean/Sorghum bicolor (L.) Moench + Brachiaria ruziziensis - rice - maize - soybean/Sorghum bicolor (L.) Moench + Brachiaria ruziziensis - soybean/Sorghum bicolor (L.) Moench + Brachiaria ruziziensis - soybean/Sorghum bicolor (L.) Moench + Brachiaria ruziziensis - maize - soybean/Sorghum bicolor (L.) Moench + Brachiaria ruziziensis; NT[6] = soybean/millet (Pennisetum glaucum Burm.) - rice - maize - soybean/millet - soybean/maize + Brachiaria ruziziensis - soybean/maize + Brachiaria ruziziensis - maize - soybean/maize.
Source: adapted from Sá et al. (2015).

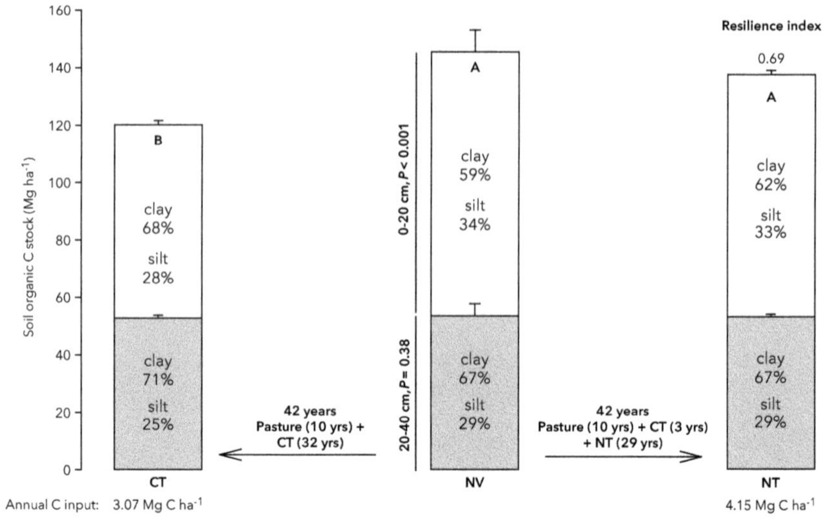

Figure 2 Effects of continuous CT and conversion from CT to NT cropping systems with contrasting C-inputs on soil organic C (Mg ha^{-1}) system at the PG site. Source: Sá et al. (2015).

biomass input (NT1, NT5, and NT6), contained on average higher SOC and N concentrations in 0–5 cm depth than those in soil under CT (23.6 to 25.2 vs. 18.3 g C kg^{-1}, and 1.2 to 1.4 vs. 0.92 g N kg^{-1}). The SOC concentration in soil under NV and cropped fields is positively correlated with the N concentration ($SOC = 19.3N + 1.25$, $R^2 = 0.90$, $n = 188$, $P < 0.001$) across all soil depths, and indicate C:N ratio of 19.3.

Soil ρ_b was the lowest in the 0–5 cm layer (1.00 ± 0.1 Mg m^{-3}) under NV, and increased slightly to 1.12 ± 0.02 Mg m^{-3} to 1-m depth (Table 3). It was higher in some croplands (NT and CT) in 0–20 cm depth, probably because of tillage (CT) and traction (seeding and harvesting) (De Oliveira Ferreira et al., 2011). Soil ρ_b under CT in 40–100 cm depth was significantly more than that under the Cerrado NV and some NT treatments ($P < 0.05$). The SOC stock to 1-m depth was 130.3 Mg ha^{-1} in Cerrado NV, of which 36% was stored in 0–20 cm, 23% in 20–40 cm, and 41% in 40–100 cm depth (Table 4). Significant losses of SOC stock occurred upon conversion to an agricultural land use and by a long-term use of CT (Fig. 3 and Table 4). Consequently, the difference in SOC stock between CT and the Cerrado NV was 14.2 and 6.0 Mg C ha^{-1} in 0–20 cm and 20–40 cm depths, respectively. The difference in SOC stock between CT and the Cerrado NV (20.1 Mg C ha^{-1}) in the 0–40 cm layer represented a 27% decrease by 23 years of CT, with an average rate of loss of 0.88 Mg C ha^{-1} yr^{-1} for the clayey typic Haplustox. In contrast, the SOC stock in the 0–20 and 20–40 cm layers under NT1, NT5, and NT6 was similar to that under the Cerrado NV (Table 4). The SOC stocks in 0–20 cm depth under

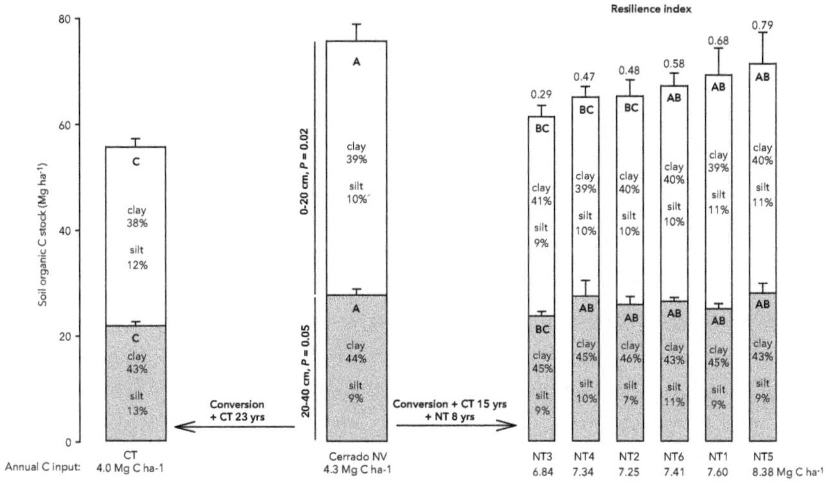

Figure 3 Effects of continuous CT and conversion from CT to NT cropping systems with contrasting C-inputs on soil organic C (Mg ha^{-1}) at the LRV site. Clay and silt %, and annual C input are given per land use. The SOC resilience index ([SOC_{NT}-SOC_{CT}]/[SOC_{NV}-SOC_{CT}]) is given for each CA for 0–40 cm depth interval, where SOC_{NT} is the C stock in the field under NT, SOC_{CT} is the C stock in the field under CT, SOC_{NV} is the C stock in the field under NV. Error bars indicate standard errors of the means. Uppercase letters indicate difference among land use and tillage treatments. The level of significance for each soil depth interval is indicated. Annual C input to soil under Cerrado NV from (Corbeels et al., 2006). Source: Sá et al. (2015).

NT, with a predominance of grasses during the dry season (NT1, NT5, NT6), were significantly higher ($P<0.05$) compared with those under CT. The difference in SOC stock to 20-cm depth between all CA and CT ranged on average from 3.9 to 10.4 Mg C ha^{-1}, representing average sequestration rates of 0.48–1.30 Mg C ha^{-1} yr^{-1}. These rates of SOC sequestration for 0–40 cm depth increased from 0.73 (NT3) to 1.98 (NT5) Mg C ha^{-1} yr^{-1}, and represented an increase of 52% more in the 20–40 cm layer, respectively (Sá et al., 2015). No significant differences were reported between any treatments in 40–100 cm depth, but SOC stocks under CT and NT were lower than those under Cerrado NV from 3.1 to 6.0 Mg C ha^{-1}.

4 Soil organic carbon (SOC) restoration and sequestration rates in response to cropping systems under Conservation Agriculture (CA)

At the PG site, concentrations of SOC and N were highly stratified with depth under NT while those under CT and MT were uniform in the 0–20 cm layer due to the mixing effect of successive plowing (Sá and Lal, 2009). The reduction of SOC in the surface layers of soil under CT occurs through a high rate of oxidation process because of soil disturbance and the illuviation of SOC into the subsoil (Osher et al., 2003). Although no significant differences were observed, soil under CT and NT stored 16.9 and 16.6 Mg ha^{-1} more SOC (Sá et al., 2015), respectively, than that under NV for 40–60, 60–80, and 80–100 cm depths (Table 4). Differences in SOC stock among tillage systems and NV at depth may be attributed to: (i) relatively more translocation of labile compounds from surface to subsoil in arable land vis-à-vis NV by increased volume of water moving through the soil profile in each rainfall event (Sollins and Radulovich, 1988); (ii) high clay content in CT soil (Table 1) contributing to a higher adsorptive capacity of organic compounds on clay minerals (Feller and Beare, 1997); and (iii) possible differences in SOC stock under NV and cropland at the time of initiating the experiment in 1989. In addition, SOC stock in subsoil may be affected by land-use changes and agricultural practices. Osher et al. (2003) reported that some of the C depleted in the topsoil under sugarcane (*Saccharum officinarum*) plantation was translocated into the subsoil, and concluded that loss of SOC upon land-use conversion can be over-estimated if gains in subsoil are not considered. Determination of ^{13}C natural abundance for NV and a tillage chronosequence, as well as the quantification of stable and labile SOC fractions reported by Sá et al. (2001), confirm these assumptions.

The rates of SOC sequestration in the present study (Sá et al., 2015; Tivet et al., 2013b) under NT cropping systems are in accord with several other studies in Brazil (Table 5). Table 5 also indicates that deeper soil layers may act as important C sink. The C accumulation in subsoil horizons of well-drained deep soils has received attention in the last years in temperate (Chabbi et al., 2009; Rumpel and Kogel-Knabner, 2011) and more recently on subtropical ecosystems (Boddey et al., 2010; Dick et al., 2013). Although the C content is lower compared to the shallow layers, the subsoil horizons contribute > 50% of the total SOC stocks in temperate ecosystems (Rumpel and Kogel-Knabner, 2011; Lal, 2018) and > 70% in subtropical ecosystems (Dick et al., 2013; De Oliveira Ferreira et al., 2016; Table 5).

At the LRV site, the SOC stocks at different depths in Cerrado NV are in accord with those of Batlle-Bayer et al. (2010). The SOC stocks in the 20-cm layer have been reported at 54 Mg ha^{-1} for a typic Haplustox (Bayer et al., 2006b), 68.1 Mg ha^{-1} for a Geri-Gibbsic Ferrasol (Corbeels et al., 2006), and 46–63 Mg ha^{-1} for some Red Latosol (Chapuis-Lardy et al., 2002).

The rate of loss of SOC stock under CT (0.88 Mg C ha^{-1} yr^{-1} in 0–40 cm depth) is in contrast with those observed by Batlle-Bayer et al. (2010), who reported the loss of SOC stock in CT between −0.3 and +0.1 Mg C ha^{-1} yr^{-1}. Zinn et al. (2005) observed no significant changes in SOC stocks for some Oxisols containing either less clay (<500 g kg^{-1}) or more clay (>500 g kg^{-1}) for contrasting land-use systems in the Cerrado region. However, Silva et al. (1994)

Table 5 Carbon sequestration rates for long- and short-term CA systems of tropical and subtropical Oxisols

Site	State	Crop Rotation (last three years) Description	Soil Type USDA-Soil Taxonomy	Soil texture (USDA)	Depth (cm)	Duration of experiment (years)	SOC Sequestration (Mg ha^{-1})	Reference
Tropical								
Lucas do Rio Verde	Mato Grosso	NT-1 (Sb/Mz+Brz - Sb/Mz+Brz - Mz - Sb/Cs)	Typic Haplustox	Clay	0–20	8	1.30	Sá et al. (2015)
Lucas do Rio Verde	Mato Grosso	NT-5 (Sb/Sg+Brz - Sb/G+Brz - M - S/Sg+Cs)	Typic Haplustox	Clay	0–20	8	1.19	Sá et al. (2015)
Lucas do Rio Verde	Mato Grosso	NT-6 (Sb/Mz+Brz - Sb/Mz+Brz - M - S/Mz)	Typic Haplustox	Clay	0–20	8	0.86	Sá et al. (2015)
Lucas do Rio Verde	Mato Grosso	NT-1 (Sb/Mz+Brz - Sb/Mz+Brz - Mz - Sb/Cs)	Typic Haplustox	Clay	0–100	8	1.89	Sá et al. (2015)
Lucas do Rio Verde	Mato Grosso	NT-5 (Sb/Sg+Brz - Sb/G+Brz - M - S/Sg+Cs)	Typic Haplustox	Clay	0–100	8	2.05	Sá et al. (2015)
Lucas do Rio Verde	Mato Grosso	NT-6 (Sb/Mz+Brz - Sb/Mz+Brz - M - S/Mz)	Typic Haplustox	Clay	0–100	8	1.83	Sá et al. (2015)
Planaltina	Brasilia	Sb-Mz/Rc/Sb-Mz/Sb	Typic Haplustox	Clay	0–30	20	0.82	Jantalia et al. (2007)
Planaltina	Brasilia	Sb-Mz/Rc/Sb-Mz/Sb	Typic Haplustox	Clay	0–100	20	1.37	Jantalia et al. (2007)
Luiziânia	Goiás	Sb-Mz/Sb-Mz/Sb-Mz	Typic Hapludox	Sand Clay Loam	0–20	8	0.30	Bayer et al. (2006b)
Montividiu	Goiás	Mt-Sb/Mz/Mt-Sb/Mz/Mt-Sb/Mz	Typic Haplortox	Clay	0–40	12	0.83	Corbeels et al. (2006)

(*Continued*)

Table 5 *(Continued)*

Site	State	Crop Rotation (last three years) Description	Soil Type USDA-Soil Taxonomy	Soil texture (USDA)	Depth (cm)	Duration of experiment (years)	SOC Sequestration (Mg ha^{-1})	Reference
Rio Verde	Goiás	Mt-Sb/Mz/Mt-Sb/Mz/Mt-Sb/Mz	Typic Haplortox	Clay	0–40	12	1.61	Blanchart et al. (2007)
Rio Verde	Goiás	Mt-Sb/Mz/Mt-Sb/Mz/Mt-Sb/Mz	Typic Haplortox	Clay	0–30	12	1.91	Maltas et al. (2007)
Rio Verde	Goiás	Mt-Sb/Mz/Mt-Sb/Mz/Mt-Sb/Mz	Typic Haplortox	Clay	0–40	13	1.29	Corbeels et al. (2006)
Rio Verde	Goiás	Mt-Sb/Mz/Mt-Sb/Mz/Mt-Sb/Mz	Typic Haplortox	Clay	0–40	13	1.48	Corbeels et al. (2006)
Rio Verde	Goiás	Mt-Sb/Mz/Mt-Sb/Mz/Mt-Sb/Mz	Typic Haplortox	Clay	0–40	21	1.61	Corbeels et al. (2006)
Dourados	Mato Grosso do Sul	(S2Psd) Sb-Pst/Sb-Pst/Sb-Pst	Haplustox	Clay	0–20	11	0.55	Salton et al. (2012)
Marcaju	Mato Grosso do Sul	(S2Psd) Sb-Pst/Sb-Pst/Sb-Pst	Haplustox	Clay	0–20	11	0.88	Salton et al. (2012)
Costa Rica	Mato Grosso do Sul	Sb-Mz/Sb-Mz/Sb-Mz	Typic Hapludox	Clay	0–20	5	0.60	Bayer et al. (2006a)
Subtropical								
Jaboticabal	São Paulo	Sb-Mz/Sb-Mz/Sb-Mz	Typic Hapludox	Sand Clay	0–30	10	0.74	Rossetti and Centurion (2015)
Londrina	Paraná	W-Sb/O-Sb/Hv-Mz	Rhodic Eutrudox	Heavy clay	0–30	20	0.80	Babujia et al. (2010)

Londrina	Paraná	W-Sb/O-Sb/Hv-Mz	Rhodic Eutrudox	Heavy clay	0–60	20	0.80	Babujia et al. (2010)
Ponta Grossa	Paraná	O-Sb/W-Sb/O-Mz	Typic Hapludox	Clay	0–20	16	0.84	Sá et al. (2014)
Ponta Grossa	Paraná	O-Sb/W-Sb/O-Mz	Typic Hapludox	Sandy clay	0–20	20	0.83	De Oliveira Ferreira et al. (2012)
Ponta Grossa	Paraná	W-Sb/O+Hv - Mz/W-Sb	Typic Hapludox	Clay	0–20	20	0.83	Sá et al. (2014)
Tibagi	Paraná	W-Sb/O-Sb/O-Mz	Typic Hapludox	Clay	0–20	22	0.81	Sá et al. (2001)
Tibagi	Paraná	W-Sb/O-Sb/O-Mz	Typic Hapludox	Clay	0–40	22	0.91	Sá et al. (2001)
Ponta Grossa	Paraná	W-Sb/O-Sb/O-Mz	Typic Hapludox	Heavy clay	0–20	20	0.76	De Oliveira Ferreira et al. (2012)
Ponta Grossa	Paraná	W-Sb/O-Sb/O-Mz	Typic Hapludox	Heavy clay	0–100	29	1.62	Sá et al. (2015)
Lages	Santa Catarina	O-Mz/W-Sb/O-Bn	Ultisol Humic	Clay	0–20	8	1.06	Bayer et al. (2000)
Campos Novos	Santa Catarina	Oat-Sb/Hv-Mz/O-B	Hapludox	Heavy clay	0–20	7	0.80	Amado et al. (2006)
Passo Fundo	Rio Grande do Sul	O-SG/Bl-Sb/HV-Mz	Typic Haplortox	Heavy clay	0–30	15	0.25	Boddey et al. (2010)
Passo Fundo	Rio Grande do Sul	O-SG/Bl-Sb/HV-Mz	Typic Haplortox	Heavy clay	0–100	15	1.11	Boddey et al. (2010)
Santo Angelo	Rio Grande do Sul	O-Sb/OR-Mz/Wh-Sb	Typic Haplortox	Heavy clay	0–30	26	0.35	Boddey et al. (2010)

(*Continued*)

Table 5 (*Continued*)

Site	State	Crop Rotation (last three years) Description	Soil Type USDA-Soil Taxonomy	Soil texture (USDA)	Depth (cm)	Duration of experiment (years)	SOC Sequestration (Mg ha^{-1})	Reference
Santo Angelo	Rio Grande do Sul	O-Sb/OR-Mz/Wh-Sb	Typic Haplortox	Heavy clay	0–100	26	0.85	Boddey et al. (2010)
Cruz Alta	Rio Grande do Sul	(R2)O+Hv- Mz/W-Sb/O-Sb	Typic Haplortox	Heavy clay	0–30	17	0.21	Boddey et al. (2010)
Cruz Alta	Rio Grande do Sul	(R2)O+Hv- Mz/W-Sb/O-Sb	Typic Haplortox	Heavy clay	0–100	17	0.52	Boddey et al. (2010)
Cruz Alta	Rio Grande do Sul	CT = O-Sb; O + HV - Mz/W-Sb/O-Sb	Hapludox	Clay	0–20	19	0.51	Amado et al. (2006)
Passo Fundo	Rio Grande do Sul	(R2) W-Sb/Hv-Mz	Typic Haplortox	Heavy clay	0–30	13	0.42	Sisti et al. (2004)
Passo Fundo	Rio Grande do Sul	(R2) W-Sb/Hv-Mz	Typic Haplortox	Heavy clay	0–100	13	1.30	Sisti et al. (2004)
Passo Fundo	Rio Grande do Sul	(R2) W-Sb/Hv-Mz	Typic Haplortox	Heavy clay	0–30	13	0.70	Sisti et al. (2004)
Passo Fundo	Rio Grande do Sul	(R3) W-Sb/O-Sb/Hv-Mz	Typic Haplortox	Heavy clay	0–100	13	1.28	Sisti et al. (2004)
Santa Maria	Rio Grande do Sul	CT = bare soil; NT = Ryegrass-Corn	Paleudalf	Sandy loam	0–20	10	0.59	Amado et al. (2006)
Santa Maria	Rio Grande do Sul	CT = bare soil; NT = VvB - Mz/ VvB-Mz	Paleudalf	Sandy loam	0–20	10	1.02	Amado et al. (2006)
Eldorado Sul	Rio Grande do Sul	CT = O-Sb; O + Hv - Mz/W-Sb/O-Sb	Paleudult	Sandy clay loam	0–20	9	0.81	Bayer et al. (2000)

Eldorado Sul	Rio Grande do Sul	CT=O-Sb; O+Hv - Mz/W-Sb/O-Sb	Paleudult	Sandy clay loam	0–30	9	1.33	Bayer et al. (2000)
Eldorado Sul	Rio Grande do Sul	Lablab+maize [L+M]	Paleudult	Sandy clay loam	0–17.5	17	1.09	Dieckow et al. (2005)
Eldorado Sul	Rio Grande do Sul	Lablab+maize [L+M]	Paleudult	Sandy clay loam	0–17.5	17	1.73	Dieckow et al. (2005)
Eldorado Sul	Rio Grande do Sul	Pigeon pea+maize [P+M]	Paleudult	Sandy clay loam	0–17.5	17	1.21	Dieckow et al. (2005)
Eldorado Sul	Rio Grande do Sul	Pigeon pea+maize [P+M]	Paleudult	Sandy clay loam	0–17.5	17	1.95	Dieckow et al. (2005)
Eldorado Sul	Rio Grande do Sul	CT=O-M; NT=O-M	Paleudult	Sandy clay loam	0–20	15	0.12	Amado et al. (2006)
Eldorado Sul	Rio Grande do Sul	CT=Oat-Mz; NT=O+Hv/ Mz+Pp	Paleudult	Sandy clay loam	0–20	15	0.50	Amado et al. (2006)

reported losses of 41% (clay soils), 76% (loamy soils), and 80% (sandy soils) of the original SOC stock after five years of heavy harrowing for cultivation of soybean. Bayer et al. (2006b) reported a loss of 15% of SOC in a sandy clay loam (350 g clay kg^{-1}) sloping land (12%) managed under CT for more than 20 years. Neto et al. (2010) observed that land-use conversion induced a total loss of around 15 Mg C ha^{-1} after 10 years of continuous CT. Similarly, Séguy et al. (2006) reported an average loss of 1.6 Mg C ha^{-1} yr^{-1} following the conversion of Cerrado NV to soybean monoculture for 10 years in a soil located in vicinity of the LRV site.

As much as 32–37% of the total SOC stock to 1-m depth was present in the 0–20-cm layer under NT cropping systems. No significant differences in SOC stock were observed among the CA, whereas a difference of almost 10.0 Mg C ha^{-1} was observed between the maximum and minimum SOC stock. In addition, more SOC stocks in 0–40 cm depth (Table 3) were observed in soil under NT which received a high amount of dry biomass annually (NT1, NT5, NT6). Trends of differences in N stock among the CA were similar to those observed for the SOC stock, with NT1, NT5, and NT6 containing relatively higher stocks than other systems (Fig. 3). The rate of SOC sequestration under some NT (i.e., NT5, NT6) in comparison with CT increased significantly with increase in soil depths. With reference to the SOC sequestration in different depths (Table 3, Sá et al., 2013b), approximately 65% of the sequestration occurred in the 0–20 cm depth and the remainder (i.e., 35%) in the 20–40 cm depth. Vegetation types, in combination with climate and mineralogy, affect the vertical distribution of SOC (Jobbagy and Jackson, 2000), and these CA have distinct imprints on the nature and depth distribution of SOC. Although additional research is warranted, SOC sequestration in subsoil could be attributed to the illuviation of dissolved organic C (Lorenz and Lal, 2005; Sá and Lal, 2009), higher SOC rhizodeposition in the soil profile by deep rooting systems of grains/forage grasses (*B. ruziziensis*, sorghum, finger millet) and legumes (*Crotalaria* sp., *C. cajan*) (Séguy et al., 2006), and enhancement of vertical distribution through bioturbation (Wilkinson et al., 2009) which influences SOC dynamics (Lavelle et al., 2006). In addition, these species enhance root/arbuscular mycorrhizal association that enhance exudates, increase soil aggregation (Rillig, 2004), and accentuate the C allocation to the whole root system (Jones et al., 1991) with a significant proportion diverted to the fungal component (Johnson et al., 2002).

Additional research, including determination of ^{13}C natural abundance for NV, CT, and NT soils, is needed to draw valid conclusions and to demonstrate that diverse biomass-C inputs under NT are efficient in terms of SOC sequestration even below 20-cm depth. Such thematic research is essential to identifying cropping systems which enhance the SOC stock, improve soil fertility, enrich the diversity and abundance of microflora and fauna communities (Hungria et al., 2009), improve porosity and hydraulic conductivity (Cassaro et al., 2011), increase aggregation (Madari et al., 2005), and strengthen nutrient cycling (Chapuis-Lardy et al., 2002). In addition, the SOC concentration decreases sharply with depth (Table 4), while the clay content follows an opposite trend. The specific surface area and thus the adsorptive capacity of minerals in the subsoil are higher than those in the surface layer. Thus, increasing nutritional reserves in subsoil is an important strategy that can be effective in increasing SOC accumulation (Briedis et al., 2016).

In an incubation experiment, Briedis et al. (2016) studied 3 Oxisols (Parana state, southern Brazil) and found that nutrient scarcity in deep soil layers; especially that of P, Ca^{2+}, and Mg^{2+}; was the driving force limiting SOC accumulation. Also in Southern Brazil (Rio Grande do Sul state), De Oliveira Ferreira et al. (2018b) studied six Oxisols (>20 yrs. under continuous no-tillage) and found that for areas with the predominance of soybean in the cropping system, higher levels of Al^{3+} and lowest of Mg^{2+} and P, the SOC recovery was smaller. On the other hand, the same authors found that in sites with medium/high cropping intensity, lower levels of Al^{3+} and higher levels of P, Ca^{2+}, M^{2+}, and K^{+} reflected in larger SOC recovery. That is, the factors that drive SOC recovery in long term no-till are proper nutrient management associated with high residue C input.

5 Conclusion

The SOC stock was drastically reduced by the conversion of NV to agro-ecosystem by a continuous use of plow-based CT leading to a depletion of 0.58 and 0.67 Mg C ha^{-1} y^{-1} in the 0- to 20-cm depth at the PG and LRV sites, respectively. Significant differences in the SOC stocks to the 20-cm depth between CT and CA cropping systems, and the rate of SOC sequestration of 0.59 Mg C ha^{-1} y^{-1} in the subtropical region and 0.48–1.30 Mg C ha^{-1} y^{-1} in the tropical Cerrado region, were observed (Sá et al., 2015).

The results support the hypothesis that SOC restoration is determined by the intensification of the NT cropping systems through the continuous input of biomass-C to maintain the C flow in the soil supplying the process of soil C transformations and driving the production systems.

6 Acronyms

CA:	Conservation Agriculture
NT:	No-till
PG:	Ponta Grossa, PR, Brazil
LRV:	Lucas do Rio Verde, MT, Brazil
CT:	Conventional tillage

SOC: Soil organic carbon
NV: Native vegetation
C: Carbon
N: Nitrogen
TOC: Total organic carbon
SOM: Soil organic matter
MAT: Mean annual temperature
MT: Minimum tillage
RI: Resilience Index
CT-S/Ct: Soybean alternating with cotton under conventional tillage
NT1: Soybean in the summer followed by maize (*Zea mays* L.) + *Brachiaria ruziziensis* as second crop
NT2: Soybean in the summer followed by finger millet (*Eleusine coracana*) or finger millet + pigeon pea (*Cajanus cajan*) as second crop
NT3: Soybean in the summer followed by finger millet + pigeon pea or finger millet + *Crotalaria spectabilis* as second crop
NT4: Soybean in the summer followed by finger millet + *C. spectabilis* or sunflower (*Helianthus annuus*) + *B. ruziziensis* as second crop
NT5: Soybean in the summer followed by sorghum (*Sorghum bicolor*) + *B. ruziziensis* as second crop
NT6: Soybean in the summer followed by millet (*Pennisetum glaucum*) or maize + *B. ruziziensis* as second crop.

7 References

Amado, T. J., Bayer, C., Conceicao, P. C., Spagnollo, E., de Campos, B. H. and da Veiga, M. 2006. Potential of carbon accumulation in no-till soils with intensive use and cover crops in southern Brazil. *Journal of Environmental Quality* 35(4), 1599–607. doi:10.2134/jeq2005.0233.

Associação de Plantio Direto no Cerrado – APDC. 2007. *Plantio direto tropical*. Brasília, DF. Disponível em: http://www.apdc.com.br (Accessed 4 June 2019).

Babujia, L., Hungria, M., Franchini, J. and Brookes, P. 2010. Microbial biomass and activity at various soil depths in a Brazilian oxisol after two decades of no-tillage and conventional tillage. *Soil Biol. and Biochem* 42, 2174-2181.

Batlle-Bayer, L., Batjes, N. H. and Bindraban, P. S. 2010. Changes in organic carbon stocks upon land use conversion in the Brazilian Cerrado: a review. *Agriculture, Ecosystems and Environment* 137(1–2), 47–58. doi:10.1016/j.agee.2010.02.003.

Bayer, C., Martin-Neto, L., Mielniczuk, J. and Ceretta, C. A. 2000. Effect of no-till cropping systems on soil organic matter in a sandy clay loam Acrisol from Southern Brazil monitored by electron spin resonance and nuclear magnetic resonance. *Soil and Tillage Research* 53(2), 95–104. doi:10.1016/S0167-1987(99)00088-4.

Bayer, C., Martin-Neto, L., Mielniczuk, J. and Pavinato, A. 2004. Carbon storage in labile fractions of soil organic matter in a tropical no-tillage Oxisol. *Pesquisa Agropecuária Brasileira* 39(7), 677–83. doi:10.1590/S0100-204X2004000700009.

Bayer, C., Martin-Neto, L., Mielniczuk, J., Pavinato, A. and Dieckow, J. 2006b. Carbon sequestration in two Brazilian Cerrado soils under no-till. *Soil and Tillage Research* 86(2), 237–45. doi:10.1016/j.still.2005.02.023.

Bayer, C., Lovato, T., Dieckow, J., Zanatta, J. A. and Mielniczuk, J. 2006a. A method for estimating coefficients of soil organic matter dynamics based on long-term experiments. *Soil and Tillage Research* 91(1-2), 217–26. doi:10.1016/j.still.2005.12.006.

Beare, M. H., Hendrix, P. F., Cabrera, M. L. and Coleman, D. C. 1994. Aggregate-protected and unprotected organic matter pools in conventional-and no-tillage soils. *Soil Science Society of America Journal* 58(3), 787–95. doi:10.2136/sssaj1994.03615995005800030021x.

Blake, G. R. and Hartge, K. H. 1986. *Bulk Density. Methods of Soil Analysis, Part I. Physical and Mineralogical Methods*. Agronomy Monograph no. 9 (2nd edn.) (Klute, A., Ed.). ASA SSSA, Madison, WI, pp. 363–75.

Blanchart, E., Bernoux, M., Sardi, X. et al. 2007. Effect of direct seeding mulch-based system on soil carbon storage in Central Brazil. *Agric. Consp. Sci.* 72, 81–87.

Boddey, R. M., Jantalia, C. P., Conceicao, P. C., Zanatta, J. A., Bayer, C., Mielniczuk, J., Dieckow, J., Dos Santos, H. P., Denardin, J. E., Aita, C., Giacomini, S. J., Alves, B. J. R. and Urquiaga, S. 2010. Carbon accumulation at depth in Ferralsols under zero-till subtropical agriculture. *Global Change Biology* 16(2), 784-95. doi:10.1111/j.1365-2486.2009.02020.x.

Briedis, C., de Moraes Sá, J. C., Lal, R., Tivet, F., De Oliveira Ferreira, A., Franchini, J. C., Schimiguel, R., Da Cruz Hartman, D. and Santos, J. Zd 2016. Can highly weathered soils under conservation agriculture be C saturated? *Catena* 147, 638–49. doi:10.1016/j.catena.2016.08.021.

Carvalho, J. L. N., Cerri, C. E. P., Feigl, B. J., Piccolo, M. C., Godinho, V. P. and Cerri, C. C. 2009. Carbon sequestration in agricultural soils in the Cerrado region of the Brazilian Amazon. *Soil and Tillage Research* 103(2), 342–9. doi:10.1016/j.still.2008.10.022.

Cassaro, F., Borkowski, A., Pires, L., Rosa, J. and Saab, S. 2011. Characterization of a Brazilian clayey soil submitted to conventional and no-tillage management practices. *Soil and Tillage Research* 111(2), 175–9.

Cerri, C. E. P., Sparovek, G., Bernoux, M., Easterling, W. E., Melillo, J. M. and Cerri, C. C. 2007. Tropical agriculture and global warming: impacts and mitigation options. *Scientia Agricola* 64(1), 83–99. doi:10.1590/S0103-90162007000100013.

Chabbi, A., Kögel-Knabner, I. and Rumpel, C. 2009. Stabilised carbon in subsoil horizons is located in spatially distinct parts of the soil profile. *Soil Biology and Biochemistry* 41(2), 256–61. doi:10.1016/j.soilbio.2008.10.033.

Chapuis-Lardy, L., Brossard, M., Assad, M. L. L. and Laurent, J. Y. 2002. Carbon and phosphorus stocks of clayey Ferralsols in Cerrado native and agroecosystems, Brazil. *Agriculture, Ecosystems and Environment* 92(2-3), 147–58. doi:10.1016/S0167-8809(01)00303-6.

Corbeels, M., Scopel, E., Cardoso, A., Bernoux, M., Douzet, J. and Neto, M. S. 2006. Soil carbon storage potential of direct seeding mulch based cropping systems in the Cerrados of Brazil. *Global Change Biology* 12(9), 1773–87. doi:10.1111/j.1365-2486.2006.01233.x.

De Oliveira Ferreira, A., Sá, J. C.d.M., Balarezo Giarola, N. F., Harms, M. G., Miara, S., Bavoso, M. A., Briedis, C. and Netto, C. Q. 2011. Variation in aggregate-tensile strength as a function of carbon content of two soils in the region of Campos Gerais. *Revista Brasileira De Ciencia Do Solo* 35, 437–46.

De Oliveira Ferreira, A., Sá, J. CdM., Harms, M. G., Miara, S., Briedis, C., Quadros Netto, C., Santos, J. Bd and Canalli, L. B. 2012. Carbon balance and crop residue management in dynamic equilibrium under a no-till system in Campos Gerais. *Revista Brasileira De Ciencia Do Solo* 36(5), 1583–90. doi:10.1590/S0100-06832012000500022.

De Oliveira Ferreira, A., Amado, T., Rice, C. W., Ruiz Diaz, D. A. R., Keller, C. and Inagaki, T. M. 2016. Can no-till grain production restore soil organic carbon to levels natural grass in a subtropical Oxisol? *Agriculture, Ecosystems and Environment* 229, 13–20. doi:10.1016/j.agee.2016.05.016.

De Oliveira Ferreira, A., Sá, J. CdM., Lal, R., Tivet, F., Briedis, C., Inagaki, T. M., Goncalves, D. R. P. and Romaniw, J. 2018a. Macroaggregation and soil organic carbon restoration in a highly weathered Brazilian Oxisol after two decades under no-till. *Science of the Total Environment* 621, 1559–67. doi:10.1016/j.scitotenv.2017.10.072.

De Oliveira Ferreira, A., Amado, T. J. C., Rice, C. W., Diaz, D. A. R., Briedis, C., Inagaki, T. M. and Goncalves, D. R. P. 2018b. Driving factors of soil carbon accumulation in oxisols in long-term no-till systems of South Brazil. *Science of the Total Environment* 622–623, 735–42.

Dick, D. P., Reis, C. S., Bayer, C. and Caldas, J. S. 2013. Carbon sequestration in subtropical Oxisol profiles: retention capacity and effect of soil management. In: *Functions of Natural Organic Matter in Changing Environment*. Springer, pp. 353–8.

Dieckow, J., Mielniczuk, J., Knicker, H., Bayer, C., Dick, D. P. and Kogel-Knabner, I. 2005. Composition of organic matter in a subtropical Acrisol as influenced by land use, cropping and N fertilization, assessed by CPMAS C-13 NMR spectroscopy. *European Journal of Soil Science* 56, 705–15.

Dieckow, J., Bayer, C., Conceicao, P. C., Zanatta, J. A., Martin-Neto, L., Milori, D. B. M., Salton, J. C., Macedo, M. M., Mielniczuk, J. and Hernani, L. C. 2009. Land use, tillage, texture and organic matter stock and composition in tropical and subtropical Brazilian soils. *European Journal of Soil Science* 60(2), 240–9. doi:10.1111/j.1365-2389.2008.01101.x.

Feller, C. and Beare, M. H. 1997. Physical control of soil organic matter dynamics in the tropics. *Geoderma* 79(1–4), 69–116. doi:10.1016/S0016-7061(97)00039-6.

Goncalves, D., Leite, W. C., Brinatti, A. M., Saab, SdC., Iarosz, K. C., Masearenhas, Y. P., Carneiro, P. I. B. and Rosa, J. A. 2008. Mineralogy of a red latosol under different management systems for twenty-four years. *Revista Brasileira De Ciencia Do Solo* 32(spe), 2647–52. doi:10.1590/S0100-06832008000700006.

Hungria, M., Franchini, J. C., Brandao-Junior, O., Kaschuk, G. and Souza, R. A. 2009. Soil microbial activity and crop sustainability in a long-term experiment with three soil-tillage and two crop-rotation systems. *Applied Soil Ecology* 42(3), 288–96. doi:10.1016/j.apsoil.2009.05.005.

Jantalia, C. P., Resck, D. V. S., Alves, B. J. R., Zotarelli, L., Urquiaga, S. and Boddey, R. M. 2007. Tillage effect on C stocks of a clayey Oxisol under a soybean-based crop rotation in the Brazilian Cerrado region. *Soil and Tillage Research* 95(1–2), 97–109. doi:10.1016/j.still.2006.11.005.

Jobbagy, E. G. and Jackson, R. B. 2000. The vertical distribution of soil organic carbon and its relation to climate and vegetation. *Ecological Applications* 10(2), 423–36. doi:10.1890/1051-0761(2000)010[0423:TVDOSO]2.0.CO;2.

Johnson, D., Leake, J. R., Ostle, N., Ineson, P. and Read, D. J. 2002. *In situ* (CO_2)-C-13 pulse-labelling of upland grassland demonstrates a rapid pathway of carbon flux from arbuscular mycorrhizal mycelia to the soil. *New Phytologist* 153(2), 327–34. doi:10.1046/j.0028-646X.2001.00316.x.

Jones, M. D., Durall, D. M. and Tinker, P. B. 1991. Fluzes of carbon and phosphorus between symbionts in willow ectomycorrhizas and their changes with time. *New Phytologist* 119(1), 99–106. doi:10.1111/j.1469-8137.1991.tb01012.x.

Lal, R. 1976. No-tillage effects on soil properties under different crops in Western Nigeria. *Soil Science Society of America Journal* 40(5), 762–8. doi:10.2136/sssaj1976.03615 995004000050039x.

Lal, R. 2004. Soil carbon sequestration to mitigate climate change. *Geoderma* 123, 1–22.

Lal, R. 2006. Enhancing crop yields in developing countries through restoration of the soil organic pool in agricultural lands. *Land. Degrad. and Develop.* 17, 197–209.

Lal, R. 2018. Digging deeper: a holistic perspective of factors affecting soil organic carbon sequestration in agroecosystems. *Global Change Biology* 24(8), 3285–301. doi:10.1111/gcb.14054.

Lavelle, P., Decaens, T., Aubert, M., Barot, S., Blouin, M., Bureau, F., Margerie, P., Mora, P. and Rossi, J.-P. 2006. Soil invertebrates and ecosystem services. *European Journal of Soil Biology* 42, S3–15. doi:10.1016/j.ejsobi.2006.10.002.

Lorenz, K. and Lal, R. 2005. The depth distribution of soil organic carbon in relation to land use and management and the potential of carbon sequestration in subsoil horizons. *Advances in Agronomy* 88, 35–66. Elsevier Academic Press Inc, San Diego. doi:10.1016/S0065-2113(05)88002-2.

Madari, B., Machado, P. L. O. A., Torres, E., de Andrade, A. G. and Valencia, L. I. O. 2005. No tillage and crop rotation effects on soil aggregation and organic carbon in a rhodic ferralsol from southern Brazil. *Soil and Tillage Research* 80(1–2), 185–200. doi:10.1016/j.still.2004.03.006.

Maltas, A., Corbeels, M., Scopel, E. et al. 2007. Long-term effects of continuous direct seeding mulch-based cropping systems on soil nitrogen supply in the Cerrado region of Brazil. *Plant and Soil* 298(1), 161–177.

Neto, M. S., Scopel, E., Corbeels, M., Cardoso, A. N., Douzet, J. M., Feller, C., Piccolo, MdC., Cerri, C. C. and Bernoux, M. 2010. Soil carbon stocks under no-tillage mulch-based cropping systems in the Brazilian Cerrado: an on-farm synchronic assessment. *Soil and Tillage Research* 110(1), 187–95. doi:10.1016/j.still.2010.07.010.

Ogle, S. M., Swan, A. and Paustian, K. 2012. No-till management impacts on crop productivity, carbon input and soil carbon sequestration. *Agriculture, Ecosystems and Environment* 149, 37–49. doi:10.1016/j.agee.2011.12.010.

Osher, L. J., Matson, P. A. and Amundson, R. 2003. Effect of land use change on soil carbon in Hawaii. *Biogeochemistry* 65(2), 213–32. doi:10.1023/A:1026048612540.

Paustian, K., Andren, O., Janzen, H. H., Lal, R., Smith, P., Tian, G., Tiessen, H., Van Noordwijk, M. and Woomer, P. L. 1997. Agricultural soils as a sink to mitigate CO_2 emissions. *Soil Use and Management* 13(s4), 230–44. doi:10.1111/j.1475-2743.1997.tb00594.x.

Rillig, M. C. 2004. Arbuscular mycorrhizae, glomalin, and soil aggregation. *Canadian Journal of Soil Science* 84(4), 355–63. doi:10.4141/S04-003.

Rossetti, K. and Centurion, J. 2015. Carbon stock and physical attributes of an oxisol in chronosequence under different management practices. *Rev. Bras. Eng. Agric.* 19(3), 252–58.

Rumpel, C. and Kogel-Knabner, I. 2011. Deep soil organic matter-a key but poorly understood component of terrestrial C cycle. *Plant and Soil* 338(1-2), 143-58. doi:10.1007/s11104-010-0391-5.

Sá, J. CdM. and De Oliveira Ferreira, A. 2018. The soil science in the evolution of no-till system in Brazil. *SBCS Bulletim (Brazilian Society of Soil Science)* 44, 54-7.

Sá, J. CdM. and Lal, R. 2009. Stratification ratio of soil organic matter pools as an indicator of carbon sequestration in a tillage chronosequence on a Brazilian Oxisol. *Soil and Tillage Research* 103(1), 46-56. doi:10.1016/j.still.2008.09.003.

Sá, J. CdM., Cerri, C. C., Dick, W. A., Lal, R., Filho, S. P. V., Piccolo, M. C. and Feigl, B. E. 2001. Organic matter dynamics and carbon sequestration rates for a tillage chronosequence in a Brazilian Oxisol. *Soil Science Society of America Journal* 65(5), 1486-99. doi:10.2136/sssaj2001.6551486x.

Sá, J. CdM., Bürkner dos Santos, J., Lal, R., de Moraes, A., Tivet, F., Machado Sá, M. F., Briedis, C., de Oliveira Ferreira, A., Eurich, G., Farias, A. and Friedrich, T. 2013. Soil-specific inventories of landscape carbon and nitrogen stocks under no-till and native vegetation to estimate carbon offset in a subtropical ecosystem. *Soil Science Society of America Journal* 77(6), 2094-110. doi:10.2136/sssaj2013.01.0007.

Sá, J. CdM., Tivet, F., Lal, R., Briedis, C., Hartman, D. C., dos Santos, J. Z. and dos Santos, J. B. 2014. Long-term tillage systems impacts on soil C dynamics, soil resilience and agronomic productivity of a Brazilian Oxisol. *Soil and Tillage Research* 136, 38-50. doi:10.1016/j.still.2013.09.010.

Sá, J. CdM., Séguy, L., Tivet, F., Lal, R., Bouzinac, S., Borszowskei, P. R., Briedis, C., dos Santos, J. B., da Cruz Hartman, D., Bertoloni, C. G., Rosa, J. and Friedrich, T. 2015. Carbon depletion by plowing and its restoration by no-till cropping systems in Oxisols of subtropical and tropical agro-ecoregions in Brazil. *Land Degradation and Development* 26(6), 531-43. doi:10.1002/ldr.2218.

Séguy, L., Bouzinac, S. and Husson, O. 2006. Direct-seeded tropical soil systems with permanent soil cover: learning from Brazilian experience. In: Uphoff, N., Ball, A. S., Fernandes, E., Herren, H., Husson, O., Laing, M., Palm, C., Pretty, J. and Sanchez, P. (Eds), *Biological Approach to Sustainable Soil Systems*. CRC Press, Taylor & Francis, pp. 323-42.

Silva, J. E., Lemainski, J. and Resck, D. V. S. 1994. Perdas de matéria orgânica e suas relações com a capacidade de troca catiônica em solos da região de cerrados do oeste baiano. *Revista Brasileira de Ciência do Solo* 18, 541-7.

Sisti, C. P. J., dos Santos, H. P., Kohhann, R., Alves, B. J. R., Urquiaga, S. and Boddey, R. M. 2004. Change in carbon and nitrogen stocks in soil under 13 years of conventional or zero tillage in southern Brazil. *Soil and Tillage Research* 76(1), 39-58. doi:10.1016/j.still.2003.08.007.

Six, J., Elliott, E. T. and Paustian, K. 2000. Soil macroaggregate turnover and microaggregate formation: a mechanism for C sequestration under no-tillage agriculture. *Soil Biology and Biochemistry* 32(14), 2099-103. doi:10.1016/S0038-0717(00)00179-6.

Six, J., Conant, R. T., Paul, E. A. and Paustian, K. 2002a. Stabilization mechanisms of soil organic matter: implications for C-saturation of soils. *Plant and Soil* 241(2), 155-76. doi:10.1023/A:1016125726789.

Six, J., Feller, C., Denef, K., Ogle, S. M., Sá, J. CdM. and Albrecht, A. 2002b. Soil organic matter, biota and aggregation in temperate and tropical soils - Effects of no-tillage. *Agronomie* 22, 755-75.

Six, J., Frey, S. D., Thiet, R. K. and Batten, K. M. 2006. Bacterial and fungal contributions to carbon sequestration in agroecosystems. *Soil Science Society of America Journal* 70(2), 555–69. doi:10.2136/sssaj2004.0347.

Sollins, P. and Radulovich, R. 1988. Effects of soil physical structure on solute transport in a weathered tropical soil. *Soil Science Society of America Journal* 52(4), 1168–73. doi:10.2136/sssaj1988.03615995005200040050x.

Tivet, F., Sá, J. C.d.M., Lal, R., Borszowskei, P. R., Briedis, C., dos Santos, J. B., Sa, M. F. M., Hartman, D. D., Eurich, G., Farias, A., Bouzinac, S. and Seguy, L. 2013a. Soil organic carbon fraction losses upon continuous plow-based tillage and its restoration by diverse biomass-C inputs under no-till in sub-tropical and tropical regions of Brazil. *Geoderma* 209–210, 214–25.

Tivet, F., de Moraes Sá, J. C., Lal, R., Briedis, C., Borszowskei, P. R., dos Santos, J. B., Farias, A., Eurich, G., Hartman, DdC., Nadolny Junior, M., Bouzinac, S. and Séguy, L. 2013b. Aggregate C depletion by plowing and its restoration by diverse biomass-C inputs under no-till in sub-tropical and tropical regions of Brazil. *Soil and Tillage Research* 126, 203–18. doi:10.1016/j.still.2012.09.004.

Virto, I., Barre, P., Burlot, A. and Chenu, C. 2012. Carbon input differences as the main factor explaining the variability in soil organic C storage in no-tilled compared to inversion tilled agrosystems. *Biogeochemistry* 108(1–3), 17–26. doi:10.1007/s10533-011-9600-4.

Wilkinson, M., Richards, P. and Humphreys, G. 2009. Breaking ground: pedological, geological and ecological implications of soil bioturbation. *Earth-Science Reviews* 97(1), 257–72.

Wright, S. F. and Upadhyaya, A. 1998. A survey of soils for aggregate stability and glomalin, a glycoprotein produced by hyphae of arbuscular mycorrhizal fungi. *Plant and Soil* 198(1), 97–107. doi:10.1023/A:1004347701584.

Zinn, Y. L., Lal, R. and Resck, D. V. S. 2005. Texture and organic carbon relations described by a profile pedotransfer function for Brazilian Cerrado soils. *Geoderma* 127(1–2), 168–73. doi:10.1016/j.geoderma.2005.02.010.

Chapter 8

Carbon management practices and benefits in Conservation Agriculture systems: soil organic carbon fraction losses and restoration

João Carlos de Moraes Sá, State University of Ponta Grossa, Brazil; Florent Tivet, CIRAD, France; Rattan Lal, The Ohio State University, USA; Ademir de Oliveira Ferreira, Federal Rural University of Pernambuco, Brazil; Clever Briedis, Brazilian Agricultural Research Corporation, Agricultural Instrumentation Center, Brazil; Thiago Massao Inagaki, Technical University of Munich, Germany; and Daniel Potma Gonçalves and Jucimare Romaniw, State University of Ponta Grossa, Brazil

1 Introduction

A better understanding of the effects of land use change and agricultural management on long-term C sequestration necessitates separation of soil organic carbon (SOC) into fractions. The SOC pool is highly diverse, composed of fractions with contrasting turnover times, and stabilized or protected against microbial decomposition through three main mechanisms: (i) spatial inaccessibility within aggregates, (ii) selective preservation through biochemical recalcitrance, and (iii) chemical protection through interactions with mineral surfaces (Lutzow et al., 2006; Tivet et al., 2013b). However, the importance of selective preservation to explain the differences between the three main SOC pools (*active*, *intermediate,* and *passive*) was largely overestimated, and the two other mechanisms are preponderant for long-term SOC stabilization.

A conceptual model with three principal types of SOC pools described by Sollins et al. (1996) and Lutzow et al. (2006) indicates the following: (i)

http://dx.doi.org/10.19103/AS.2019.0049.15

active or labile pool has a short residence time (i.e. a few months to a decade); is easily impacted by soil use and management; and is comprised of microbial biomass C (MBC), plant residues, roots, and fungal hyphae at different decomposition stages; and is stabilized by selective preservation; (ii) *slow or intermediate pool* has a half-life of 10–100 years; is comprised of decomposed residues, is stabilized by organic matter (OM) occlusion and through organo-mineral interactions; and is less impacted by land use and management; and (iii) *passive pool* has half-life of >100 years; is comprised of stable or recalcitrant and mineral-associated organic C (MAOC), is stabilized through encapsulation within stable micro-aggregates, in clay micro-structures and interactions with minerals; and is least impacted by land use and management. The *passive pool* is rather heterogeneous in terms of the turnover time and composition, hence represents a mixture of intermediate and passive SOC fractions.

Changes in SOC pool generally occur in fractions comprising of polysaccharides, water soluble SOC, and to a lesser extent in the stable or mineral-associated fractions. In contrast, Conservation Agriculture (CA) enhance soil aggregate stability, reduce the rate of decomposition of soil organic matter (SOM), and accentuate physical protection of SOM within stable aggregates (Tivet et al., 2013b; De Oliveira Ferreira et al., 2018a). Conversion of CT to CA increases storage of SOC (Sá et al., 2015; De Oliveira Ferreira et al., 2016, 2018b).

At the process level, the SOC stabilization in aggregates is the principal mechanism of long-term sequestration of C in SOM, because the degree and stability of aggregates protect the enzymatic and microbial attacks, and reduce the rate of mineralization in temperate (Six et al., 1999; Balesdent et al., 2000) and subtropical environments (Bayer et al., 2006a,b). While quantitative estimates of changes in SOC have been widely reported for contrasting conditions, qualitative changes upon conversion of native vegetation (NV) to CA have not been widely studied for subtropical and tropical Brazilian environments. Yet, the research information is essential toward a better understanding of the SOM dynamics, and to predict changes in SOC storage among contrasting land uses and management.

Milori et al. (2006) reported a good applicability of Laser-Induced Fluorescence Spectroscopy (LIFS) in characterizing the SOM and assessing its humification degree on the bulk soil samples collected from different depths in three long-term experiments from tropical agro-ecoregion in Brazil. Milori et al. reported that LIFS can be used as a fast and nondestructive tool for analyzing SOM, and reported a high correlation between the data from LIFS and those from conventional fluorescence spectroscopic analysis of the humic acids extracted from the same soil samples. The LIFS have been widely used in conjunction with other analytical techniques (e.g. EPR and/or

NRM). Gonzalez-Perez et al. (2007) and Segnini et al. (2010) reported a close relationship among humification degree obtained by LIFS and semiquinone-type-free radical levels determined by EPR, and also with the degree of aromaticity determined by VACP/MAS ^{13}C NMR spectra. These relationships were observed for contrasting conditions such as Brazilian Oxisols and Wetlands from the high Andes. Other studies, including those by Milori et al. (2006) and Dieckow et al. (2009), indicated an increase in the aromatic components of SOM upon conversion of NV to CT and, to an opposite trend when CT is converted to an NT-based cropping system. In addition, Fourier Transform Infra-Red spectroscopy (FTIR) has been widely used for the characterization of organic compounds (Stevenson, 1994; Barancikova et al., 1997) to study the vibrational characteristics of their structural chemical bonds. Ellerbrock et al. (1999) and Kaiser et al. (2007) analyzed the changes in functional groups of humic acids with FTIR for different land uses and management in temperate conditions.

Several researchers have focused their analyses on specific organic compounds (C-H, C=O, C=C), which reflect SOM characteristics such as its cation exchange capacity (Kaiser et al., 2007), soil wettability (Capriel et al., 1995), and organo-minerals interactions (Lehmann et al., 2007). In addition, partial least square models have been applied to the FTIR data to predict the concentration of functional groups (Ludwig et al., 2008), SOC pools (Bornemann et al., 2010), decomposition status of SOM on bulk samples from different horizons Haberhauer et al. (2000) aggregation Madari et al. (2006), and as a tool for assessing soil quality (Artz et al., 2008). Therefore, more emphasis has to be given to the qualitative changes in functional organic C groups, the stabilization mechanism, and contributions of land use to SOC fraction restoration (Tivet et al., 2013a,c).

2 Soil organic carbon fraction losses and restoration: a case study

2.1 Context

CA is a C-conservative system because it reduces biological SOC oxidation, decreases soil temperature, increases water content of soil, slows the turnover of macroaggregates, and prevents erosion (Golchin et al., 1994; Jastrow et al., 1996; Six et al., 1999). The continued deposition of organic residues on the soil surface combined with minimum soil disturbance enhances biological activities that stimulate bio-physico-chemical C stabilization (De Oliveira Ferreira et al., 2018a).

An array of physical and chemical (i.e. oxidation, hydrolysis) methods is used to isolate different functional fractions involved in diverse processes in soil including, among others, biological activity, C distribution and stabilization, and soil structural stability. Physical fractionation is a useful tool to provide a rough differentiation between active, intermediate, and passive SOC pools, and also to assess the impact of soil management on dynamics (Cambardella and Elliott, 1994; Sá et al., 2001) and qualitative changes (Bayer et al., 2000). In addition, chemical methods are also used to isolate (i) hot water extractable organic C (HWEO-C) representing the microbial C (Ghani et al., 2010) involved in aggregate stability (Haynes and Swift, 1990), (ii) acid hydrolyzable total polysaccharides (TPS) which are the main binding agents involved in soil aggregation (Haynes and Francis, 1993; Tisdall and Oades, 1982), and (iii) a refractory C fraction obtained through oxidation with H_2O_2 which oxidizes labile organic compounds but retains chemically stabilized organic C (CSO-C) (Jagadamma and Lal, 2010). A combination of physical and chemical methods can provide valuable information to fingerprint quantitative and qualitative changes in SOC pool due to land use and agricultural management.

Sá et al. (2015) assessed changes in total SOC stocks upon conversion of NV to CT, and from CT to NT in subtropical and tropical regions of Brazil. Relatively rapid changes in SOC stock with intensive CA systems (e.g., diversity of cover/relay crops and high annual biomass input) indicate a large potential to reverse the process of soil degradation and of SOC depletion through conversion to CA based on complex rotations and cover cropping (Séguy et al., 2006; Sá et al., 2015; De Oliveira Ferreira et al., 2016, 2018a,b).

In this case study, the specific objectives were as follows: (i) the magnitude of changes in SOC fractions (HWEO-C, TPS, CSO-C, particulate, and MAOC) by conversion of the NV into agricultural land and CT, (ii) the spatial distribution of SOC fractions among land use and management practices, (iii) the humification degree of SOM, and the relative composition and changes of functional organic C groups throughout the soil profile and on aggregate size fractions, and (iv) the rate of recovery of SOC fractions and the resilience index (RI) of CA systems for different inputs of biomass-C.

2.2 Experimental procedures

2.2.1 Site description, land uses management

Field experiments were conducted at two research sites: (i) near the town of Ponta Grossa, in Paraná State, southern Brazil, at the experimental station of IAPAR (Instituto Agronômico do Paraná), and hereafter called the PG site, and (ii) near the city of Lucas do Rio Verde, in Mato Grosso State, center western Brazil, at the experimental station of Fundação Rio Verde, and hereafter called the LRV site (Table 1).

Table 1 Sites description: Location, soil type, parent material, climate, land use and management, duration of experiment, sampling depth, chemical properties, and clay content

Description	Ponta Grossa, PR (PG site)				Lucas do Rio Verde, MT (LRV site)		
Location-coordinates	25°09'S-50°09'W				13°00'S-55°58'S		
Altitude	865 m				380 m		
Soil type (FAO)	Red Latosol, Oxisol				Red Yellow Latosol, Oxisol		
Soil type (Soil taxonomy)	Rhodic Hapludox				Typic Haplustox		
Parent material	Shale				Shale and Sandstone		
Climate–type	Mesothermic, summer and winter wet, cold winter (Cfa)				Humid tropic, summer hot and very wet, winter hot and dry (Aw)		
Mean annual temperature (MAT)	18.5°C				25.2°C		
Mean annual rainfall (MAR)	1545 mm				1950 mm		
Land use[a]	NV, CT, MT, and NT				NV, CT and NT1 to NT6		
Years of experiment	29				8		
Sampling depth (cm)	0–5, 5–10, 10–20, 20–40, 40–60, 60–80, 80–100				0–5, 5–10, 10–20, 20–40, 40–60, 60–80, 80–100		
Soil properties	**NV**	**CT**	**MT**	**NT**	**NV**	**CT**	**NT[b]**
Chemical (0–40 cm)							
pH ($CaCl_2$, 0.01 M)	4.01	4.54	4.64	4.47	3.95	4.80	4.57
Al^{3+} ($cmol_c\ dm^{-3}$)	2.12	0.48	0.35	0.50	1.00	0.17	0.31
H + Al ($cmol_c\ dm^{-3}$)	14.34	9.77	8.59	9.47	8.26	5.45	6.15
Ca^{2+} ($cmol_c\ dm^{-3}$)	2.50	3.40	3.35	3.06	0.33	1.62	1.83
Mg^{2+} ($cmol_c\ dm^{-3}$)	0.90	1.27	1.51	1.31	0.26	0.63	0.68
K^+ ($cmol_c\ dm^{-3}$)	0.16	0.25	0.19	0.19	0.09	0.19	0.19
$P_{available}$ ($mg\ dm^{-3}$)	4.16	4.46	3.29	4.56	0.47	18.96	8.10
Clay content ($g\ kg^{-1}$)							
0–20 cm	593	681	637	619	393	381	398
20–40 cm	673	713	690	667	437	425	444
40–100 cm	717	731	719	692	529	455	490

[a] NV: native vegetation of forest (PG) and Cerrado (LRV), CT: conventional plow-based tillage, MT: minimum tillage, NT: no-tillage.
[b] Mean values of the six no-till systems were used for the quantification of the chemical properties and clay content.
Source: adapted from Sá et al. (2015).

2.2.2 The Ponta Grossa (PG) site

The present experiment was initiated in 1981, and the experimental design consisted of three tillage treatments involving conventional plow tillage (CT), minimum tillage (MT), and no-till (NT) laid out as whole plots. The CT consisted of plowing with a 70-cm disk after summer harvest and one after winter harvest to 20-cm depth followed by two 60-cm harrowing. The MT comprised of one chisel plowing to 25-cm depth and one 60-cm narrow disking. In contrast to CT and MT, there was no pre-sowing soil disturbance in NT. Each plot was 100 × 100 m for the NT, and 50 × 140 m for the CT and MT treatments with six subplots (33 × 50 m for NT, and 25 × 46 m for MT and CT). A 3-year cropping sequence was used for all tillage treatments over the 10-year period, involving two crops per year (see Table 2) with soybean (*Glycine max,* L. Merril) grown in six and maize in four summers alternating with oats (*Avena strigosa* Schreb), wheat (*Triticum aestivum* L.), and vetch (*Vicia sativa* L.) in the winter.

An area under NV and adjacent to the experimental plots was selected as a baseline to assess the management-induced changes in SOC stock. Six subplots were demarcated for soil sampling on an area of 100 × 140 m under NV.

2.2.3 The Lucas do Rio Verde (LRV) site

The NV was converted to cropland in 1986 by clearing, burning, and plowing. Immediately after conversion of NV between 1988 and 2000, soil acidity was managed by incorporating 4.5 Mg ha^{-1} of dolomite lime to 20-cm depth. The soil was cultivated for rice for 1 year and soybean as monoculture for 10 years under CT involving a disk harrow to 20 cm depth followed by a fallow during the winter. A uniform area of 10.9 ha was selected in 2001 to initiate the present experiment.

The experiment was designed to compare the standard tillage management of the region represented by one crop per year in the summer by alternating soybean with cotton (*Gossypium hirsutum* L.) under plow-based CT, compared with cropping systems involving different biomass-C input under NT. Tillage treatments were assigned as whole plots (see Table 2) and cropping sequences consisted of (i) soybean alternating with cotton under plow-based CT designated as CT-S/Ct [C-input = 4.01 Mg ha^{-1} yr^{-1}]; (ii) NT1–soybean in the summer followed by maize (*Zea mays* L.) + *Brachiaria ruziziensis* as second crop [C-input = 7.60 Mg ha^{-1} yr^{-1}]; NT2–soybean in the summer followed by finger millet (*Eleusine coracana*) or finger millet + pigeon pea (*Cajanus cajan*) as second crop [C-input = 7.25 Mg ha^{-1} yr^{-1}]; NT3–soybean in the summer followed by finger millet + pigeon pea or finger millet + *Crotalaria spectabilis* as second crop [C-input = 6.84 Mg ha^{-1} yr^{-1}]; NT4–soybean in the summer followed by finger millet + *C. spectabilis* or sunflower (*Helianthus annuus*) + *B. ruziziensis* as second crop [C-input = 7.34 Mg ha^{-1} yr^{-1}]; NT5–soybean in the summer followed by sorghum (*Sorghum bicolor*) + *B. ruziziensis* as second crop [C-input = 8.38 Mg ha^{-1} yr^{-1}]; and NT6–soybean in the summer followed by millet (*Pennisetum glaucum*) or maize + *B. ruziziensis* as second crop [C-input = 7.41 Mg ha^{-1} yr^{-1}].

Table 2 Site, tillage systems, crop sequence, cumulative, and annual biomass-C input in the 29 years experiment period at the PG site and 8 years experiment period at the LRV site

			Biomass-C input[d] (Mg ha^{-1})	
Site	Tillage systems[a]	Crop sequence[b,c]	Cumulative	Annual
PG	CT	S/O - M/O - S/W - S/O + V - M/O - S/O - M/O - S/O - M/R - S/V	86.1	3.07
	MT	S/O - M/O - S/W - S/O + V - M/O - S/O - M/O - S/O - M/R - S/V	83.7	2.99
	NT	S/O - M/O - S/W - S/O + V - M/O - S/O - M/O - S/O - M/R - S/V	116.1	4.15
LRV	CT	S/M - S/M - S/Ct - S/Ct - S/Ct - S/ Ct-S/Ct-S/Ct	32.1	4.01
	NT_1	S/M + Brz - Rc - M - S/M + Brz - S/M + Brz - S/M + Brz - M - S/Cs	60.8	7.60
	NT_2	S/Fm - Rc - M - S/Fm - S/Fm + Pp - S/Fm + Pp - M - S/M + Cs	58.0	7.25
	NT_3	S/Fm + Pp - Rc - M - S/Fm + Pp - S/Fm + Cs - S/Fm + Cs - M - S/G + Cs	54.7	6.84
	NT_4	S/Fm + Cs - Rc - M - S/Fm + Cs - S/G + Brz - S/G + Brz - M - S/ Sg + Cs	58.7	7.34
	NT_5	S/Sg + Brz - Rc - M - S/Sg + Brz - S/Sg + Brz - S/Sg + Brz - M -S/ Sg + Brz	67.0	8.38
	NT_6	S/Mt - Rc - M - S/Mt - S/M + Brz - S/M + Brz - M - S/M	59.3	7.41

[a] Tillage systems: CT = conventional tillage, MT = minimum tillage, NT = no-tillage.
[b] crop sequence for the last 10 years at the PG site: S = soybean (*Glycine max* (L.) Merr.), O = black oats (*Avena strigosa* Schreb), M = maize (*Zea mays* L.), W = wheat (*Triticum aestivum* L.), V = vetch (*Vicia sativa*).
[c] Crop sequence and Tillage systems for the 8 years experiment at the LRV site: Ct = cotton (*Gossypium hirsutum* L.), Brz = *Brachiaria ruziziensis* cv. ruzi, Rc = rice (*Oryza sativa*); Cs = *Crotalaria spectabilis* Roth, Fm = finger millet (*Eleusine coracana* (L.) Gaertn), Pp = pigeon pea (*Cajanus cajan* (L.) Millsp.), G = sunflower (*Helianthus annuus* L.), Sg = *Sorghum bicolor* (L.) Moench, Mt = millet (*Pennisetum typhoides* Burm.).
[d] C-input by the crop residues: cumulative = sum of C-input along the 29 years of experiment at PG site and 8 years at LRV site; Annual C-input = Cumulative C-input/29 and 8 years, for PG site and LRV site, respectively.
Source: adapted from Sá et al. (2015).

2.2.4 Soil sampling

Soil samples at the end of the 29th year (September, 2009) for the PG and 8th year (October, 2009) of experimentation for the LRV site were collected from seven depths: 0–5, 5–10, 10–20, 20–40, 40–60, 60–80, and 80–100 cm. Bulk soil samples for each subplot were obtained for 0–5, 5–10, and 10–20 cm depths by digging 15 × 15 cm trenches. Soil bulk density (ρ_b) was measured by the core method (Blake and Hartge, 1986) using steel cylinders of 5 × 5 cm.

2.2.5 Soil organic carbon pools extraction and analysis

Different SOC fractions were isolated by (i) chemical methods to obtain the HWEO-C, the acid-extracted polysaccharides (TPS), and the CSO-C extracted by H_2O_2 oxidation; and (ii) physical fractionation to separate the particulate organic carbon (POC), wet-sieving of the whole sample isolating the particle size fraction between 53 and 2000 μm (POC) and the slow and passive pool of MAOC which is <53 μm. Details of the fractionation methods are described in the following section (Table 3).

Table 3 Description of methods of SOC fractions

SOC fractions	Description	Method (reference)
Hot water extractable organic carbon (HWEO-C)	The HWEO-C is representative of the microbial biomass, containing more microbially derived than acid hydrolysable carbohydrates (Haynes and Francis, 1993). Given the strong correlations between HWEO-C and other biochemical measurements (microbial biomass, mineralizable N), HWEO-C could be used as an integrated measure of soil quality (Ghani et al., 2003).	Ghani et al. (2003)
Acid-extracted total polysaccharides (TPS)	The C-TPS refer to labile organic C pool mostly involved in diverse processes in soil including the biological activity (Degens et al., 2000), C distribution and stabilization Kaiser et al. (2007), and aggregation (Puget et al., 1998), which physically protect the organic matter.	Lowe (1993)
Chemically stabilized organic C (CSO-C)	The CSO-C constitutes a stable, refractory, pool largely enriched in alkyl C (Jagadamma et al., 2010).	Jagadamma et al. (2010)
Particulate organic carbon (POC) and mineral-associated organic carbon (MAOC)	The POC represents a labile fraction (Cambardella and Elliott, 1994) largely involved as a main source of nutrients for soil microorganisms, in plant nutrition, and in soil structural stability on short-term basis. Meanwhile, the OM associated with finer fractions (e.g. fine silt and clay size), and with lower turnover rates, is considered as the most stabilized pool (Balesdent et al., 2000).	Feller (1994) and adapted by Sá et al. (2001)
Water-stable aggregate	The following classification was used herein: macroaggregates (2-4 to 8-19 mm), mesoaggregate (0.25-0.5 to 1-2 mm) and microaggregate (0.053-0.25 mm).	Castro Filho et al. (2002)

Table 4 Indexes as soil quality indicators

Index	Description	Formula	Reference
Resilience	The resilience index is used to evaluate the recovery of SOC for different tillage systems (SOC management). It takes into consideration the native vegetation as the upper limit, and the conventional tillage as the lower limit for SOC levels	$RI = \frac{SOC_{management} - SOC_{CT}}{SOC_{NV} - SOC_{CT}}$	Herrick and Wander (1997)
Conversion	The conversion index is used to estimate the amount of SOC converted from the biomass input	$C - CR = \frac{\Delta SOC}{Biomass\,C\,input}$	Sá et al. (2014)
Humification	The humification index (H_{LIF}) is used to estimate the degree of SOC humification. It takes into consideration data recorded by Laser-Induced Fluorescence Spectroscopy	$HLIF = \frac{FCA}{SOC}$	Milori et al. (2006)

SOC = soil organic carbon stock (Mg ha^{-1}); CT = Conventional Tillage, NV = Native Vegetation; ΔSOC = Final SOC-Initial SOC stock (Mg ha^{-1}); FCA = Fluorescence curve area.

2.2.6 Indexes used as soil quality indicators

We have used three indexes as soil quality indicators to test the efficiency of the soil management systems: Resilience, Conversion, and Humification index. Table 4 describes the characteristics of each one.

3 Key results

3.1 Chemically stabilized organic C (CSO-C), hot water extractable organic carbon (HWEO-C), and acid-extracted total polysaccharides (TPS)

Concentration of CSO-C differed significantly among land uses and management practices along the soil profile (Fig. 1). The amounts of C removed by the H_2O_2 treatment were significantly ($P < 0.05$) higher in soil under NV (64.4 g kg^{-1}) and NT (53.5 g kg^{-1}) than those under CT, where the amounts of labile C removed by H_2O_2 was 27.4 g kg^{-1} at 0–5 cm depth. A similar CSO-C concentration was observed among CT, MT, and NT treatments in 5–10 cm until 60 cm depth, despite contrasting TOC concentration. In contrast, CSO-C concentration in soil under NV was significantly higher (Fig. 5) than that in cropped fields for 0–5 cm depth and for subsoil layers (60–100 cm). The percentage of CSO-C in TOC increased with increase in depth, and ranged on average among cropped fields from 16% to 35% in 0–5 cm and in 80–100 cm

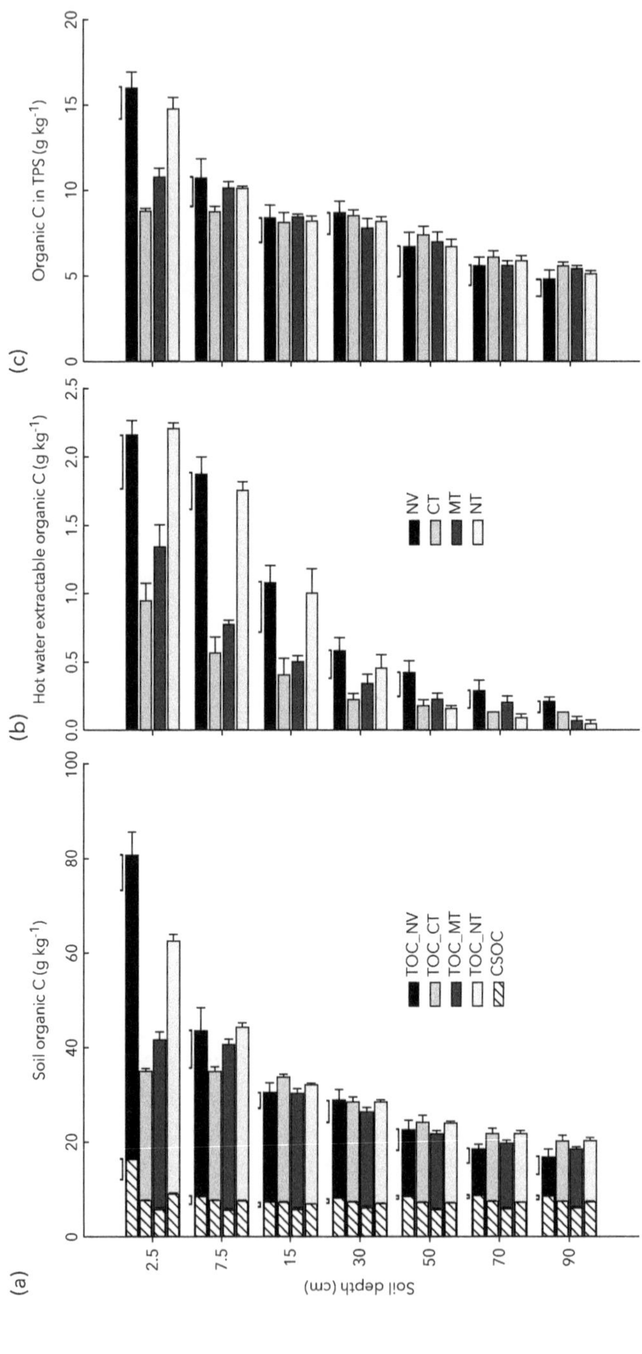

Figure 1 Soil organic carbon concentration (g kg^{-1}) to 100 cm depth managed under either no-till (NT), conventional tillage (CT), minimum tillage (MT), and under the neighboring native vegetation (NV), at the PG site: (a) total soil organic C (TOC), chemically stabilized organic C (CSO-C), (b) hot water extractable organic C (HWEO-C), and (c) organic C in total polysaccharides (C in TPS). Error bars indicate standard error of the mean ($n = 6$) and the horizontal bars denote significance ($P = 0.05$) among land use. Source: adapted from Tivet et al. (2013a).

depths, respectively. The proportion of the SOC left after H_2O_2 treatment ranged under NV from 20% (16.3 g kg^{-1}) in 0–5 cm depth to 51% (8.5 g kg^{-1}) in 80–100 cm depth.

The HWEO-C and C in the TPS exhibited the same pattern as TOC in 0–5 cm depth. HWEO-C concentration accounted for a small percentage of the TOC, and varied in the soil profile according to the variation in TOC and CTPS fractions (Fig. 1). This trend was also confirmed by the positive correlation between HWEO-C and TOC across all soil layers and for each land use.

The HWEO-C concentration decreased significantly with increase in depth independent of land uses and tillage treatments. In soil under NV, the average concentration of HWEO-C represented 2.7% of TOC in 0–5 cm depth and it decreased progressively to 1.5% in 80–100 cm depth. The average proportion of CTPS in soil under NV represented 19.8% of TOC (15.99 g kg^{-1}) in 0–5 cm depth and increased with soil depth to 28.6% (4.8 g kg^{-1}) in 80–100 cm depth (Tivet et al., 2013a). This proportion is similar under cropped fields with values ranging from 25.3% (NT, 5.1 g kg^{-1}) to 29.3% (MT, 5.4 g kg^{-1}) in subsoil layers. The HWEO-C and CTPS concentrations were significantly affected by the land use. However, the concentration of HWEO-C was more impacted by land use changes than those of TOC and CTPS. The concentration of HWEO-C and CTPS decreased from 2.16 g kg^{-1} and 15.99 g kg^{-1} under NV to 0.95 g kg^{-1} and 8.78 g kg^{-1} under CT in 0–5 cm depth, a decrease of 56% and 45% of HWEO-C and CTPS, respectively. The ratio HWEO-C:TOC decreased sharply under CT from 0.027 (0–5 cm) to 0.012 (10–20 cm). In contrast, values close to or even higher than those for NV were observed in soil under NT for 0–20 cm depth (Tivet et al., 2013a). As already observed in the case of TOC, the CTPS concentrations were also uniformly distributed in the surface layers (0–20 cm) under CT due to the mixing effect of plowing and incorporation of crop residues in the 0–20 cm layer.

Concentrations of HWEO-C reported in this study are in accord with those reported by several others (i.e. Leinweber et al., 1995; Haynes, 2005), who observed HWEO-C concentrations of 1–5% of the TOC). Soil carbohydrates constitute a significant part of the labile SOC pool indicating a large contribution of non-cellulosic and cellulosic polysaccharides in the soil. Soil under NT (0–20 cm) had HWEO-C concentration close to or even higher than those under NV at the PG site, leading to a clear formation of a gradient from surface to subsoil layers, which is a characteristic of an NV (Tivet et al., 2013a). These trends suggest that biomass input under NT provided higher and diversified amount of labile organic compounds supporting interdependent food web, enhancing soil microbial communities, and contributing to HWEO-C and CTPS pools. These trends were less evident at the LRV site, but higher values of HWEO-C in 0–5 cm depth were observed under NTs when compared with CT.

Several studies (i.e. Ball et al., 1996; Beare et al., 1994) showed an enrichment of microbial-derived carbohydrates under NT versus CT. A greater proportion of microbial-derived compared with plant-derived carbohydrates under NT compared with CT may be due to the relatively higher fungal biomass under NT (Frey et al., 1999). In addition, incorporation of forage species into crop rotations (NT1, NT5, and NT6) seems to increase HWEO-C (Fig. 2) in 0–5 cm depth, probably due to higher root inputs and the stimulation of microbial activity that follows. In addition, HWEO-C is associated with high concentration of soluble carbohydrates, which is more likely to be present in systems with a high annual C-input. In contrast, a constant proportion of SOC present as acid hydrolyzable carbohydrates was observed across differing CA at the LRV site. Since HWEO-C are considered more microbially derived than acid hydrolyzable carbohydrates, the microbially derived carbohydrates seem to be more affected by tillage (at both locations) and by NT cropping sequences (LRV) than plant-derived carbohydrates. Thus, a large variation in the HWEO-C concentration observed among land uses and management corroborates the importance of HWEO-C as a soil quality indicator to characterize contrasting land uses and tillage treatments (Tivet et al., 2013a).

3.2 Changes in POC and MAOC concentrations

In both sites, land uses with higher TOC concentration exhibited greater amount of SOC lost during fractionation, emphasizing greater water-soluble C under NV and CA. Metay et al. (2007), reporting the concentration of all SOC fractions, concluded that water-soluble SOC, obtained through fractionation, ranged from 8.0% to 14.6% of the TOC of bulk soil for a clayey Oxisol of the Cerrado region. Chivenge et al. (2007) reported losses of SOC during fractionation from 1.7 g kg^{-1} to 3.0 g kg^{-1} on a red clay soil, representing 8–18% of TOC in the bulk soil (Tivet et al., 2013a).

A close correlation was observed between HWEO-C, and TPS, with the MAOC, and more so than the correlation observed with the POC (data not shown) at both sites. These trends indicate the importance of microbial carbohydrates from HWEO-C and TPS in MAOC fraction (Puget et al., 1998). The MAOC is largely enriched in polysaccharides and CSO-C in subsoil layers. Averaged across land uses, CTPS and CSO-C in 80–100 cm depth at the PG site accounted approximately for 28% and 41% of MAOC, respectively (Fig. 3). For the same soil depth interval, CTPS and CSO-C represented 14% and 36% of MAOC at the LRV site, respectively. Similarly, Wattel-Koekkoek et al. (2001) indicated that kaolinite-associated SOC is enriched in polysaccharide products for kaolinitic soils from Brazil, Kenya, Mali, and Mozambique. Thus, the presence of easily degradable substances such as polysaccharides in this fraction in subsoil layers emphasized that physical protection provided by the mineral

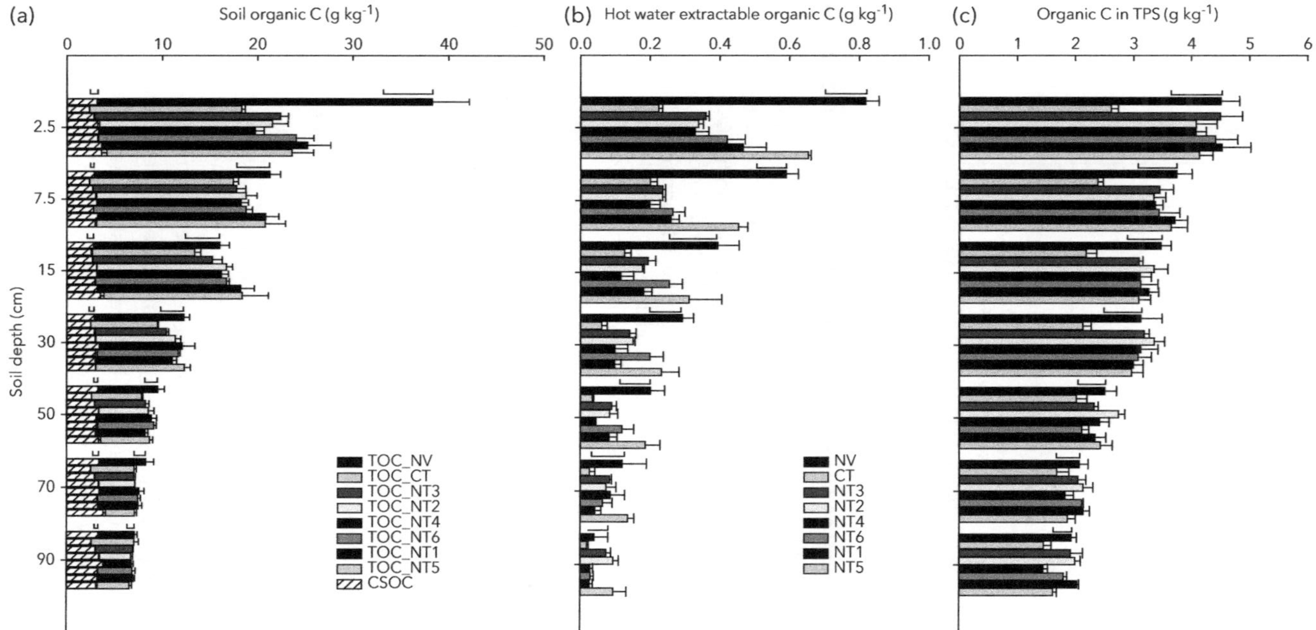

Figure 2 Soil organic carbon concentration (g kg^{-1}) to 100 cm depth under no-till systems (NT1-6), conventional tillage (CT), and under the neighboring native vegetation of Cerrado (NV), at the LRV site: (a) total organic C (TOC), chemically stabilized organic C (CSO-C), (b) hot-water extractable organic C (HWEO-C), and (c) organic C in total polysaccharides (C in TPS). Error bars indicate standard error of the mean ($n = 3$) and the horizontal bars denote significance ($P = 0.05$) among land use. Source: adapted from Tivet et al. (2013a).

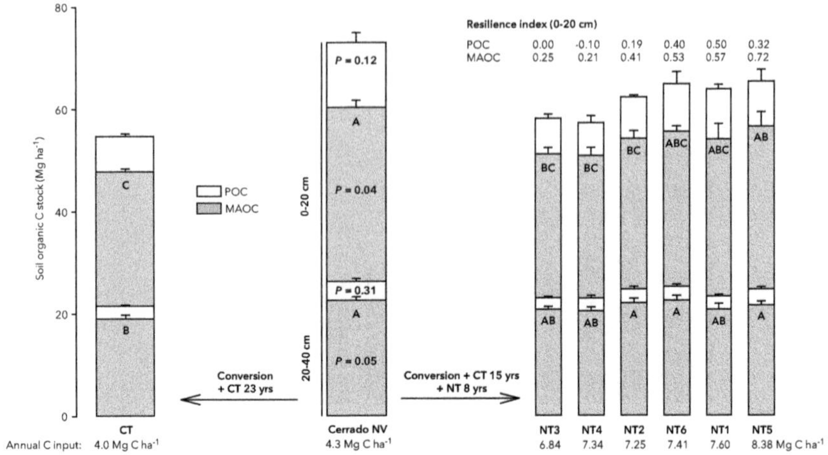

Figure 3 Particulate organic C (POC) and mineral-associated organic C (MAOC) pools (Mg C ha^{-1}) in 0–20 cm and 20–40 cm depths under native vegetation (Cerrado-NV), conventional tillage (CT), and no-till (NT 1–6) systems at the LRV site. Error bars indicate standard error of the mean ($n = 3$). MAOC fraction followed by different uppercase letters is significantly different across land uses based on LSD-Student tests. The level of significance for each soil depth interval is indicated for both SOC fractions. The SOC resilience index ([SOC_{NT}-SOC_{CT}]/[SOC_{NV}-SOC_{CT}]) is given for both SOC fractions for 0–20 cm depth interval, where SOC_{NT} is the C stock in the field under NT, SOC_{CT} is the C stock in the field under CT, and SOC_{NV} is the C stock in the field under NV. The proportion of each SOC fractions is indicated for each soil depth interval. Annual C input (above- and below-ground) of tillage treatments are from Sá et al. (2013) and of Cerrado NV from Corbeels et al. (2006). Source: adapted from Tivet et al. (2013a).

matrix plays a crucial role in SOC stabilization (Rumpel and Kogel-Knabner, 2011).

The SOC fractions are negatively affected by CT in soil surface layers of these two clayey soils, principally in 0–5 cm depth for POC, and 0–5 and 5–10 cm depths for MAOC.

This result is consistent with a previous study (i.e. Feller and Beare, 1997), which also showed that, under continuously cultivated soils, the decrease in SOC is primarily due to a loss of POC in sandy soils and of clay associated-C in clayey soils. At the LRV site, much of the variation in SOC among CT and CA is due to the silt + clay associated-C, indicating that a significant proportion of the SOC associated with minerals is relatively labile. Furthermore, physical stabilization of SOC plays a significant role in the short-term SOC accumulation process for the clayey soil under harsh climate. However, the difference recorded in clay and silt contents (Sá et al., 2013) among land use and management practices, considered being key variables affecting SOC (Feller and Beare, 1997), may interact with these rates of SOC depletion and restoration. Lower clay contents were observed under CT in 0–5 cm, 5–10 cm, and below 40 cm depths. In contrast, higher clay contents were observed in CT soil at the PG site when compared with soils under NV and NT land uses (Sá et al., 2015).

3.3 Soil organic carbon (SOC) pools of different fractions

The CTPS and CSO-C accounted for 54% of TOC up to 1-m depth under cropped fields emphasizing the importance of both fractions. The HWEO-C in 0–20 cm depth accounted for only a small proportion of TOC pool but represented over 3.3% of the TOC under NV, 3.5% under NT, and only 1.7% and 2.1% under CT and MT, respectively. Difference in SOC pools among NV and CT was mainly observed in the 0–20 cm layer, while significant differences were observed in labile pools such as HWEO-C and POC to 1-m depth. Similarly, SOC pools in NT in 0–20 cm depth were close to NV, and also significantly higher under NT than that under CT. In addition, HWEO-C was significantly higher under NT than that under CT when considering the entire profile to 1-m depth.

The POC pool (10.2 Mg C ha^{-1}) represented 56% of total POC pool in the soil profile, while that of the 'silt + clay associated-C' (71.0 Mg C ha^{-1}) accounted for 34% of the total MAOC under NV in 0–20 cm depth. Higher contribution in the 0–20 cm layer was observed for POC under NT with 78% of the total POC pool, and MAOC accounting for 29%. In contrast, these proportions were 58% and 25% under CT, and 64% and 28% under MT. The POC and MAOC pools in 20 cm depth represented 1.9 times (4.74 Mg C ha^{-1}) and 1.3 times (15.10 Mg C ha^{-1}) more C under NV than under CT, respectively. Both POC and MAOC pools in the 0–20 cm soil depth were significantly higher in soils under NT compared with that under CT. At the end of the 29 year, POC (5.5 vs. 8.8 Mg C ha^{-1}, $P < 0.01$) and MAOC (55.9 vs. 63.6 Mg C ha^{-1}, $P < 0.001$) pools in 0–20 cm depth were consistently lower under CT than those under NT (Fig. 4). In contrast, higher POC stocks were recorded under CT in 20–40 and 40–100 cm depths. Higher RI were observed for HWEO-C, and CTPS, followed by POC and MAOC in NT soil. The RI in 0–20 cm depth increased among SOC pools in the order HWEO-C > CTPS > POC > MAOC.

3.4 Distribution of aggregate size classes, soil organic carbon (SOC) concentrations, and stocks

The data of aggregate size distribution in Fig. 4 show that soil under NT had significantly larger macroaggregate (8–19 mm) than that under NV for 0–5 and 5–10 cm depths. Consequently, the soil under NT had significantly lower amounts of 4–8 and 2–4 mm aggregate size fractions in 0–10 cm depth. The results show that, after wet-sieving, a maximum of 59%, 43%, and 80% of the

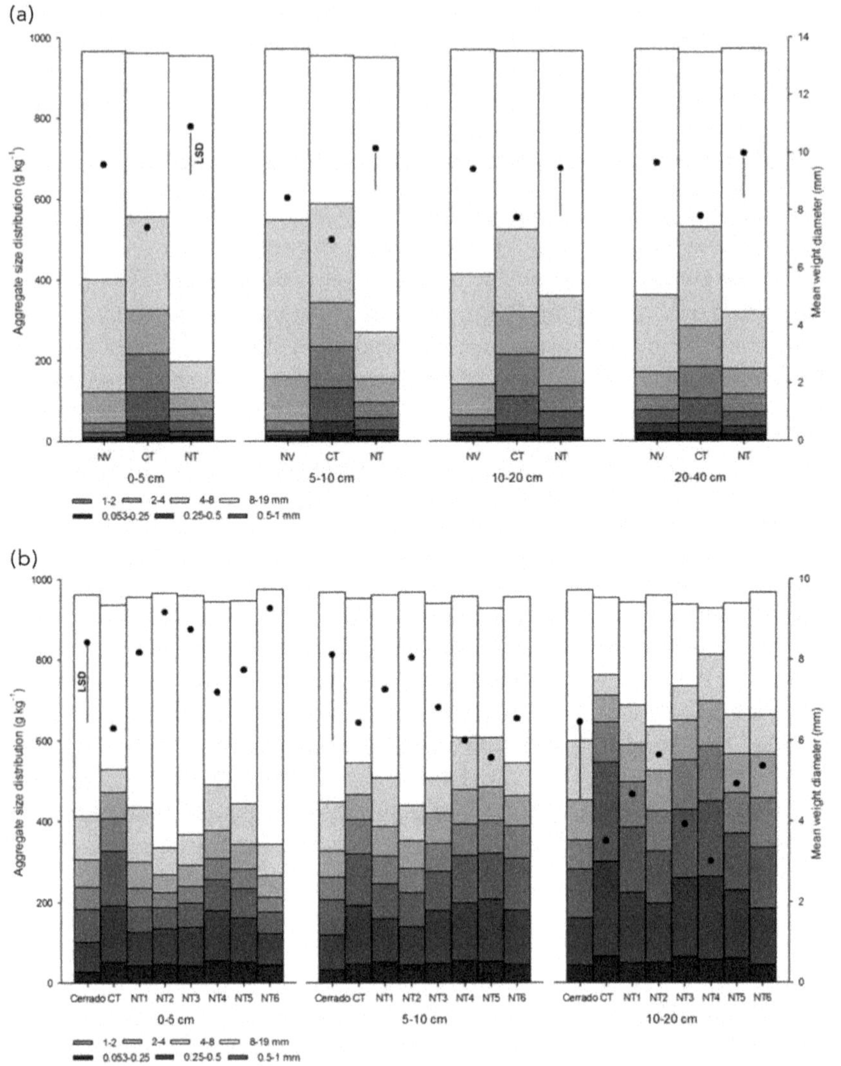

Figure 4 Land use effect on the distribution of aggregate size classes (g soil in aggregate fraction kg^{-1} soil) and on the mean weight diameter (MWD) at the PG (a) and LRV (b) sites at the different depths. Bars represent LSD at $P \leq 0.05$ for MWD. Source: adapted from Tivet et al. (2013b).

large macroaggregates (8–19 mm) were stable to wet-sieving in NV, CT, and NT for 0–5 cm depth, respectively (Fig. 4).

On average, and across all soil depths, the proportion of 8–19 mm aggregate size fraction was 56%, 44%, and 71%, in NV, CT, and NT, respectively. For 0–20 cm depth, soil under NV had significantly ($P < 0.001$) lower amounts of small macroaggregates (<2 mm) and microaggregates than that under CT.

Significant differences in aggregates among soils under CT and NV treatments were observed for all depths, and for almost all aggregate size classes, indicating that the disruptive effect of plowing, the changes in oxidation rate, and microbial environment into the soil profile were observed to 20–40 cm depth. Because of the decrease in 8–19 mm aggregate size fraction, soil under CT had almost three times the amount of 0.5–4 mm aggregate size, 2.5 times the amount of small macroaggregates (0.25–0.5 mm), and 1.5 times the amount of microaggregates than that under NT treatment for 0–5 cm depth.

The SOC concentration in each aggregate size class under NV was significantly higher than that observed under cropped fields (CT and NT) in 0–10 cm depth (Fig. 5). Even the microaggregate size class of the Rhodic Hapludox was strongly affected by the land use conversion, and SOC concentration under CT represents less than 50% of that under NV. This trend of a high SOC concentration under NV was also observed in 10–20 cm depth. The SOC concentration under NT was significantly more than that under CT in all aggregate size classes but especially in 0–5 cm depth. Although no significant differences existed, the highest SOC concentration was still observed for 5–10 cm depth under NT when compared with that under CT. In comparison with CT, higher SOC stock of the 8–19 mm aggregate size fraction was observed under NV and NT for all depths, while those for NV and NT differed significantly for the 0–5 cm and 5–10 cm depths ($P < 0.01$).

The SOC stock in aggregates was severely depleted upon land conversion and continuous CT, and represented a 23% (34.1 Mg ha^{-1}) decline in 0–40 cm depth with 49 years of cultivation ($P < 0.001$). The 8–19 mm aggregate size fraction stored 58%, 44%, and 70% of the total SOC stock in the 0–40 cm depth interval under NV, CT, and NT, respectively. The second largest macroaggregate class (4–8 mm) was also significantly affected by land use change, and followed the order NV > CT > NT in 0–5 and 5–10 cm depths. Consequently, the sum of the SOC stocks of the small macroaggregates (<1 mm) and microaggregates represented ~3%, 12%, and 6% of the total SOC stock among aggregate classes for NV, CT, and NT, respectively. These trends signify the importance of large macroaggregates in protecting and stabilizing SOC within stable microaggregates.

Another aspect which has contributed to the SOC accumulation and recovery within macroaggregates is the C addition through crop residues over time. De Oliveira Ferreira et al. (2018a) studying two soils (clay loam Oxisol and sandy clay Oxisol) near the site of the present study (PG site) observed that the sandy clay Oxisol with the highest soybean frequency (soybean/maize ratio of ≈3:1), and with total input of 119.6 Mg ha^{-1} of biomass-C input over 20 years (5.9 Mg ha^{-1} year^{-1}), had a lower SOC recovery (58.3%) comparing to clay loam Oxisol (73.1%) with higher frequency of maize in the rotation (soybean-maize ratio of ≈1:1) and with total input of 140.4 Mg ha^{-1} of biomass-C over 20 years

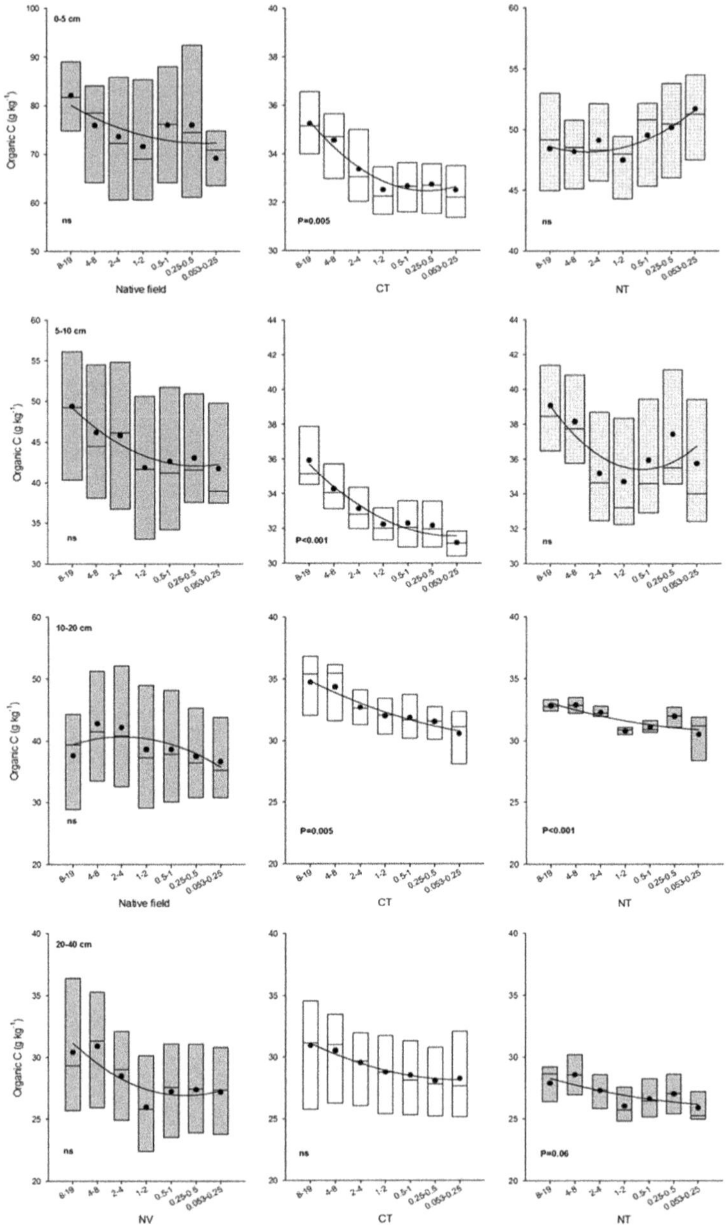

Figure 5 Land use effect on aggregate-associated C (g C in aggregate fraction kg^{-1} soil in aggregate fraction) among aggregate size classes at the PG site at the different depths. The lower and upper limits of each bar represent the first and second quartile; the line within each bar represents the median values and the close circle the mean of C concentration. *P* values: comparisons between aggregate size classes; ns: not significant at $P \leq 0.05$. Source: adapted from Tivet et al. (2013b).

(7 Mg ha^{-1} $year^{-1}$). These results reflect on the importance of crop intensification in the recovery of SOC at the level of extra-large macroaggregates (8–19 mm) in subtropical soils.

At the LRV site, a significant effect of the land use × aggregate size interaction was observed for the distribution of aggregate size classes, SOC concentration, and stock for different depths. In 0–20 cm depth, the proportion of WSA with diameters of 8–19 mm decreased from 50% in soil under Cerrado NV to 35% under CT.

In soils under NT, the proportion of the 8–19 mm size represented 33% (NT4) to 51% (NT2) of the total aggregate distribution in 0–20 cm depth, and greater proportion was observed under cropping systems with a predominance of *Poaceae* + legume (NT2) or only *Poaceae* (NT1, NT6). Furthermore, soil under NV and some NT (e.g. NT2, NT3, and NT6) had significantly ($P < 0.05$) more amounts of large macroaggregates (8–19 mm) than those under CT for 0-5 cm depth. Consequently, soil under CT had higher amounts of small macroaggregates of 1-2 mm ($P < 0.05$, with the exception of that under Cerrado NV and NT4), 0.5–1 mm ($P < 0.05$), and 0.25–0.5 mm size fractions ($P < 0.05$, except in NT4 and NT5). Above all, the SOC concentration decreased significantly ($P < 0.01$) in smaller (<0.5 mm) vis-a-vis larger aggregate size classes for all land use and tillage systems (Fig. 6). Thus, SOC concentration in the 0.053-0.25 mm aggregate size class under the Cerrado NV represents on average <40% of that in 8–19 mm aggregate size classes.

Irrespective of land use, SOC concentration in 0–5 cm and 5–10 cm depths was the highest in 1-2 mm macroaggregate, and it strongly decreased with reduction in aggregate size to 1-2 mm and also in microaggregate size class. The SOC concentration in the 1-2 mm aggregate class ranged from 48.7 g kg^{-1} under NV, 25.8 g kg^{-1} on average under NT, and 21.8 g kg^{-1} under CT in 0–5 cm depth. The SOC concentration decreased by almost 50% as the aggregate size decreased to <0.5 mm. In the microaggregate (0.053–0.25 mm) fraction, the SOC concentration decreased further by 30% when compared with that in 0.25–0.5 mm fraction under NV.

The largest macroaggregate (8–19 mm) stored 52%, 37%, and 41% of the total SOC stock across all aggregate size classes in the 0–20 cm depth interval under NV, CT, and NT (range from 30% to 49%), respectively (Fig. 7). In contrast, the sum of the SOC stocks for the small macroaggregates (<1 mm) and microaggregates amounted to ~19%, 36%, and 27% of the total SOC in aggregates for NV, CT, and NT, respectively. Overall, aggregates SOC stock was severely depleted upon conversion from NV and represented a 36% (17.7 Mg ha^{-1}) decrease over 21 years of CT in the 20-cm layer ($P < 0.001$), compared with a 77% decrease in the 8–19 mm aggregate size fractions. Significant differences among land uses were observed in the SOC stock of the 8–19 mm ($P < 0.01$), 4–8 mm ($P < 0.01$), 2–4 mm size ($P < 0.05$), and 0.5–1 mm size classes

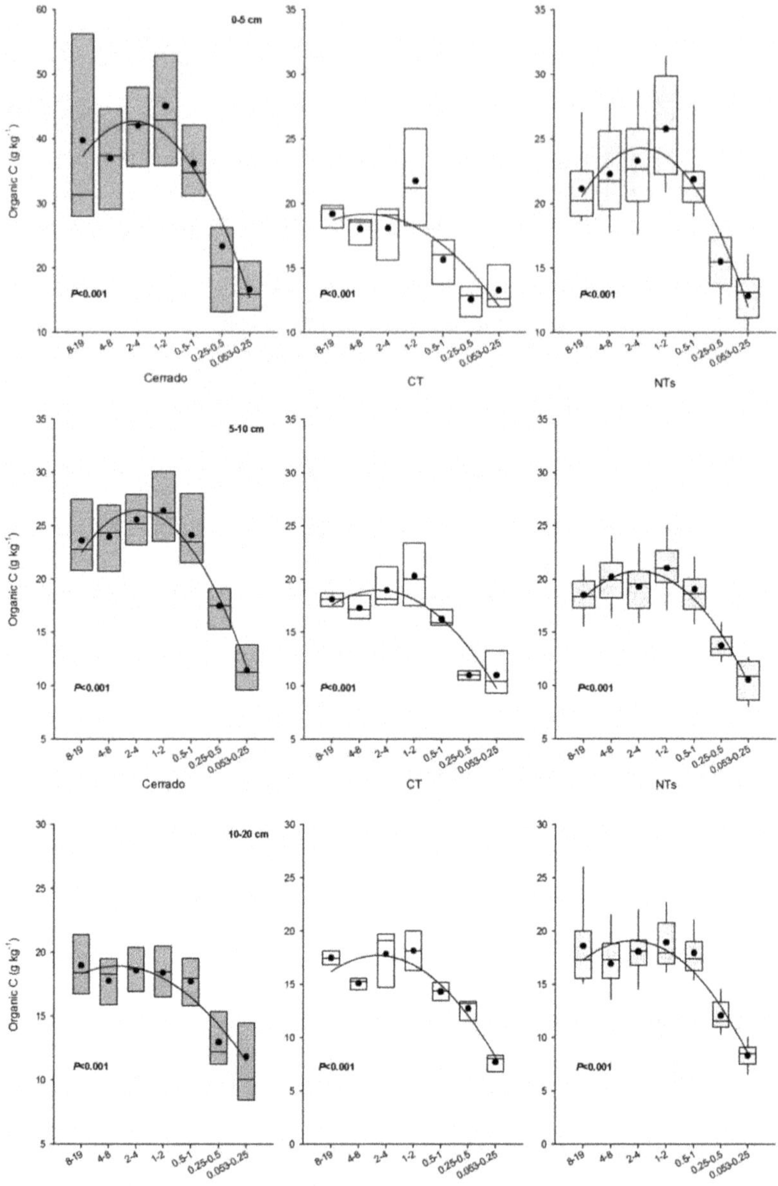

Figure 6 Land use effect on aggregate-associated C (g C in aggregate fraction kg^{-1} soil in aggregate fraction) among aggregate size classes at the LRV site at the different depths. The lower and upper limits of each bar represent the first and second quartile; the line within each bar represents the median values and the close circle the mean of C concentration. The six biomass-C input cropping sequences managed under NT were combined in one group. *P* values: comparisons between aggregate size classes; ns: not significant at $P \leq 0.05$. Source: adapted from Tivet et al. (2013b).

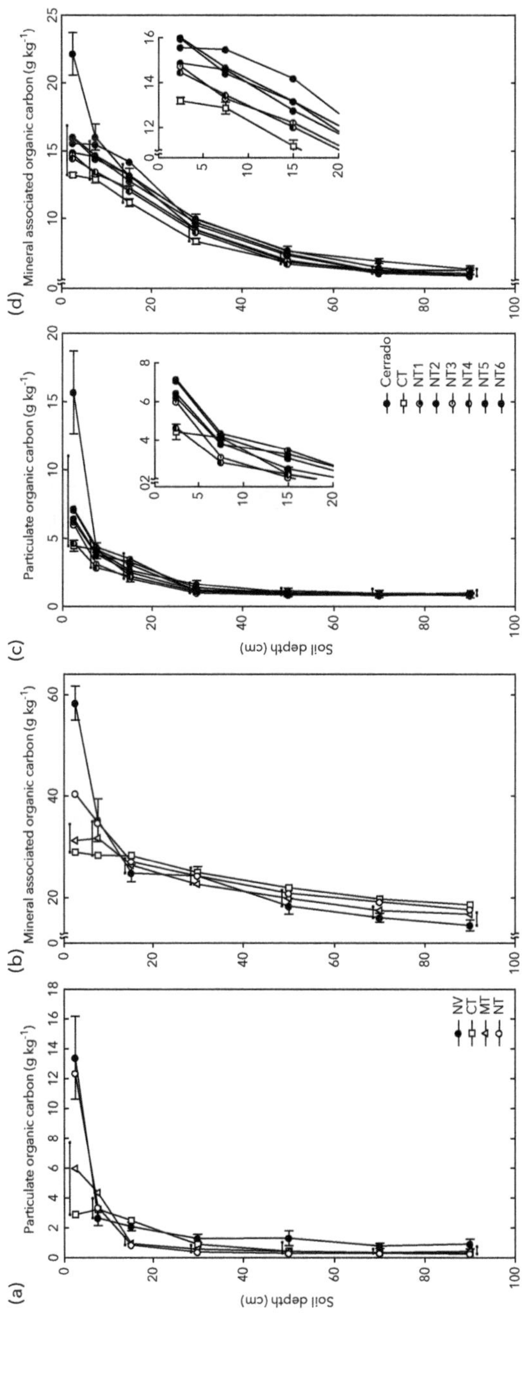

Figure 7 Particulate and mineral associated organic C (g kg^{-1}) to 100 cm depth under NT, CT, MT, and under the neighboring native vegetation (NV) at the PG site (a, b), and under CT, CA, and Cerrado NV at the LRV site (c, d). Error bars for NV (PG and LRV) and CT (LRV) indicate standard error of the mean, the horizontal bars refer to the least significant difference (LSD-Student, $P=0.05$) between land use. Source: adapted from Tivet et al. (2013a).

($P<0.05$) in 0–5 cm depth. The soil under Cerrado NV had a higher SOC stock for the first three aggregate classes mentioned previously, while no significant difference was observed when compared with CT in the 0.5–1 mm size class. The soil under NT had a higher SOC stock than that under CT for the 8–19 mm and 4–8 mm aggregate size classes in 0–5 cm depth.

3.5 Particulate and mineral-associated organic carbon concentrations

On average, the sum of SOC concentration in both fractions was lower than that of the SOC in bulk soil by 8.4% (CT) to 12.9% (NT). Averaged across soil layers, the SOC depleted during fractionation amounted to 3.94, 2.43, 2.86, and 4.49 g kg^{-1} under NV, CT, MT, and NT, respectively. The silt + clay associated-C on the whole represented on average 90.1% (NF) to 93.8% (CT) of the soil mass and 92.0% (NV) to 96.0% (CT) of the SOC fraction mass. The POC, averaged across all soil depths, accounted for 9.9%, 7.4%, 7.0%, and 7.1% of the total soil mass under NV, CT, MT, and NT, respectively, and SOC represented 3.22 (8.0%), 1.71 (4.0%), 1.87 (5.8%), and 2.55 g kg^{-1} (5.9%) under NV, CT, MT, and NT, respectively. Independent of land uses and tillage treatments, the proportion of MAOC concentration increased slightly with increase in depth in the profile, and represented 90% of TOC under cropped fields and 86% under NV in 80–100 cm depth.

Both SOC fractions were markedly affected by land use conversion and tillage treatments. The adoption of CT reduced TOC concentration to 45.7 g C kg^{-1}, compared with decline by 82% in POC (10.5 g C kg^{-1}) and 49% in MAOC (29.3 g C kg^{-1}) in 0–5 cm depth (Fig. 7). Most of the differences in SOC concentration in both aggregate fractions between NT and CT were observed in 0–5, 5–10, 10–20, and 20–40 cm depths. Higher POC concentrations were recorded under CT at 10–20 and 20–40 cm depths when compared with NT ($P<0.05$) resulting from the redistribution of SOC by successive plowing. In contrast, higher MAOC concentrations ($P<0.05$) were recorded under NT at 0–5 and 5–10 cm depths. Similar values of MAOC concentration were observed in 40–100 cm depth under NT and CT. In addition, higher MAOC concentrations ($P<0.05$) were observed in soil under CT and NT than those under NV in 60–100 cm depth. At the end of 29 years, POC (5.5 vs. 8.8 Mg C ha^{-1}, $P<0.01$) and MAOC (55.9 vs. 63.6 Mg C ha^{-1}, $P<0.001$) stocks in the 0–20 cm depth were consistently lower under CT than those under NT (Fig. 8).

3.6 Comparison among subtropical and tropical sites

The MAOC (210.7 Mg ha^{-1}) and CSO-C (84.4 Mg ha^{-1}) pools in the subtropical Rhodic Hapludox under NV were more than twice as high as the respective pools in the Cerrado Typic Haplustox (MAOC = 102.2 Mg ha^{-1}; CSO-C = 34.8 Mg

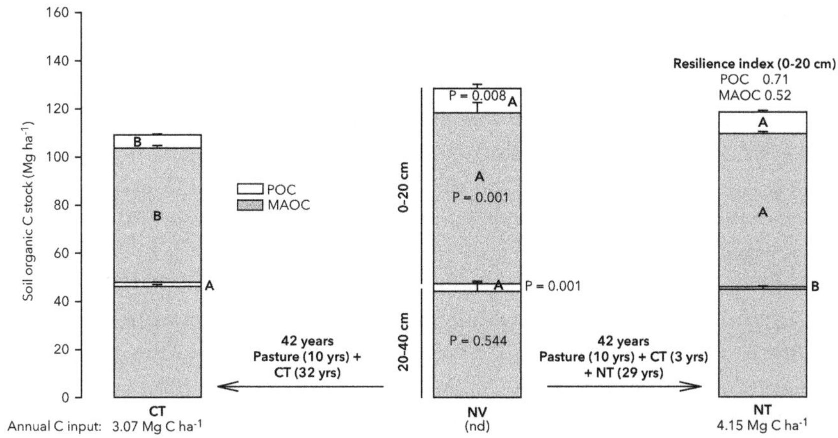

Figure 8 Particulate organic C (POC) and mineral-associated organic C (MAOC) pools (Mg C ha^{-1}) in 0–20 cm and 20–40 cm depths under native vegetation (NV), conventional tillage (CT), and no-till (NT) systems at the PG site. Error bars indicate standard error of the mean ($n = 6$). SOC fractions followed by different uppercase letters are significantly different across land uses based on LSD-Student. The level of significance for each soil depth interval is indicated for both SOC fractions. The SOC resilience index ([SOC_{NT}-SOC_{CT}]/[SOC_{NV}-SOC_{CT}]) is given for both SOC fractions for 0–20 cm depth interval, where SOC_{NT} is the C stock in the field under NT, SOC_{CT} is the C stock in the field under CT, and SOC_{NV} is the C stock in the field under NV. Annual C-input (above- and below-ground) of tillage treatments from Sá et al. (2013). Source: adapted from Tivet et al. (2013a).

ha^{-1}), while the POC pool was comparable but higher at the LRV compared with the PG site: 22.4 versus 18.3 Mg ha^{-1}. The high clay content and Al and Fe-sesquioxides at the PG site (Goncalves et al., 2008) increase the specific surface area, micropores, and the potential stabilization of OM against biological processes, and are the principal determinants of a higher MAOC and CSO-C pools under this subtropical climate. In addition, several studies, including those by Shang and Tiessen (1998) and Barthes et al. (2008), have highlighted the role of both crystalline and non-crystalline Fe and Al-sesquioxides in stabilizing the SOC pool through the formation of organo-mineral complexes. The POC pool ranged from 4.1% (CT) to 8.0% (NV) of TOC pool at the PG site, compared with 14.1–18.0% (NV) in the 0–100 cm depth at the LRV site. Thus, relatively more POC could be mineralized in the tropical Typic Haplustox due to the higher proportion of this fraction.

Jagadamma and Lal (2010) conducted an analysis to assess the relevance of chemical and physical methods of isolating the stable SOC pool. For a soil in the temperate climate, chemical oxidation of bulk soil with H_2O_2 and physical fractionation plus chemical oxidation with H_2O_2 were efficient and similar in isolating the stable SOC fractions in the 30–45 cm layer. These results support the assumption that 100% of the CSO-C was included in the MAOC fraction in

60–100 cm depth. The CSO-C fraction is lower than the MAOC, indicating that part of the MAOC is susceptible to chemical oxidation by H_2O_2. The proportion of the nonchemically stabilized C from MAOC decreases with increase in soil depths, and shows a similar trend among subtropical and tropical regions. Under NV, the nonchemically stabilized C accounted for 54% (9.8 g kg^{-1}) and 57% (4.3 g kg^{-1}) of the MAOC in 40–60 cm depth at the PG and LRV sites, and decreased to 41% (6.0 g kg^{-1}) and 47% (2.9 g kg^{-1}) in 80–100 cm depth, respectively.

The RI was 0.69 under CA at the PG site and ranged from 0.29 (NT3) to 0.79 (NT5) at the LRV site in accord with the increase in dry-matter input among the CA ($RI = 0.18$ C-input–0.76, $R^2 = 0.88$, $P < 0.001$) and indicating the potential of biomass-C input under the CA to restore SOC depleted by the conversion of the NV cropland managed by CT (Sá et al., 2015).

3.7 Soil aggregation indices

The soil under NV was well aggregated and characterized by a high ASI of 98.9% and 94.9% of the total soil mass in macroaggregates at both PG and LRV sites, respectively (Tables 5 and 6). At the PG site, soils under NT and NV had significantly larger ($P < 0.01$) MWD than those under CT for each depth, and that under NT had significantly larger values than those under NV for 0–5 and 5–10 cm depths. A decline in MWD under continuous CT management was due to disruption of large macroaggregates (8–19 mm) by 24% in the top 40-cm layer which resulted in an increase in the proportion of smaller macroaggregates and microaggregates by cultivation. At the LRV site, the soil under Cerrado NV had significantly larger MWD than that under CT in 0–5 and 10–20 cm depths. In addition, similar values of MWD were observed between Cerrado NV and NT in 0–5 cm depth. This effect was primarily due to the improvement of macro aggregation under NT leading to an increase in the largest aggregates.

3.8 Aggregate hierarchy and soil organic carbon (SOC) stabilization

The identification of an aggregate hierarchy in these soils is based on two sets of observations (Six et al., 2004a). One, an increase in SOC concentration followed by an increase in aggregate size class. Two, a higher concentration of SOC labile moieties occurs in macroaggregates than in microaggregates. The first observation is true at both locations with an exception under NT in 0–5 cm depth at the PG site where SOC concentration is on average higher in microaggregates. The data on H_{LIF} assessed through LIFS at the LRV site (Tivet et al., 2013c) supported the second hypothesis, indicating that the SOC associated with macroaggregates is less humified than that associated with

Table 5 Total soil organic carbon concentration (g kg^{-1}) and humification index (HLIF = fluorescence curve area/TOC) managed under either no-till (NT) or conventional tillage (CT), and under the neighboring native vegetation (NV), at the PG site

Soil depth/	Aggregate size (mm)													
Land uses	8–19		4–8		2–4		1–2		0.5–1		0.25–0.5		0.053–0.25	
	TOC concentration (g kg^{-1})													
0–5 cm														
NV	82.1	A ns	75.9	A	73.6	A	71.6	A	76.0	A	76.0	A	69.2	A
CT	35.3	Ca	34.6	Cab	33.4	Cbc	32.5	Cc	32.7	Cc	32.7	Cc	32.5	Cc
NT	48.4	B ns	48.2	B	49.1	B	47.5	B	49.5	B	50.2	B	51.7	B
5–10 cm														
NV	49.4	A ns	46.2	A	45.8	A	41.9	A	42.6	A	43.1	A	41.8	A
CT	35.9	Ba	34.3	Bab	33.1	Bbc	32.2	Bcd	32.3	Bcd	32.2	Bcd	31.2	Cd
NT	39.1	AB ns	38.2	B	35.2	B	34.7	AB	35.9	AB	37.4	AB	35.8	BC
	H_{LIF} (a.u.)													
0–5 cm														
NV	3.8	C ns	4.0	C	3.8	C	4.1	C	3.7	C	3.8	C	3.7	C
CT	9.4	Aa	9.0	Aab	8.5	Ab	9.0	Aab	8.9	Aab	8.7	Ab	8.8	Aab
NT	6.5	B ns	6.7	B	5.6	B	5.8	B	6.3	B	5.9	B	5.4	B
5–10 cm														
NV	6.9	B ns	7.2	B	6.4	B	7.5	A	7.4	A	7.2	A	6.9	B
CT	9.3	Aab	9.5	Aa	8.5	Ac	8.9	Abc	9.1	Aab	8.7	Abc	9.3	Aab
NT	8.4	A ns	8.7	AB	8.2	A	8.4	A	9.1	A	8.5	A	8.0	AB

Table 6 Total soil organic carbon concentration (g kg^{-1}) and humification index (H_{LIF} = fluorescence curve area/TOC) managed under either no-till (NT) or conventional tillage (CT), and under the neighboring native vegetation (NV), at the LRV site

Soil depth/ land uses[a]	Aggregate size (mm)													
	19–8		8–4		4–2		2–1		1–0.5		0.5–0.25		0.25–0.053	
	TOC concentration (g kg^{-1})[b]													
0–5 cm														
Cerrado NV	41.2	Aab	37.8	Aab	41.3	Aab	48.7	Aa	36.0	Aab	25.1	Aab	15.6	ns b
CT	19.2	Bab	18.0	Bbc	18.1	Bb	21.8	Ba	15.6	Ccd	12.6	Bd	13.3	d
NT1	24.4	ABab	24.6	Bab	25.5	Ba	27.8	Ba	21.2	BCb	15.6	Bc	14.4	c
NT4	19.7	Ba	20.3	Ba	20.1	Ba	22.7	Ba	20.8	BCa	15.7	Bb	10.4	c
NT5	22.2	Ba	24.8	Ba	24.3	Ba	27.4	Ba	24.9	ABa	14.9	Bb	12.8	b
NT6	20.5	Bbc	25.3	Bab	23.8	Bab	27.7	Ba	23.3	BCab	16.0	Bcd	14.5	d
5–10 cm														
Cerrado NV	20.4	ns[c] ab	23.4	Aab	22.8	ns ab	25.0	ns a	23.6	ns ab	18.2	Ab	10.2	ns c
CT	18.1	abc	17.3	Bbc	19.0	ab	20.3	a	16.3	c	11.0	Cd	11.0	d
NT1	19.8	bc	21.1	ABab	21.0	abc	23.0	a	18.0	c	14.1	Bd	12.3	d
NT4	16.5	ab	16.9	Ba	17.6	a	17.9	a	17.1	a	13.4	BCb	9.6	c
NT5	19.5	ab	23.4	Aa	20.7	ab	21.9	ab	22.9	a	14.9	Bb	9.8	c
NT6	18.4	b	19.8	ABab	19.6	ab	21.8	a	20.1	ab	13.9	Bc	9.9	d

	H_{FIL} (a.u.)													
0–5 cm														
Cerrado NV	8.9	ns b	8.8	Bb	8.2	Bb	7.6	ns b	10.2	Bb	14.4	ns ab	24.5	ns a
CT	12.5	c	14.4	Abc	14.5	Abc	12.0	c	16.6	Aab	19.1	a	18.4	a
NT1	8.5	b	7.8	Bb	7.8	Bb	7.6	b	9.6	Bb	15.4	a	17.8	a
NT4	9.9	b	10.1	Bb	9.9	Bb	8.4	b	10.2	Bb	14.4	a	25.8	a
NT5	9.4	b	8.7	Bb	8.4	Bb	7.2	b	8.7	Bb	14.7	a	20.1	a
NT6	9.7	c	8.2	Bc	8.5	Bc	7.3	c	9.2	Bc	13.6	b	17.4	a
5–10 cm														
Cerrado NV	17.3	Ab	15.0	Ab	15.5	Ab	13.9	Ab	16.2	Ab	20.5	ABb	42.1	Aa
CT	13.6	Bb	15.1	Ab	13.3	ABb	12.6	ABb	15.1	ABb	22.0	Aa	22.3	Ba
NT1	10.1	Bcd	9.2	Bcd	9.6	Bcd	8.9	BCd	11.2	ABCc	16.0	BCb	21.4	Ba
NT4	11.7	Bc	11.3	Bc	10.5	Bc	10.2	BCc	12.0	ABCc	16.9	BCb	27.6	Ba
NT5	10.4	Bbc	8.6	Bc	9.4	Bbc	8.8	Cc	9.4	Cbc	15.0	Cb	28.8	Ba
NT6	10.5	Bc	9.3	Bc	9.6	Bc	8.9	BCc	10.7	BCc	14.7	Cb	26.7	Ba

NT1, NT4, NT5, and NT6: no-till systems with contrasted biomass-C input.
NT[1] = soybean (Glycine max (L.) Merr.)/maize (Zea mays L.) + *Brachiaria ruziziensis* cv. ruzi - rice (*Oryza sativa* L.) - maize - soybean/maize + *Brachiaria ruziziensis* - soybean/maize + Brachiaria - soybean/maize + Brachiaria - maize - soybean/*Crotalaria spectabilis* Roth; NT[4] = soybean/finger millet + *Crotalaria spectabilis* Roth - rice - maize - soybean/finger millet + *Crotalaria spectabilis* Roth - soybean/sunflower + *Brachiaria ruziziensis* - soybean/sunflower + *Brachiaria ruziziensis* - maize - soybean/*Sorghum bicolor* (L.) Moench + *Crotalaria spectabilis* Roth; NT[5] = soybean/*Sorghum bicolor* (L.) Moench + *Brachiaria ruziziensis* - rice - maize - soybean/*Sorghum bicolor* (L.) Moench + *Brachiaria ruziziensis* - soybean/*Sorghum bicolor* (L.) Moench + *Brachiaria ruziziensis* - soybean/*Sorghum bicolor* (L.) Moench + *Brachiaria ruziziensis* - maize - soybean/*Sorghum bicolor* (L.) Moench + *Brachiaria ruziziensis*; NT[6] = soybean/millet (*Pennisetum glaucum* Burm.) - rice - maize - soybean/millet - soybean/maize + *Brachiaria ruziziensis* - soybean/maize + *Brachiaria ruziziensis* - maize - soybean/maize.
Source: adapted from Tivet et al. (2013c).

microaggregates. In addition, higher values of H_{LIF} in 0–5 and 5–10 cm depths at the LRV site were observed under CT than under NV and NT for any aggregate size (except the microaggregates). Independent of land use, no significant differences in H_{LIF} were observed among aggregate size classes at the PG site. However, and similarly to the LRV site, H_{LIF} of aggregate-associated SOC in soil under CT was significantly higher than those under NV (2.3 times) and NT (1.6 times) for any aggregate size in 0–5 cm depth, indicating a release of labile SOC under CT as a result of the continuous and strong disruption of aggregates. The latter leads to a rapid mineralization of SOC and a preferential loss of labile moieties. In contrast, the enhanced protection of SOC by aggregates in soil under NV and NT enhances labile SOC in each aggregate size fraction. Although not significant, a similar trend was observed in 5–10 cm depth with higher values of H_{LIF} in soil under CT than those under NV and NT management (Tivet et al., 2013c).

3.9 Humification index assessed through fluorescence spectroscopy

In accord with observations by Milori et al. (2006), the ratio between the area of the fluorescence curve and the TOC concentration is used to estimate the humification degree of the bulk soil samples. The H_{LIF} in bulk soils for different depths is shown in Tables 5 and 6. A strong vertical distribution of H_{LIF} degree with depths was observed for both locations ($P < 0.001$). The H_{LIF} increased from the surface to the subsoil layers, which is consistent with the deposition of labile SOM from plant residues on the soil surface resulting in a less aromatic and humified compounds. In the subsoil, however, the decomposition of humic substances continues, increasing the degree of aromaticity and complexation. The H_{LIF} was more than five times greater under the NV of Cerrado than in the forest soil at the PG site for 40–100 cm depth. The difference in H_{LIF} among sites is a function of the TOC concentration, the morphology and chemical structure of SOM, the chemical and physical nature of the mineral fraction, and the architecture of the matrix (Baldock and Skjemstad, 2000). High clay content, and Al and Fe-sesquioxides at the PG site, which enhance the potential adsorptive and protective capacities of the mineral fractions, might be the principal reason for the lowest H_{LIF} under subtropical conditions where larger physical protection mechanisms operate in contrast with highly weathered soils of the tropics.

At PG, the H_{LIF} is significantly higher under CT (Table 5) than NT in 0–10 cm depth ($P < 0.001$), while no significant difference was observed among land use and management systems for 10–60 cm depth. In subsoil layers, the H_{LIF} is more under NV than those in cropped fields. Yet, the same humification gradient is observed between NT and NV soil for 0–60 cm depth, while CT homogenized H_{LIF} in the 0–20 cm depth. Favoretto et al. (2008) and Martins et al. (2011) also reported a similar trend for the PG site and observed that H_{LIF} of humic acids was significantly lower under NT than CT for the 0–20 cm layer. Using fluorescence spectroscopy and electron pin resonance, Bayer et al. (2002) reported that the lowest concentration of stable semiquinone-type free radicals of humic acids samples was observed under NT for a typic Paleudeult in Southern Brazil for the 0–25 cm layer. Several studies, including those by Gonzalez-Perez et al. (2007) and Dieckow et al. (2005), have reported similar results for soils of subtropical and tropical regions of Brazil when comparing CT and CA. A low humification degree under NT was attributed to a higher input of metabolizable organic compounds, a lower turnover time of the labile pools of SOC, a lower oxidation rate (Bayer et al., 2006a), and a higher C stabilization by incorporation within microaggregates and interactions with surface minerals (Sollins et al., 1996; Six et al., 2004b). At the LRV site, H_{LIF} index is significantly higher under CT in 0–5 cm depth, while a higher index was observed under NV soil for 60–100 cm depth (Table 6). In contrast with the subtropical Rhodic Hapludox, H_{LIF} is almost constant under NTs for the 0–20 cm depth.

3.10 Fourier-transform infra-red (FTIR) spectroscopy

3.10.1 General observations of Fourier-transform infra-red (FTIR) spectra

The FTIR spectra of the forest Rhodic Hapludox and the Cerrado Typic Haplustox displayed a number of characteristics absorbance peaks (Fig. 9). In general, the spectra of both locations were characterized by the same peak pattern except at the regions of 3400–3700 cm^{-1} and 1200–900 cm^{-1} that refer mainly to inorganic components of clay minerals and quartz. The predominant clay minerals in the subtropical Rhodic Hapludox are gibbsite, kaolinite, halloysite, and hematite. Goncalves et al. (2008) also reported no significant difference in the mineralogy of the CT and NT soils for this subtropical Rhodic Hapludox. The Typic Haplustox of the Cerrado biome has kaolinite, gibbsite, hematite, and goethite as principal constituents of the mineral matrix. Thus, well-resolved peaks are identified at both locations. Additional well-resolved peaks of clay minerals and quartz were observed for Cerrado soil at 1006 (Si-O-Si in plane stretching modes), 1092 (quartz and O-H stretch of kaolinite or halloysite), and 3646 cm^{-1} (O-H stretch on kaolinite surface). Bands in the regions of 1632 and 1404 cm^{-1} represented coupled vibrations of complex organic compounds. Thus, it is not possible to group them under one specific functional group. The peak at 1632 cm^{-1} was termed as 'aromatic-C', the peak at 1404 cm^{-1} as 'aliphatic-C_1', and the cumulative bands at 2852 and 2922 cm^{-1} as 'aliphatic-C_2'. Band from 3300 to 3700 cm^{-1} due to hydroxyl stretching vibrations associated with clay minerals is included also in alcohols and phenols (Ellerbrock et al.,

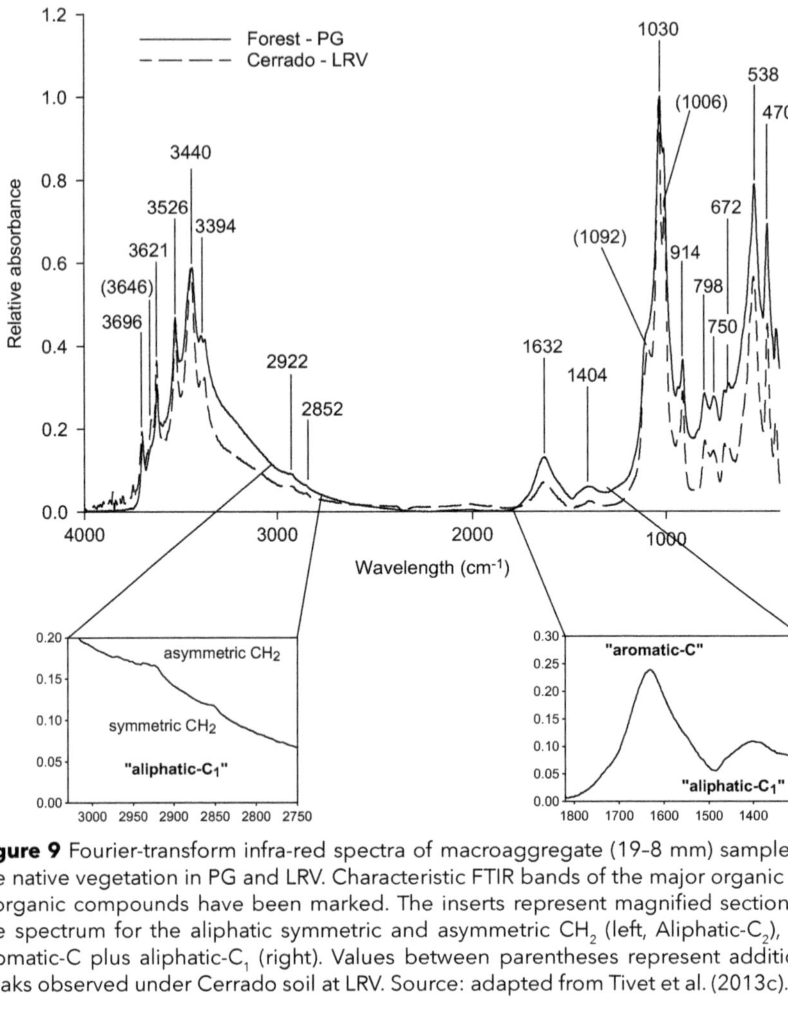

Figure 9 Fourier-transform infra-red spectra of macroaggregate (19–8 mm) samples of the native vegetation in PG and LRV. Characteristic FTIR bands of the major organic and inorganic compounds have been marked. The inserts represent magnified sections of the spectrum for the aliphatic symmetric and asymmetric CH_2 (left, Aliphatic-C_2), and aromatic-C plus aliphatic-C_1 (right). Values between parentheses represent additional peaks observed under Cerrado soil at LRV. Source: adapted from Tivet et al. (2013c).

1999). The same characteristic is observed between 1030 and 1080 cm^{-1} with the superposition of Si-O-Si of quartz, Si-O stretching of clay, and C-O stretching of polysaccharides. It is due to this superposition that these peaks were not classified as functional C groups, and are categorized as inorganic peaks.

4 Conclusion

4.1 Soil organic carbon (SOC) fraction losses and restoration

The conversion to cultivated field under CT reduced both labile (HWEO-C, TPS, OXC, and POC) and stable (MAOC, <53 μm) pools. In contrast, only small changes were observed in the refractory (CSO-C) pool, which is highly stable and is less impacted by changes in land use and management (Tivet et al., 2013a).

Intensive CA systems (e.g. diversity of cover/relay crops and high annual biomass input) at the tropical LRV increased HWEO-C, C in the TPS, and enhanced the short-term process of SOC stabilization in the mineral-associated fraction. In addition, relatively higher resilience indices of POC and MAOC fractions were observed under CA systems with high biomass input at the LRV site.

The results presented emphasize the high potential of CA systems to restore SOC fractions previously depleted by CT, contributing to a high farm productivity and sustainability.

4.2 Aggregate carbon (C) depletion and restoration

Large macroaggregates (8–19 mm) dominated aggregate size distribution under NV and NT at both sites, which in turn was positively associated with the SOC concentration. CA systems restore aggregation and the proportion of SOC stock occluded in 8–19 mm aggregate size to a value similar to that under NV (Tivet et al., 2013b).

The difference in SOC concentration between CT and CA at both locations is largely attributed to SOC storage in the largest macroaggregate size classes, indicating that macroaggregates are crucial to the storage and stabilization of SOC within macroaggregate-occluded microaggregates.

4.3 Assessing humification and organic carbon (C) compounds

The degree of aromaticity was relatively higher in the top soil layers under CT than under NT and NV, indicating that physical protection by aggregation was not sufficient to protect the most labile SOM fraction in CT. In addition, a general depletion of functional C groups was observed under CT. On average, a higher concentration of TOC in the whole soil sample, and in aggregate size classes, as well as greater intensity of functional C groups were observed under CA cropping systems (Tivet et al., 2013c). In addition, lower H_{LIF} under CA suggests that spatial inaccessibility by aggregates plays a major role in enhancing SOC accumulation rather than the selective preservation by aromaticity of SOM. These processes are essential to promote long-term carbon sequestration in subtropical and tropical soils.

5 Acronyms

CA: Conservation Agriculture
NT: No-till
PG: Ponta Grossa, PR, Brazil

LRV: Lucas do Rio Verde, MT, Brazil
CT: Conventional tillage
SOC: Soil organic carbon
NV: Native vegetation
HWEO-C: Hot water extractable organic carbon
C: Carbon
N: Nitrogen
TPS: Acid-extracted total polysaccharides
POC: Particulate organic carbon
MAOC: Mineral-associated organic carbon
CSO-C: Chemically stabilized organic carbon
TOC: Total organic carbon
NPP: Net primary production
SOM: Soil organic matter
LIFS: Laser-Induced Fluorescence Spectroscopy
FTIR: Fourier-Transform Infra-Red spectroscopy
MAT: Mean annual temperature
MT: Minimum tillage
ρ_b: Soil bulk density
WSA: Water-stable aggregates
MWD: Mean weight diameter (mm)
MGD: Mean geometric diameter (mm)
ASI: Aggregate Stability Index
FCA: Fluorescence curve area
RI: Resilience Index
C-CR: Conversion index
HLIF: Humification Index
CT-S/Ct: Soybean alternating with cotton under conventional tillage
NT1: Soybean in the summer followed by maize (*Zea mays* L.) + *Brachiaria ruziziensis* as second crop
NT2: Soybean in the summer followed by finger millet (*Eleusine coracana*) or finger millet + pigeon pea (*Cajanus cajan*) as second crop
NT3: Soybean in the summer followed by finger millet + pigeon pea or finger millet + *Crotalaria spectabilis* as second crop
NT4: Soybean in the summer followed by finger millet + *C. spectabilis* or sunflower (*Helianthus annuus*) + *B. ruziziensis* as second crop
NT5: Soybean in the summer followed by sorghum (*Sorghum bicolor*) + *B. ruziziensis* as second crop
NT6: Soybean in the summer followed by millet (*Pennisetum glaucum*) or maize + *B. ruziziensis* as second crop

6 References

Artz, R. R. E., Chapman, S. J., Robertson, A. H. J., Potts, J. M., Laggoun-Defarge, F., Gogo, S., Comont, L., Disnar, J. R. and Francez, A. J. 2008. FTIR spectroscopy can be used as a screening tool for organic matter quality in regenerating cutover peatlands. *Soil Biology and Biochemistry* 40(2), 515–27. doi:10.1016/j.soilbio.2007.09.019.

Baldock, J. and Skjemstad. J. 2000. Role of soil matrix and minerals in protecting natural organic materials against biological attack. *Organic Geochemistry* 31(7), 697–710.

Balesdent, J., Chenu, C. and Balabane, M. 2000. Relationship of soil organic matter dynamics to physical protection and tillage. *Soil and Tillage Research* 53(3–4), 215–30. doi:10.1016/S0167-1987(99)00107-5.

Ball, B. C., Cheshire, M. V., Robertson, E. A. G. and Hunter, E. A. 1996. Carbohydrate composition in relation to structural stability, compactibility and plasticity of two soils in a long-term experiment. *Soil and Tillage Research* 39(3–4), 143–60. doi:10.1016/S0167-1987(96)01067-7.

Barancikova, G., Senesi, N. and Brunetti, G. 1997. Chemical and spectroscopic characterization of humic acids isolated from different Slovak soil types. *Geoderma* 78(3–4), 251–66. doi:10.1016/S0016-7061(97)00033-5.

Barthes, B. G., Kouakoua, E., Larre-Larrouy, M. C., Razafimbelo, T. M., de Luca, E. F., Azontonde, A., Neves, C. S. V. J., de Freitas, P. L. and Feller, C. L. 2008. Texture and sesquioxide effects on water-stable aggregates and organic matter in some tropical soils. *Geoderma* 143(1–2), 14–25. doi:10.1016/j.geoderma.2007.10.003.

Bayer, C., Martin-Neto, L., Mielniczuk, J. and Ceretta, C. A. 2000. Effect of no-till cropping systems on soil organic matter in a sandy clay loam Acrisol from Southern Brazil monitored by electron spin resonance and nuclear magnetic resonance. *Soil and Tillage Research* 53(2), 95–104. doi:10.1016/S0167-1987(99)00088-4.

Bayer, C., Mielniczuk, J., Martin-Neto, L. and Ernani, P. R. 2002. Stocks and humification degree of organic matter fractions as affected by no-tillage on a subtropical soil. *Plant and Soil* 238(1), 133–40. doi:10.1023/A:1014284329618.

Bayer, C., Lovato, T., Dieckow, J., Zanatta, J. A. and Mielniczuk, J. 2006a. A method for estimating coefficients of soil organic matter dynamics based on long-term experiments. *Soil and Tillage Research* 91(1–2), 217–26. doi:10.1016/j.still.2005.12.006.

Bayer, C., Martin-Neto, L., Mielniczuk, J., Pavinato, A. and Dieckow, J. 2006b. Carbon sequestration in two Brazilian Cerrado soils under no-till. *Soil and Tillage Research* 86(2), 237–45. doi:10.1016/j.still.2005.02.023.

Beare, M. H., Hendrix, P. F. and Coleman, D. C. 1994. Water-stable aggregates and organic matter fractions in conventional-and no-tillage soils. *Soil Science Society of America Journal* 58(3), 777–86. doi:10.2136/sssaj1994.03615995005800030020x.

Blake, G. R. and Hartge, K. H. 1986. *Bulk Density. Methods of Soil Analysis, Part I. Physical and Mineralogical Methods*. Agronomy Monograph no. 9 (2nd ed.) (Klute, A., Ed.). ASA SSSA, Madison, WI, pp. 363–75.

Bornemann, L., Welp, G. and Amelung, W. 2010. Particulate organic matter at the field scale: rapid acquisition using mid-infrared spectroscopy. *Soil Science Society of America Journal* 74(4), 1147–56. doi:10.2136/sssaj2009.0195.

Cambardella, C. A. and Elliott, E. T. 1994. Carbon and nitrogen dynamics of soil organic matter fractions from cultivated grassland soils. *Soil Science Society of America Journal* 58(1), 123-30. doi:10.2136/sssaj1994.03615995005800010017x.

Capriel, P., Beck, T., Borchert, H., Gronholz, J. and Zachmann, G. 1995. Hydrophobicity of the organic-matter in arable soils. *Soil Biology and Biochemistry* 27(11), 1453–8. doi:10.1016/0038-0717(95)00068-P.

Castro Filho, C., Lourenco, A., Guimaraes, M.d.F. and Fonseca, I. C. B. 2002. Aggregate stability under different soil management systems in a red latosol in the state of Parana, Brazil. *Soil and Tillage Research* 65(1), 45–51. doi:10.1016/S0167-1987(01)00275-6.

Chivenge, P. P., Murwira, H. K., Giller, K. E., Mapfumo, P. and Six, J. 2007. Long-term impact of reduced tillage and residue management on soil carbon stabilization: implications for conservation agriculture on contrasting soils. *Soil and Tillage Research* 94(2), 328–37. doi:10.1016/j.still.2006.08.006.

Corbeels, M., Scopel, E., Cardoso, A., Bernoux, M., Douzet, J. and Neto, M. S. 2006. Soil carbon storage potential of direct seeding mulch based cropping systems in the Cerrados of Brazil. *Global Change Biology* 12(9), 1773–87. doi:10.1111/j.1365-2486.2006.01233.x.

Degens, B., Schipper, L., Sparling, G. and Vojvodic-Vukovic, M. 2000. Decreases in organic C reserves in soils can reduce the catabolic diversity of soil microbial communities. *Soil Biology and Biochemistry* 32, 189–96.

De Oliveira Ferreira, A., Amado, T., Rice, C. W., Ruiz Diaz, D. A. R., Keller, C. and Inagaki, T. M. 2016. Can no-till grain production restore soil organic carbon to levels natural grass in a subtropical oxisol? *Agriculture, Ecosystems and Environment* 229, 13–20. doi:10.1016/j.agee.2016.05.016.

De Oliveira Ferreira, A., Sá, J. C. M., Lal, R., Tivet, F., Briedis, C., Inagaki, T. M., Goncalves, D. R. P. and Romaniw, J. 2018a. Macroaggregation and soil organic carbon restoration in a highly weathered Brazilian oxisol after two decades under no-till. *Science of the Total Environment* 621, 1559–67. doi:10.1016/j.scitotenv.2017.10.072.

De Oliveira Ferreira, A., Amado, T. J. C., Rice, C. W., Diaz, D. A. R., Briedis, C., Inagaki, T. M. and Goncalves, D. R. P. 2018b. Driving factors of soil carbon accumulation in oxisols in long-term no-till systems of South Brazil. *Science of the Total Environment* 622–623, 735–42.

Dieckow, J., Mielniczuk, J., Knicker, H., Bayer, C., Dick, D. P. and Kogel-Knabner, I. 2005. Organic N forms of a subtropical Acrisol under no-till cropping systems as assessed by acid hydrolysis and solid-state NMR spectroscopy. *Biology and Fertility of Soils* 42(2), 153–8. doi:10.1007/s00374-005-0002-4.

Dieckow, J., Bayer, C., Conceicao, P. C., Zanatta, J. A., Martin-Neto, L., Milori, D. B. M., Salton, J. C., Macedo, M. M., Mielniczuk, J. and Hernani, L. C. 2009. Land use, tillage, texture and organic matter stock and composition in tropical and subtropical Brazilian soils. *European Journal of Soil Science* 60(2), 240–9. doi:10.1111/j.1365-2389.2008.01101.x.

Ellerbrock, R. H., Hohn, A. and Gerke, H. H. 1999. Characterization of soil organic matter from a sandy soil in relation to management practice using FT-IR spectroscopy. *Plant and Soil* 213(1/2), 55–61. doi:10.1023/A:1004511714538.

Favoretto, C. M., Goncalves, D., Milori, D. M. B. P., Rosa, J. A., Leite, W. C., Brinatti, A. M. and Saab, SdC. 2008. Determination of humification degree of organic matter of an oxisol and of its organo-mineral fractions. *Quimica Nova* 31(8), 1994–6. doi:10.1590/S0100-40422008000800015.

Feller, C. 1994. La matière organique dans les sols tropicaux. Thèse Doctorat ès Sciences (in French). University of Strasbourg (ULP), Strasbourg, France.

Feller, C. and Beare, M. H. 1997. Physical control of soil organic matter dynamics in the tropics. *Geoderma* 79(1–4), 69–116. doi:10.1016/S0016-7061(97)00039-6.

Frey, S. D., Elliott, E. T. and Paustian, K. 1999. Bacterial and fungal abundance and biomass in conventional and no-tillage agroecosystems along two climatic gradients. *Soil Biology and Biochemistry* 31(4), 573–85. doi:10.1016/S0038-0717(98)00161-8.

Golchin, A., Oades, J. M., Skjemstad, J. O. and Clarke, P. 1994. Soil structure and carbon cycling. *Soil Research* 32(5), 1043–68. doi:10.1071/SR9941043.

Ghani, A., Dexter, M. and Perrott, K. W. 2003. Hot-water extractable carbon in soils: a sensitive measurement for determining impacts of fertilisation, grazing and cultivation. *Soil Biology and Biochemistry* 35(9), 1231–43. doi:10.1016/S0038-0717(03)00186-X.

Ghani, A., Muller, K., Dodd, M. and Mackay, A. 2010. Dissolved organic matter leaching in some contrasting New Zealand pasture soils. *European Journal of Soil Science* 61(4), 525–38. doi:10.1111/j.1365-2389.2010.01246.x.

Goncalves, D., Leite, W. C., Brinatti, A. M., Saab, SdC., Iarosz, K. C., Masearenhas, Y. P., Carneiro, P. I. B. and Rosa, J. A. 2008. Mineralogy of a red latosol under different management systems for twenty-four years. *Revista Brasileira De Ciencia Do Solo* 32(spe), 2647–52. doi:10.1590/S0100-06832008000700006.

Gonzalez-Perez, M., Milori, D. M. B. P., Colnago, L. A., Martin-Neto, L. and Melo, W. J. 2007. A laser-induced fluorescence spectroscopic study of organic matter in a Brazilian oxisol under different tillage systems. *Geoderma* 138(1–2), 20–4. doi:10.1016/j.geoderma.2006.10.010.

Haberhauer, G., Feigl, B., Gerzabek, M. H. and Cerri, C. 2000. FT-IR spectroscopy of organic matter in tropical soils: changes induced through deforestation. *Applied Spectroscopy* 54(2), 221–4. doi:10.1366/0003702001949131.

Haynes, R. J. 2005. Labile organic matter fractions as central components of the quality of agricultural soils: an overview. *Advances in Agronomy* 85, 221–68. doi:10.1016/S0065-2113(04)85005-3.

Haynes, R. J. and Francis, G. S. 1993. Changes in microbial biomass-C, soil carbohydrate-composition and aggregate stability induced by growth of selected crop and forage species under field conditions. *Journal of Soil Science* 44(4), 665–75. doi:10.1111/j.1365-2389.1993.tb02331.x.

Haynes, R. J. and Swift, R. S. 1990. Stability of soil aggregates in relation to organic-constituents and soil-water content. *Journal of Soil Science* 41(1), 73–83. doi:10.1111/j.1365-2389.1990.tb00046.x.

Herrick, J. E. and Wander, M. M. 1997. *Relationships between Soil Organic Carbon and Soil Quality in Cropped and Rangeland Soils: the Importance of Distribution, Composition, and Soil Biological Activity*. CRC Press, Boca Raton, FL.

Jagadamma, S. and Lal, R. 2010. Integrating physical and chemical methods for isolating stable soil organic carbon. *Geoderma* 158(3–4), 322–30. doi:10.1016/j.geoderma.2010.05.014.

Jagadamma, S., Lal, R., Ussiri, D. A. N., Trumbore, S. E. and Mestelan, S. 2010. Evaluation of structural chemistry and isotopic signatures of refractory soil organic carbon fraction isolated by wet oxidation methods. *Biogeochemistry* 98(1–3), 29–44. doi:10.1007/s10533-009-9374-0.

Jastrow, J. D., Miller, R. M. and Boutton, T. W. 1996. Carbon dynamics of aggregate associated organic matter estimated by carbon-13 natural abundance. *Soil Science*

Society of America Journal 60(3), 801–7. doi:10.2136/sssaj1996.03615995006000030017x.

Kaiser, M., Ellerbrock, R. H. and Gerke, H. H. 2007. Long-term effects of crop rotation and fertilization on soil organic matter composition. *European Journal of Soil Science* 58(6), 1460–70. doi:10.1111/j.1365-2389.2007.00950.x.

Lehmann, J., Kinyangi, J. and Solomon, D. 2007. Organic matter stabilization in soil microaggregates: implications from spatial heterogeneity of organic carbon contents and carbon forms. *Biogeochemistry* 85(1), 45–57. doi:10.1007/s10533-007-9105-3.

Leinweber, P., Schulten, H. -R. and Korschens, M. 1995. Hot-water extracted organic-matter R – chemical-composition and temporal variations in a long-term field experiment. *Biology and Fertility of Soils* 20(1), 17–23. doi:10.1007/BF00307836.

Lowe, L. 1993. Total and labile polysaccharide analysis of soils. *Soil Sampling and Methods of Analysis*, 373–6.

Ludwig, B., Nitschke, R., Terhoeven-Urselmans, T., Michel, K. and Flessa, H. 2008. Use of mid-infrared spectroscopy in the diffuse-reflectance mode for the prediction of the composition of organic matter in soil and litter. *Journal of Plant Nutrition and Soil Science* 171(3), 384–91. doi:10.1002/jpln.200700022.

Lutzow, M. v., Kogel-Knabner, I., Ekschmitt, K., Matzner, E., Guggenberger, G., Marschner, B. and Flessa, H. 2006. Stabilization of organic matter in temperate soils: mechanisms and their relevance under different soil conditions – a review. *European Journal of Soil Science* 57(4), 426–45. doi:10.1111/j.1365-2389.2006.00809.x.

Madari, B. E., Reeves, J. B., Machado, P. L. O. A., Guimaraes, C. M., Torres, E. and McCarty, G. W. 2006. Mid- and near-infrared spectroscopic assessment of soil compositional parameters and structural indices in two Ferralsols. *Geoderma* 136(1–2), 245–59. doi:10.1016/j.geoderma.2006.03.026.

Martins, T., Saab, S. C., Milori, D. M. B. P., Brinatti, A. M., Rosa, J. A., Cassaro, F. A. M. and Pires, L. F. 2011. Soil organic matter humification under different tillage managements evaluated by Laser Induced Fluorescence (LIF) and C/N ratio. *Soil and Tillage Research* 111(2), 231–5. doi:10.1016/j.still.2010.10.009.

Metay, A., Oliver, R., Scopel, E., Douzet, J.-M., Moreira, J. A. A., Maraux, F., Feigl, B. J. and Feller, C. 2007. N_2O and CH_4 emissions from soils under conventional and no-till management practices in Goiânia (Cerrados, Brazil). *Geoderma* 141(1–2), 78–88. doi:10.1016/j.geoderma.2007.05.010.

Milori, D. M. B. P., Galeti, H. V. A., Martin-Neto, L., Dieckow, J., Gonzalez-Perez, M., Bayer, C. and Salton, J. 2006. Organic matter study of whole soil samples using laser-induced fluorescence spectroscopy. *Soil Science Society of America Journal* 70(1), 57–63. doi:10.2136/sssaj2004.0270.

Puget, P., Angers, D. A. and Chenu, C. 1998. Nature of carbohydrates associated with water-stable aggregates of two cultivated soils. *Soil Biology and Biochemistry* 31(1), 55–63. doi:10.1016/S0038-0717(98)00103-5.

Rumpel, C. and Kogel-Knabner, I. 2011. Deep soil organic matter-a key but poorly understood component of terrestrial C cycle. *Plant and Soil* 338(1–2), 143–58. doi:10.1007/s11104-010-0391-5.

Sá, J. CdM., Cerri, C. C., Dick, W. A., Lal, R., Filho, S. P. V., Piccolo, M. C. and Feigl, B. E. 2001. Organic matter dynamics and carbon sequestration rates for a tillage chronosequence in a Brazilian Oxisol. *Soil Science Society of America Journal* 65(5), 1486–99. doi:10.2136/sssaj2001.6551486x.

Sá, J. CdM., Bürkner dos Santos, J., Lal, R., de Moraes, A., Tivet, F., Machado Sá, M. F., Briedis, C., de Oliveira Ferreira, A., Eurich, G., Farias, A. and Friedrich, T. 2013. Soil-specific inventories of landscape carbon and nitrogen stocks under no-till and native vegetation to estimate carbon offset in a subtropical ecosystem. *Soil Science Society of America Journal* 77(6), 2094–110. doi:10.2136/sssaj2013.01.0007.

Sá, J. CdM., Tivet, F., Lal, R., Briedis, C., Hartman, D. C., dos Santos, J. Z. and dos Santos, J. B. 2014. Long-term tillage systems impacts on soil C dynamics, soil resilience and agronomic productivity of a Brazilian Oxisol. *Soil and Tillage Research* 136, 38–50. doi:10.1016/j.still.2013.09.010.

Sá, J. CdM., Séguy, L., Tivet, F., Lal, R., Bouzinac, S., Borszowskei, P. R., Briedis, C., dos Santos, J. B., da Cruz Hartman, D., Bertoloni, C. G., Rosa, J. and Friedrich, T. 2015. Carbon depletion by plowing and its restoration by no-till cropping systems in Oxisols of subtropical and tropical agro-ecoregions in Brazil. *Land Degradation and Development* 26(6), 531–43. doi:10.1002/ldr.2218.

Segnini, A., Posadas, A., Quiroz, R., Milori, D. M. B. P., Saab, S. C., Neto, L. M. and Vaz, C. M. P. 2010. Spectroscopic assessment of soil organic matter in wetlands from the high Andes. *Soil Science Society of America Journal* 74(6), 2246–53. doi:10.2136/sssaj2009.0445.

Séguy, L., Bouzinac, S. and Husson, O. 2006. *Direct-Seeded Tropical Soil Systems with Permanent Soil Cover: Learning from Brazilian Experience. Biological Approach to Sustainable Soil Systems* (Uphoff, N., Ball, A. S., Fernandes, E., Herren, H., Husson, O., Laing, M., Palm, C., Pretty, J. and Sanchez, P., Eds). CRC Press, Taylor & Francis, pp. 323–42.

Shang, C. and Tiessen, H. 1998. Organic matter stabilization in two semiarid tropical soils: size, density, and magnetic separations. *Soil Science Society of America Journal* 62(5), 1247–57. doi:10.2136/sssaj1998.03615995006200050015x.

Six, J., Elliott, E. T. and Paustian, K. 1999. Aggregate and soil organic matter dynamics under conventional and no-tillage systems. *Soil Science Society of America Journal* 63(5), 1350–8. doi:10.2136/sssaj1999.6351350x.

Six, J., Bossuyt, H., Degryze, S. and Denef, K. 2004a. A history of research on the link between (micro)aggregates, soil biota, and soil organic matter dynamics. *Soil and Tillage Research* 79(1), 7–31. doi:10.1016/j.still.2004.03.008.

Six, J., Ogle, S. M., Breidt, F. J., Conant, R. T., Mosier, A. R. and Paustian, K. 2004b. The potential to mitigate global warming with no-tillage management is only realized when practised in the long term. *Global Change Biology* 10(2), 155–60. doi:10.1111/j.1529-8817.2003.00730.x.

Sollins, P., Homann, P. and Caldwell, B. A. 1996. Stabilization and destabilization of soil organic matter: mechanisms and controls. *Geoderma* 74(1–2), 65–105. doi:10.1016/S0016-7061(96)00036-5.

Stevenson, F. J. 1994. Humus chemistry. *Wiley, New-York*.

Tivet, F., Sá, J. C.d.M., Lal, R., Borszowskei, P. R., Briedis, C., dos Santos, J. B., Sa, M. F. M., Hartman, D. D., Eurich, G., Farias, A., Bouzinac, S. and Seguy, L. 2013a. Soil organic carbon fraction losses upon continuous plow-based tillage and its restoration by diverse biomass-C inputs under no-till in sub-tropical and tropical regions of Brazil. *Geoderma* 209-210, 214–25.

Tivet, F., de Moraes Sá, J. C., Lal, R., Briedis, C., Borszowskei, P. R., dos Santos, J. B., Farias, A., Eurich, G., Hartman, DdC., Nadolny Junior, M., Bouzinac, S. and Séguy, L. 2013b. Aggregate C depletion by plowing and its restoration by diverse biomass-C inputs under no-till in sub-tropical and tropical regions of Brazil. *Soil and Tillage Research* 126, 203–18. doi:10.1016/j.still.2012.09.004.

Tivet, F., Sá, J. C. M., Lal, R., Milori, D. B. M., Briedis, C., Letourmy, P., Pinheiro, L. A., Borszowskei, P. R. and Hartman, D. D. C. 2013c. Assessing humification and organic C fractions by laser-induced fluorescence and FTIR spectroscopies in conventional and no-till methods in Brazil's sub-tropical and tropical agro-ecoregions. *Geoderma* 207–208, 71–81.

Tisdall, J. M. and Oades, J. M. 1982. Organic-matter and water-stable aggregates in soils. *Journal of Soil Science* 33(2), 141–63. doi:10.1111/j.1365-2389.1982.tb01755.x.

Wattel-Koekkoek, E. J. W., van Genuchten, P. P. L., Buurman, P. and van Lagen, B. 2001. Amount and composition of clay-associated soil organic matter in a range of kaolinitic and smectitic soils. *Geoderma* 99(1–2), 27–49. doi:10.1016/S0016-7061(00)00062-8.

Chapter 9

Biodiversity management practices and benefits in Conservation Agriculture systems

Scott Day, Treelane Farms Ltd, Canada; Ademir Calegari, Agricultural Research Institute of Paraná State (IAPAR), Brazil; Alessandra Santos, Marcus Cremonesi, Lilianne Maia and Wilian Demetrio, Federal University of Paraná, Brazil; and Marie L. C. Bartz, Coimbra University, Portugal

1 Introduction

Soil biodiversity is vitally important because soil is one of the most complex habitats on earth and hosts an astounding diversity of life. It is estimated that 25–30% of all species on Earth live in soils for all or part of their lives (Orgiazzi et al., 2016). Biodiversity in soils and its functioning is critical to sustaining life on Earth, including human life. Ecosystem services provided by soil organisms include supporting agricultural production, controlling nutrient loads in soils and water, and regulating greenhouse gas cycling and climate change (Bardgett and van der Putten, 2014). For many years, the concept of biodiversity was associated only with natural systems, especially due to low aboveground diversity in agricultural lands (i.e. low diversity of crops) and the constant soil disturbance practices used (House and Parmelee, 1985). However, the adoption of conservation practices in agriculture has been promoting the development of new habitats for several organisms, mainly soil inhabitants, which are absents in conventional tillage systems (Barros et al., 2003; Bartz et al., 2013). These new agricultural models were developed in order to improve the sustainability

http://dx.doi.org/10.19103/AS.2019.0049.09

of agriculture and reduce the degradation of agroecosystems caused by poor agricultural practices.

Conservation Agriculture (CA) management practices such as no or minimum soil disturbance, soil mulch cover, plant diversity, and crop rotation/association directly affect biodiversity by increasing system complexity and promoting the establishment of a high number of species, particularly those associated with soil (Brown et al., 2018). This increase in the biodiversity in CA systems promotes not only optimum functioning of agroecosystems but also increases crop productivity due to the large number of beneficial organisms such as plant growth-promoting bacteria and invertebrates such as earthworms. Besides the high diversity of organisms in CA systems, there is also an increase in water infiltration and storage, soil organic matter, nutrients cycling, and pest control (Brown et al., 2018). This chapter reviews the effects of CA systems and practices on biodiversity, especially soil biodiversity, and the benefits of promoting biodiversity.

2 Soil microorganisms and their importance

The soil is home to a large variety of microorganisms and invertebrates known as soil biota which are responsible for numerous processes in the soil (Brown et al., 2018; Correia and Oliveira, 2000). Among the many organisms that are part of the soil biota, the soil fauna is exclusively composed of invertebrates. They can be classified according to the length of the organism's body: organisms larger than 4.0 mm are classified as macrofauna, organisms between 0.2 mm and 4.0 mm are mesofauna, and microfauna are composed of small organisms whose dimensions can only be visualized with the aid of optical microscopy (Ruiz et al., 2008).

Soil macrofauna includes more than 20 taxonomic groups: millipedes (Diplopoda), centipedes (Chilopoda), termites (Isoptera), spiders (Araneae), crickets (Orthoptera), cockroaches (Blattodea), beetles (Coleoptera), isopods (Isopoda), earthworms (Oligochaeta), ants (Formicidae), among many others. These organisms have diversified eating habits, feeding on wood (xylophages), plant remains and litter (detritus), live plants or roots (rhizophagus, phytophagous), other organisms (predators, necrophages), organic matter (humivores) and soil (geophages), as well as those that feed on many different sources (omnivores; Brown et al., 2008).

Earthworms, ants, termites, and beetle larvae are also considered ecosystem engineers, as they influence soil structure through bioturbation (producing nests, galleries, and aggregates), and thus affect the availability of resources to other organisms (Lavelle et al., 1997). Van Groenigen et al. (2014) showed that earthworms contributed to a 23% increase in belowground biomass and a 25% increase in harvest yields. Earthworms are also used for the biocontrol of pests and diseases, stimulating symbionts that promote the production of regulatory substances for plant growth (Bertrand et al., 2015; Blouin et al., 2013; Lavelle et al., 2016). Ants have the capacity to survive in different environments and play a role in the maintenance of soil quality (Correia, 2002). Beetles (*Carabidae*) are sensitive to soil management practices such as the use of pesticides in agroecosystems and the contamination of soils with heavy metals (Butovsky, 2011).

The relationship between soil attributes and the biodiversity of the soil organisms makes it possible to identify the functional groups most vulnerable to the land use system and which can be used as bioindicators to evaluate soil quality and functioning (Abreu et al., 2004). The decrease in the soil organism diversity due to the intensive use of the soil associated with poor management practices leads to changes which modify the diversity and structure of soil fauna communities, affecting ecosystem services and causing environment degradation (Rousseau et al., 2010, 2013; Peña-Peña and Irmler, 2016).

3 Effects of cropping practices on soil biodiversity and ecosystem functioning

The relationships between soil biodiversity and the functioning of the soil ecosystem have been extensively studied in the last decades. The functioning of the soil ecosystem can be measured through provision of environmental services such as organic matter decomposition, carbon sequestration, and plant production (Brown et al., 2018). Several studies reveal the potential value of the services rendered each year by soil biota worldwide, possibly exceeding $1.5 trillion (Pimentel et al., 1997; van der Putten et al., 2004).

Soil invertebrates and soil microbial diversity are part of the biological resources of the agroecosystems and should be considered in soil management decisions. Soil biodiversity is influenced by climate and management. These factors influence soil processes (carbon, nutrients, and water cycles) and services (nutrient and water-use efficiency and carbon sequestration; Fig. 1).

Agricultural practices can have a positive, negative, or no effect on soil macrofauna communities. The soil fauna is sensitive to disturbances in the agricultural environment caused by crop and soil management and is also affected by plant cover and quantity and quality of organic matter, pH, temperature, humidity, and soil texture (Baretta et al., 2006; Benazzi et al., 2013). For example, plowing decreases earthworm and termite populations (Stork and Eggleton, 1992), while pastures (native or planted) may increase their populations (Barros et al., 2003; Decaëns et al., 1994).

The management of agroecosystems can take place at several levels. Swift (1999) proposed a number of potential 'starting points' or management practices that may affect the soil, its biological processes, biodiversity, and the

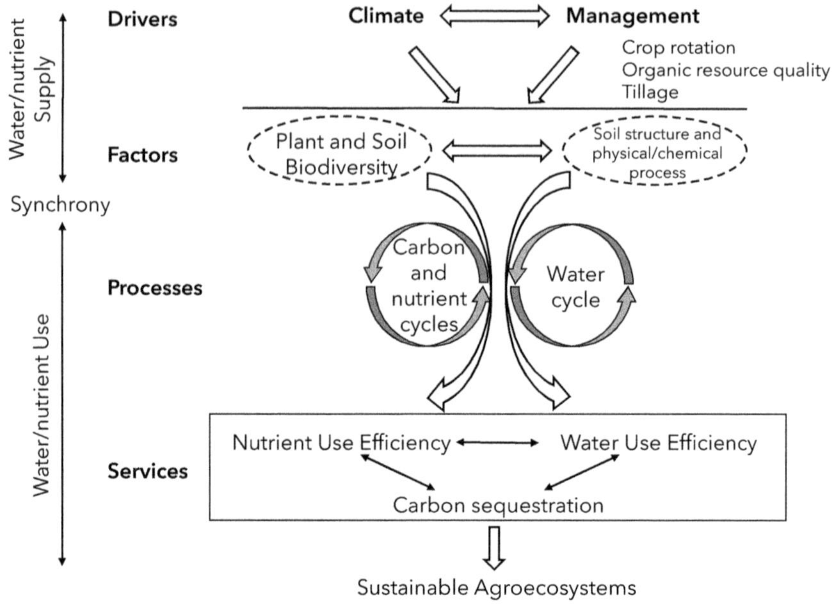

Figure 1 Conceptual diagram on the relationships between management, plant and soil biodiversity, soil structure, and nutrient and water ecosystem use efficiencies. Source: adapted from Brussaard et al. (2007).

contribution of soil biota to agricultural sustainability (Fig. 2). Each of these points can influence soil biota positively or negatively, directly or indirectly. Benefits for soil biota (activity and diversity), particularly in the long term, may come from the appropriate choice of crop rotation (e.g. grass and legume plants), trees, and their spatial and temporal distribution in the ecosystem, increasing the natural resistance to pests and diseases of the plants chosen (Altieri et al., 2007), improving the quality of straw produced, and management of organic matter and other external inputs (e.g. fertilizers) in the system.

The practice of minimal soil disturbance and of permanent soil surface cover in CA systems (e.g. crop litter, straw, stubble, or live cover crops) greatly benefit the food webs and belowground processes compared to conventional agriculture fields (House and Parmelee, 1985; Brown et al., 2002). The adoption of these practices can help to increase the complexity of soil organism communities due to the creation of a more suitable environment for their activities (Fries and Aita, 1999).

3.1 Positive effects of CA systems on soil biodiversity

The soil, as a living, open, and dynamic system, undergoes constant changes of different intensities and origins, whether chemical, physical or biological, natural

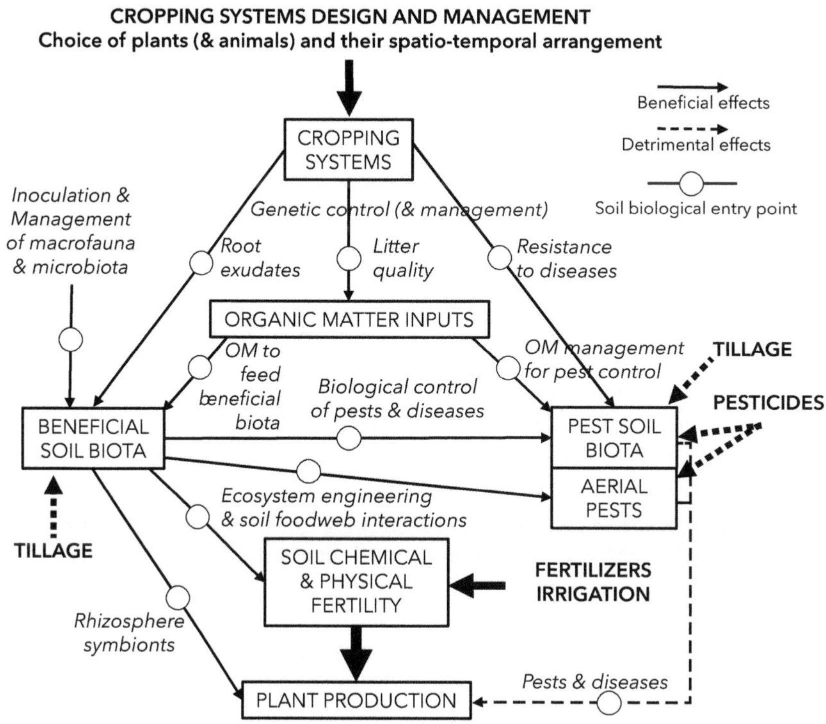

Figure 2 Potential starting points for the biological management of beneficial organisms and soil pests, cropping systems, organic matter inputs and soil fertility, affecting production. Source: adapted from Swift (1999); modified by Susilo et al. (2004).

or anthropic, and imposes such changes on its inhabitants. This change alters the population dynamics and therefore the biodiversity of the soil. Processes caused by soil fauna are also constant and of different intensities, which are reflected again in the soil habitat making up the cycle, where macro-effects alter and modify the micro environment, just as the micro-effects change and modify the macroenvironment. Doran and Parkin (1994) have shown that the effects of the interaction between macro and micro-agents can be considered as beneficial when they improve the system or harmful when resulting in negative impacts, both modifying the production and the quality of the soil. When dealing with the soil as an open system, human intervention can alter such effects in order to mitigate the negative effects or intensify the positive effects.

Agricultural systems, in general, tend to have a close relationship between complexity and biodiversity. More complex systems (e.g. with high aboveground plant diversity) produce high edaphic biodiversity, especially in the upper soil layers, where the biological component represented mostly by microorganisms (microbial biomass) can reach up 10% of soil organic content (Jenkinson and Ladd, 1981; De Araújo and Monteiro, 2007). Soil organisms have significant effects through their functioning, such as the mineralization of organic matter solubilizing and releasing nutrients, assimilation and degradation of potentially toxic substances, biological control of pathogens, soil aggregation due to exudate release, and beneficial symbiotic associations with roots (Assad, 1997; Kennedy and Doran, 2002; Decaëns et al., 2003; Baretta et al., 2003; Brown et al., 2003).

Conservation soil management systems, such as the no-tillage CA system, act through less soil disturbance, the use of cover crops/green manure, the adoption of crop rotation and decrease in the use of synthetic chemical molecules In doing so, they promote high soil biodiversity, with higher volumes of microbial biomass and invertebrate populations (Winter et al., 1990; Kladivko, 2001; Alves et al., 2006; Baretta et al., 2006; Paul et al., 2013; Balota et al., 2014). The increase in soil biodiversity through CA systems can have mitigating effects on climate change through carbon sequestration and also improve the cycling and inputs of nutrients in agricultural environments (Gregorich et al., 1994; Hu et al., 1997). In addition, soil organisms can improve the physical and chemical soil attributes in CA systems when compared to conventional management practices (e.g. monoculture and excessive soil preparation; Baretta et al., 2011; Baini et al., 2012; Ashford et al., 2013; Martínez et al., 2013).

Integrated crop-livestock systems, associated with no-tillage management, especially with grass/legume species associations, provide more favorable conditions for the development of soil ecosystem engineers such as earthworms (Da Silva et al., 2006; Marchão et al, 2009a,b). As an example, in Brazil (Cerrado biome), crop-livestock systems associated with no-till CA practices showed higher diversity of ecosystem engineers than continuous pasture fields and tilled arable fields, emphasizing the potential of CA systems to enhance invertebrate diversity in agriculture fields (Aquino et al., 2008; Marchão et al., 2009a,b). Consequently, the earthworm has become the symbol of the Brazilian No-Tillage Farmers Federation (Brown et al., 2003; Bartz et al., 2013), because of the high population of these organisms in agricultural lands following the adoption of no-tillage CA management. Earthworms are universal indicators of soil health and quality (Paoletti, 1999; Huerta et al., 2009). Several studies in south Brazil, the region where the adoption of no-tillage practice began in South America, have shown the significant numbers of earthworms, especially native species which are more sensitive to land-use changes. Most of these native species are new undescribed species, and the results confirm the capacity of the no-tillage CA system to preserve or enhance soil biodiversity (Santos et al., 2008; Bartz and Brown, 2011; Bartz et al., 2011, 2013, 2014; Demetrio et al., 2018).

3.2 Importance of plant diversity to biodiversity and soil health

The successful experiences and results obtained during the last two decades in Paraná State, and in other parts of Brazil and Latin America countries, have shown that cover crops and crop rotation in properly designed no-tillage CA systems are both economically feasible and ecologically sustainable. Such systems not only provide higher long-term crop yields, but also conserve, maintain and/or regenerate soil health and fertility, sustain greater biological activity in the soil, and decrease the effects of insect pests and/or diseases, and nematodes. They represent a harmonized and dynamic way of managing soils and the crops sustainably. In 2018, no-tillage CA systems occupied more than 6 million ha in Paraná and estimates show that the no-tillage CA system covers more than 33 million ha in Brazil (Febrapdp, 2019).

Research internationally shows strong evidence of climate change leading to greater variations in the distribution and amount of precipitation, tending in many regions to diminish the number of events (when rain falls) but to increase rainfall intensity. This will lead to greater risk of runoff and soil erosion and the consequent loss of soil particles, water and nutrients and, thus, soil fertility. Consequently, it is essential that the soil surface must be kept covered with crop residues, and managed with minimum soil disturbance as possible (e.g., no-tillage and direct drilling the crops), so that the soil profile favors water infiltration and has higher water-holding capacity. Integrated soil and water conservation methods include cover crops/green manure, crop rotation, terraces, strip crops, watershed ways, and so forth (Bolliger et al., 2006; Derpsch et al., 1986; Calegari et al., 2013).

Continuous no-tillage system management that includes the use of summer/winter cover crops results in the greatest amount of soil organic matter in the surface soil. No-tillage CA systems with winter cover crops store greater amounts of soil organic C, sustaining in particular the productivity of Oxisols in tropical and subtropical regions of the world (Calegari et al., 2008).

Experiences in some temperate climates also show favorable results when no-tillage CA systems are implemented. The use of cover crops, returning of crop biomass as soil cover, crop rotation, and no soil disturbance has promoted soil recovery and enhancement of soil productive capacity in Brazil, the United States, Australia, Russia, among other countries.

Many different cover crop species have been used successfully in all these countries. These include *Cajanus cajan* (normal, dwarf), crotalarias (*C. juncea*, *C. spectabilis*, *C. ochroleuca*, *C. breviflora*), *Mucuna aterrima* (grey, black, dwarf), *Clitoria ternatea*, buckwheat (*Fagopirum sculetum*), millet (pearl, finger), sudangrass, *Stylosanthes* sp., *Arachis pintoi*, *Neonotonia wightii*, *Brachiaria* sp., *Paspalum* sp. (including fall/winter species in South Brazil), *Avena strigosa* (black oat), *Raphanus sativus* (Radish), *Vicia villosa* (hairy vetch), field pea (*Pisum sativum* subsp. Arvense), lupin (*Lupinus albus*), rye (*Secale cereale*), *Melilotus* sp., ryegrass, clovers, and serradella.

Beyond the use of individual species, some mixtures of cover crops (cocktails) have been studied and recommended over the last 20 years (Calegari, 2010). These cover crop mixtures include combining two, three, four, six, or more species in the field, such as:

- oat + vetch, radish + black oat + vetch;
- buckwheat + radish + pear millet, crotalaria + pearl millet; and
- cajanus + pear millet ADR-300 + crotalaria, pear millet ADR-300 + crotalaria + buckwheat.

These mixtures can be used in annual and perennial cropping systems. Some species can also be undersown in standing crops such as soybean, maize, and beans around 20–30 days before the grain harvest. The cover crop seeds are broadcast by plane over the crops on large farms or by hand on small farms. After harvesting the grains, the cover crops achieve growth of around 20–30 cm in height. Many cover crop species can increase the abundance of predators or entomopathogenic organisms which decrease nematode populations as well as promoting plant bioactivation. Some biological products such as *Bacillus subtilis,* and other species of *Bacillus*, *Trichoderma*, and so forth, can help to enhance soil health in different CA cropping systems.

The understanding of how crop residues work on different soil attributes is important in the development of soil health and fertility. Their influence on nutrient cycling and soil chemical properties, and the integration of residue management strategies into different cropping systems, are one of the keys to the development of good soil fertility management (Hargrove, 1991; Calegari, 2002a,b; Calegari et al., 2013).

Continuous monocropping succession systems in Brazil and other parts of the world, such as soybean–maize, rice–rice, cotton–wheat, soybean–wheat, and so forth, have promoted pests, diseases, nematodes, weed infestation, soil compaction, degradation, and decrease in crop yields. This contrasts with the effects of no-tillage systems when properly combined with different cover crops, both as individual species or mixes/cocktails of species (two, three, six, or more species together) and other soil conservation practices to minimize soil degradation and promote soil health.

3.3 Crop and livestock integration in CA system

Integration of crops and livestock is a common practice with farmers in Chile, Uruguay, Argentina, Brazil, and Paraguay (Rodríguez et al., 2016). It is an important management practice in CA systems that increases system biodiversity. CA systems can also solve potential conflict between herders and farmers, particularly in agropastoral areas. In individual farming systems, the potential conflict is in the use of plant residues to meet the double goals of animal feeding and soil protection. In Africa, agroforestry species have been shown to provide animal feed alternatives to crop residues and cover crops (Mbow et al., 2014). In countries in Latin America (Brazil, Argentina, Chile), there are good examples of integrated crop-livestock CA systems (Moraes et al., 2014).

According to Bolliger et al. (2006), in the seasonally dry Cerrado region of Brazil, it is common to have ley systems where forage land is set aside and grazed for 3–4 years. This land is grazed until the unfertilized pasture becomes nitrogen deficient, the soil shows some compaction and there is an increase in *Pratylenchus brachyurus*. This decrease in pasture quality is reversed by growing legumes or fertilized crops, and a new parcel of land is set aside for grazing. *Brachiaria*, *Stylosanthes,* and pearl millet species have been shown to be both efficient in recycling nutrients and good forage species in the Cerrado (Scopel et al., 2004). If grown as cover crops toward the end of the rainy season, they can be grazed immediately at the beginning of the following dry season.

Due to their production of large amounts of biomass, sequences of *Brachiaria*, *Stylosanthes,* and pearl millet, and occasionally forage sorghum species, combined with commercial crops, provide a low-cost way of building SOM levels and rehabilitating degraded pastures (Scopel et al., 2004). Broch et al. (1997) found that such systems can support soybean yield increases over a number of years. Kluthcouski et al. (2000), Kluthcouski and Aidar (2003), and Kluthcouski and Stone (2003) describe the 'Santa Fé' cropping system in the Cerrado, which mixes a maize crop and a *Brachiaria* pasture. The *Brachiaria* is made to germinate and emerge later than the maize. During the whole maize cycle, the *Brachiaria* is shaded by maize plants and remains minimally competitive to the maize, but at maize harvest, the pasture is already in place, and grows very quickly over the maize residues. The *Brachiaria* is then killed with a desiccant herbicide, and soybean is planted directly into the dead/dying *Brachiaria* sward, using a zero-till planter to break through the superficial compaction layer (8–10 cm) of the pasture (Landers, 1998). Similar types of systems have been devised in southern Brazil, including a rotation of ryegrass used as pasture during winter, followed by a soybean crop planted directly on the chemically desiccated ryegrass (Scopel et al., 2004). This tight integration between forage and grain crops generally leads to a better use of total farmland and a more intensive use of pastures, with shorter turnover and less pasture degradation.

Research conducted by Alves et al. (2006) showed that integrated animal and grain crop systems have enabled the production of high-quality milk and meat at highly competitive costs. A combination of perennial and annual forage species, pasture fertilization, and animals with high genetic potential has allowed the production of high-quality animal products with high market value. In stocker operations, the system has made it possible for animals to be ready for slaughter at less than 20 months of age, finished on pasture and with a high-quality carcass. In dairy enterprises, work has been conducted on pasture production with and without supplementation, resulting in milk yields ranging from 15 to 20 L animal^{-1} day^{-1}, at low cost of production, demonstrating that high profitability is possible in this system.

Legumes are less used in extensive pastures in Brazil. Reasons for this situation include lack of information, the availability and cost of seeds, the poor persistence of legume plants under grazing, low palatability, and high fertility requirements. Legumes can be used in integrated animal and grain crop systems, because a crop rotation with pasture achieves higher levels of fertilization. It would be useful for researchers to evaluate leguminous pastures in this integrated system and their potential in extensive systems. There are already many examples in different regions of Brazil that show the favorable effects of integrated systems on soil attributes (chemical, physical, and biological), crop yield and farm net income when the integration of crop and livestock is managed effectively.

4 Effectiveness of diversified CA cropping systems

In no-tillage CA systems, Da Silva et al. (2006) showed that the effects of biomass mulch on soil surface avoided soil warming compared to other systems (Fig. 3), demonstrating the role of mulching in reducing the maximum soil temperature. This effect is due to low thermal conductivity and high sunlight reflectivity of straw (Hillel, 1998). Other studies have also shown a reduction in soil temperature down to 20 cm depth through oat straw cover throughout the year compared

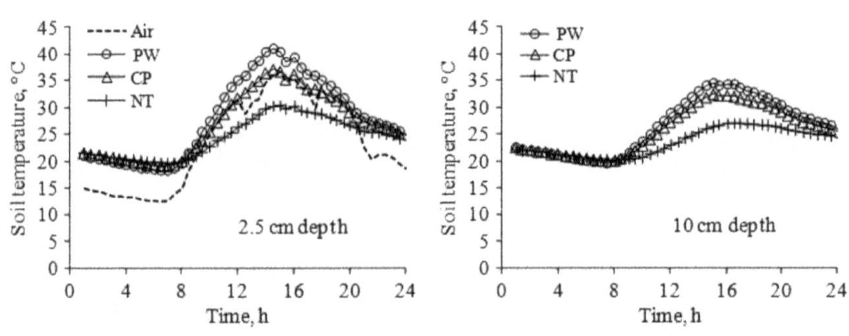

Figure 3 Air and soil temperature in the three soil management systems on 2 December 2001. NT = No-Tillage, PW = No-Tillage followed by plowing and harrowing; CP = No-Tillage followed by chisel plowing. Source: adapted from Silva et al. (2006).

with soil without straw cover (Trevisan et al., 2002). Research suggests this effect is magnified in the order: conventional tillage > no-tillage > permanent cover (Sidiras and Pavan, 1986). A lower portion of surface energy balance will be used as latent heat in the system, reducing evaporation of water from soil.

Adding organic carbon to the soil through crop growth, as well as keeping crop residues on the soil surface, are important measures to maintain and improve organic matter in soil. Plants used as cover crops both produce vegetal biomass (shoot and roots) and have important direct and indirect effects on the soil-water-plant system, playing a fundamental role in enhancing soil health and productivity.

The long-term efficiency of CA has been reviewed by Calegari and Alexander (1998), Calegari (2002a,b), Calegari et al. (2008, 2012, 2013), and Balota et al. (2014). They conducted long-term experiments over periods ranging from 12 to 23 years. The soil at the experimental site is very acid Oxisol (Rhodic Hapludox), with a high clay content (72% clay, 14% silt, and 14% sand), located in Southwestern Paraná State, Brazil (52° 41′ W, 26° 07′ S and 700 m altitude). The site is situated in a subhumid tropical zone (with no dry season, rainy summers and the average hottest month lower than 22°C).

Research by Calegari and Alexander (1998) showed that, after 12 years of cultivation with cover crops and crop rotation in a no-tillage CA system, there was a significant improvement in soil fertility and crop yields. Small differences were found in the nutrient content in the soil profile with maize crops regardless of whether mineral nitrogen was applied or not. Because of this, the nutrient results presented are only for the no-nitrogen application treatment in the maize crop. The experiment results showed that, at 0–5 cm depth, the accumulation of organic material on the soil surface and the absence of plowing in a no-tillage system play an important role in maintaining higher levels of organic carbon and phosphorus than in a conventional tillage system (Fig. 4).

A later study by Calegari et al. (2008) concluded that no-tillage CA system management sequestered 6.84 Mg ha^{-1} more organic carbon compared to conventional tillage system management (64.6%) at a soil depth of 0–10 cm and 29.4% more at a soil depth of 0–20 cm, but there were equivalent amounts of organic carbon as conventional tillage at a soil depth of 20–40 cm. Higher amounts of soil organic carbon were found at a depth of 0–20 cm (i.e., the moldboard plow layer). Results also showed that, when winter cover crops were used in no-tillage CA system, the organic carbon sequestered was generally increased.

More recently, Calegari et al. (2012) have reviewed the impact over 23 years of using cover crops and crop rotations in a no-tillage CA system. They showed that the phosphorus (P) stored in the topsoil (0–10 cm depth) and soil microbial biomass were higher in the no-tillage CA system than in a conventional tillage system, while the topsoil to a depth of 5 cm under the

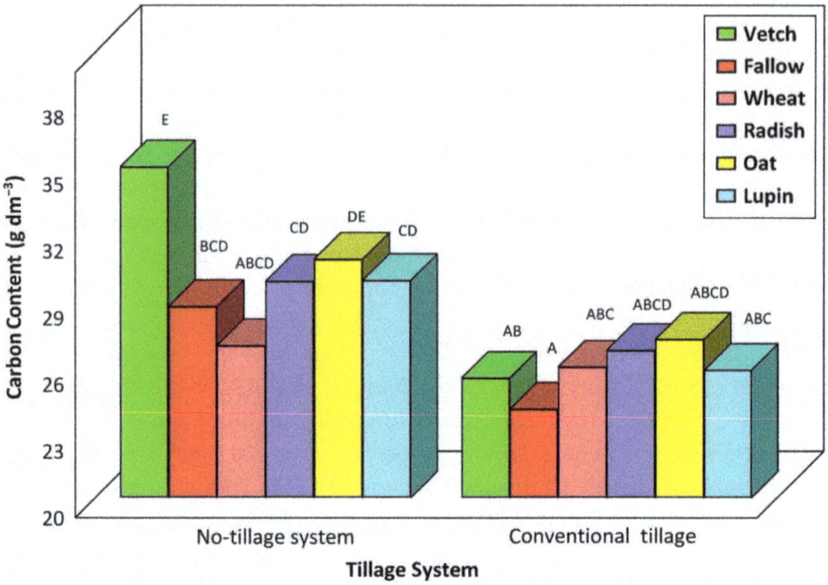

Figure 4 Effects of the different winter cover crops in the tillage systems (no-tillage and conventional tillage) on the carbon content (0–5 cm) in the soil profile. (LSD test, $p < 0.15$). Source: adapted from Calegari and Alexander (1998).

no-tillage system showed a higher content of organic phosphorus, higher total phosphorus, a higher content of total soil organic carbon, and an increase in the activity of acid phosphatase enzyme, compared to conventional tillage. In the 0–5 cm soil layer, labile organic P was higher in soil under a no-tillage CA system than in conventional tillage, but only in rotation with oat, vetch, lupin, and wheat. The total phosphorus content in the 10–30 cm soil layer in conventional tillage system was 10% higher than that found in a no-tillage system. Regardless of soil management, the winter crops had a higher P stored in the soil microbial biomass compared to the fallow in all soil layers, particularly lupin. In the soil under no-tillage, the winter crop increased the content of labile organic P extracted by $NaHCO_3$ up to a depth of 0–5 cm, and acid phosphatase enzyme activity up to a depth of 10 cm. At a depth of 0–5 cm under no-tillage, the highest value of acid phosphatase enzyme activity among the winter crops was obtained with vetch treatment (1625 $\mu g\ g^{-1}\ h^{-1}$), and the highest level of labile organic P was obtained with the cultivation of black oat and blue lupin. Moreover, crop production in winter also increased the total organic carbon content in both soil tillage systems, especially in the 0–5 cm and 5–10 cm soil layers. Black oat and blue lupin were found to be more efficient in phosphorus cycling, increasing the soil-available phosphorus content especially in the surface soil under no-tillage (Calegari et al., 2013). The lower amount of biomass produced over time when no cover crops were used in the winter period resulted in lower phosphorus and potassium availability in the soil, showing the important role of growing winter species to maintain or increase soil fertility.

The different cover crops used had significant effects on organic carbon content at differing soil depths (0–5 and 5–20 cm depth). The concentration of organic material on the surface contributed to an increase in organic carbon values (0–5 cm depth) compared to fallow. The mean values of all winter crops were greater than fallow in both depths. This showed clearly the beneficial effect of winter cover crop in increasing the organic carbon content in the soil profile.

The results also showed difference among different cover crops regime (0–5 cm) for calcium, magnesium, and potassium no-tillage CA systems; however, in the lower layer (5–20 cm), no significant difference was observed for phosphorus and organic carbon (ploughed system) and variable values were found for other nutrients and pH (Calegari and Alexander, 1998; Calegari, 2002a,b). Growing winter crop species rotated with summer crops, especially under no-tillage system, resulted in a lower requirement for phosphate fertilizers.

Winter cover crop species as a part of a no-tillage system are an important practice for enhancing soil microbial quality and soil organic carbon stocks. The inclusion of legumes in crop rotation is important for nitrogen balance in the soil-plant system and provides a key driver to enhance soil organic carbon.

Balota et al. (2014) have also reviewed the efficiency of the CA system with cover crops over a 23-year period. They highlight the importance of soil microbial biomass and activity, not only as indicators but because they have an important role in the supply of nutrients to plants. The microbial biomass immobilizes significant amounts of nutrients, decreasing the pools of free nutrients, with a consequential reduction in nutrient losses by leaching or denitrification. One of the benefits of the use of cover crop in CA systems is to utilize the microbial biomass as a substantial reserve of nutrients for crops. Although no-tillage system had higher microbial biomass than conventional tillage, the qCO_2 was lower; a decrease of about 39% at 0–5 cm layer and 24% at 5–10 cm layer.

According to Balota et al. (2014), the change of microbial properties due to different winter crops may be because considerable residue is returned to the soil annually. The amount of aboveground dry-matter yield produced in 23 years of fallow treatment (weedy biomass) was about 31.0 Mg ha^{-1} in conventional tillage and 42.5 Mg ha^{-1} in the no-tillage system. The winter crop treatments produced from 68 Mg ha^{-1} (wheat) to 88 Mg ha^{-1} (oat) in conventional tillage and 75 Mg ha^{-1} (wheat) to 100 Mg ha^{-1} (oat) in the no-tillage system (Table 1).

Significant effects were observed in different winter crops on organic carbon and microbial activity (Balota et al., 2014). The C:N ratio of winter cover

Table 1 Total aboveground dry biomass yield over 23 years in different winter cover crops in conventional (plow) tillage (CT) and no-tillage (NT)

Winter crops	Winter cover crops[a]		Summer crop residues[b]		Total		Annual mean	
	CT	NT	CT	NT	CT	NT	CT	NT
	Mg ha^{-1}							
Fallow[c]	31.0	42.5	91.7	93.1	122.7	135.5	5.3	5.9
Oat	87.5	99.9	93.6	95.4	181.1	195.3	7.9	8.5
Wheat	68.6	75.0	86.0	89.9	154.6	164.9	6.7	7.2
Radish	69.5	84.9	97.1	99.0	166.5	183.9	7.2	8.0
Lupin	76.0	87.5	91.1	97.2	167.1	184.7	7.3	8.0
Vetch	73.3	85.6	91.0	97.7	164.3	183.3	7.1	8.0
Mean	405.9	475.4	550.5	572.3	956.3	1047.6	41.5	45.6

[a] Values are the sum of biomass yield of the winter cover crop used as treatments in 1986–1990, 1992, 1994, 1999–2001, and 2005; plus black oat biomass yield in 1991, 1995, 1996, and 1998; plus black oat + radish biomass yield in 1997, 2002–2004.
[b] Values are the sum of crop residues produced by maize cultivated in 1986–1988, 1992, 1994, 1996, 1999, and 2003 and by soybean cultivated in 1989–1991, 1993, 1995, 1997, 1998, 2000–2002, and 2004.
[c] The fallow biomass yield consisted of weed biomass.
Source: adapted from Balota et al. (2014).

crops affects the level of their decomposition. The C:N ratio is 26 for oat, 43 for wheat, 23 for radish, 16 for lupin, and 13 for vetch. Generally, low C:N ratios promote faster decomposition in legumes (lupin and vetch) and high C:N ratios promote slower decomposition rate in grasses (oat and wheat). The amount of above- and belowground biomass produced by winter cover crops also has a fundamental role in soil carbon sequestration, which is an important sink for C storage, because soil contains three times more C than vegetation and twice that of the atmosphere (Kumar et al., 2006).

Winter crops, such as vetch, increased microbial biomass which, in turn, increased enzyme activity (phosphatase and arylsulfatase) and polysaccharide content. Winter crops had a higher microbial biomass than fallow treatment and the qCO_2 was lower, a decrease of up to 60% under conventional tillage and no-tillage system. As expected, the fallow treatment created soil conditions that decreased the efficiency of C substrate use by the soil microbial community.

Some winter crops species enhanced the percentage of the total C as MBC (MBC:Corg) up to 178% in conventional tillage and 85% in no-tillage systems, in the 0–10 cm layer. However, the MBC:Corg varied from 0.66% to 1.77% and increased soil organic carbon over the 23-year period. The highest accumulation of organic C occurred in lupin, which increased by 62% in CT and 126% in NT, with, respectively, a MBC:Corg ratio of 0.93% and 1.43%, followed by black oat. These results suggest that the MBC:Corg relationship in tropical/

subtropical conditions may have a unique threshold for carbon accumulation compared to temperate regions (Balota and Chaves, 2011).

The use of legume species in crop rotations is particularly important, mainly in tropical/subtropical soils, due to the potential for biological N_2 fixation (BNF), which contributes to soil fertility and the balance of the organic carbon pool. Introducing lupin as a cover crop improved N inputs due to BNF, which replaced N exports by grain removal, and may provide a key driver for preserving the soil organic C stock (Balota et al., 2014). Another fact that should be considered is the large amount of root mass that is produced and recycled annually by the groundcover. For example, oat can produce up to 2.7 t ha^{-1} of root dry matter, while radish and vetch produce 1.4 and 0.34 t ha^{-1} of root dry matter, respectively (Sanchez, 2012). Beyond the large quantity of roots in the soil profile, these different species also release root exudates, different organic acids, and other metabolites that have significant effects on soil microbial activity. The patterns of root exudation, which vary quantitatively and qualitatively among plant species, influence the soil microbial community and the biological processes that the community regulates in different ways and intensities. From 5% to 21% of all photosynthetically fixed carbon by plant is transferred to the rhizosphere through root exudates, which range from 20% to 50% of plant biomass (Kumar et al., 2006).

5 Case study: biodiversity management practices and benefits in CA systems in South-West Manitoba (Canada)

This section is written by Scott Day. In 1989 I began farming with my family and also working as a Provincial Government Extension Agronomist and then a Research Farm Manager in Southwest Manitoba. My work took me throughout the Southwest corner of Manitoba and beyond to other parts of the World. In 2012, I left the Manitoba Government to help start a farmland investment and management company based in the USA. We invest in farmland across the US with a focus on soil conservation in our subsequent management of the properties. Our company also invests in new Ag Technology. However, I never stopped farming and I continue to farm with my family to this day. I spend the winter months in Silicon Valley working out of our head office there and I spend the summer months back in Manitoba near the family farm during the growing and harvest season.

Our farm is located approximately in the middle of Southwest region of Manitoba: 25 kms North of the North Dakota border and about 60 kms East from the Saskatchewan border. Our home farming region is located near the exact geographical centre of North America, making this region one of the farthest on the continent from the moderating effects of any ocean. As a result, it is also a region that sees some of the most extreme weather. In our farming region we can easily see −40 C and plus 40 C in a 6-month period. We can see 40 C temperature changes in a few days during the spring and fall. It can freeze one day and be plus 30 C the next day. Our number 1 cause of crop failure is drought, and our number 2 is flooding, our number 3 is too much heat and our number 4 is too much cold. With these incredible variations it is hard to compare one year to the next and somewhat impossible to define what a normal year represents.

With our farm location in the Eastern Canadian Prairies we grow a wide variety of crops depending on weather, agronomy and market conditions. Over the years our farm has produced sunflowers, flax, peas, dry beans, soybeans, barley, winter wheat, spring wheat, and all types of canola and rapeseed. Other crops that are common in my region now include grain corn, hemp, grass seed, alfalfa, lentils, oats, and rye. The main crops on our farm going into the 2019 planting season are canola, hard red spring (HRS) wheat, malt barley, soybeans and dry beans. Our farm has seen virtually no widespread tillage for almost 30 years. All crops are seeded and fertilized in one pass with ammonia, with the other nutrients being applied as dry products with our air seeder. The seed is placed in a very narrow band generally at 2.5 cm in depth, while the fertilizer is applied about 4 cm to the side and 4 cm below the seed in a separate narrow band (generally 1.25 cm wide). Most seeding is now done in a similar fashion on our Northern Plains.

Early in my career both as a farmer and as an Extension Agronomist, I was very fortunate to have been mentored by some of the pioneering no-till farmers in Western Canada and the Northern USA. These incredible visionaries encouraged me to get involved in the Manitoba-North Dakota No-Till Farmers Association along with various other Conservation Agriculture initiatives. The Man-Dak No-Till group was one of the first formed in the early 1980s and grew rapidly in members and influence in the early 1990s. It was a great experience and one I feel incredibly fortunate to have been involved with at the time. I got to be a part of the tremendous change in agriculture on the Northern Plains where, in a relatively short period of time, most farmland switched from being tilled to no longer being tilled. In my school years up until the late 1980's I remember when strong spring winds would often whip up dust storms by blowing top soil from 100's of km away, almost blocking out the sun at times. Since the advent of no-till in the 1990's I have never seen another one of these large dust storms in my region.

Trying to evaluate the changes or the impact of no-till on biodiversity is difficult in such a variable environment as the region in which I farm, but I will share some observations from my almost 30 years of working with and promoting CA systems and practices in my region of Central North America. However, before I begin, I want to share a section from the FAO Book 'The State of the World's Biodiversity for Food and Agriculture' (FAO, 2019, p 87):

> *'Links between biodiversity, biodiversity loss and disease risk:*
> *An increasing number of emerging infectious diseases in humans, animals and plants have been reported over recent decades (Anderson et al., 2004; Fisher et al., 2012; Jones et al., 2008). This has been linked to rapid habitat changes caused by urbanization and agriculture intensification (Hassell et al., 2017). Empirical studies show that high levels of biodiversity are associated with high levels of pathogen diversity (Morand and Lajaunie, 2017). However, increases in epidemics and the risk of disease emergence are associated with decreased biodiversity (ibid.), as deforestation and agricultural intensification increase contacts between wildlife, domestic animals and humans, favouring the spread of zoonotic diseases (Keesing et al., 2010). Empirical studies have also shown that species-rich host communities contribute to reducing the transmission of infectious diseases, a phenomenon known as the "dilution effect". The effect has been observed in studies on several vector-borne and zoonotic diseases. A meta-analysis of 90 studies on various diseases that affect humans, wildlife, livestock or plants concluded that they provide broad support for a negative effect of diversity on disease transmission (Johnson et al., 2015). The magnitude of the effect, however, appears to be related to the structure of species communities and not only to species diversity per se (Civitello et al., 2015). Similar effects operate at the intraspecies (genetic) level. Species that have high genetic diversity may sustain a high diversity of pathogens, but with each pathogen showing low transmission and rarely causing an epidemic. In contrast, while species that have low genetic diversity may sustain fewer pathogens, these pathogens may have high transmissibility and the potential to cause dramatic epidemics (Heesterbeek et al., 2015; Karvonen et al., 2016; King and Lively, 2012).'*

One of the issues that I remember clearly was the emergence of orange blossom wheat midge as a significant problem in spring wheat during the mid-1990's across the Northern Plains. This tiny, fragile midge has no mouth and simply exists to lay eggs on the flowering heads of cereal crops - mainly spring wheat. These midge eggs hatch into larvae that feed on the growing kernels in the wheat head causing yield reductions and quality problems. Before the wheat heads fully mature the well-fed larvae drop to the ground where they remain over winter, sometimes even for years, before they pupate in the summer into the tiny midges that fly up into the wheat canopy again to lay their eggs and start the cycle over again.

Many farmers and some scientists attributed its increased presence to the rise of no-till system at that time as they thought the undisturbed soil was a perfect harbor for the midge larvae/eggs. Though that was just an anecdotal observation, this assumed connection did gain considerable acceptance at the time. However, farmers spraying indiscriminately, regardless of the midge number threshold, seemed to need to spray the following year regardless of their tillage practice. However, farmers that did not spray indiscriminately often did not need to spray the following year. Tilled ground did not seem to make much difference in providing a less hospitable environment for the larvae to complete their life cycle. What some farmers found was that small ground-based predators consumed the larvae much faster if the soil was not disturbed and the field had not been sprayed before·

Dr. John Gavloski – Manitoba Agriculture's Provincial Entomologist – provided me with the following comments when I asked him recently about this wheat midge issue:

> *'I have never seen a peer-reviewed article linking increases in wheat midge levels to zero-till. So, this is anecdotal at best. There has been some research on ground-based predators feeding on wheat midge larvae. Ground beetles (Carabidae) prey on wheat midge larvae, and a study in Saskatchewan found 14 species of ground beetles feeding on wheat midge larvae (Floate et al., 1990. Environmental Entomology. 1503–1511). Four of these species were tested in lab studies with no alternate prey (just wheat midge larvae) and consumed 9, 15, 43, and 48 midge larvae per beetle per day respectively. I doubt they consume these levels in the field, where searching time and alternate prey would be factored in, but it does show that when wheat midge levels are high predation by ground beetles can certainly help reduce levels. The authors of this study estimated that daily predation in the field from all species of ground beetles combined ranged from < 1 to 86 midge larvae/m2. Most of this predation would occur during June and August.*
>
> *I agree that farmers can get themselves in a trap (where they actually inflate long-term pressure from wheat midge) by applying insecticides for wheat midge when not needed. Likely a large part of what is happening is that in addition to predators they are damaging populations of a natural egg-larval parasitoid of wheat midge called Macroglenes penetrans. Research in Saskatchewan found they controlled an average of 31.5% of wheat midge across Saskatchewan (Olfert et al., 2009). Research in North Dakota found parasitism rates averaged 22% and ranged from 0 to 100% over the 12-year period.*
>
> *Entomologists in Alberta have noted high levels of Macroglenes penetrans when sweeping canola fields that were wheat the previous season. They would be emerging in the current year's crop (in this case canola) and searching for wheat fields and wheat midge to parasitize. So unnecessary insecticide applications to canola could not only affect predators and parasitoids of the canola pests, but in this case (where there was wheat the previous year) may impact regional levels of Macroglenes penetrans. This is another example of why using economic thresholds and minimizing unnecessary insecticide applications can be the most economical strategy for a farmer in the long-term.'*

It is my observation that we do not really know why orange blossom wheat midge became such a significant problem 25 years ago. It may have just been a coincidence that it happened at the same time no-till farming was taking off. However, it seems that once we got a few years into the no-till system there was enough biodiversity in the soil to help keep the insect pest in check. I personally have not sprayed for wheat midge in 24 years – but back then we sprayed 4 or 5 years in a row. There are areas where wheat midge continues to remain a problem and there are now more effective genetic and chemical methods for control of the pest, but I believe maintaining undisturbed topsoil helps significantly in the bio-control of the insect as well. These natural controls have helped keep midge at overall levels which are much less significant than they were 25 years ago.

At that same time during the advent of no-till on the Canadian Prairies, crucifer flea beetles were already a significant problem in canola – this was even though very potent insecticides were being used at that time with the seed. To this day virtually no canola is seeded on the Prairies without an insecticide, primarily because of this insect, but some of the most effective insecticides are no longer in use. It is my observation that no-till may have helped stabilize flea beetle populations. It is unreasonable to say that no-till on its own will control flea beetles because that just is not the case. On my farm and others in the region the pest is still a serious concern despite many years without tillage, and applying insecticides while seeding is still very important. However, we have only needed to spray for the insect once in the 25 years since we stopped tilling the land, while before then the need to spray was a regular occurrence. This could be attributed to more effective seed treatments in recent years too but, regardless of the other reasons for the insect's control, the elimination of tillage in the farming system certainly did not make the situation worse. This is the opposite of what many felt would happen 25 years ago when tillage was taken out of the seeding process. It was felt then that undisturbed soil would simply be a better environment for the flea beetle to exist and proliferate. Thankfully that does not seem to have been the situation:

The following are observations recently lifted directly from the Canola Council's Official website:

> *'Tillage: Flea beetles prefer bright relatively warm conditions. Direct seeding provides a microclimate that is less ideal for flea beetles than tillage before seeding. Planting into stubble may provide a microclimate that is cooler, moister and less favourable to flea beetles. Injury may be reduced due to the cooler microenvironment created by stubble shading on surface soil. Cooler temperatures at the soil surface slow flea beetle activity reducing damage. However, crop development can also be slowed by cooler temperatures, so frequent crop scouting is still essential to ensure damage thresholds are not exceeded.*
>
> *In Alberta, zero till canola production can reduce flea beetle damage to levels equivalent to those found with the use of insecticidal seed treatments. This seeding method produces large plants early and may reduce the grower's dependence on seed treatments and foliar insecticides, except under conditions of intense flea beetle pressure.'*

With the indiscriminate spraying of insects, we are often upsetting a balance or equilibrium that we do not fully understand. This is different to the indiscriminate spraying to control weeds or disease where the main risk is simply eventual resistance of that plant or disease. It is vital to preserving a healthy and biodiverse ecosystem in your fields that you only spray insecticides when economic population thresholds have been exceeded (Fig. 5).

Figure 5 Flea Beetle damage to canola seedlings.

If we look specifically at the economic benefits of not having to spray for those insects because they are already being controlled to some degree by the biodiversity found within the soil, we can assume the following savings: Total costs for wheat midge or flea beetle control are generally Canadian$8 per acre for the product and $9 for the application cost which equals about $17 an acre for treatment. There are often 45 million acres of wheat and canola planted in Canada each year so, without the biological reduction of these two insects (and other insects as well) from the biodiversity in Prairie soils, there would be hundreds of millions of dollars in extra production costs required to grow these crops each year. This extra spraying would also likely come with unintended negative impacts to the general insect community. Keep in mind that this simple and singular example is just for Canada; the implications worldwide for these crops would likely be in the billions$ of extra costs not to mention the environmental impacts of that excessive need for insecticides!

In the northern Regions of North America, the last glacier was only 8000–10 000 years ago. Almost the entire Northern Plains was covered in a sheet of ice up to 2 km thick. This glacier essentially killed whatever earthworms were in the soil at the time. It has been determined that most of the earthworms we see in a handful of topsoil in Western Canada can be now attributed to earthworm eggs being distributed across the Plains on the roots of trees being provided for new farm yard shelterbelts. Much of the seed stock for these shelterbelt trees came from Eastern Europe over 100 years ago and that is probably where most of our current earthworms have originated. There are countless articles that talk about the importance of earthworms in a healthy no-till CA system and how CA contributes to the increase in earthworm numbers in the soil.

I can say with certainty that earthworm numbers have increased dramatically since we stopped tilling the land on our farm. There are times when earthworm numbers are so high, they appear to be now 'dripping' off the seeding openers of our air seeder whenever we lift the unit out of the ground. In my youth when we tilled our fields the only place you would find many earthworms on our farm was in the garden in the spring. Earthworms improve the internal drainage of the soil, they improve the nutrient cycling within the soil, they contribute to the organic matter, they consume crop residues, and the list goes on. Tillage is the enemy of the earthworm and like many other farms we have seen their increase since we stopped tilling.

I would like to mention we have also used exclusively anhydrous ammonia (NH_3) as our nitrogen source on our farm over the past 40 years. The $NH_3$is banded 4 cm to the side and 4 cm below the seed row while seeding and the seeding rows are spaced 25 cm apart. While the soil biota will be disrupted in that narrow band of NH_3 every 25cm - there is obviously plenty of unaffected soil between the seed rows that allow the soil to return to equilibrium quickly and the high earthworm population to remain relatively stable. Seeding with NH_3 is very common in my local region but is not common across all the Prairies. The lack of local infrastructure for supporting the NH_3 is part of the reason it is no longer used extensively (Figs. 6 and 7).

The Northern Plains of North America are sometimes called the 'Duck Factory' of the continent. Migrating waterfowl of all types return to this large region after overwintering in the South to nest amongst and near the millions of potholes and temporary wetlands that dot the Prairie landscape after the

Figure 6 Earthworms have become abundant in our no-till fields.

Figure 7 Our air seeder with NH_3.

snow melts each spring. The advent of no-till has provided a much better habitat for these ground nesting birds compared to when fields were tilled. This improvement in habitat with no-till has motivated Ducks Unlimited (DU) - the waterfowl conservation company - to focus much of its activity on the promotion and perfection of no-till farming systems. DU, along with several government and farmer initiatives focusing on no-till research and extension, was instrumental in the adoption of no-till on the Prairies going back over 30 years. There are many studies done by the biologists that work with DU that supports their work in no-till promotion as an excellent tool in their goals of improving the habitat and increasing the populations of Prairie waterfowl, upland birds and shore birds.

It appears that the use of no-till along with fall seeded crops provides the best nesting environments in crop production agriculture for many of these migrating birds. DU biologists in Southern Alberta have seen that even finicky shore birds like the long-billed curlew have been nesting in no-till winter wheat crops at levels similar to their native prairie nesting densities. It is the practice of combining winter crops such as winter wheat, fall rye and winter triticale with a no-till farming system that provides the best habitat for ground nesting birds. This is primarily because there is basically no activity on that field when the birds are nesting in the spring because the crop is already growing. With spring seeded no-till fields the quality of the nesting habitat is almost as good but the seeding process itself will often destroy many of the nests.

In cultivated fields the nesting of birds is often not as successful because with so little residue left on the soil surface it is hard to disguise their nest from predators. Also, the color, look and feel of a cultivated field is so foreign to these birds compared to the native grasslands where they once nested. Winter crops planted in a no-till system is the closest they are going to get to a native grassland environment but still be nesting on fields that produce crops and food. Winter crops used in a conjunction with a spring seeded crops provide excellent diversity to the Prairie cropping system. It is like having a cover crop that you get to harvest for income as well. However, the value and poor economics of those winter crops at this time have made them much less attractive than they were even 5 or 6 years ago. Hopefully that cycle will rebound, and winter cereal crops will once again be popular in Western Canada (Fig. 8).

A few years ago, while seeding canola into thick and tall undisturbed wheat stubble, I came across this nest of the ground nesting 'marsh hawk' (also known as the 'northern harrier'). This is something I had never seen on cultivated land before as these hawks generally only nest in tall grass near marshes and swamps. Their presence is appreciated by farmers and others for their control of field rodents and even grasshoppers and other pests. I carefully

Figure 8 Mallard Duck nest in barley stubble that has been seeded to canola - our farm yard is in the background.

seeded around this nest, so it was left undisturbed and hopefully those chicks are now adults feeding on various vermin in my fields when they return each spring. This is an example of a beneficial bird that would not have nested in a tilled field (Fig. 9).

In addition to ducks and hawks we also find a myriad of other prairie birds nesting in no-till fields. Some of the most common are killdeer, sandpipers, and plovers. Once again, their nesting densities and successes seem to be much higher on our no-till fields compared to when we used to till the land. Other farmers in my region have observed these same results.

Along with the birds, earthworms, beneficial ground beetles and many other soil organisms that seem to now be flourishing in our no-till fields the other phenomenon I have observed in recent years is the return of many large mammals to our landscape. This cannot be attributed to no-till on its own, but our reduction or elimination of tillage that started 25 years ago certainly has not deterred wildlife becoming more prevalent on the Canadian Prairie landscape.

On our farm virtually all the trees have been planted by my family or other humans. When our family first took over the farm in the mid-1930s, you could not see a tree on the farm as it was simply some farm building sticking out amongst a vast sea of grass. Despite the few trees and the large swaths of open Prairie, I have seen over the past few years many animals on or near our farm that one would associate with more forested areas: moose, elk, whitetail deer, coyotes, foxes, wolves, and even black bears and mountain lions have been observed (and causing problems) only a few kilometers from our farm.

Figure 9 Marsh Hawk nest.

However, the really exciting recent observations for me are the open Prairie mammals that have returned to our farm: pronghorn antelope (which have not been officially seen in our region for 130 years) and mule deer (not officially seen in our region for 60 years) have now been on our farm in the past few years. If you were to come to our small community, outside of winter, I could almost guarantee I would be able to show you many of the animals I have mentioned (except the bear and mountain lion) within a 24-hour period. There is now a more concerted effort to provide proper habitat and there are many programmes that are helping to support wildlife on the Prairies such as the recent initiative from Ducks Unlimited. However, another reason we may be seeing this upswing in wildlife is because no-till has been implemented at the same time we have simply less people living on the land. Farms are much larger now and many farmyards have become abandoned. These small diverse woodlands where farm families once lived can provide excellent habitat for some of these larger animals. In addition to these old yards we now have large stretches of relatively undisturbed (no-till) farmland in between all these pockets of wooded habitat. These uncultivated fields provide predators with a better chance of finding small mammals like mice, gophers and those nesting birds and eggs that have been already mentioned. You essentially have, with no-till CA landscape, a better source of food in between the patches of protective habitat compared to when fields were always tilled. When fields are left untilled it is like having more restaurants on the route between your destinations (Figs. 10–13).

Tillage does not exist in nature. Therefore, it stands to reason that nature has evolved to appreciate/thrive in a soil that is left as undisturbed as possible. These undisturbed conditions provide the best possible environment for biodiversity to reach its full potential in any cropping system. Adding diversity to that no-till CA cropping system with winter and summer crops, using different crop species, utilizing cover cropping where appropriate, and including forages and grazing where possible greatly enhances this environment for the soil biology and the overall biology of a region. However, it is the elimination of tillage that is key to getting this all started and providing the base to build a biologically diverse farming system above and below the soil surface.

The following section is from the FAO publication 'The State of the World's Biodiversity for Food and Agriculture' (FAO, 2019, p 448):

> *'Diversification and management practices at production level:*
>
> *The use of a number of diversification strategies in food and agricultural production systems seems to be increasing. Evidence indicates that agroforestry is becoming more widespread in all regions of the world. Priorities in terms of strengthening the contributions of agroforestry to sustainable development include addressing problems in germplasm supply, improving the provision of marketing advice and developing a better understanding of gender-related implications. Home*

Figure 10 Moose in my Canola.

Figure 11 Coyote running across no-till wheat stubble after hunting for mice.

gardens are major reservoirs of biodiversity for food and agriculture (BFA) in many parts of the world. However, knowledge of the status and trends of these systems is limited. In the case of diversification in aquaculture, while traditional extensive diversified systems are tending to decline as a consequence of resource constraints, innovative polyculture approaches are creating opportunities to increase efficiency

Figure 12 Pronghorn on our farm.

Figure 13 Moose in my wheat.

and tackle problems related to fish health and effluent discharge. Integrated crop-livestock systems remain widespread globally. There is need for research into how complementarities between crop and livestock production can be enhanced in the context of limited availability of land and other resources, including research into the significance of within-species genetic diversity. The use of many management practices believed to help promote the conservation of BFA, or that utilize BFA in a sustainable way, is reportedly increasing, as is awareness of the benefits of such practices among consumers, producers, governments and international agencies. This appears to be the case, for example, for organic agriculture, low external input agriculture, management practices implemented with the aim of preserving and enhancing soil biodiversity, conservation agriculture, integrated plant nutrient management, integrated pest management, pollination management and sustainable forest management practices. Nevertheless, the availability of global data on the levels of implementation of many of these practices remains limited, and knowledge of their impacts on BFA and the supply of ecosystem services needs to be improved. Biodiversity-based and biodiversity-friendly management practices generally require detailed knowledge of local production systems and ecosystems and are often relatively labour intensive. Consequently, their implementation tends to require the active participation of producers and their organizations, as well as the presence of effective extension services. Management interventions often need to extend beyond farm boundaries into the broader landscape or seascape. Attention needs to be paid to maintaining or restoring ecosystems that deliver services to food and agriculture and conserving the species and genetic diversity that will allow adaptation to changing conditions'.

6 Where to look for further information

A good source of information on Biodiversity management practices and benefits in Conservation Agriculture systems can be found at FAO website (see Sustainable agriculture for biodiversity – Biodiversity for sustainable agriculture; http://www.fao.org/3/a-i6602e).

Several resources, including published articles, can be found at websites of the following centres of expertise include:

- The Conservation Agriculture Group (http://conservationagriculture.mannlib.cornell.edu);
- Brazilian No-Tillage Farmers Federation (FEBRAPDP) (https://febrapdp.org.br/);
- European Conservation Agriculture Federation (ECAF) (http://www.ecaf.org/); and
- Agricultural Research Institute of Paraná State (IAPAR) (http://www.iapar.br).

7 References

Abreu, S. L., Reichert, J. M. and Reinert, D. J. 2004. Mechanical and biological chiseling to reduce compaction of a sandy loam alfisol under No-Tillage age. *Revista Brasileira de Ciência do Solo* 28(3), 519–31. doi:10.1590/S0100-06832004000300013.

Altieri, A. M., Ponti, L. and Nicholls, C. I. 2007. Melhorando o manejo de pragas através da saúde do solo: direcionando uma estratégia de manejo do habitat solo. In: *Controle biológico de pragas através do manejo de agroecossistemas. Secretaria da Agricultura Familiar*. Ministério do Desenvolvimento Agrário, Brasília, DF, pp. 17–31.

Alves, M. V., Baretta, D. and Cardoso, E. J. B. N. 2006. Fauna edáfica em diferentes sistemas de cultivo no estado de São Paulo. *Revista de Ciências Agroveterinárias, Lages* 5(1), 33–43.

Anderson, P. K., Cunningham, A. A., Patel, N. G., Morales, F. J., Epstein, P. R. and Daszak, P. 2004. Emerging infectious diseases of plants: pathogen pollution, climate change and agrotechnology drivers. *Trends in Ecology and Evolution* 19(10), 535–44. doi:10.1016/j.tree.2004.07.021.

Aquino, A. M., Correia, M. E. F., Guimarães, M. F., Lavelle, P., Mercante, F. M. and Silva, R. F. 2008. Invertebrate soil macrofauna under different ground cover plants in the No-Tillage system in the Cerrado. *European Journal of Soil Biology* 44, 191–7.

Ashford, O. S., Foster, W. A., Turner, B. L., Sayer, E. J., Sutcliffe, L. and Tanner, E. V. J. 2013. Litter manipulation and the soil arthropod community in a lowland tropical rainforest. *Soil Biology and Biochemistry* 62, 5–12. doi:10.1016/j.soilbio.2013.03.001.

Assad, M. L. L. 1997. Fauna do solo. In: Vargas, M. A. T. and Hungria, M. (Eds), *Biologia Dos Solos Dos Cerrados*. EMBRAPA-CPAC, Planaltina, pp. 363–443.

Baini, F., Pitzalis, M., Taiti, S., Taglianti, A. V., Zapparoli, M. and Bologna, M. A. 2012. Effects of reforestation with *Quercus* species on selected arthropod assemblages (Isopoda Oniscidea, Chilopoda, Coleoptera Carabidae) in a Mediterranean area. *Forest Ecology and Management* 286, 183–91. doi:10.1016/j.foreco.2012.08.042.

Balota, E. L. and Chaves, J. C. D. 2011. Microbial activity in soil cultivated with different summer legumes in coffee crop. *Brazilian Archives of Biology and Technology* 54(1), 35–44. doi:10.1590/S1516-89132011000100005.

Balota, E. L., Calegari, A., Nakatani, A. S. and Coyne, M. S. 2014. Benefits of winter cover crops and No-Tillage for microbial parameters in a Brazilian Oxisol: a long-term study. *Agriculture, Ecosystems and Environment* 197, 31–40. doi:10.1016/j.agee.2014.07.010.

Bardgett, R. D. and van der Putten, W. H. 2014. Belowground biodiversity and ecosystem functioning. *Nature* 515(7528), 505–11. doi:10.1038/nature13855.

Baretta, D., Santos, J. C. P., Mafra, Á. L., Widner, L. P. and Miquelluti, D. J. 2003. Fauna edáfica avaliada por armadilhas e catação manual afetada pelo manejo do solo na região oeste catarinense. *Revista de Ciências Agroveterinárias* 2, 97–106.

Baretta, D., Santos, J. C. P., Bertol, I., Alves, M. V., Manfoi, F. and Baretta, C. R. D. M. 2006. Efeito do cultivo do solo sobre a diversidade da fauna edáfica no planalto sul catarinense. *Revista de Ciências Agroveterinárias* 5(2), 108–117.

Baretta, D., Santos, J. C. P., Segat, J. C., Geremia, E. V., Oliveira Filho, L. C. I. and Alves, M. V. 2011. Fauna edáfica e qualidade do solo. *Tópicos Ciência do Solo* 8, 119–70.

Barros, E., Neves, A., Blanchart, E., Fernandes, E. C. M., Wandelli, E. and Lavelle, P. 2003. Development of the soil macrofauna community under silvopastoral and agrosilvicultural systems in Amazonia. *Pedobiologia* 47(3), 273–80. doi:10.1078/0031-4056-00190.

Bartz, M. L. C. and Brown, G. G. 2011. As minhocas e o Sistema Plantio Direto na Palha. *A Ganja (Porto Alegre)* 67, 65–7.

Bartz, M. L. C., Brown, G. G., Rosa, M. G., Locatelli, M., James, S. W. and Baretta, D. 2011. Minhocas Urobenus sp.: das matas para as áreas sob plantio direto. *Revista Plantio Direto* 1, 6–7.

Bartz, M. L. C., Pasini, A. and Brown, G. G. 2013. Earthworms as soil quality indicators in Brazilian No-Tillage systems. *Applied Soil Ecology* 69, 39–48. doi:10.1016/j.apsoil.2013.01.011.

Bartz, M. L. C., Brown, G. G., Da Rosa, M. G., Filho, O. K., James, S. W., Decaëns, T. and Baretta, D. 2014. Earthworm richness in land-use systems in Santa Catarina, Brazil. *Applied Soil Ecology* 83, 59-70. doi:10.1016/j.apsoil.2014.03.003.

Benazzi, E. S., Bianchi, M. O., Fernandes, C. M. E., Lima, E. and Zonta, E. 2013. Impacts of harvesting methods of sugar cane on the soil macrofauna in production area in Espírito Santo – Brazil. *Semina: Ciências Agrárias* 34(6), 3425–42.

Bertrand, M., Barot, S., Blouin, M., Whalen, J. K., de Oliveira, T. and Roger-Estrada, J. 2015. Earthworm services for cropping systems. A review. *Agronomy for Sustainable Development* 35(2), 553–67. doi:10.1007/s13593-014-0269-7.

Blouin, M., Hodson, M. E., Delgado, E. A., Baker, G., Brussaard, L., Butt, K. R., Dai, J., Dendooven, L., Peres, G., Tondoh., J. E., Cluzeau, D. and Brun, J. J. 2013. A review of earthworm impact on soil function and ecosystem services. *European Journal of Soil Science* 64, 161–82. doi:10.1111/ejss.12025.

Bolliger, A., Magid, J., Amado, J. C. T., Skóra Neto, F., Ribeiro, MdFdS., Calegari, A., Ralisch, R. and de Neergaard, A. 2006. Taking stock of the Brazilian "zero till revolution": a review of landmark research and farmers' practice. *Advances in Agronomy* 91, 47–110. doi:10.1016/S0065-2113(06)91002-5.

Broch, D. L., Pitol, C. and Borges, E. P. 1997. Integração Agricultura-Pecuária: plantio direto da soja sobre pastagem na integração agropecuária. Fundação MS para Pesquisa e Difusão de Tecnologias Agropecuárias. Informativo Técnico, Maracaju-MS, p. 24.

Brown, G. G., Pasini, A., Benito, N. P., de Aquino, A. M. and Correia, M. E. F. 2002. Diversity and functional role of soil macrofauna comunities in Brazilian no-tillage agroecosystems. In: *International Symposium on Managing Biodiversity in Agricultural Ecosystems*. United Nations University/Convention on Biological Diversity, Montreal, Canada, pp. 1–20.

Brown, G. G., Benito, N. P., Pasini, A., Sautter, K. D., de, F., Guimarães, M. and Torres, E. 2003. No-Tillage greatly increases earthworm populations in Paraná state, Brazil: The 7th International Symposium on Earthworm Ecology·Cardiff·Wales 2002. *Pedobiologia* 47(5–6), 764–71.

Brown, G. G., Pasini, A., Oliveira, L. J., Korasaki, V., Sautter, K. D., Pereira, S. D. and Torres, E. 2008. A vida do solo no sistema plantio direto macro, micro e mesofauna, suas funções e importância. In: Canally and Cury. *Resumos. Encontro nacional de plantio direto na palha*, 11, Londrina: Febrapdp, pp. 97–105.

Brown, G. G., da Silva, E., Thomazini, M. J., Niva, C. C., Decaëns, T., Cunha, L. F. N., Nadolny, H., Demetrio, W. C., Santos, A., Ferreira, T., Maia, L., Conrado, A. C., Segalla, R. F., Ferreira, A. C., Pasini, A., Bartz, M. J. C., Sautter, K. D., James, S., Baretta, D., Antoniolli, Z. I., Briones, M. J. I., Sousa, J. P., Römbke, J. and Lavelle, P. 2018. The role of soil fauna in soil health and delivery of ecosystem services. In: Reicosky, D. (Ed.), *Managing Soil Health for Sustainable Agriculture Volume 1: Fundamentals*. Burleigh Dodds Science Publishing, Cambridge, UK, pp. 197–241.

Brussaard, L., Kuyper, Th. W., Didden, W. A. M., de Goede, R. G. M. and Bloem, J. 2004 Biological soil quality from biomass to biodiversity - importance and resilience to management stress and disturbance. In: Schjønning, P., Emholt, S. and Christensen, B. T. (Eds), *Managing Soil Quality - Challenges in Modern Agriculture*. CAB International, Wallingford, UK, pp. 139-61.

Brussaard, L., de Ruiter, P. C. and Brown, G. G. 2007. Soil biodiversity for agricultural sustainability. *Agriculture, Ecosystem and Environment* 121(3), 233–44. doi:10.1016/j.agee.2006.12.013.

Butovsky, R. O. 2011. Heavy metals in carabids (Coleoptera, Carabidae). *ZooKeys* 100(100), 215–22. doi:10.3897/zookeys.100.1529.

Calegari, A. 2002a. The spread and benefits of No-Tillage agriculture in Paraná State, Brazil. In: Uphoff, N. (Ed.), *Agroecological Innovations: Increasing Food Production with Participatory Development*. Earthscan Publications, London, pp. 187–202.

Calegari, A. 2002b. Conservation Agriculture: the Brazilian experience of No-Till including crop rotation and cover crops towards Sustainable Agriculture. In: *International Workshop on Conservation Agriculture for Sustainable Wheat Production in Rotation with Cotton in Limited Water Resource Areas*. Proceedings. FAO, Ministry of Agriculture. October, Uzbekistan, pp. 12-18.

Calegari, A. 2010. Rotação de culturas/Culturas de cobertura. In: *12o. Encontro nacional de Plantio direto na Palha. Plantio direto Tecnologia que mudou a visão do produtor*. Resumos/Anais. Federação Brasileira Plantio direto na Palha; Editado por Lutécia Canalli, Ponta Grosa, Pr: FEBRAPDP, Foz do Iguaçu, 23 a 25 de junho de 2010, pp. 165-71.

Calegari, A. and Alexander, I. 1998. The effect of tillage on some chemical properties of an oxisol and summer crop yields in Southwestern Paraná, Brazil. *Advances in Geoecology. Reiskirchen, Germany* 31, 1239-46.

Calegari, A., Hargrove, W. L., Rheinheimer, D. D. S., Ralisch, R., Tessier, D., de Tourdonnet, S. and de Fatima Guimarães, M. 2008. Impact of long-term no-tillage and cropping system management on soil organic carbon in an Oxisol: a model for sustainability. *Agronomy Journal* 100(4), 1013-9. doi:10.2134/agronj2007.0121.

Calegari, A., Tiecher, T., Rheinheimer, D., Bender, M. A., Piccin, R., Gubiani, E., Sartori Bellinaso, R. J. and Casali, C. A. 2012. Soil management systems and how winter crops affect soil organic phosphorus cycle. In: *The 3rd International Conference on Conservation Agriculture in Southeast Asia. Conservation Agriculture and Sustainable Upland Livelihoods. Innovations for, with and by Farmers to Adapt to Local and Global Changes*. Proceedings. Hanoi, Vietnam, 10-15 December 2012. Printed by: Steps Media Co, Ltd. Hanoi Vietnam. Cirad, Nomafsi, University of Queensland, pp. 232-3. ISBNCIRAD: 978-2-87614-687-7. Ean Cirad: 9782876146877.2012.

Calegari, A., Araújo, A. G., Costa, A., Lanillo, R. F., Casão Junior, R. and Rheinheimer, D. S. 2013. Conservation Agriculture in Brazil. In: Jat, R. A., Sahrawat, K. L. and Kassam, A. H. (Eds), *Conservation Agriculture - Global Prospects and Challenges*. CAB International, Oxfordshire, pp. 54-88.

Civitello, D. J., Cohen, J., Fatima, H., Halstead, N. T., Liriano, J., McMahon, T. A., Ortega, C. N., Sauer, E. L., Sehgal, T., Young, S. and Rohr, J. R. 2015. Biodiversity inhibits parasites: broad evidence for the dilution effect. *Proceedings of the National Academy of Sciences of the United States of America* 112(28), 8667-71. doi:10.1073/pnas.1506279112.

Correia, M. E. F. 2002. *Relações entre a Diversidade da Fauna de Solo e o Processo de Decomposição e seus reflexos sobre a estabilidade dos Ecossistemas*. Embrapa Agrobiologia, Seropédica, p. 33.

Correia, M. E. F. and Oliveira, L. C. M. de. 2000. *Fauna de solo: Aspectos gerais e metodológicos*. Documentos No 112. Embrapa Agrobiologia, Rio de Janeiro, 46pp.

Da Silva, R. F., Aquino, A. M., Mercante, F. M. and Guimarães, M. F. 2006. Macrofauna invertebrada do solo sob diferentes sistemas de produção em Latossolo da Região do Cerrado. *Pesquisa Agropecuária Brasileira* 41(4), 697-704. doi:10.1590/S0100-204X2006000400022.

De Araújo, A. S. F. and Monteiro, R. T. R. 2007. Indicadores biológicos de qualidade do solo. *Bioscience Journal* 23(3), 66-75.

Decaëns, T., Lavelle, P., Jiineiiez, J. J. J., Escobar, G. and Rippstein, G. 1994. Impact of land management on soil macrofauna in the Oriental Llanos of Colombia. *European Journal of Soil Biology* 30(4), 157-68.

Decaëns, T., Lavelle, P., Jiménez, J. J., Escobar, G., Rippstein, G., Schneidmadl, J., Sanz, J. I., Hoyos, P. and Thomas, R. J. 2003. Impacto del uso de la tierra en la macrofauna del suelo de los Llanos Orientales de Colombia. In: Jiménez, J. J. and Thomas, R. J. (Eds), *El arado natural: las comunidades de macroinvertebrados del suelo en las savanas neotropicales de Colombia*. Centro Internacional de Agricultura Tropical, Cali, Colombia, pp. 21–45 (Publicación CIAT, 336).

Demetrio, W. C., Santos, A., Ferreira, T., Nadolny, H., Cardoso, G. B. X., Torre, J. L. M., Dudas, R., De Oliveira, V., Barreto, J. O., James, S. W., Da Silva, E., Brown, G. G. and Bartz, M. L. C. 2018. Earthworm species in various land use systems in the Campos Gerais region of Lapa, Paraná, Brazil. *Zootaxa* 4496(1), 503–16. doi:10.11646/zootaxa.4496.1.39.

Derpsch, R., Sidiras, N. and Roth, C. H. 1986. Results of studies made from 1977 to 1984. To control erosion by cover crops and zero tillage techniques in Paraná, Brazil. *Soil and Tillage Research* 8, 253–63. doi:10.1016/0167-1987(86)90338-7.

Doran, J. W. and Parkin, T. B. 1994. Defining and assessing soil quality. In: Doran, J. W., Coleman, D. C., Bezdicek, D. F., Stewart, B. A (Eds), *Defining Soil Quality for a Sustainable Environment*. Special Publication 35. Soil Science Society of America and American Society of Agronomy, Madison, pp. 3–21.

FAO. 2019. *The State of the Word's Biodiversity for Food and Agriculture*. Food and Agriculture Organization of the United Nations, Rome, Italy, p. 547.

Febrapdp. 2019. Federação Brasileira de Plantio Direto e Irrigação. Área sob Plantio Direto no Brasil e por estado brasileiro. Available at: https://febrapdp.org.br/area-de-pd (accessed on 15 January 2019).

Fisher, M. C., Henk, D. A., Briggs, C. J., Brownstein, J. S., Madoff, L. C., McCraw, S. L. and Gurr, S. J. 2012. Emerging fungal threats to animal, plant and ecosystem health. *Nature* 484(7393), 186–94. doi:10.1038/nature10947.

Floate, K. D., Doane, J. F. and Gillott, C. 1990. Carabid predators of the wheat midge (Diptera: Cecidomyiidae) in Saskatchewan. *Environmental Entomology* 19, 1503–11.

Fries, M. R. and Aita, C. 1999. Aspectos básicos sobre a importância dos microorganismos em plantio direto. In: *Curso sobre aspectos básicos de fertilidade e microbiologia do solo sob plantio direto*, 3., 1999, Cruz Alta. Fertilidade do solo em plantio direto: resumos de palestras. Aldeia Norte, Passo Fundo, pp. 90–110.

Gregorich, E. G., Monreal, C. M., Carter, M. R., Angers, D. A. and Ellert, B. H. 1994. Towards a minimum data set to assess soil organic-matter quality in agricultural soils. *Canadian Journal of Soil Science* 74(4), 367–85. doi:10.4141/cjss94-051.

Hargrove, W. L. 1991. Cover crops for clean water. *Proceedings of an International Conference*. West Tennessee Experiment Station, 9–11 April 1991, Jackson, TN. Soil and Water Conservation Society, Ankeny, p. 198.

Hassell, J. M., Begon, M., Ward, M. J. and Fèvre, E. M. 2017. Urbanization and disease emergence: dynamics at the wildlife–livestock–human interface. *Trends in Ecology and Evolution* 32(1), 55–67. doi:10.1016/j.tree.2016.09.012.

Heesterbeek, H., Anderson, R. M., Andreasen, V., Bansal, S., De Angelis, D., Dye, C., Eames, K. T., Edmunds, W. J., Frost, S. D., Funk, S., Hollingsworth, T. D., House, T., Isham, V., Klepac, P., Lessler, J., Lloyd-Smith, J. O., Metcalf, C. J., Mollison, D., Pellis, L., Pulliam, J. R., Roberts, M. G., Viboud, C. and Isaac Newton Institute IDD Collaboration. 2015. Modeling infectious disease dynamics in the complex landscape of global health. *Science* 347(6227), aaa4339. doi:10.1126/science.aaa4339.

Hillel, D. 1998. Flow of water in unsaturated soil. In: Hillel, D. (Ed.), *Environmental Soil Physics*. Academic Press, New York, NY, pp. 203–41.

House, G. J. and Parmelee, R. W. 1985. Comparison of soil arthropods and earthworms from conventional and No-Tillage agroecosystems. *Soil and Tillage Research* 5(4), 351–60. doi:10.1016/S0167-1987(85)80003-9.

Hu, F., Li, H. X. and Wu, S. M. 1997. Differentiation of soil fauna populations in conventional tillage and No-Tillage soil ecosystems. *Pedosphere* 7, 339–48.

Huerta, E., Kampichler, C., Geissen, V., Ochoa-Gaona, S., Jond, Bd and Hernández-Daumás, S. 2009. Towards an ecological index for tropical soil quality based on soil macrofauna. *Pesquisa Agropecuaria Brasileira* 44(8), 1056–62. doi:10.1590/S0100-204X2009000800039.

Jenkinson, D. S. and Ladd, J. N. 1981. Microbial biomass in soil: measurement and turnover. In: Paul, E. A. and Ladd, J. N. (Org.), *Soil Biochemistry*. Marcel Dekker, New York, NY, pp. 415–71.

Johnson, P. T. J., Ostfeld, R. S. and Keesing, F. 2015. Frontiers in research on biodiversity and disease. *Ecology Letters* 18(10), 1119–33. doi:10.1111/ele.12479.

Jones, K. E., Patel, N. G., Levy, M. A., Storeygard, A., Balk, D., Gittleman, J. L. and Daszak, P. 2008. Global trends in emerging infectious diseases. *Nature* 451(7181), 990–3. doi:10.1038/nature06536.

Karvonen, A., Aalto-Araneda, M., Virtala, A. M., Kortet, R., Koski, P. and Hyvärinen, P. 2016. Enriched rearing environment and wild genetic background can enhance survival and disease resistance of salmonid fishes during parasite epidemics. *Journal of Applied Ecology* 53(1), 213–21. doi:10.1111/1365-2664.12568.

Keesing, F., Belden, L. K., Daszak, P., Dobson, A., Harvell, C. D., Holt, R. D., Hudson, P., Jolles, A., Jones, K. E., Mitchell, C. E., Myers, S. S., Bogich, T. and Ostfeld, R. S. 2010. Impacts of biodiversity on the emergence and transmission of infectious diseases. *Nature* 468(7324), 647–52. doi:10.1038/nature09575.

Kennedy, A. and Doran, J. 2002. Sustainable agriculture: role of microorganisms. In: Bitton, G. (Org.), *Encyclopedia of Environmental Microbiology*. John Wiley & Sons, New York, pp. 3116–26.

King, K. C. and Lively, C. M. 2012. Does genetic diversity limit disease spread in natural host populations? *Heredity (Edinb.)* 109(4), 199–203. doi:10.1038/hdy.2012.33.

Kladivko, E. J. 2001. Tillage systems and soil ecology. *Soil and Tillage Research* 61(1–2), 61–76. doi:10.1016/S0167-1987(01)00179-9.

Kluthcouski, J. and Aidar, H. 2003. Sistema Santa Fé. In: Kluthcouski, J., Stone, L. F. and Aidar, H. (Eds), *Integração lavoura-pecuária*. Embrapa Arroz e Feijão, Santo Antônio de Goiás, pp. 405–41.

Kluthcouski, J. and Stone, L. F. 2003. Desempenho de culturas anuais sobre palhada de braquiária. In: Kluthcouski, J., Stone, L. F. and Aidar, H. (Eds), *Integração lavoura pecuária*. Embrapa Arroz e Feijão, Santo Antônio de Goiás, pp. 499–522.

Kluthcouski, J., Cobucci, T., Aidar, H., Yokoyama, L. P., Oliveira, I. P., Costa, J. L. S., Silva, J. G., Vilela, L., Barcellos, A. O. and Magnabosco, C. U. 2000. *Integração lavoura/ pecuária pelo consorcio de culturas anuais com forrageiras, em áreas de lavoura, nos sistemas direto e convencional*. Circular Técnica No 28. Embrapa Arroz e Feijão, p. 28.

Kumar, R., Pandey, S. and Pandey, A. 2006. Plant roots and carbon sequestration. *Current Science* 91, 885–90.

Landers, J. N.. 1998. Zero tillage development in tropical Brazil. In: *Conservation Tillage for Sustainable Agriculture*. Proceedings of the International Workshop, Harare,

Zimbabwe, 22-27 June 1998. Annexe III: Background Papers (International). Benites, J., Chuma, E., Fowler, R. J., Kienzle, K., Molapong, I., Manu, N., Steiner, K. and van Veenhuizen, R. (Eds). GTZ, Eschborn, pp. 247–62.

Lavelle, P., Bignell, D., Lepage, M., Wolters, V., Roger, P., Ineson, P., Heal, O. W. and Dhillion, S. 1997. Soil function in a changing world: the role of invertebrate ecosystem engineers. *European Journal of Soil Biology* 33, 159–93.

Lavelle, P., Spain, A., Blouin, M., Brown, G., Decaens, T., Grimaldi, M., Jimenez, J. J., McKey, D., Mathieu, J., Velasquez, E. and Zangerle, A. 2016. Ecosystem engineers in a self-organized soil: a review of concepts and future research questions. *Soil Science* 181(3–4), 91–109.

Marchão, R. L., Lavelle, P., Celini, L., Balbino, L. C., Vilela, L. and Becquer, T. 2009a. Soil macrofauna under integrated crop - livestock systems in a Brazilian cerrado ferralsol. *Pesquisa Agropecuária Brasileira* 44(8), 1011–20. doi:10.1590/S0100-204X2009000800033.

Marchão, R. L., Lavelle, P., Celini, L., Balbino, L. C., Vilela, L. and Becquer, T. 2009b. Soil macrofauna under integrated crop-livestock systems in a Brazilian Cerrado Ferralsol. *Pesquisa Agropecuária Brasileira* 44(8), 1011–20. doi:10.1590/S0100-204X2009000800033.

Martínez, E., Fuentes, J. P., Pino, V., Silva, P. and Acevedo, E. 2013. Chemical and biological properties as affected by No-Tillage and conventional tillage systems in an irrigated Haploxeroll of Central Chile. *Soil and Tillage Research* 126, 238–45. doi:10.1016/j.still.2012.07.014.

Mbow, C., Van Noordwijk, M., Luedeling, E., Neyfeldt, H., Minang, P. A. and Kowero, G. 2014. Agroforestry solution to address food security and climate challenges in Africa. *Current Opinion in Environmental Sustainability* 6, 61–7. doi:10.1016/j.cosust.2013.10.014.

Moraes, A., Carvalho, P. C. F., Lustosa, S. B. C., Lang, C. R. and Deiss, L. 2014. Research on integrated crop-livestock systems in Brazil. *Revista Ciência Agronômica* 45(5spe), 1024–31. doi:10.1590/S1806-66902014000500018.

Morand, S. and Lajaunie, C. 2017. *Biodiversity and Health. Linking Life, Ecosystems and Societies*. Elsevier & ISTE Press, London.

Orgiazzi, A., Bardgett, R. D., Barrios, E., Behan-Pelletier, V., Briones, M. J. I., Chotte, J.-L., De Deyn, G. B., Eggleton, P., Fierer, N., Fraser, T., Hedlund, K., Jeffery, S., Johnson, N. C., Jones, A., Kandeler, E., Kaneko, N., Lavelle, P., Lemanceau, P., Miko, L., Montanarella, L., Moreira, F. M. S., Ramirez, K. S., Scheu, S., Singh, B. K., Six, J., van der Putten, W. H. and Wall, D. H. (Eds). 2016. *Global Soil Biodiversity Atlas*. European Commission, Publications Office of the European Union, Luxembourg, p. 176.

Olfert, O., Elliot, R. H. and Hartley, S. 2009. Non-native insects in agriculture: strategies to manage the economic and environmental impact of wheat midge, *Sitodiplosis mosellana*, in Saskatchewan. *Biological Invasions* 11, 127–33.

Paoletti, M. G. 1999. Using bioindicators based on biodiversity to assess landscape sustainability. *Agriculture, Ecosystems and Environment* 74(1–3), 1–18. doi:10.1016/S0167-8809(99)00027-4.

Paul, B. K., Vanlauwe, B., Ayuke, F., Gassner, A., Hoogmoed, M., Hurisso, T. T., Koala, S., Lelei, D., Ndabamenye, T., Six, J. and Pulleman, M. M. 2013. Medium-term impact of tillage and residue management on soil aggregate stability, soil carbon and crop productivity. *Agriculture, Ecosystems and Environment* 164, 14–22. doi:10.1016/j.agee.2012.10.003.

Peña-Peña, K. and Irmler, U. 2016. Moisture seasonality, soil fauna, litter quality and land use as drivers of decomposition in Cerrado soils in SE-Mato Grosso, Brazil. *Applied Soil Ecology* 107, 124–33. doi:10.1016/j.apsoil.2016.05.007.

Pimentel, D. C., Wilson, C., McCullum, C., Huang, R., Dwen, P., Flack, J., Tran, Q., Saltman, T. and Cliff, B. 1997. Economic and environmental benefits of biodiversity. *BioScience* 47(11), 747–57. doi:10.2307/1313097.

Rodríguez, D. I., Anríquez, G. and Riveros, J. L. 2016. Food security and livestock: the case of Latin America and the Caribbean. *Ciencia e Investigación Agraria* 43(1), 5–15.

Rousseau, G. X., Silva, P. R. dos S. and Carvalho, C. J. R. de. 2010. Earthworm ants and other arthropods as soil health indicators in traditional and no fire agriecosystems from Eastern Brazilian Amazonia. *Acta Zoologicica Mexicana* 2, 117–34.

Rousseau, L., Fonte, S. J., Téllez, O., Hoek, R. V. D. and Lavelle, P. 2013. Soil macrofauna as indicators of soil quality and land use impacts in smallholder agroecosystems of western Nicaragua. *Ecological Indicators* 27, 71–82. doi:10.1016/j.ecolind.2012.11.020.

Ruiz, N., Lavelle, P. and Jiménez, J. 2008. *Soil Macrofauna Field Manual 2008*. Food and Agriculture Organization of the United Nations, Rome, Italy, pp. 1–132.

Sanchez, E. 2012. Physical properties of soil and soybean productivity in succession of winter cover. MSc Dissertation. Universidade Estadual do Centro-Oeste, Unicentro, PR, Brazil. (in: Portuguese, abstract in English), p. 48.

Santos, G. de A., da Silva, L. S., Canellhas, L. P. and Camargo, F. de O. 2008. *Fundamentos da matéria orgânica do solo: ecossistemas tropicais & subtropicais*. Metrópole, Porto Alegre, 636p.

Scopel, E., Triomphe, B., Ribeiro, M. F. S., Séguy, L., Denardin, J. E. and Kochann, R. A.. 2004. Direct seeding mulch-based cropping systems (DMC) in Latin America. In: *New Directions for a Diverse Planet: Proceedings for the 4th International Crop Science Congress*, Brisbane, Australia, 26 September – 1 October 2004. Fischer, T., Turner, N., Angus, J., McIntyre, L., Robertsen, M., Borrell, A. and Llyod, D. (Eds). Available online at: www.cropscience.org.au.

Sidiras, N. and Pavan, M. A. 1986. Influência do sistema de manejo na temperatura do solo. *Revista Brasileira de Ciência do Solo* 10, 181–4.

Silva, V. R., da Reichert, J. M. and Reinert, D. J. 2006. Soil temperature variation in three different systems of soil management in blackbeans crop. *Revista Brasileira de Ciência do Solo* 30(3), 391–9.

Stork, N. E. and Eggleton, P. 1992. Invertebrates as determinants and indicators of soil quality. *American Journal of Alternative Agriculture* 7(1-2), 38–47. doi:10.1017/S0889189300004446.

Susilo, F. X., Neutel, A. M., Van Noordwijk, M., Hairiah, K., Brown, G. G. and Swift, M. J. 2004. Soil biodiversity and food webs. In: van Noordwijk, M., Cadisch, G. and Ong, C. K. (Eds), *Below-Ground Interactions in Tropical Agroecosystems: Concepts and Models with Multiple Plant Components*. CAB International, Wallingford, UK, pp. 285-307.

Swift, M. J. 1999. Towards the second paradigm: integrated biological management of soil. In: Siqueira, J. O., Moreira, F. M. S., Lopes, A. S., Guilherme, L. R. G., Faquin, V., Furtani, A. E. and Carvalho, J. G. (Eds), *Inter-relação fertilidade, biologia do solo e nutrição de plantas*. Universidade Federal de Lavras, Lavras, Brazil, pp. 11-24.

Trevisan, R., Herter, F. G. and Pereira, I. S. 2002. Variação da amplitude térmica do solo em pomar de pessegueiro cultivado com aveia preta (*Avena* sp.) e em sistema convencional. *Revista Brasileira de Agrociência* 8, 155–7.

Van der Putten, W. H., Anderson, J. M., Bardgett, R. D., Behan-Pelletier, V., Bignell, D. E., Brown, G. G., Brown, V. K., Brussaard, L., Hunt, H. W., Ineson, P., Jones, T. H., Lavelle, P., Paul, E. A., St. John, M., Wardle, D. A., Wojtowicz, T. and Wall, D. H. 2004. The sustainable delivery of goods and services provided by soil biota. In: Wall, D. H. (Ed.), Sustaining biodiversity and ecosystem services in soil and sediments. Island Press, Washington, UK, pp. 15–43

Van Groenigen, J. W., Lubbers, I. M., Vos, H. M. J., Brown, G. G., De Deyn, G. B. and Van Groenigen, K. J. 2014. Earthworms increase plant production: a meta-analysis. *Scientific Reports* 4, 6365. doi:10.1038/srep06365.

Winter, J. P., Voroney, R. P. and Ainsworth, D. A. 1990. Soil microarthropods in long-term No-Tillage and conventional tillage corn production. *Canadian Journal of Soil Science* 70(4), 641–53. doi:10.4141/cjss90-067.

Chapter 10

Conservation Agriculture: climate change mitigation and adaptation benefits

Emilio J. Gonzalez Sanchez, Universidad de Córdoba, Spain, European Conservation Agriculture Federation (ECAF), Belgium and Asociación Española Agricultura de Conservación. Suelos Vivos (AEAC.SV), Spain; Oscar Veroz-Gonzalez, Asociación Española Agricultura de Conservación. Suelos Vivos (AEAC.SV), Spain; Manuel Morena-Garcia and Rafaela Ordoñez-Fernandez, IFAPA Centro Alameda del Obispo, Spain; Jesus A. Gil-Ribes and Julio Roman-Vazquez, Universidad de Córdoba, Spain; Antonio Holgado-Cabrera, IFAPA Centro Alameda del Obispo, Spain; Amir Kassam, University of Reading, UK; Gordon Conway, Imperial College London, UK; Saidi Mkomwa, African Conservation Tillage Network, Kenya; Paula Triviño-Tarradas, Antonio Miranda-Fuentes and Francisco Marquez-Garcia, Universidad de Córdoba, Spain; and Rosa M. Carbonell-Bojollo, IFAPA Centro Alameda del Obispo, Spain

1 Introduction

If there is any productive activity that depends directly on the climate and its variability, it is undoubtedly agriculture. A change in temperature and rainfall patterns or an increased concentration of atmospheric CO_2 will significantly affect crop development. For good reason it is estimated that, globally, climate variability is responsible for between 32% and 39% of variability in yields (Ray et al., 2015).

Although some aspects of climate change, such as longer growing seasons and a rise in temperature, can be beneficial, the lack of water availability and the more frequent occurrence of extreme weather phenomena will in

http://dx.doi.org/10.19103/AS.2019.0049.10

turn have negative and adverse effects on agriculture. Based on the above, if there is no adaptation over time and measures are not taken to mitigate the effects of global warming, there could be considerable economic, social and environmental consequences. This becomes even more significant when considering forecasts for growth in the demand for agricultural products over the coming decades due to growth in world population.

In agricultural systems, one of the most relevant natural resources for combating climate change is soil. Its potential for capturing CO_2 from the atmosphere and incorporating it in the form of organic carbon (OC) makes it a powerful mitigation tool. Proof of this is that soil is the greatest reserve of carbon (C) in terrestrial ecosystems (Lal, 2008) and the second in the world behind the oceans, accumulating three times more C than the atmosphere (Smith, 2004) and aerial biomass (Sommer and Bossio, 2014).

Accordingly, soil management systems have a great deal of mitigation potential in terms of crop management. Agricultural systems based on tillage can lead to a reduced OC content. There is a general agreement in the literature that soil disturbance from tillage is one of the major causes of reduced OC in soil (Balesdent et al., 1990; Six et al., 2000; Olson et al., 2005). Thus, some studies confirm that intensive farming contributed to a loss of between 30% and 50% of OC in soil in the last two decades of the twentieth century (Reicosky, 2011). Taking into account soil's storage capacity of C and the ongoing systematic loss of C for decades, it could be thought that any strategy aimed at increasing the OC content of soil, however small those increases may be, will have a positive impact on mitigating climate change. At this point, it is worth bearing in mind the soil's capacity to store C is limited and that a point comes when a balance is reached between the carbon captured and the carbon released through decomposition of organic matter. It can be unequivocally stated that this balance in agricultural ecosystems is a long way from being reached, according to currently available carbon loss figures, so farming practices that help increase the sink effect clearly still have a large potential contribution.

On the other hand, tillage has a direct influence on emissions of CO_2 from the soil into the atmosphere both in the short term (immediately after ploughing) and in the long term (crop, tillage management). This is because tillage stimulates the production and accumulation of CO_2 in the soil's porous structure from the processes of organic matter (OM) mineralization. The mechanical action of tillage includes breaking down soil aggregates making occluded OM available to microbes, with the consequent release of the CO_2 trapped within them, and which is then emitted into the atmosphere.

Furthermore, the energy consumption associated with different farming practices (soil tillage, fertilizers and amendments, irrigation, pesticides etc.) is essentially based on the use of fossil fuels, particularly fuel oil, which means inevitable emissions into the atmosphere of greenhouse gases (GHGs).

Soil can also play a relevant role in relation to the adaptation of agrarian ecosystems to climate change. The Intergovernmental Panel on Climate Change (IPCC), in its fifth evaluation report (IPCC, 2014a), warns of the increase in the occurrence of episodes of extreme rainfall, the appearance of new pests and diseases and the lower availability of water resources for crops. For achieving a more resilient ecosystem, a soil with higher quality is required, which means soils with higher contents of OM, better structured, with a greater capacity for water retention and with a greater degree of biodiversity.

There is a broad consensus that conventional farming systems, based on intensive tillage, have adverse environmental effects, with the potential to put natural ecosystems at risk (Duru et al., 2015). Some of the main risks of such systems are depletion of water sources, pollution of soil and water resources, air pollution and GHG emissions, depletion of soil OC, erosion of the fertile soil layers by wind and water and soil salinization (Horrigan et al., 2002).

Based on all these considerations, mitigation measures in the agricultural sector involve fixing the C found inside the oxidized compound in the soil while reducing GHG emissions in general. In turn, if the measures adopted cannot only mitigate climate change but also improve water balance and soil quality, as well as increasing biodiversity, it will be possible to confirm that these measures will help crops adapt to scenarios involving a lower availability of water resources, a higher incidence of extreme weather conditions that increase the risk of erosion and the incidence of new pests and diseases.

Conservation Agriculture (CA) can be considered a practice that mitigates climate change and contributes to the adaptation of crops to the effects of global warming. This chapter sets out the reasons that make CA a suitable global response to address the challenges posed by climate change to agricultural ecosystems, in terms of both mitigation and adaptation.

2 Fundamentals of climate change mitigation

According to the Food and Agricultural Organization of the United Nations (FAO, 2018), CA is a farming system that promotes continuous no or minimum soil disturbance (i.e. no-tillage), maintenance of a permanent soil mulch cover and diversification of plant species. It enhances biodiversity and natural biological processes above and below the ground surface and in so contributing to increased water- and nutrient-use efficiency and productivity, to more resilient cropping systems, and to improved and sustained crop production. CA is based on the practical application of three interlinked principles:

1 Avoiding or minimizing mechanical soil disturbance involving seeding or planting directly into untilled soil, eliminating tillage altogether

once the soil has been brought to good condition and keeping soil disturbance from cultural operations to the minimum possible.

2 Maintaining year-round biomass mulch cover over the soil, including specially introduced cover crops and intercrops and/or the mulch provided by retained biomass and stubble from the previous crop.
3 Diversifying crop rotations, sequences and associations, adapted to local environmental and socio-economic conditions, and including appropriate nitrogen fixing legumes. Such rotations and associations contribute to maintaining biodiversity above and in the soil, add biologically fixed nitrogen to the soil-plant system and help avoid build-up of pest populations. In CA, the sequences and rotations of crops encourage agrobiodiversity as each crop will attract different overlapping spectra of microorganisms and natural enemies of pests.

In its Fifth Assessment Report, the IPCC defines climate change mitigation as 'a human intervention to reduce the sources or enhance the sinks of greenhouse gases' (IPCC, 2014b). In this context, CA could be considered a mitigation measure, by fulfilling the double premise set out by the IPCC of reducing sources and enhancing sinks. Therefore, a direct consequence of changes brought about by CA in the dynamics of C in the soil is an increase of carbon in the soil, making these techniques a carbon sink activity. On the other hand, the drastic reduction in soil tillage and the non-mechanical alteration of the soil mean lower CO_2 emissions resulting from the energy saved and the reduction of OM mineralization processes (Fig. 1).

2.1 Carbon sink

Soil management in CA leads to alterations in the dynamics of carbon in the soil, increasing in both amount and concentration. There are two reasons for this. First, by leaving harvest stubble and crop biomass on the ground and root biomass in the soil, CA produces a dynamic in the OM that is similar to how natural ecosystems behave. The result of these practices is an increase in OM accumulation in the vertical dimension. This accumulation is used as the recovery index in the quality of agricultural soil degraded by tillage (Franzluebbers, 2002; Moreno et al., 2005). A significant part of this humidified surface OM is incorporated into the soil by worms, the populations of which are given a highly significant boost by CA (Cantero-Martínez et al., 2004; Bescansa et al., 2005). Secondly, the less the soil is tilled, the more carbon it absorbs, and as a result synthesizes more OM, which in the long term increases its productive capacity, while also reducing the CO_2 released into the atmosphere.

Tebruegge (2001) stated that through the processes of microbiological oxidation that occur in the soil, 3.7 tonnes of CO_2 are generated from 1 tonne

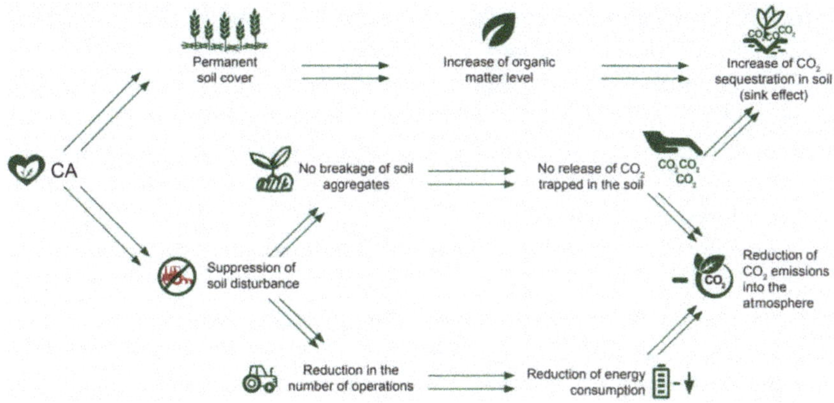

Figure 1 Climate change mitigation mechanisms used in Conservation Agriculture. Source: adapted from González-Sánchez et al. (2017).

of C decomposition, such that based upon the increases in OM observed in CA management systems in comparison with conventional tillage management systems, it is possible to determine, based on OC, what these increases mean in terms of CO_2 amounts, thus making it possible to provide a range of values on the potential of these techniques with regard to atmospheric carbon fixation.

Without losing sight of this evidence, it is important to take into account that the magnitude of response from CA systems to C capture varies considerably depending on the soil and weather conditions of the area, the amount of crop biomass being returned and how long the CA system has been in place. There are studies that suggest that the OC content of the soil rapidly increases over the first 10 years of CA practice, after which it slows down until reaching almost zero growth, as the soil reaches a new equilibrium (Yang and Wander, 1999; Franzluebbers and Arshad, 1996; Puget and Lal, 2005).

In any event, as a result of these two processes, CA promotes sequestration of CO_2 into the soil due to the presence of stubble and root biomass, and reduces CO_2 release due to less tillage, which in turn translates into an increased sink effect. Results all over the world show how CA effectively boosts the soil's capacity to capture atmospheric carbon. In a worldwide meta-analysis, Govaerts et al. (2009) found that the OC content in CA soils was greater than that in soils managed through conventional tillage in 51% of the cases studied.

In more recent meta-analyses, Ranaivoson et al. (2017) found that the OC in soils under CA increased by 0.38 t ha^{-1} yr^{-1}. Along the same lines, in a meta-analysis studying the C capture capacity of different crop management systems in areas around the world with a Mediterranean climate, Aguilera et al. (2013) obtained C capture rates of 0.44 t ha^{-1} $year^{-1}$ in the case of no-till systems in herbaceous crops, and 0.27 t ha^{-1} yr^{-1} in the case of groundcovers in woody crops, an increase in soil carbon content of 11.4% and 10%, respectively, compared with conventional farming methods.

The results of worldwide studies are in line with those obtained in research carried out on different continents. In Europe (EU-15), the increase in soil organic carbon (SOC) sequestration of no-till systems in comparison with conventional tillage is 0.4 t ha^{-1} yr^{-1} (Freibauer et al., 2004; Smith et al., 2005). However, there is no general data on this scale for groundcovers. The closest approximation is provided by Freibauer et al. (2004), which indicates an increase of 0.3–0.8 t ha^{-1} yr^{-1}. This information, however, refers to cover crops, which are crops that are not aimed for harvest, but protect the soil from erosion and nutrient loss and to build biological forms of crop nutrient sources especially when legumes are used as cover crops. Groundcovers are permanent grass or grass-legume cover crop running between rows of woody crops. In this case, there is only information at the European level for the Mediterranean biogeographical region. In particular, Vicente-Vicente et al. (2016) used meta-analysis to determine that groundcovers improve the SOC contents, 1.1 t ha^{-1} yr^{-1} in olive groves, 0.78 t ha^{-1} yr^{-1} in vineyards and 2.0 t ha^{-1} yr^{-1} in land planted with almond trees.

In a study conducted by the European Conservation Agriculture Federation (ECAF) on the potential of CA to sequester carbon in Europe, González-Sánchez et al. (2017) analysed the carbon sequestration rates of CA in the different biogeographic regions of Europe and estimated that the current capacity of carbon sequestration in agricultural soils due to CA is about 1.5 million tonnes per year in the case of no-tillage, whereas 2.8 million tonnes for the case of groundcovers in woody crops. If CA was more broadly applied in Europe, the potential becomes more than 37.4 million tonnes of C per year in the case of no-tillage and 14.2 million tonnes in the case of groundcovers in woody crops (Figs. 2 and 3).

Carbon sink rates for C in no-till systems in the Americas are highly variable, due to the continent's significant diversity in both soils and climate. However, CA soils generally show an OC gain compared with conventional tillage systems. Thus, there are C capture increase rates that range from 0.64 to 1.17 t ha^{-1} yr^{-1} in the south of Brazil (Sá et al., 2013), from 0.4 to 1.7 t ha^{-1} yr^{-1} in the centre of Brazil (Blanchart et al., 2007), from 0.43 to 0.9 in the United States (Causarano et al., 2006; Hollinger et al., 2005) and from 0.31 to 0.44 t ha^{-1} yr^{-1} in Canada (West and Post, 2002).

In Africa there are several studies on each of the continent's major bioclimatic areas. Thus, in an equatorial climate, Barthès et al. (2004) obtained mean rates of carbon capture in no-till CA soils of 1.50 t ha^{-1} yr^{-1}. In a tropical climate, mean C increase rates in no-till CA soils range from 0.33 t ha^{-1} yr^{-1} obtained on farms in Kenya (Okeyo et al., 2016) to 2.76 t ha^{-1} yr^{-1} obtained on farms located in Zimbabwe (Gwenzi et al., 2009).

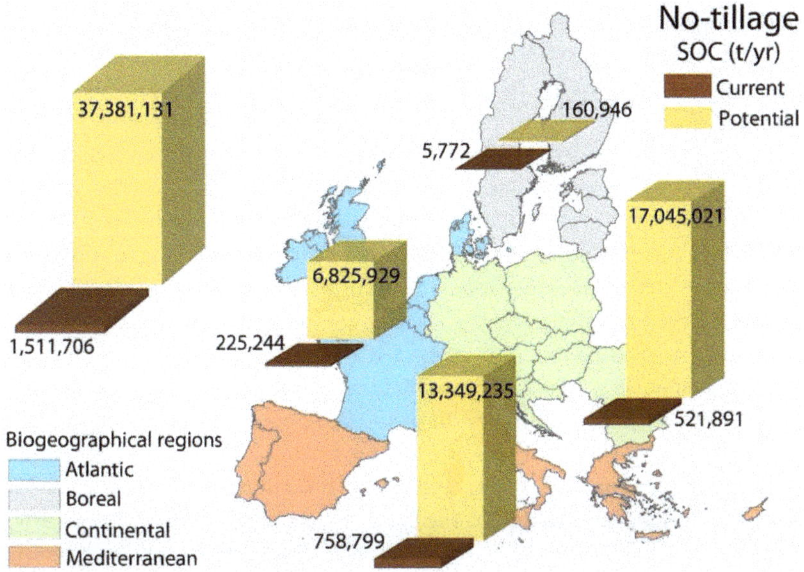

Figure 2 Current and potential SOC fixed by no-till compared to systems based on soil tillage in EU-28 and in the different biogeographical regions. Source: adapted from González-Sánchez et al. (2017).

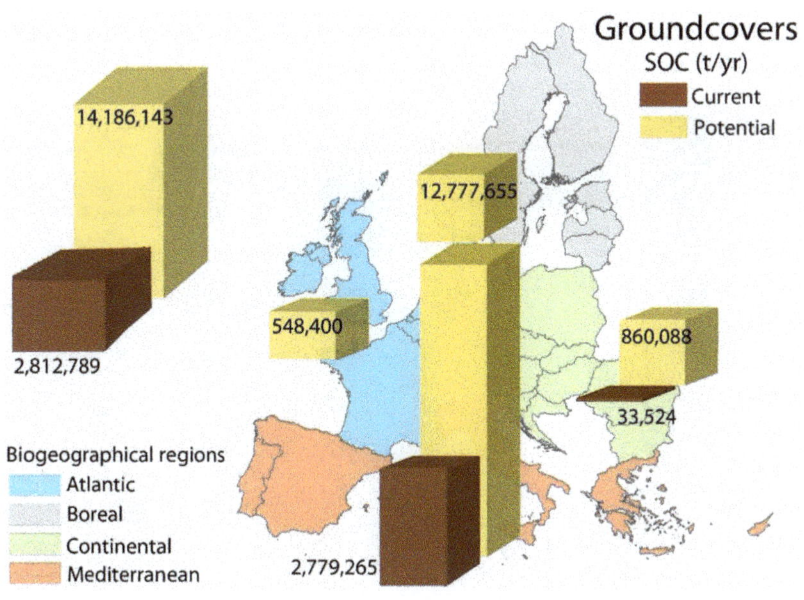

Figure 3 Current and potential SOC fixed by groundcovers compared to systems based on soil tillage in EU-28 and in the different biogeographical regions. Source: adapted from González-Sánchez et al. (2017).

In a report published by the African Conservation Tillage Network (ACT) and the ECAF (González-Sánchez et al., 2019), based on the results of almost 30 peer reviewed papers, the potential for annual SOC fixation in Africa was estimated. The assessment was based upon the average rates of carbon sequestration by CA systems in agricultural soils for the major climatic zones in Africa. The review concluded that currently, agricultural soils in Africa have the potential of capturing more than 1.5 million tonnes of carbon per year. In the same review, it is concluded that, if CA was broadly adopted across Africa, the sequestration potential would reach more than 145 million tonnes of SOC per year (Fig. 4).

In Asia, studies have been carried out both in China and on the Indo-Gangetic plains. In China, a meta-analysis conducted by Du et al. (2017) showed that no-till agriculture increased C soil content by an average of 3.8% compared with soil in conventional systems. Although *a priori* this figure is relevant, this same study highlighted that the results are not conclusive, and that more analyses are needed to reduce the variability of existing data among the studies conducted. In the case of the Indo-Gangetic plains, in rotations of

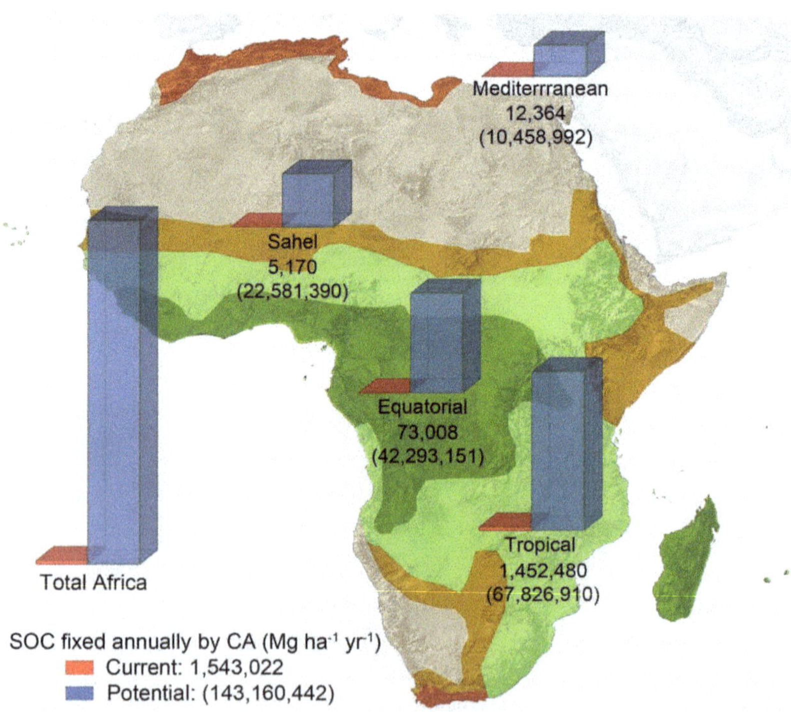

Figure 4 Potential soil organic carbon (SOC) fixed annually by CA cropland systems compared to tillage-based agricultural production systems in Africa. Authors diagram. Source: adapted from Gonzalez-Sanchez et al. (2019).

irrigated wheat and rice, after 20 years of study Grace et al. (2003) found that no-till soils increased OC content by an average of 1.8 t ha^{-1} yr^{-1}, in contrast with 0.37 t ha^{-1} yr^{-1} obtained by Powlson et al. (2016) in the same region.

Finally, in a study carried out in different agroclimatic zones of Australia, which applied the GHG emissions accounting method described by the IPCC (IPCC, 2006), Grace et al. (2010) obtained sequestration values for no-till soils that ranged from 0.08 t ha^{-1} yr^{-1} to 0.64 t ha^{-1} yr^{-1}.

At a global level, there are more general studies that have included CA practices (i.e. no-tillage, soil mulch and crop diversification) as a recommendation to achieve the carbon sequestration rate potential in agricultural soils. Together with other complementary practices, that potential was estimated between 0.4 and 1.2 trillion tonnes per year (US-EPA, 2005, 2006; Smith, 2004; Smith et al., 2000, 2005, 2008).

2.2 Reducing greenhouse gas (GHG) emissions

Adopting CA means a drastic reduction or complete elimination of any mechanical actions that alter the soil profile. This reduction affects the volume of CO_2 emissions produced, due on the one hand to the lessened fracture of soil aggregates and the subsequent gas exchange that occurs following tillage, and on the other, to the reduced consumption of fuel required for the land to be tilled.

With regard to the first of the processes mentioned, there is ample scientific evidence that tilled soils emit a higher amount of CO_2 than CA soils. In research carried out in the United States, Sainju et al. (2008) observed that tillage-based practices increased the CO_2 emissions in comparison with no-till systems from 62% to 118%. Reicosky and Archer (2007) assessed the short-term effects on CO_2 emissions of two soil management systems, one based on the use of mouldboard ploughing and the other using no-till methods. The result was higher emission levels from tilled plots than no-till plots, in both the short and medium term, with values ranging from emissions 3.8 times higher for shallow tillage (10 cm) to 10.3 times higher for deep tillage (28 cm) compared with no-till plots. The research was also able to quantify the emissions associated with the different types of equipment used in each of the assessed soil management systems (Fig. 5).

These results are in agreement with other studies in Spain by Carbonell-Bojollo et al. (2011), which studied not only the reduction in CO_2 emissions in soil when changing the management system, but also the behaviour of these emissions after each operation. In all of these operations, an emission peak was observed between 2 and 4 h after tillage, with the maximum difference between the two management systems (no-till vs conventional tillage) occurring during that period. This difference was greater the more deeply the soil was altered

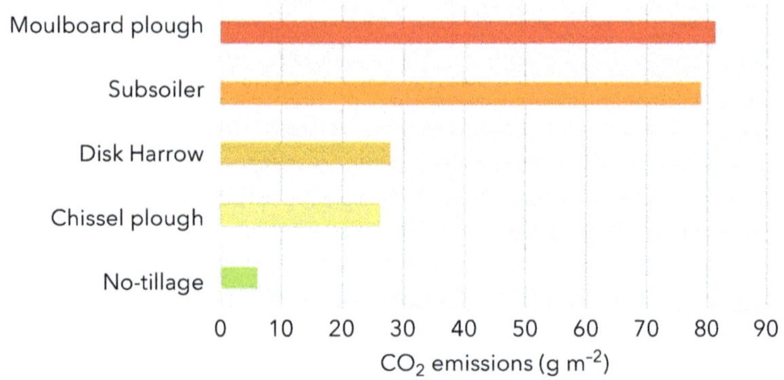

Figure 5 Cumulative CO_2 loss after conservation tillage tools for the first 5 h after tillage. Source: adapted from Reicosky (1997).

in the operation, indicating how tillage intensity has a direct relationship with the level of CO_2 emissions, although other edaphoclimatic factors such as soil humidity, temperature and incidence of rainfall can also have an influence.

Another mechanism responsible for reduced CO_2 emissions in CA systems is the reduction in energy used in its application. Energy consumption is converted into CO_2 emissions into the atmosphere, such that lower energy consumption means lower CO_2 emissions. By eliminating tillage in CA, these systems require a lower amount of energy than conventional tillage, where the larger amounts of fuel consumed in preparing the land is responsible for this difference. As with the other production factors, the higher the number of passes with the plough and harrows, the higher the fuel consumption.

Several studies have now been carried out worldwide on the effect the energy consumption in operations to prepare the soil for sowing has in terms of equivalent C emissions. Some studies have estimated C emissions equivalent to 35.3 kg ha^{-1} in conventional tillage, 7.9 kg ha^{-1} in minimum tillage using a chisel plough and 5.8 kg ha^{-1} in no-till farming, making for a reduction of 84% of emissions in comparison with conventional agriculture (Lal, 2004). Similarly, in studies conducted over 20 years in an irrigated wheat and rice rotation on the Indo-Gangetic plains, reductions of CO_2 emissions linked to lower fuel consumption due to no-till farming reached 67% a year (Grace et al., 2010).

3 Fundamentals of adapting to climate change

The term 'adapting' refers here to all the adjustments that are needed to produce a system (in this case, a farming system) that can respond to real or predicted changes that may occur as a consequence of climate change, and in so doing reduce their vulnerability and make the most of the opportunities provided by the new climatic scenarios. The IPCC defines it as 'the process of adjustment to actual or expected climate and its effects. In human systems, adaptation seeks to moderate or avoid harm or exploit beneficial opportunities. In some natural systems, human intervention may facilitate adjustment to expected climate and its effects' (IPCC, 2014a).

In their daily work, farmers have always had to make decisions to adapt their crops and system to the changing meteorological conditions that occur naturally from one year to another, such as the need to modify the sowing date due to rainfall variation, applying pesticides due to the particular virulence of pests or disease, or adjusting the crop mix in the cropping system and so forth. Such decisions are circumstantial in nature, and until recently farmers have not felt climate incidents to have been persistent enough to warrant a major change in behaviour patterns. However, the expected scenario in the medium and long term due to climate change, in terms of rising temperatures, changes in rainfall patterns and increases in extreme weather events (drought, intense rainfall etc.), will mean that normal crop management routines will need to change substantially.

In this regard, Delgado et al. (2011) identify a series of applicable principles for the agricultural sector to mitigate and adapt to climate change These principles include implementing practices aimed at conserving soil and water, which includes CA. Findings by Lal (2010) agree with this diagnosis, stating that CA is a good strategy not just for mitigating climate change, but also in helping to adapt crops to its effects, by helping to increase the resilience of agricultural ecosystems. This is due in large part to the improvement of many of the soil's functions and properties, which are boosted as a result of carbon capture and which are related to crop productivity (Delgado et al., 2011). However, CA also offers other benefits that help crops and cropping systems adapt to climate change (Table 1).

3.1 Improved water balance

Implementing and operating CA practices brings a series of benefits in the water resource within agricultural ecosystems, leading to increased availability of water for the crop and an improvement in its quality. This increased availability and quality are the result of a reduction in run-off, an increase in water infiltration capacity and a reduction in evapotranspiration. This brings an increased efficiency in water use, as indicated by Wang et al. (2018) in a meta-analysis carried out in China, which suggests that crops sowed using no-till methods with high rates of evapotranspiration will produce better yields than crops planted using conventional practices (Fig. 6).

Furthermore, the higher OM content in soils as a result of these practices also helps increase water storage capacity. Thus, Thierfelder et al. (2012, 2013)

Table 1 Contribution of Conservation Agriculture in agricultural systems adapting to climate change

Consequences of climate change	Needs of agricultural ecosystem	Contribution of Conservation Agriculture
Reduced availability of water for crops and increased evapotranspiration	Improved water balance	Reduced run-off Increased infiltration Reduced water evaporation
Degradation of soil due to extreme weather events (drought, intense rainfall)	Improved soil quality and structure	Reduced erosion Increase in the natural fertility of soil and OM
Higher incidence of pests and diseases	Improved pest and disease control	Increased biodiversity (increase in natural predators and competition) Crop rotation (controlling diseases and pests by changing plant species)

Source: compiled by authors.

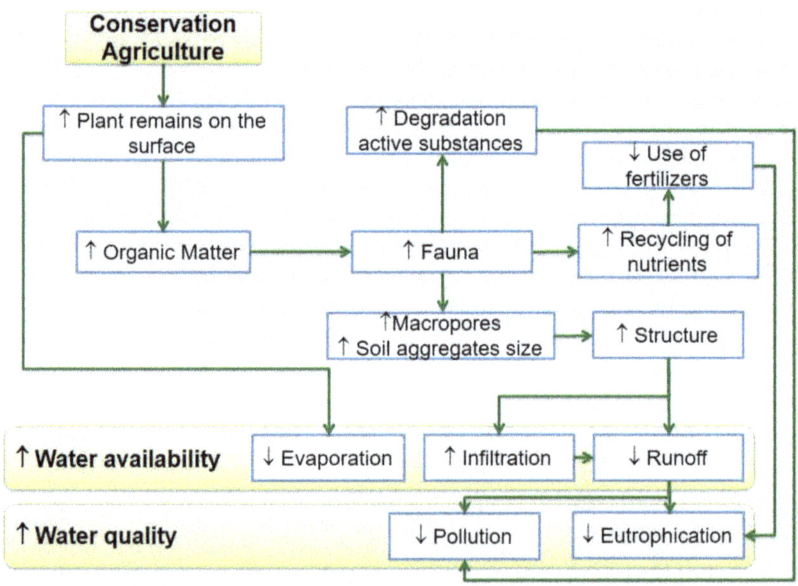

Figure 6 Processes in Conservation Agriculture with regard to benefits of water. Source: authors' diagram.

found that soil moisture content in no-till crops was significantly higher than in crops farmed using conventional tillage.

This improvement in soil moisture content also has a positive impact on irrigation, as indicated by Grassini et al. (2011) and Gathala et al. (2013). These studies showed that, with a reduction of 20–50% in irrigation water, no-till crops obtained a similar yield to crops irrigated with the same irrigation amounts. This means that irrigated no-till crops will adapt better to future scenarios of lower water availability as a consequence of climate change.

3.1.1 Reduced run-off

Significantly less run-off is produced in CA, due on the one hand to increased water infiltration as a result of structural improvements in the soil, and on the other to the lower run-off flow speeds as a consequence of the crop residue and mulch cover on the soil surface, increasing the opportunity for infiltration to occur. Furthermore, crop residue protects the soil from the direct impact of raindrops, which are responsible for breaking up aggregates and producing a sealing effect on the surface (resulting in crusting), a phenomenon that limits infiltration and increases run-off. All this succeeds in improving the soil's water balance by reducing water discharge.

Several worldwide studies confirm the reduced run-off that occurs in CA systems, with a reduction being approximately 67% in no-till systems (Kertész et al., 2010) and 43% in groundcovers (Márquez et al., 2010a). In a meta-analysis to quantify the effectiveness of no-tillage in reducing surface run-off, Sun et al. (2015) concluded that CA significantly reduced run-off by 22% and 27% in comparison with minimum tillage and conventional tillage, respectively. The percentage of soil surface covered, as well as the intensity of rainfall, influences the rate of decrease in run-off. That is the reason of the differences between estimations given by the different studies.

3.1.2 Increased infiltration rate

The higher infiltration rate in CA soils means that groundwater recharge in these soils after periods of rain is greater than soils managed using a tillage system. In this regard, several authors have studied the effects of soil management on water dynamics and conservation, with some concluding that the volumetric content of the first 20 cm of soil is higher in CA soils than in conventionally tilled soils. Along the same lines, in a comparison of conventional tillage and CA systems with monoculture durum wheat, Troccoli et al. (2009) reported an average soil moisture content that was significantly higher in no-till systems (13% dry basis) than in conventional tillage (10%), equating to a water depth saving of approximately 22 mm.

Studies conducted in Spain by Muriel et al. (2005) concluded that not only did CA allow for greater water retention in the soil profile (particularly in the initial 30 cm of depth), they also slowed the rate of water discharge, which has a positive effect on crop development in spring-summer, where the limiting factor for production is undoubtedly water availability.

3.1.3 Reduced evaporation

CA reduces water evaporation, as it prevents the direct incidence of radiation on humid soil and reduces the turbulent transfer of vapour into the atmosphere. As a result, dryland crops are able to cope better with stress conditions, as found in certain studies carried out on dryland crops in Andalusia (Spain) (Moreno et al.,1997; Murillo et al., 1998), where spring and summer temperatures are high, with this positive effect being particularly notable in dry years.

3.2 Improved soil quality and structure

One of the key factors in the increased resilience of agricultural ecosystems as a result of implementing CA is the substantial improvement of the physicochemical properties of soils under CA management. Crop residues and their decomposition have both direct and indirect impacts favouring soil quality and structure (Fig. 7). Soils with a better structure and which are less susceptible to erosion will respond better to intense rainfall events. Furthermore, soils with a higher OM content and higher natural fertility are more resilient to adverse weather situations that could play a role in their degradation.

3.2.1 Reduced erosion

The IPCC, in its fifth evaluation report (IPCC, 2014a), warns of the increase in the occurrence of episodes of extreme rainfall as a consequence of climate change. These episodes affect the soil in an important way, as the intensity of

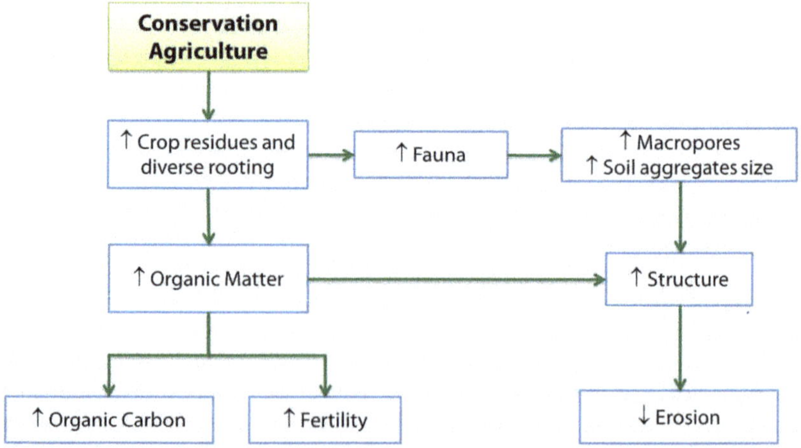

Figure 7 Processes in Conservation Agriculture with regard to benefits of soil. Source: authors' diagram.

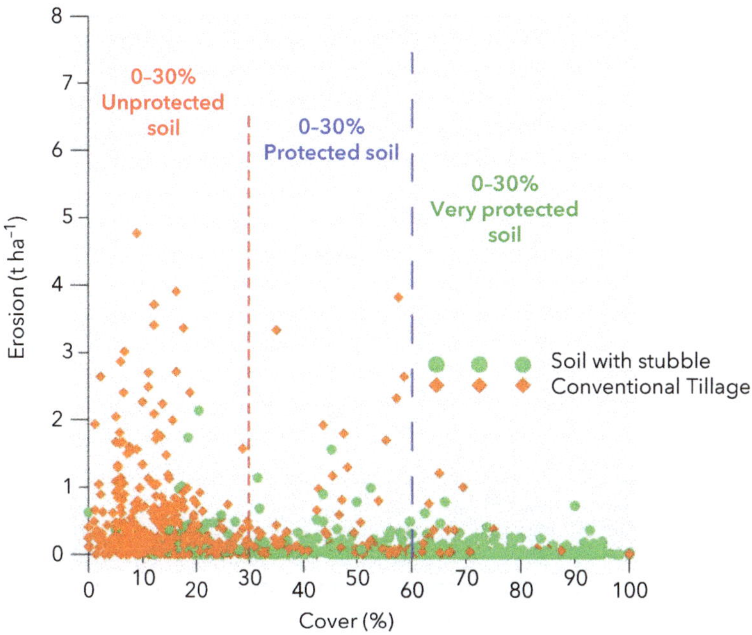

Figure 8 Relationship between percentage of groundcover and soil loss. Source: adapted from Márquez et al. (2010b).

rainfall is a factor directly linked to erosion. In these circumstances, having more protected soils against these phenomena is a strategy of adaptation.

As a result of the groundcover that CA provides for the soil, the direct impact of raindrops is minimized, helping to increase infiltration and reduce run-off, as the water loses its erosive power. This reduction is more effective the greater the groundcover, and therefore the less residue is buried through tillage. Studies carried out along these lines show that with 30% or more groundcover, erosion levels fall, and practically disappear at 60% or more (Fig. 8). Recent studies show that erosion can be reduced by as much as 98% in no-till systems (Kertész et al., 2010).

Maintaining permanent soil cover also plays an important role in reducing wind erosion. According to results obtained by Fryrear (1985), in soil with a surface covered with 20% crop stubble, soil loss was reduced by 57%. In soils with a stubble cover of 50%, erosion was reduced by 95% (Fig. 9).

3.2.2 Increase in the natural fertility of soil and organic matter (OM)

The increase in OM content in soils under CA means an improvement in soil quality, as it releases nutrients to the plants, increases chemical and physical

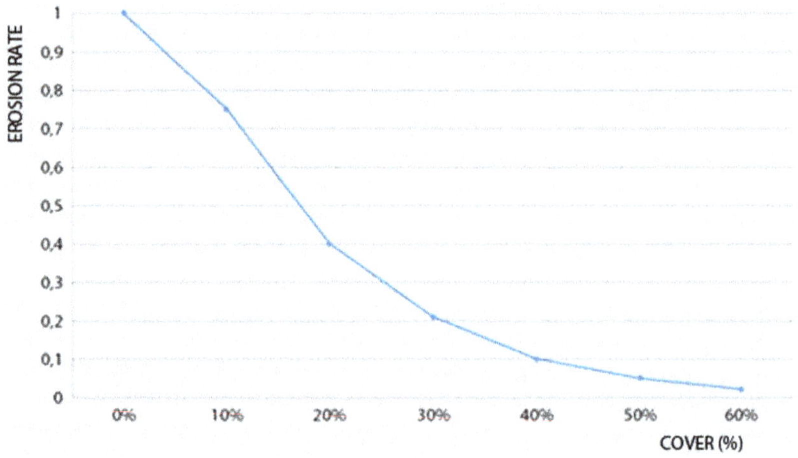

Figure 9 Reduction of wind erosion rate according to the percentage of crop residue. Source: adapted from Fryrear (1985).

fertility and helps develop structure or aggregates, thus increasing the soil's resistance to erosion and helping water to infiltrate. Further, because of the capacity of the humus to retain cations and adsorb heavy and noxious elements, the OM acts as a filter for the water, improving its quality.

Furthermore, the improvement in the moisture regime during the growing season and improved soil storage of water and nutrients, together with the use of leguminous crop rotation, as well as the presence of a groundcover of stubble, all help to ensure that in the medium to long term, crops grown using CA methods require less fertilizers and pesticides (Lafond et al., 2008; Crabtree, 2010; Lindwall and Sonntag, 2010). Plant groundcover also 'buffers' extreme temperatures on the soil's surface, which could otherwise damage plant tissue in the soil/atmosphere interface, thus minimizing one possible cause of limited yields (Kassam et al., 2012).

3.3 Improved pest and disease control

3.3.1 Increased biodiversity

One of the effects of climate change will be new pests and diseases for crops in all regions (IPCC, 2014a). In this sense, an agricultural ecosystem with a greater degree of biodiversity will have more tools to minimize the impact of these pests and diseases. The existence of a greater number of natural predators in the trophic chain helps maintain a balance in the ecosystem, thus increasing its resilience.

One of the environmental benefits of implementing CA for agricultural ecosystems is the improvement in biodiversity in general, and in soil in particular (Fig. 10). Specifically, under CA practices the soil biota is enriched, with improved nutrient recycling and helping to control pests and diseases (Holland, 2004).

One of the populations that benefits from CA are microorganisms from a diverse range of groups (bacteria, fungi, protozoa, nematodes etc.) found in no-till soils. In research carried out in semi-arid regions with a Mediterranean climate (Muñoz et al., 2007), significant differences were found in the amount of microorganisms from the beginning to the end of the study in several management systems, and always in favour of CA systems. According to this study, the soil managed using the no-till system had an average of 50% greater microbial biomass than conventionally tilled soil.

Another population benefited by implementing CA, and the activity of which spells an improvement in soil fertility and structural stability, is that of earthworms. These creatures are hugely significant, particularly in productive ecosystems, because of their influence on decomposing OM, developing soil structure and nutrient cycles. Furthermore, worm activity helps reduce apparent soil density and increases water infiltration with the subsequent advantages indicated previously of improving the soil's moisture content. It has been demonstrated that no-till agriculture increases worm activity, due to less mechanical alteration of the soil and increased OM. Studies carried out in

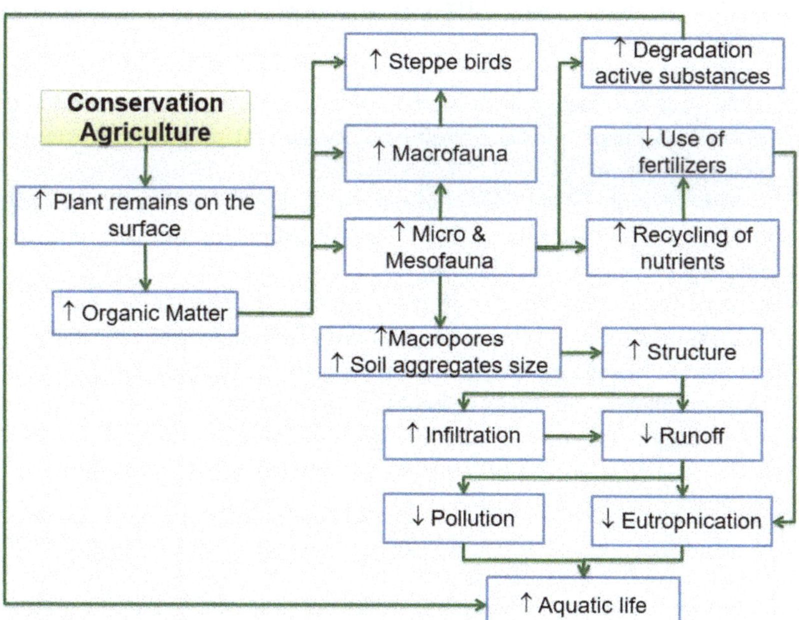

Figure 10 Processes in Conservation Agriculture with regard to benefits of biodiversity. Source: authors' diagram.

France (Piron et al., 2010) compared three management systems (no till, surface till and conventional till) with two fertilization strategies (mineral and organic fertilizer) and found that after 7 years in all cases the worm population was higher in CA than in conventional tillage treatments.

3.3.2 Crop rotation/diversification

One of the main elements on which CA is based is crop rotation or diversification in annual crops. Pests and diseases are better controlled, pest and disease cycles that remain in monoculture farming are broken, crop diversity can help improve the soil's natural fertility and biodiversity.

Further, crop rotation not only brings benefits in optimized water management and soil moisture, it has other advantages that help the agricultural ecosystem be more and better prepared for climate scenarios resulting in increased resilience and sustainability. In this regard, crop rotation has the following advantages:

- With roots exploring different edaphic horizons and having different water needs, rotated crops boost synergies between them, and in general help produce larger yields in the medium and long term.
- Rotation is used to reduce pests and diseases in the crop system and to control weeds.
- Rotations can also bring benefits in terms of improved soil quality (deeper roots, root exudates), improved nutrient distribution in the soil profile (deep-root crops mobilize nutrients located at depth) and increased biological activity.
- Periods of high labour demand can be reduced, with work distributed throughout the year, if sowing and harvest times are different in the different crops used in rotation.
- Risks posed by extreme weather events such as drought or flooding can be lessened, as their incidence does not affect all crops in the same way. Rotation is therefore a way of diversifying and mitigating risk.
- Crop rotations can balance stubble production by alternating crops that produce very little and/or easily degradable stubble with crops that produce large amounts of or highly durable stubble.

Similar advantages are present with CA systems in which crop diversification is achieved through associations. As a result, crop rotations and associations boosted by CA methods increase the resilience of the agricultural ecosystem, generally improving soil properties while increasing the crop's potential for greater yields.

4 Case study: the LIFE+ Agricarbon Project

As indicated throughout this chapter, CA is a practice with a double response to climate change by both mitigating the driving phenomena behind it and helping crops adapt to its effects. These agricultural practices are not exempt from being used along with other techniques, the synergies of which boost the benefits that they can both have in combating impacts of climate change.

This is the focus of the European project LIFE+ Agricarbon, an initiative that, over 4 years, has shown the potential for mitigation and adaptation of agricultural systems implemented under CA with support from precision farming techniques in the Mediterranean regions.

To that end, a network of demo farms was set up in the Guadalquivir valley (Spain), with crop rotation implemented using CA and precision agriculture practices. These farms were monitored for reduced GHG emissions, the soil's increased sink effect, crop production and soil moisture. In total, 90 ha of crops were established, where conventional tillage-based techniques were compared with the methods proposed by the project (Figs. 11 and 12).

Moving from conventional agriculture to the adoption of CA increased SOC by an average of 30% and in some cases as much as a 56% increase (Fig. 13).

At the same time, with CA and precision agriculture, CO_2 emissions from the soil were reduced by 19% compared with conventional agriculture. Regarding energy use, the joint implementation of CA and precision agriculture reduced energy consumption by 12% in the wheat crop, 26% in the sunflower crop and 18% in the legume crop. This means that in a growing season using no-till methods, the wheat crop emitted 199 kg CO_2 ha^{-1} less, the sunflower crop 64 kg CO_2 ha^{-1} less and the legume crop 107 kg CO_2 ha^{-1} less than in conventional tillage (Fig. 14). The surface area covered by CA and precision agriculture crops

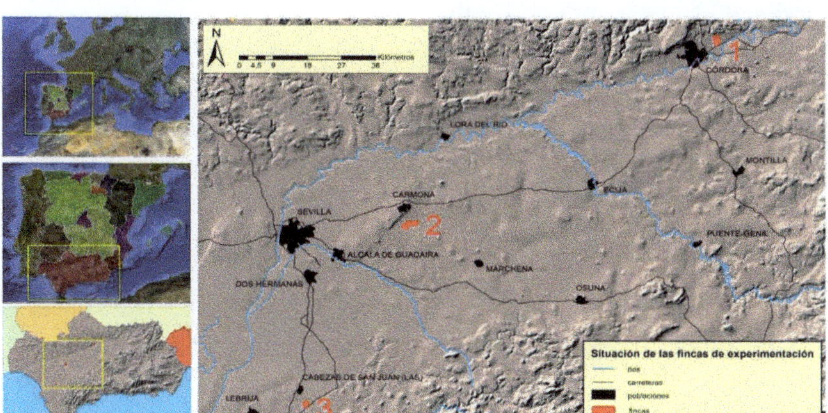

Figure 11 Location of the demo farms.

Figure 12 Layout of the demo farms.

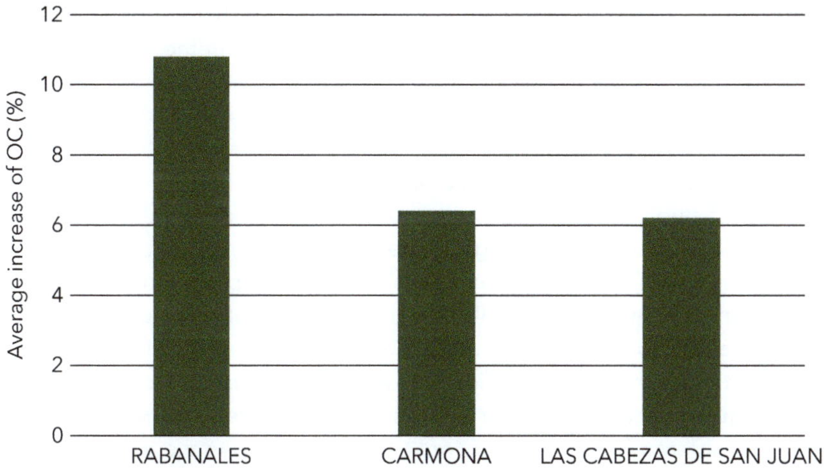

Figure 13 Average increase in organic carbon (OC) in the soil in CA compared with CT in three demo farms after 4 years.

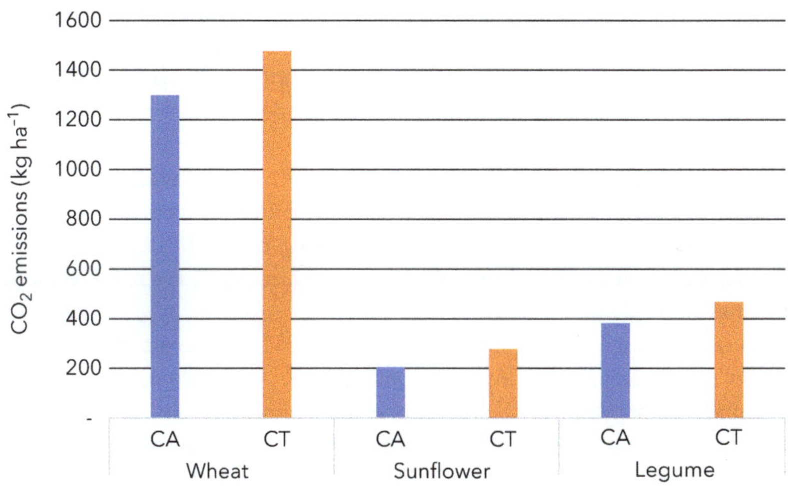

Figure 14 CO_2 emissions for each crop as a consequence of the energy consumption of the operations carried out (CA: Conservation Agriculture, CT: Conventional Tillage).

project's network of demo farms resulted in 1296 t more CO_2 sequestered and 20 t less CO_2 emitted into the atmosphere.

5 Conclusion

Climate change is undoubtedly one of the major challenges faced by the agricultural sector today. Indeed, this challenge is even more relevant in the regions where in the coming years, it is foreseeable for there to be an increase in average temperatures, erratic rainfall regimes and extreme weather events (drought, torrential rain), placing agricultural ecosystems and the model of agricultural production as we know it today at serious risk. In this scenario, it would be necessary to implement good agricultural practices that would both mitigate climate change and adapt crops to its effects.

Most of studies carried out throughout the world confirm that soils under CA increase SOC, thus mitigating climate change. The main factors that make possible a change in the dynamics of the C in the soil, promoting the increase of this element in the soil's profile, are the presence of crop residues on the soil surface and root biomass in the soil and the suppression or avoidance of tillage. Data found in the scientific literature show a great variability regarding the C sequestration rates under CA. This fact is related to the agroclimatic region considered (0.08–2.76 t ha^{-1} yr^{-1}), but also linked to the amount of crop residues returned after harvest.

Due to the minimum soil disturbance principle, CA reduces CO_2 emissions into the atmosphere. On the one hand, by not mechanically altering the soil the mineralization of OM is considerably reduced, which is the main CO_2 generation process in the soil. On the other hand, the reduction in the number of field passes leads to a lower energy consumption, of up to 25% in some cases, which in the end results in a similar reduction in CO_2 emissions linked to energy consumption.

CA promotes more resilient agricultural ecosystems, due to the environmental benefits that these practices bring to the soil and water resources and also due to the increase in biodiversity. Thus, soils under CA are less erodible and, therefore, with a better capacity to respond to extreme climatic events.

CA also improves the structure of the soil, increases the infiltration of water in the soil profile and reduces run-off. All this leads to increase in the availability of water for cultivation, a fundamental issue in those regions where due to climate change, an increase in temperature is expected and, as a consequence of evapotranspiration, a reduction in the availability of water resources.

The increase in biodiversity is another benefit that will contribute to the adaptation of agrarian ecosystems to climate change. The existence of an abundant and diverse biota increases the natural control of pests and diseases, helping to maintain balance. In this sense, the studies show how the practice of CA increases the biodiversity of the micro and meso fauna fundamentally. Moreover, not only would residues benefit soil fauna, but they also offer shelter to birds and small animals.

For all this, CA can be considered as a holistic response to climate change and the challenges it involves.

6 Future trends

Despite the advances in knowledge that have occurred in recent years with regard to CA in general, and most particularly to everything related to these practices and climate change, there are still unanswered questions on this matter, and new lines of investigation remain to be explored.

With regard to climate change mitigation, many authors have studied the capacity that techniques such as no-till farming or groundcover in woody crops have for capturing atmospheric carbon, but it is no less true that these studies represent only one part of the capacity that these practices have in mitigating climate change, given that a global balance is not being struck between sources of emission and carbon sinks. Furthermore, reviews such as those by Smith et al. (2005), Powlson et al. (2016) and Du et al. (2017) question the mitigating role that CA plays, due to the heterogeneous nature of the data between the studies conducted, the lack of unification of criteria about what CA really is (three interlinked principles by FAO) or when approaching research aimed at studying the sink capacity of CA, and the lack of consideration given to certain OC values at particular soil depths. In this regard, comprehensive studies that combine research into carbon capture and emission of the main GHGs (CO_2, N_2O and CH_4) in these management systems would shed light on the true potential of CA in mitigating climate change.

Another line of research that has not long been under study, particularly with the launch of the '4 per 1000' initiative, is that of OC content in soils and the relevance of this knowledge in determining their capture capacity. The '4 per 1000' initiative proposes improving content OM by increasing C in soils by 0.4%, by implementing agricultural practices that will help achieve this increase. In this regard, knowledge of the existing C stock in soils is very relevant, and several studies are being carried out across the world using modelling techniques (Batjes, 2016; Aksoy et al., 2016; De Brogniez et al., 2015; Lugato et al., 2014). Integrating variables into these models that take into account the management system, such as CA for example, would provide more global data on the capacity of these practices for soil carbon capture in different agroecologies.

Although studies based on the experimental research and demonstration of certain agricultural practices in mitigating climate change have been carried out for some time, research focussing on adaptation strategies is more recent, taking into account the fact that the initial international focus was on finding ways to alleviate the driving forces behind climate change. In this regard, existing scientific evidence in terms of CA and its capacity to help crops adapt to the effects of climate change is the indirect result of studies with objectives that did not include studying adaptation as a goal, but rather in the provision of services within the ecosystems of said practices (Palm et al., 2014; Hobbs et al.,

2008; Grace et al., 2003). In this regard, and taking into account the fact that the models used to predict the impacts of climate change as ENSEMBLESS (van der Linden and Mitchell, 2009) do not explicitly consider management systems as variables from the outset, new opportunities arise in the field of modelling by including the mitigation and adaptability benefits brought by CA in those models.

Another line of research to consider in the future is the study of the benefits that the joint use of a range of agricultural practices might have with regard to climate change. Some reviews already identify a series of good agricultural practices for combating climate change (Smith et al., 2008), but they analyse their benefits individually. The synergies that can occur between the joint application of CA and other agricultural practices may be highly beneficial for climate change due to the amount of ecosystem services that are generated. Studies such as those by Pinto et al. (2017) and Villamil et al. (2006) are pioneering in America, but are much less frequent in Europe.

Finally, what is important is to maximize the contribution that the locally adapted CA systems can make to both mitigation and adaptation objectives at the farm, landscape and regional levels for producers and their communities, for society at large and for the environment. Research has already shown enough evidence that increase in soil organic matter is related to both mitigation and adaptability, and consequently CA systems offer both these benefits simultaneously. Research should focus on how to optimize CA's contribution to both mitigation and adaptability in the context of contributing to achieving the Sustainable Development Goals.

7 Where to look for further information

Existing information on CA and climate change is plentiful and has increased considerably in recent years. There are many research groups and both public and private bodies that have begun to show an interest in these issues, and have therefore generated new documentation, adding to existing knowledge.

For all matters relating to climate change and its impact on the agricultural sector, the website of the IPCC (www.ipcc.ch) provides ample information on the subject. The current Fifth Assessment Report includes data and figures on impacts, adaptation and vulnerability, and mechanisms of mitigation in the report carried out by Working Group II and Working Group III, respectively, with a section devoted to agriculture.

Additionally, to discover existing options in the agricultural sector for mitigating climate change, a wide-reaching review was carried out by Smith et al. (2008), which considers the different agricultural practices currently available for mitigating climate change, and the possibilities of them being implemented in agricultural ecosystems.

For the specific case of CA and climate change, a report from the ECAF condenses current knowledge on mitigation and adaptation mechanisms that have been implemented using these techniques and the results reported by the scientific literature. This report, titled 'Conservation Agriculture: Making Climate Change Mitigation and Adaptation Real in Europe' (González-Sánchez et al., 2017), is available in the ECAF website (www.ecaf.org). The ECAF also has a large database on CA (conservationagriculturedatabase.eu), with documents ranging from scientific articles to technical reports, as well as video archives that include aspects relating to the benefits regarding climate changes reported by these techniques.

As well as the information published, there are several international events and international institutions that either directly or indirectly cover issues of climate change and CA. These include:

- World Congress on Conservation Agriculture: an international event held every 3 years, with CA as the central issue.
- Eurosoil: an event organized at the European level by the European Confederation of Soil Science Societies, held every 4 years. With soil as the main issue, information can be found on the functions of soil as a mitigating agent and which can help crops adapt to climate change.
- World Congress of Soil Sciences: an international event organized by the International Union of Soil Sciences (IUSS), held every 4 years, which draws together all knowledge relating to all the fields of sciences involved in the study of soil, which naturally includes those that focus on research into resilience and the sink effect.
- FAO website about Conservation Agriculture (www.fao.org/conservation-agriculture).
- Cornwell University website of Conservation Agriculture resources (http://conservationagriculture.mannlib.cornell.edu/).

8 References

Aguilera, E., Lassaletta, L., Gattinger, A. and Gimeno, B. 2013. Managing soil carbon for climate change mitigation and adaptation in Mediterranean cropping systems: a meta-analysis. *Agriculture, Ecosystems and Environment* 168, 25–36.

Aksoy, E., Yigini, Y. and Montanarella, L. 2016. Combining soil databases for topsoil organic carbon mapping in Europe. *PLoS ONE* 11(3), e0152098. doi:10.1371/journal.pone.0152098.

Balesdent, J., Mariotti, A. and Boisgontier, D. 1990. Effects of tillage on soil organic carbon mineralization estimated from 13C abundance in maize fields. *Journal of Soil Science* 41(4), 587–96. doi:10.1111/j.1365-2389.1990.tb00228.x.

Barthès, B., Azontonde, A., Blanchart, E., Girardin, C., Villenave, C., Lesaint, S., Oliver, R. and Feller, C. 2004. Effect of a legume cover crop (*Mucuna pruriens* var. utilis) on soil carbon in an Ultisol under maize cultivation in southern Benin. *Soil Use and Management* 20, 231–9.

Batjes, N. H. 2016. Harmonized soil property values for broad-scale modelling (WISE30sec) with estimates of global soil carbon stocks. *Geoderma* 269, 61–8. doi:10.1016/j.geoderma.2016.01.034.

Bescansa, P., Imaz, M. J., Virto, I., Enrique, A. and Briones, M. J. I. 2005. Influencia de laboreo de conservación y manejo de residuos en el desarrollo de poblaciones de lombrices en suelos semiáridos. *En Congreso Internacional sobre Agricultura de Conservación: El reto de la agricultura, el medio ambiente, la energía y la nueva Política Agraria Común*. Libro de Actas: 285–290. AEAC/SV, ECAF, Diputación de Córdoba, Córdoba.

Blanchart, E., Bernoux, M., Sarda, X., Siqueira Neto, M., Cerri, C. C., Piccolo, M., Douzet, J., Scopel, E. and Feller, C. 2007. Effect of direct seeding mulch-based systems on soil carbon storage and macrofauna in Central Brazil. *Agriculturae Conspectus Scientificus* 72, 81–7.

Cantero-Martínez, C., Ojeda, J., Angás, P. and Santiveri, P. 2004. Efectos sobre las poblaciones de lombrices: técnicas de laboreo en suelo en zonas de secano semiárido. *Agricultura* 866, 724–8.

Carbonell-Bojollo, R., González-Sánchez, E. J., Veroz-González, O. and Ordóñez-Fernández, R. 2011. Soil management systems and short term CO_2 emissions in a clayey soil in southern Spain. *Science of the Total Environment* 409(15), 2929–35. doi:10.1016/j.scitotenv.2011.04.003.

Causarano, H. J., Franzluebbers, A. J., Reeves, D. W. and Shaw, J. N. 2006. Soil organic carbon sequestration in cotton production systems of the Southeastern United States: a review. *Journal of Environmental Quality* 35(4), 1374–83. doi:10.2134/jeq2005.0150.

De Brogniez, D., Ballabio, C., Stevens, A., Jones, R. J. A., Montanarella, L. and van Wesemael, B. 2015. A map of the topsoil organic carbon content of Europe generated by a generalized additive model. *European Journal of Soil Science* 66(1), 121–34. doi:10.1111/ejss.12193.

Delgado, J. A., Groffman, P. M., Nearing, M. A., Goddard, T., Reicosky, D., Lal, R., Kitchen, N. R., Rice, C. W., Towery, D. and Salon, P. 2011. Conservation practices to mitigate and adapt to climate change. *Journal of Soil and Water Conservation* 66(4), 118A–29A. doi:10.2489/jswc.66.4.118A.

Du, Z., Angers, D. A., Ren, T., Zhang, Q. and Li, G. 2017. The effect of no-till on organic C storage in Chinese soils should not be overemphasized: a meta-analysis. *Agriculture, Ecosystems and Environment* 236, 1–11. doi:10.1016/j.agee.2016.11.007.

Duru, M., Therond, O., Martin, G., Martin-Clouaire, R., Magne, M. A., Justes, E., Journet, E. P., Aubertot, J. N., Savary, S., Bergez, J. E. and Sarthou, J. P. 2015. How to implement biodiversity-based agriculture to enhance ecosystem services: a review. *Agronomy for Sustainable Development* 35(4), 1259–81. doi:10.1007/s13593-015-0306-1.

Crabtree, B. 2010. *Search for Sustainability with No-Till Bill in Dryland Agriculture*. Crabtree Agricultural Consulting, Australia.

FAO. 2018. Conservation Agriculture website. Available at: http://www.fao.org/conservation-agriculture/en/ (accessed on 19 June 2018).

Franzluebbers, A. J. 2002. Soil organic matter stratification ratio as an indicator of soil quality. *Soil and Tillage Research* 66(2), 95–106. doi:10.1016/S0167-1987(02)00018-1.

Franzluebbers, A. J. and Arshad, M. A. 1996. Soil organic matter pools with conventional and zero tillage in a cold, semiarid climate. *Soil and Tillage Research* 39(1-2), 1-11. doi:10.1016/S0167-1987(96)01055-0.

Freibauer, A., Rounsevell, M. D. A., Smith, P. and Verhagen, J. 2004. Carbon sequestration in the agricultural soils of Europe. *Geoderma* 122(1), 1-23. doi:10.1016/j.geoderma.2004.01.021.

Fryrear, D. W. 1985. Soil cover and wind erosion. *Transactions of the ASAE* 28(3), 781-4.

Gathala, M. K., Kumara, V., Sharmac, P. C., Saharawata, Y. S., Jat, H. S., Singh, M., Kumar, A., Jat, M. L., Humphreys, E., Sharmac, D. K., Sharmaa, S. and Ladhaa, J. K. 2013. Optimizing intensive cereal-based cropping systems addressing current and future drivers of agricultural change in the northwestern Indo-Gangetic Plains of India. *Agriculture, Ecosystems and Environment* 177, 85-97. doi:10.1016/j.agee.2013.06.002.

González-Sánchez, E. J., Moreno-García, M., Kassam, A., Holgado-Cabrera, A., Triviño-Tarradas, P., Carbonell-Bojollo, R., Pisante, M., Veroz-González, O. and Basch, G. 2017. *Conservation Agriculture: Making Climate Change Mitigation and Adaptation Real in Europe*. European Conservation Agriculture Federation (ECAF). Available at: http://www.ecaf.org/downloads/books/23-conservation-agriculture-climate-change-report/file.

González-Sánchez, E. J., Veroz-Gonzalez, O., Conway, G., Moreno-Garcia, M., Kassam, A., Mkomwa, S., Ordoñez-Fernandez, R., Triviño-Tarradas, P. and Carbonell-Bojollo, R. 2019. Meta-analysis on carbon sequestration through Conservation Agriculture in Africa. *Soil and Tillage Research* 190, 22-30. doi:10.1016/j.still.2019.02.020.

Govaerts, B., Verhulst, N., Castellanos-Navarrete, A., Sayre, K. D., Dixon, J. and Dendooven, L. 2009. Conservation agriculture and soil carbon sequestration: between myth and farmer reality. *Critical Reviews in Plant Sciences* 28(3), 97-122, doi:10.1080/07352680902776358.

Grace, P. R., Harrington, L., Jain, M. C. and Robertson, G. P. 2003. Long-term sustainability of the tropical and subtropical rice-wheat system: an environmental perspective. In: Ladha, J. K., Hill, J., Gupta, R. K., Duxbury, J. and Buresh, R. J. (Eds), *Improving the Productivity and Sustainability of Rice-Wheat Systems: Issues and Impact*. ASA Special Publications 65. Australian society of anaesthetists, Madison, WI, pp. 27-43. chapter 7.

Grace, P. R., Antle, J., Ogle, S., Paustian, K. and Basso, B. 2010. Soil carbon sequestration rates and associated economic costs for farming systems of south-eastern Australia. *Soil Research* 48(8), 720-9. doi:10.1071/SR10063.

Grassini, P., Yang, H., Irmak, S., Thorburn, J., Burr, C. and Cassmann, K. G. 2011. High-yield irrigated maize in the Western U.S. Corn Belt: II. Irrigation management and crop water productivity. *Field Crops Research* 120(1), 133-41. doi:10.1016/j.fcr.2010.09.013.

Gwenzi, W., Gotosa, J., Chakanetsa, S. and Mutema, Z. 2009. Effects of tillage systems on soil organic carbon dynamics, structural stability and crop yields in irrigated wheat (*Triticum aestivum* L.)-cotton (*Gossypium hirsutum* L.) rotation in semi-arid Zimbabwe. *Nutrient Cycling in Agroecosystems* 83(3), 211-21. doi:10.1007/s10705-008-9211-1.

Hobbs, P. R., Sayre, K. and Gupta, R. 2008. The role of conservation agriculture in sustainable agriculture. *Philosophical Transactions of the Royal Society of London, Series B* 363, 543-55.

Holland, J. M. 2004. The environmental consequences of adopting conservation tillage in Europe: reviewing the evidence. *Agriculture, Ecosystems and Environment* 103(1), 1–25. doi:10.1016/j.agee.2003.12.018.

Hollinger, S. E., Bernacchi, C. J. and Meyers, T. P. 2005. Carbon budget of mature no-till ecosystem in North Central Region of the United States. *Agricultural and Forest Meteorology* 130(1–2), 59–69. doi:10.1016/j.agrformet.2005.01.005.

Horrigan, L., Lawrence, R. S. and Walker, P. 2002. How sustainable agriculture can address the environmental and human health harms of industrial agriculture. *Environmental Health Perspectives* 110(5), 445–56. doi:10.1289/ehp.02110445.

IPCC. 2006. *2006 IPCC Guidelines for National Greenhouse Gas Inventories*. Eggleston, H. S., Buendia, L., Miwa, K., Ngara, T. and Tanabe, K. (Eds). IPCC/IGES, Hyama, Japan.

IPCC. 2014a. *Climate Change 2014: Impacts, Adaptation, and Vulnerability. Part B: Regional Aspects. Contribution of Working Group II to the Fifth Assessment Report of the Intergovernmental Panel on Climate Change*. Barros, V. R., Field, C. B., Dokken, D. J., Mastrandrea, M. D., Mach, K. J., Bilir, T. E., Chatterjee, M., Ebi, K. L., Estrada, Y. O., Genova, R. C., Girma, B., Kissel, E. S., Levy, A. N., MacCracken, S., Mastrandrea, P. R. and White, L. L. (Eds). Cambridge University Press, Cambridge, United Kingdom and New York, NY, p. 688.

IPCC. 2014b. *Climate Change 2014: Mitigation of Climate Change. Contribution of Working Group III to the Fifth Assessment Report of the Intergovernmental Panel on Climate Change*. Edenhofer, O., Pichs-Madruga, R., Sokona, Y., Farahani, E., Kadner, S., Seyboth, K., Adler, A., Baum, I., Brunner, S., Eickemeier, P., Kriemann, B., Savolainen, J., Schlömer, S., von Stechow, C., Zwickel, T. and Minx, J. C. (Eds). Cambridge University Press, Cambridge, United Kingdom and New York, NY.

Kassam, A., Friedrich, T., Derpschs, R., Lanmar, R., Mrabet, R., Basch, G., González-Sánchez, E. J. and Serraj, R. 2012. Conservation agriculture in the dry Mediterranean climate. *Field Crops Research* 132, 7–17. doi:10.1016/j.fcr.2012.02.023.

Kertész, A., Madarász, B., Csepinszky, B. and Bádonyi, K. 2010. Conservation agriculture as a tool against soil erosion and for improving biodiversity. In: *Proceedings of the European Congress on Conservation Agriculture*. Ministerio de Medio Ambiente y Medio Rural y Marino. Asociación Española Agricultura de Conservación/Suelos Vivos, Madrid, pp. 501–6.

Lafond, G. P., Walley, H., Schoenau, J., May, W. E., Holzapfel, C. B., Mckell, J. and Halford, J. 2008. Long-term vs. short-term conservation tillage. In: *Proceedings of the 20th Annual Meeting and Conference of the Saskatchewan*. Soil Conservation Association, Regina, pp. 28–43.

Lal, R. 2004. Soil carbon sequestration impacts on global climate change and food security. *Science* 304(5677), 1623–7. doi:10.1126/science.1097396.

Lal, R. 2008. Carbon sequestration. *Philosophical Transactions of the Royal Society B: Biological Sciences* 363(1492), 815–30. doi:10.1098/rstb.2007.2185.

Lal, R. 2010. Managing soils and ecosystems for mitigating anthropogenic carbon emissions and advancing global food security. *BioScience* 60(9), 708–21. doi:10.1525/bio.2010.60.9.8.

Lindwall, C. W. and Sonntag, B. 2010. *Landscape Transformed: the History of Conservation Tillage and Direct Seeding. Knowledge Impact in Society*. University of Saskatchewan, Sakskatoon, Canada.

Lugato, E., Panagos, P., Bampa, F., Jones, A. and Montanarella, L. 2014. A new baseline of organic carbon stock in European agricultural soils using a modelling approach. *Global Change Biology* 20(1), 313–26, doi:10.1111/gcb.12292.

Márquez, F., Ordóñez, R., Gil Ribes, J., González, E. and Gómez, M. 2010a. Implantación de cubiertas vegetales en olivar como sistema de conservación y mejora del agua. In: *Proceedings of the European Congress on Conservation Agriculture*. Ministerio de Medio Ambiente y Medio Rural y Marino. Asociación Española Agricultura de Conservación/Suelos Vivos, Madrid, pp. 545–54.

Márquez, F., Ordóñez, R., Gil Ribes, J., González, E. and Gómez, M. 2010b. Implantación de cubiertas vegetales en olivar como sistema de conservación y mejora del suelo. In: *Proceedings of the European Congress on Conservation Agriculture*. Ministerio de Medio Ambiente y Medio Rural y Marino. Asociación Española Agricultura de Conservación/Suelos Vivos, Madrid, pp. 555–63.

Moreno, F., Pelegrin, F., Fernandez, J. E. and Murillo, J. M. 1997. Soil physical properties, water depletion and crop development under traditional and conservation tillage in southern Spain. *Soil and Tillage Research* 41(1–2), 25–42. doi:10.1016/S0167-1987(96)01083-5.

Moreno, F., Murillo, J. M., Madejón, E., Girón, I. F. and Pelegrín, F. 2005. Laboreo de conservación: efectos a largo plazo sobre la calidad del suelo. *En Congreso Internacional sobre Agricultura de Conservación: El reto de la agricultura, el medio ambiente, la energía y la nueva Política Agraria Común. Libro de Actas*: 515–520. AEAC/SV, ECAF, Diputación de Córdoba, Córdoba.

Muñoz, A., López-Piñeiro, A. and Ramírez, M. 2007. Soil quality attributes of conservation management regimes in a semi-arid region of south western Spain. *Soil and Tillage Research* 95(1–2), 255–65. doi:10.1016/j.still.2007.01.009.

Muriel, J. L., Vanderlinden, K., Perea, F., Jiménez, J. A., García-Tejero, I. and Pérez, J. J. 2005. Régimen hídrico en suelos arcillosos de campiña sometidos a distintos sistemas de manejo. In: En Veroz González, O., Holgado Cabrera, A., Ordóñez Fernández, R., González Sánchez, E. and Gil Ribes, J. (Coord.), *Actas Congreso Internacional sobre Agricultura de Conservación*. Asociación Española Agricultura de Conservación/Suelos Vivos, Córdoba, pp. 537–42.

Murillo, J. M., Moreno, F., Pelegrín, F. and Fernández, J. E. 1998. Responses of sunflower to traditional and conservation tillage under rainfed conditions in southern Spain. *Soil and Tillage Research* 49(3), 233–41. doi:10.1016/S0167-1987(98)00177-9.

Okeyo, J. M., Norton, J., Koala, S., Waswa, B., Kihara, J. and Bationo, A. 2016. Impact of reduced tillage and crop residue management on soil properties and crop yields in a long-term trial in western Kenya. *Soil Research* 54(6), 719–29. doi:10.1071/SR15074.

Olson, K. R., Lang, J. M. and Ebelhar, S. A. 2005. Soil organic carbon changes after 12 years of no tillage and tillage of Grantsburg soils in southern Illinois. *Soil and Tillage Research* 81(2), 217–25. doi:10.1016/j.still.2004.09.009.

Palm, C., Blanco-Canqui, H., DeClerck, F., Gatere, L. and Grace, P. 2014. Conservation agriculture and ecosystem services: an overview. *Agriculture, Ecosystems and Environment* 187, 87–105. doi:10.1016/j.agee.2013.10.010.

Pinto, P., Fernández Long, M. E. and Piñeiro, G. 2017. Including cover crops during fallow periods for increasing ecosystem services: is it possible in croplands of Southern South America? *Agriculture, Ecosystems and Environment* 248, 48–57. doi:10.1016/j.agee.2017.07.028.

Piron, D., Peres, G. and Cluzeau, D. 2010. The evaluation and improvement of biodiversity in conservation agriculture systems. In: *Proceedings of the European Congress on Conservation Agriculture*. Ministerio de Medio Ambiente y Medio Rural y Marino. Asociación Española Agricultura de Conservación/Suelos Vivos, Madrid, pp. 155–62.

Powlson, D. S., Stirling, C. M., Thierfelder, C., White, R. P. and Jat, M. L. 2016. Does conservation agriculture deliver climate change mitigation through soil carbon sequestration in tropical agro-ecosystems? *Agriculture, Ecosystems and Environment* 220, 164–74. doi:10.1016/j.agee.2016.01.005.

Puget, P. and Lal, R. 2005. Soil organic carbon and nitrogen in a Mollisol in central Ohio as affected by tillage and land use. *Soil and Tillage Research* 80(1–2), 201–13. doi:10.1016/j.still.2004.03.018.

Ranaivoson, L., Naudin, K., Ripoche, A., Affholder, F., Rabeharisoa, L. and Corbeels, M. 2017. Agro-ecological functions of crop residues under conservation agriculture. A review. *Agronomy for Sustainable Development* 37(4), 26. doi:10.1007/s13593-017-0432-z.

Ray, D. K., Gerber, J. S., MacDonald, G. K. and West, P. C. 2015. Climate variation explains a third of global crop yield variability. *Nature Communications* 6, 5989 doi:10.1038/ncomms6989.

Reicosky, D. C. 1997. Tillage-induced CO_2 emission from soil. *Nutrient Cycling in Agroecosystems* 49(1/3), 273–85. doi:10.1023/A:1009766510274.

Reicosky, D. C. 2011. Conservation agriculture: global environmental benefits of soil carbon management. In: *Fifth World Congress on Conservation Agriculture* (vol. 1). Australian Center for International Agricultural Research, Canberra, pp. 3–12.

Reicosky, D. C. and Archer, D. W. 2007. Moldboard plow tillage depth and short-term carbon dioxide release. *Soil and Tillage Research* 94(1), 109–21. doi:10.1016/j.still.2006.07.004.

Sá, J. C. D. M., Bürkner dos Santos, J., Lal, R., de Moraes, A., Tivet, F., Machado Sá, M. F., Briedis, C., de Oliveira Ferreira, A., Eurich, G., Farias, A. and Friedrich, T. 2013. Soil-specific inventories of landscape carbon and nitrogen stocks under no-till and native vegetation to estimate carbon offset in a subtropical ecosystem. *Soil Science Society of America Journal* 77(6), 2094–110. doi:10.2136/sssaj2013.01.0007.

Sainju, U. M., Jabro, J. D. and Stevens, W. B. 2008. Soil carbon dioxide emissions and carbon contents as affected by irrigation, tillage, cropping system and nitrogen fertilization. *Journal of Environmental Quality* 37(1), 98–106. doi:10.2134/jeq2006.0392.

Six, J., Elliott, E. T. and Paustian, K. 2000. Soil macroaggregate turnover and microaggregate formation: a mechanism for C sequestration under no-tillage agriculture. *Soil Biology and Biochemistry* 32(14), 2099–103. doi:10.1016/S0038-0717(00)00179-6.

Smith, P. 2004. Carbon sequestration in croplands: the potential in Europe and the global context. *European Journal of Agronomy* 20(3), 229–36. doi:10.1016/j.eja.2003.08.002.

Smith, P., Powlson, D. S., Smith, J. U., Falloon, P. D. and Coleman, K. 2000. Meeting Europe's climate change commitments: quantitative estimates of the potential for carbon mitigation by agriculture. *Global Change Biology* 6, 525–39.

Smith, P., Andrén, O., Karlsson, T., Perälä, P., Regina, K., Rounsevell, M. and Wesemael, B. 2005. Carbon sequestration potential in European croplands has been overestimated. *Global Change Biology* 11(12), 2153–63. doi:10.1111/j.1365-2486.2005.01052.x.

Smith, P., Martino, D., Cai, Z., Gwary, D., Janzen, H., Kumar, P., McCarl, B., Ogle, S., O'Mara, F., Rice, C., Scholes, B., Sirotenko, O., Howden, M., McAllister, T., Pan, G., Romanenkov, V., Schneider, U., Towprayoon, S., Wattenbach, M. and Smith, J. 2008. Greenhouse gas mitigation in agriculture. *Philosophical Transactions of the Royal Society of London, Series B* 363, 789–813.

Sommer, R. and Bossio, D. 2014. Dynamics and climate change mitigation potential of soil organic carbon sequestration. *Journal of Environmental Management* 144, 83–7. doi:10.1016/j.jenvman.2014.05.017.

Sun, Y., Zeng, Y., Shi, Q., Pan, X. and Huang, S. 2015. No-tillage controls on runoff: a meta-analysis. *Soil and Tillage Research* 153, 1–6. doi:10.1016/j.still.2015.04.007.

Tebruegge, F. 2001. No-tillage visions- Protection of soil, water and climate and influence on management and farm income. En García-Torres, L., Benites, J. and Martínez-Vilela, A. (Eds), *I World Congress on Conservation Agriculture: Conservation Agriculture, A Worldwide Challenge. Volume I*. FAO, ECAF, Córdoba, Spain, pp. 303–16.

Thierfelder, C., Cheeseman, S. and Rusinamhodzi, L. 2012. Benefits and challenges ofcrop rotations in maize-based conservation agriculture (CA) cropping systems of southern Africa. *International Journal of Agricultural Sustainability* 11, 108–24.

Thierfelder, C., Mwila, M. and Rusinamhodzi, L. 2013. Conservation agriculture in eastern and southern provinces of Zambia: long-term effects on soil quality and maize productivity. *Soil and Tillage Research* 126, 246–58. doi:10.1016/j.still.2012.09.002.

Troccoli, A., Colecchia, S. A., Russo, M., Cattivelli, L. and Gallo, A. 2009. Agricoltura conservativa ideale per il grano duro al Sud. *L'Informatore Agrario* 34, 37–41.

US-EPA. 2005. Greenhouse gas mitigation potential in US forestry and agriculture. EPA 430-R-05-006. US Environmental Protection Agency, Washington DC.

US-EPA. 2006. Global mitigation of non-CO_2 greenhouse gases. EPA 430-R-06-005. US Environmental Protection Agency, Washingdon DC.

van der Linden, P. and Mitchell, J. F. B. (Eds). 2009. *ENSEMBLES: Climate Change and its Impacts: Summary of Research and Results from the ENSEMBLES Project*. Met Office Hadley Centre, Exeter, UK, 160p.

Vicente-Vicente, J. L., García-Ruiz, R., Francaviglia, R., Aguilera, E. and Smith, P. 2016. Soil carbon sequestration rates under Mediterranean woody crops using recommended management practices: a meta-analysis. *Agriculture, Ecosystems and Environment* 235, 204–14. doi:10.1016/j.agee.2016.10.024.

Villamil, M. B., Bollero, G. A., Darmody, R. G., Simmons, F. W. and Bullock, D. G. 2006. No-till corn/soybean systems including winter cover crops. *Soil Science Society of America Journal* 70(6), 1936–44. doi:10.2136/sssaj2005.0350.

Wang, Y., Zhang, Y., Zhou, S. and Wang, Z. 2018. Meta-analysis of no-tillage effect on wheat and maize water use efficiency in China. *Science of the Total Environment* 635, 1372–82. doi:10.1016/j.scitotenv.2018.04.202.

West, T. O. and Post, W. M. 2002. Soil organic carbon sequestration rates by tillage and crop rotation: a global data analysis. *Soil Science Society of America Journal* 66(6), 1930–46. doi:10.2136/sssaj2002.1930.

Yang, X. M. and Wander, M. M. 1999. Tillage effects on soil organic carbon distribution and storage in a silt loam soil in Illinois. *Soil and Tillage Research* 52(1–2), 1–9. doi:10.1016/S0167-1987(99)00051-3.

Chapter 11

Benefits of Conservation Agriculture to farmers and society

Patrick Wall, Independent Consultant – Sustainable Agricultural Systems, Mexico; Christian Thierfelder, International Maize and Wheat Improvement Center (CIMMYT), Zimbabwe; Peter Hobbs, Cornell University, USA; Jon Hellin, International Rice Research Institute (IRRI), The Philippines; and Bram Govaerts, International Maize and Wheat Improvement Center (CIMMYT), Mexico

1 Introduction

Conservation Agriculture (CA) is not a technology, but rather an improved management system for producing agricultural crops while doing away with many of the harmful effects on the environment of conventional, tillage-based agriculture. It endeavours to bring commercial agriculture as close as possible to natural vegetation systems that are populated by diverse plant species, are not tilled, and where dead leaves and plant residues remain on the soil surface. The adoption of CA, however, does not imply doing away with most of the advances in crop science and production agronomy that lead to productive agricultural systems, such as improved crop varieties, adequate crop nutrition and integrated pest, disease and weed management – these are just as much a part of CA systems as they are of tilled systems (Thierfelder et al., 2018). However, CA omits the three components of conventional production systems that are the cause of soil and land degradation – soil tillage, removal of most of

http://dx.doi.org/10.19103/AS.2019.0049.11

the organic soil cover, and monoculture (i.e. continuous production of the same crop or crops on the same field).

CA should not be seen as simply another alternative method of crop production that can be practised in tandem or rotation with conventional, tilled agriculture, but rather as an imperative in the fight to feed nine billion people by 2050 without further damaging the environment and jeopardizing future generations. Alarming levels[1] of soil erosion and degradation in agricultural lands around the world (Lowdermilk, 1948; Oldeman, 1994; Montgomery, 2007a; Anderson and Johnson, 2016) show that an alternative to tilled agriculture is urgent. However, the objective of most farmers is to provide a livelihood for themselves and their families, and they therefore require the promise of increased benefits from a new or modified system before they leave their known and tried farming system. CA potentially provides major benefits both on and off the farm - the extent to which these benefits are expressed under any given situation depends on the local biophysical and socio-economic conditions, the resource-base of the farmer and therefore the methodology he or she uses to apply the principles of CA. Society in general benefits from the off-farm effects including reduced damage to infrastructure from water run-off and erosion, less dust and smoke pollution, less greenhouse gas (GHG) generation, and greater biodiversity in the rural landscape. However, in most places, farmers are not rewarded for these off-farm benefits or ecosystem services (Ehrlich and Ehrlich, 1981), thereby limiting the profitability of CA to the farmer and reducing the incentives for CA adoption. As a result, soil and land degradation caused by tillage-based agriculture continue unabated in many regions of the world.

In the following sections we describe the principal benefits of CA, as well as problems and difficulties associated with the successful management of CA systems, describe the adoption of CA, and argue for a more equitable sharing of costs and benefits from the widespread adoption of the system.

2 Farm benefits of Conservation Agriculture

Well-managed CA systems result in a number of changes in soil physical, chemical, and biological conditions that affect the farming operation (Verhulst et al., 2010). Most of these changes are positive, while some are negative and require specific management practices to overcome them. There are also benefits to the farmer simply from stopping doing unnecessary farm operations in terms of both the time and the energy involved in tillage. The benefits of CA on soil quality have

1 Soil erosion worldwide was estimated to be 35,900 Mt in 2012 (Borrelli et al., 2017). Although most but not all of this erosion comes from agricultural land, it is equivalent to 7.3 t/ha/y for each of the approximately 4,900 M ha of global agricultural (crop and grazing) land, ten times greater than the average rate of annual soil formation of 700 kg/ha (Wakatsuki and Azwar, 1992).

been amply described by Verhulst et al. (2010), while Hobbs (2007), Hobbs et al. (2008), Triplett and Dick (2008), Derpsch et al. (2010), and Kassam and Friedrich (2011) have covered many of the agronomic and societal benefits of CA. In the following sections we summarize these benefits, followed by a section on overcoming problems or difficulties often associated with a change to CA.

2.1 Energy savings

One of the most immediate and important benefits emanating from the adoption of CA is energy savings. The extent and type of energy saved depend on the particular type of agriculture: fuel savings are most important in mechanized agricultural systems, whereas savings in manual labour and drudgery are most important in systems relying on human and animal energy (Baudron et al., 2015a).

2.1.1 Fuel and machinery cost savings

In mechanized farming systems, fuel savings of 35–80% (Corsi et al., 2012) are usually an immediate benefit that accrues in the first season of a change to CA. The fuel savings in any one situation are obviously linked to the extent of the reduction in tillage passes, as well as to the draft requirements of the tillage operations. In many farming situations, such as Brazil, Paraguay, and Bolivia, where farmers were conducting intensive tillage prior to the move to CA, fuel savings of 65–80% are common (Rocco, 1997). In Bangladesh, Haque et al. (2011) found an 82% reduction in fuel use with CA strip-tillage systems compared to the conventional tillage system, and surveys in Pakistan and north-west India showed that because of the multiple tillage passes that farmers use in conventional systems to get a good seed-bed in degraded soils, farmers save 60–80 litre/ha of fuel with zero-tillage wheat after puddled rice (Hobbs and Gupta, 2004). In other situations, for instance the northern steppes of Kazakhstan where markedly reduced tillage systems have been in place for decades to overcome wind erosion, the reduction in fuel consumption may only be in the order of 25–30%. Ploughing and subsoiling generally require more draft energy than seeding or spraying operations – the principal mechanized activities in CA systems – and so reductions in fuel use are not directly proportional to the reduction in tractor passes. Where fuel subsidies are in place, direct benefits to the farmer from reduced fuel use are decreased, and benefits to society are increased.

2.1.2 Labour savings

Labour savings may result in some cost savings in large-scale agriculture (Triplett and Dick, 2008), but for smallholder farmers, who often rely on manual labour and/or animal traction for their farming operations, reduced labour requirements for tillage may be the key to increasing total farm productivity and enhancing farm family livelihoods. Labour availability, commonly from family members, is often a limiting factor in smallholder systems, and increases in productivity are better expressed in output per unit of labour expended. For smallholder farmers, savings in one enterprise may free up labour that can then be used in other activities and enterprises, both on and off the farm (Byerlee et al., 1989; Ekboir et al., 2002). The greater opportunity for off-farm employment provided by labour savings on the farm may benefit not only household income, but also the adoption of technologies that require increased investment (Mwangi and Kariuki, 2015). However, Fouzai et al. (2018) found that increased off-farm income was correlated with lower CA adoption levels in the Siliana Governate of Tunisia. Importantly in some instances, farmers also report that freeing up family labour has meant that children can spend more time at school (Ekboir, 2002).

Again the labour savings offered by a change to CA depend on the intensity of tillage in the present system. In ploughing for and seeding a wheat crop, farmers using animal traction in the Andes and Ethiopia commonly walk behind their bullocks and an ard-type plough for 100 km for each hectare they seed – 12 days of work to seed one hectare (Wall et al., 2001). With a small zero-tillage animal traction seeder, this can be reduced to 17 km/ha and completed in 2 days (Wall et al., 2001), a major saving in time and energy. The preparation of the traditional ridges on one hectare of land with a hand hoe in Malawi takes approximately 30 person days (Thierfelder et al., 2015a) and involves moving an estimated 600 t of soil – an incredible amount of work and drudgery that can be obviated using CA. The labour savings for farmers in Pakistan and north-west India using animal traction to prepare a seed-bed for wheat after rice are very high, because of the difficulty of achieving an 'adequate seed-bed' after puddled rice (Byerlee et al., 1989). However, labour savings depend not only on tillage, but also on how weeds are controlled in the system (see below) – the benefits of reduced labour for tillage can be more than offset by increased labour for weed control if only manual weed control is used (Siziba, 2008).

In Southern Africa, Oldrieve (1993) developed a system of manually dug planting basins that restrict tillage to only about 15% of the field area, a system that is the most widely used form of CA in Zambia and Zimbabwe. Although labour requirements for digging the basins are high, these should be prepared in the dry season (when labour demand for farm activities is low but the activity may still conflict with off-farm employment opportunities) and then can be seeded immediately with the first planting rain, saving valuable moisture (Haggblade and Tembo, 2003). Inputs are concentrated in the basins, ideally dug in the same place each year, which also capture water making this an interesting example of precision agriculture for smallholders, but farmers state that, even though the system increases crop yield, the labour requirement for digging the basins is the main reason for not adopting or adopting on only part of the farm (Umar et al., 2012).

2.2 Improved soil water balance

One important short-term benefit commonly experienced after the adoption of a CA system is an improved crop water balance, principally from the combination of three processes: increased water infiltration into the soil, reduced evaporation from the soil surface (Thierfelder and Wall, 2009; Verhulst et al., 2010), and obviating the moisture loss occasioned by tillage itself (Piggin and Devlin, 2012; quoted by Haddad et al., 2014). In a tilled soil, unprotected by surface cover with crop residues, raindrops break down the aggregates, already weakened by tillage, into their constituent soil particles, which flow with the rainwater and block soil pores, restricting water from entering the soil. The resulting surface layer of fine, disaggregated soil particles also restricts the diffusion of oxygen into the soil, and, when it dries, forms a surface crust that may restrict seedling emergence, leading to poor plant populations. In an untilled soil covered by residues, aggregates, unloosened by tillage, are protected from breakdown by the surface residues, which absorb the impact of the raindrops. As they are not blocked by disaggregated soil particles, pores remain open and water is able to enter the soil.

The second component of the increased water balance is a reduction in evaporation. Surface cover reduces the radiation reaching the soil and thus restricts evaporation from the soil surface – the greater the amount of soil cover, the greater the reduction in evaporation (Hatfield et al., 2001). As little as 30% ground cover can reduce evaporation by 30%, while 80% ground cover reduces evaporation by over 40% (Schertz, 1994). The combination of increased infiltration of water into the soil profile and decreased moisture loss through evaporation results in greater water availability to the crop in situations where water is a limiting factor (Verhulst et al., 2011a), and it is common to see these benefits in the first or second season after a switch to CA from conventional tillage (Thierfelder and Wall, 2009). We believe that this fact, together with the effects on soil erosion described below, largely explains why most of the area where CA has been adopted to date is in rainfed situations, often on sloping land and with periods of moisture stress (Baudron et al., 2015a).

Although water infiltration at the soil surface depends on a combination of residue protection and lack of loose soil, infiltration at depth depends more on the maintenance of soil porosity and pore continuity. In a tilled soil, porosity may be high after inversion tillage before the soil mass slumps after rainfall, after which there is little continuity of the pores in the tilled horizon. In an untilled soil, pores and channels made by roots and soil fauna and flora are not disturbed by tillage and so pore continuity is maintained, facilitating water infiltration and gas diffusion - allowing carbon dioxide out and oxygen into the soil.

The improved soil water balance under CA is especially important under water-limited conditions, and yield benefits with CA are common in drier areas and regions with seasonal droughts and may be observed even in the first year, especially if sufficient surface mulch is retained. In Syria, Piggin and Devlin (2012) (quoted by Haddad et al., 2014) found that farmers using CA were able to produce 2–3 t/ha of wheat with 300 mm rainfall compared to only1 t/ha with conventional tillage. In contrast, under conditions of abundant rainfall, there is little benefit to increased soil moisture, and, as stated earlier, if there is some impediment to drainage (e.g. a plough pan), increased soil moisture may prove detrimental to crop yield. Under irrigated conditions, the benefits of the improved soil water balance may be masked by irrigation. However, when irrigation is practised efficiently, savings in irrigation water with CA are common. Sikka et al. (2005) found 30% water savings to produce a wheat crop and, because of the omission of puddling which demands a large amount of water, 72% savings of the water to produce a rice crop. In both cases crop yields were higher without tillage resulting in 62% and 446% increases in water efficiency (kg grain/mm water) in wheat and rice, respectively. In many irrigated areas there is a shortage of irrigation water and increasing competition for water between agriculture, industry and urban populations. Thus, savings in irrigation water are an important social benefit of CA, permitting either increased agricultural (and food) production or the liberation of water for other uses. Unfortunately, water subsidies in place in many irrigated conditions reduce the need for efficiency (e.g. the north-west of India, many irrigation districts in Mexico) and therefore the need to adopt more efficient and sustainable production systems.

2.3 Reduction in soil erosion (by wind and water)

CA reduces soil erosion by both wind and water, often by over 90% (Montgomery, 2007b). The erodibility of untilled soil is far less than that of a recently tilled soil and surface residues absorb much of the kinetic energy of raindrops, thereby protecting the surface aggregates. With respect to water erosion, CA reduces the erosivity of rainfall (and overhead irrigation) by increasing water infiltration and therefore reducing the amount of run-off. At the same time surface residues impede the flow of water, reduce the velocity of water running off the field and allow soil particles to sediment out - even when water does run off a CA field, it is cleaner and holds less sediment than water running off a tilled field. Therefore, it is common to see that reductions in soil erosion of well over 80% with CA (see summaries in Thierfelder and Wall, 2009; Wall et al., 2014) are greater than the reductions in water run-off.

Crop residues are also the key to reducing wind erosion. However, whereas residues lying on the soil surface are most effective for reducing water erosion, standing residues are more effective in the reduction of wind erosion (Fryrear and Skidmore, 1985) especially under limited residue conditions. Complete control of wind erosion can be achieved with far less standing residues (the actual amounts depend on wind speeds, soil erodibility, stalk diameter and height of the residues) than with flattened residues but large quantities of flattened residues will also give complete wind erosion control (Siddoway et al., 1965). Taller standing residues are more effective than short stubble, and many thinner stalks (e.g. wheat and barley) are more effective in reducing wind erosion than are fewer, thicker stalks (e.g. maize or sorghum). Interestingly, in cold climates, standing residues also play an important role in capturing and protecting snow from being blown off the field. On the steppes of northern Kazakhstan, where approximately 70% of the annual precipitation (<350 mm/yr.) falls as snow, snow capture is extremely important as is protection of the untilled soil by surface residues to protect the soil from water erosion during spring snow-melt.

The surface soil is generally the most fertile and holds higher concentrations of soil organic matter, nutrients, soil microbes, soil fauna, and flora, as well as recently applied fertilizer and seed, than soil deeper in the profile. Erosion losses therefore also imply considerable losses of soil organic matter and nutrients. In many cases heavy rainfall soon after seeding results in the loss of all the pulverized soil above the tillage pan, and with it all of the applied fertilizer and seed. Re-seeding is a relatively common practice in conventionally tilled systems in areas with heavy rainfall during the seeding period (e.g. Brazil, Paraguay), whereas in CA systems it is seldom necessary.

2.4 Saving of time

Removing tillage from crop production practices provides time savings that can have important effects on the yield and yield stability of crops. In the rice–wheat system of north-western India and Punjab of Pakistan, where yields of wheat decline for each day of delay of seeding after mid-November, the time savings of doing away with tillage after the rice crop was an important initial driver of the development of CA systems (Hobbs et al., 2019; Vol. 1, Ch. 6 in this series). Early seeding of wheat also helps avoid heat stress during the grain-filling period, resulting in higher yields. Likewise in Zimbabwe, Nyagumbo (2008) found that one of the major benefits of CA for smallholder farmers was being able to seed more of the maize crop on time – delaying maize seeding leads to yield reductions of 5% for each week delayed.

In many rainfed systems, for instance in the Andes and the highlands of Mexico, the ground is hard prior to the onset of the rains, making ploughing difficult and expensive. Farmers wait until the first rains to be able to plough and prepare the land, and then seed. In this process they have lost several weeks of the cropping season, often missing the optimum seeding time, as well as losing precious soil moisture that evaporates when the moist soil is tilled. Doing away with tillage allows farmers to seed before or with the first rains, capturing all of the rainfall for use by the crop and essentially increasing the season length. However, tillage after the onset of the rains can be an effective weed control practice, and so in CA systems other weed management methods are required.

Untilled soils provide better trafficability than tilled soils - in most soils, a tractor with a seeder can enter an untilled field within 2–3 days after a good seeding rainfall, whereas it is common to have to wait a week before similar equipment can get into a tilled field with loosened soil. Also, because of the reduced moisture evaporation with residue cover, the surface soil remains moist longer in CA systems. The combination of these factors provides a longer seeding opportunity in untilled situations and increases the area that can be seeded after one rainfall event, often allowing more of the crop to be seeded at the optimum time.

2.5 Increased machinery productivity

In mechanized farming systems there are savings in machinery costs associated with a change to CA over and above the savings in fuel costs. While the initial outlay to purchase a zero-tillage seeder can be considerable, the total value of equipment on the farm can be reduced. Because of the reduction in operations, a single tractor is able to cover a greater area in CA systems than in conventional systems. This is an important aspect at the agricultural frontier in developing countries, such as in lowland Bolivia in the 1990s, where farmers could expand their agricultural operations without purchasing new tractors: Rocco (1997) showed that one tractor was sufficient for double the area that could be managed by the same tractor under a conventionally tilled system. Another aspect of the increased area that can be managed by one tractor is found in north-west India, where larger farmers purchased zero-tillage seed drills and found that they still had the capacity to seed more land after finishing seeding their own fields (Hobbs et al., 2019). This allowed them to provide a service to smaller farmers who could not afford, nor had sufficient land to warrant, the purchase of a zero-tillage drill, resulting in areas of CA adoption around the larger farmers. For smallholder farmers, the availability of contract services for seeding and spraying operations is likely to be an important component of CA adoption patterns because of the economy of scale that it offers to the service provider that can be passed on to the service user. Lower machinery costs per hectare make service provision a very viable option, even though the initial cost of the CA seeding equipment is higher.

Tractor maintenance and equipment wear and tear are also reduced by a change to CA situations (Rocco, 1997; Triplett and Dick, 2008) as the tractor works only in seeding and spraying operations that require relatively low draft, avoiding the strenuous, heavy draft activities of ploughing and subsoiling. Tractors and equipment also work in a cleaner environment under CA compared to the dusty situations common in conventional agriculture, again reducing the maintenance costs of machinery and giving a longer useful lifespan to the equipment.

2.6 Increases in soil organic matter

There has been considerable discussion over claims that soil organic matter (SOM) increases under CA systems, and a meta-analysis of published results by Govaerts et al. (2009a) showed that SOM increases were not universal. This no doubt reflects the different environments where research on CA has been conducted – different residue retention levels, soil textures and cropping systems. In favourable environments, with well-managed CA systems, it appears that steady increases in SOM are possible. Six et al. (2002) found a general increase of 325 kg C/ha/y without tillage compared to tilled soils in both tropical and temperate soils: very similar to the results of Vagen et al. (2005) in sub-Saharan Africa and Loss et al. (2015) in the Middle East. Changes in soil organic matter levels will depend on how much organic matter is returned to the system and the rate of breakdown of this organic matter – SOM increases are unlikely where few residues are returned – and will be slower in tropical situations with high temperatures and soil moisture (Cheesman et al., 2016). At the same time tillage leads to more rapid oxidation of SOM and decomposition of crop residues, especially when residues are incorporated into moist, aerated soil conducive to increased short-term biological activity. Once the residues are decomposed, populations of soil fauna and flora die off releasing nitrogen, thereby creating the flush of nitrogen that commonly coincides with early crop growth in tilled agriculture – one reason for more intensive tillage in many subsistence farming situations. The breakdown of the residues is generally much slower under untilled conditions where they are left on the soil surface – and occurs principally at the soil/residue interface. The increased longevity of the crop residues, and the more stable microbial populations resulting from this, often mean that early crop growth is more limited by nitrogen in CA systems due to the lack of the 'nitrogen flush' common in tilled soils: the greater soil microbial biomass in CA systems effectively 'ties-up' more of the nitrogen in the system (Govaerts et al., 2006). Thus, after a conversion to a CA system, a little more nitrogen needs to be applied in the early stages of the crop (commonly about 20 kg/ha N) until the system stabilizes and a new equilibrium of SOM and N cycling are achieved. We have observed that over time soil biological activity in soils in CA systems increases and that, therefore, the rate of decomposition of surface residues increases.

Reicosky and Lindstrom (1993) measured CO_2 emissions from fields after different tillage operations, providing an interesting example of the carbon dynamics with and without tillage. The same levels of crop residues were returned to all situations – 185 g C/m^2 or about 3.7 t/ha of wheat residues – and were either left on the surface in plots without tillage or incorporated into the soil with tillage. After 19 days, 250 g C/m^2 had been emitted from the mouldboard ploughed treatment – nearly 35% more than the carbon returned and incorporated with the crop residues. However, from the untilled plots only 50 g C/m^2 were emitted in the 19 days. Emissions of CO_2 from the mouldboard ploughed treatment were reduced when ploughing was followed by two passes of a disc harrow to reduce the roughness and, therefore, gas diffusion, but even so, 70% of the incorporated C was lost in only 19 days.

Even though results are ambiguous with respect to C sequestration by no-till systems, increased SOM levels in the surface soil should result in benefits due to the improved surface soil structure and fertility (Verhulst et al., 2010; Powlson et al., 2016). Associated with the stabilization or increase in SOM in CA systems, available nutrient levels are generally higher in soils that have been under CA for a number of years. Sidiras and Pavan (1985) have shown increases in available phosphorus (P), potassium (K), calcium (Ca) and magnesium (Mg) throughout the principal crop rooting zone under CA systems. Results from one of the pioneer CA farmers in Brazil, Mr. Franke Dijkstra, show that between 1977 and 1998 maize and soybean yields on his farm increased by 100% and 60%, respectively, and this with 30% less fertilizer in the case of maize and 50% less fertilizer in the case of soybean (Derpsch, personal communication). While part of this increased fertilizer efficiency is no doubt due to reduced losses through soil erosion, improved varieties and pest control, the soils are undoubtedly more fertile after 20 years of CA as a result of improved nutrient recycling due to increased microbial activity.

2.7 Improved soil structure

Soil structure is damaged by soil tillage, both physically and by oxidation of SOM (humus) that binds together soil particles into soil aggregates (Six et al., 2000). Soil aggregates are the key element of soil structure and soil quality, and aggregate stability, a measure of the strength of the aggregates, is commonly used as a measure of soil structure (Bronick and Lal, 2005). Soil porosity and water-holding capacity are directly linked to aggregate size and stability. Aggregate stability is generally considerably higher under CA – a change that can be observed within one or two seasons of starting a CA system (Romero et al., 1993; Govaerts et al., 2007a, 2009b; Thierfelder et al., 2015b). Strong soil aggregates are more resilient to physical pressure and so withstand compression and compaction – although it is unwise to allow tractors, machinery and animals onto agricultural fields while the soil is wet, soils under CA are better able to withstand the compacting effects of traffic.

2.8 Increased soil biological activity

Soils with little or no biological activity are infertile. Soil organisms are responsible for organic matter decomposition and distribution and therefore define soil structural improvement and nutrient cycling (Kladivko, 2001). Populations, diversity and activity of soil fauna and flora are all reduced by soil tillage (Kladivko, 2001), which therefore reduces soil fertility. CA, on the other hand, due to the reduction in soil disturbance and the provision of a continuous supply of residues, rapidly leads to increased populations, diversity and activity of soil fauna and flora (Govaerts et al., 2008). The increased microbial populations are also likely to result in increased populations of microorganisms antagonistic to pathogens (Cook, 1990) and crop pests.

Soil macro- and mesofauna can be divided into two functional groups – litter transformers and ecosystem engineers (Lavelle, 1997). As these terms suggest, the former group comprising large arthropods and mesofauna fragment plant material and have little effect on soil structure and quality, whereas the ecosystem engineers (earthworms and termites) ingest plant material and soil particles and incorporate organic material into the soil profile. Earthworm populations commonly increase markedly after only a few years of CA (Verhulst et al., 2010; Thierfelder and Wall, 2010), whereas the effects on termite numbers are variable (Thierfelder et al., 2013a, 2015b), probably as a result of species and spatial variability. However, termite numbers were reduced when velvet bean (*Mucuna pruriens* L.) was included as a green manure cover crop in the CA system (Thierfelder et al., 2015b). In general termites have a positive effect on soil structure, but variable direct effects on crop plants, due to different species composition – some of which feed on live plants while some only feed on dead tissue. Visible crop damage often overshadows the benefits provided by termites in soil structure, especially in the eyes of farmers.

The effects of tillage on soil-borne pests (soil fauna detrimental to the crop) are extremely variable. In a review of tillage effects on 45 soil pests, Stinner and House (1990) found that reducing tillage reduced populations of 43% of the species, whereas 29% were not significantly affected, and populations of 28% increased. Govaerts et al. (2007b) found that both parasitic and non-parasitic nematode populations were higher in maize under zero-tillage conditions, but not in wheat. The effects of many crops on soil nematodes has been well documented (e.g. Howard et al., 1998; Bagayoko et al., 2000; Govaerts et al., 2007b) but interestingly a particular crop may affect different nematode species differently - maize increased nematode populations in central Mexico (Govaerts et al., 2007b) but reduced soybean cyst nematodes (*Heterodera glycines*) under zero tillage in the USA (Howard et al., 1998). Green manure cover crops, important components of CA systems in Brazil (see Vol. 1, Ch. 6 in this series), have differential effects on parasitic nematodes. The use of pearl millet (*Pennisetum glaucum*) has increased with the finding that it reduces populations of three important nematodes, whereas the use of dolichos (*Lablab purpureum*) has declined with the finding that it results in higher populations of some nematodes.

2.9 Soil health

Doran and Zeiss (2000) defined soil health as 'the capacity of soil to function as a vital living system, ... to sustain plant and animal productivity, maintain or enhance water and air quality, and promote plant and animal health'. As such soil health includes chemical, physical and, importantly, biological components, and the assessment of soil health is used as a measure of the sustainability of current agricultural practices. As discussed previously, CA, as compared to tillage-based agricultural systems, has positive effects on all three components of soil health - largely mediated by soil organic matter dynamics as affected by lack of tillage, crop residue retention, and crop rotation - and therefore markedly enhances soil health.

2.10 Crop yield

Many of the effects of CA noted above do not affect crop yield, so it is not surprising that in many cases yield is not affected initially by a switch to a CA system. In general terms, CA is promoted more to increase crop and faming system productivity through reduced costs of production and increased sustainability than through increased yields. However, in conditions where moisture is a limiting factor for crop production and soils are not very degraded, yield benefits with CA may be observed from the first year due to the beneficial effects of CA on water infiltration and evaporation and the potential for more optimal seeding dates. In areas with highly degraded soils, crop yield may be lower for a few years after initiating a CA system (Derpsch et al., 2016) - it takes several years for natural processes to increase the physical and biological fertility of a structureless, inert soil with little porosity to a state where crop growth and development is not adversely affected.

Reports summarizing the effects of CA on crop yields can be somewhat confusing. Pittlekow et al. (2015) conducted a global meta-analysis of the effects of no-till on crop yields, which included 678 sets of published data from 63 countries and 50 crops, and showed that in general there was a small yield penalty to the use of no-till. Yields were unaffected by no-till in cotton, oilseeds, and legumes, slightly reduced (-2.6%) in wheat, while the greatest yield reductions were found in rice (-7.5%) and maize (-7.6%). Surprisingly in a review of the effects of CA on maize yields in eastern and southern Africa, Wall et al. (2014) found that maize yields were generally higher under CA than under conventional tillage, especially when the crop was adequately fertilized. Interestingly only one of the 24 data sets used in the Wall et al. (2014) review formed part of the Pittlekow et al. (2015) meta-analysis because the latter included only results from peer-reviewed journals, whereas, especially in developing countries, there is a wealth of information in the 'grey' literature, which is generally more accessible to local scientists.

Several studies show that although yield is little affected during the first years after installation of a CA system (in some studies there is a small increase in yield, in others a small decrease), after 2–3 years, yields under CA are consistently better than under conventional systems (Farooq et al., 2011; Brouder and Gomez-Macpherson, 2014; Corbeels et al., 2014b; Thierfelder et al., 2015c). Much of this effect is likely due to progressive improvements in soil conditions leading to increasing yield benefits in CA systems compared to conventionally tilled systems. However, research results are often unintentionally biased against CA: researchers frequently establish trials before they have adapted a CA system under local conditions, prejudicing results in the first years until they know how to manage the system properly (Wall et al., 2002; Laborte et al., 2012; Derpsch et al., 2014). To be able to compare a management system to current management practices that have been adapted over many years, and even over decades, it is obvious that the new system needs first to be adapted to local conditions before a valid comparison can be made. Another common error made by researchers is to overlook system effects – for instance even though CA treatments can often be seeded earlier than conventionally tilled treatments, researchers delay seeding until all treatments can be seeded at the same time, thereby negating one of the benefits of the CA system.

In a meta-analysis of cereal crop yield data from 61 cereal experiments, largely on maize, in sub-Saharan Africa, Corbeels et al. (2014b) were able to discriminate between effects of zero tillage with and without residue retention, and a full CA package that included zero tillage, residue retention and crop rotation. Zero tillage without residues gave, overall, a slight yield increase in the first one or two years, after which yields declined relative to the conventionally tilled treatment. However, when residues were retained, there was a 0.3 t/ha yield benefit initially, which increased to 0.5 t/ha over time. A full CA package gave yield benefits of approximately 0.2 t/ha compared to the conventionally tilled treatments, and importantly yield was more stable across seasons than zero tillage without rotations. However, on degraded soils, including those in Bolivia (Wall, 1999), Mexico (Govaerts et al., 2005), and southern Africa

(Thierfelder et al., 2013b), zero-tillage treatments without surface residues yielded considerably less than the conventionally tilled treatments whereas when surface residues were retained, yields without tillage were higher than the tilled treatments.

Risk is an important deterrent to technology adoption. Importantly the water savings due to the adoption of CA practices lead to reduced risk by increasing water use efficiency and yield in drier sites and years (Ngwira et al., 2013), an important consideration in many rainfed agricultural situations. However, yield may be reduced in wetter than normal years, especially in soils with limited drainage (Rusinamhodzi et al., 2011). In a study of a large number of farmers in Zimbabwe using the basin system of CA referred to earlier, Michler et al. (2019) found that basins had little effect on yield during years of average rainfall, but a positive effect on maize, sorghum, and cowpea yields in dry years.

3 Difficulties with Conservation Agriculture

3.1 Weeds

As Edward Faulkner wrote in his book 'Ploughman's Folly' (referring to the role of soil tillage as the cause of the US Dust Bowl), 'The truth is that no-one has ever advanced a scientific reason for ploughing' (Faulkner, 1943). However, there are many practical reasons why farmers plough the soil, largely to make it easier to produce crops. Probably the most important of these reasons is to control weeds, and, therefore, once one does away with tillage in CA systems, weed control becomes a major issue. The traditional milpa system in Mexico and Central America is, in effect, a CA system, but one which involves considerable labour in weed control. CA systems may facilitate the control of some weeds (see the control of *Phalaris minor* in Hobbs, Vol. 1, Ch. 6 in this series) but, in general, weed control is more difficult and requires an efficient, integrated, weed control strategy (Nichols et al., 2015). The development of herbicides that allowed farmers to desiccate the weeds present in the field at the time of seeding, including, first, paraquat, and later glyphosate, provided an initial impetus for the development of no-till and CA systems. This latter herbicide has become the backbone (and therefore a weak link) of CA systems in many/most regions, and the spread of CA in many regions in the late 1980s and early 1990s appears to be linked to the reduction in the price of glyphosate herbicide. Later the development of soybeans, cotton, and maize resistant to glyphosate has increased the reliance on this herbicide, a substantial risk as the continuous use of glyphosate herbicide increases the possibility of selecting resistant weed biotypes leading to severe weed control problems. Herbicide resistance, not just to glyphosate but to other herbicide groups as well, is today a major problem in CA systems (especially in Australia) and some farmers find they have to till the soil occasionally to control resistant weeds, as well as employing multiple other strategies to restrict weed seed infestation.

Given the problems that have arisen with herbicide resistance through excessive and continuous use of the same or similar herbicides, it is imperative that farmers incorporate an integrated weed control programme into the farming system. The aim of an integrated weed control strategy in CA systems is to reduce the seed bank in the soil profile (initially rich with seed deposits from previous conventional tillage), limit as far as possible new weed seed production and entry into the field, and control future weed populations using as little chemical weed control as possible. Crop rotation plays an important role in a weed control strategy in CA systems, as a rotation of crops also implies a rotation of herbicides and herbicide types. Crops with different sowing dates, or grown in different seasons favour, and therefore permit the control of, different weed species and types (Nichols et al., 2015). Herbicide use may be higher in the initial stages of implementation of a CA system but, if an effective weed control strategy is pursued, herbicide usage can later be reduced (Nichols et al., 2015). Ground cover with crop residues or, especially, cover crops helps reduce weed populations. Full weed control with crop residues can only be attained with very high levels of residues, generally not attainable with commercial crops in rainfed conditions, but certainly possible with productive green manure cover crops (Kliewer et al., 2000) and irrigated crops. Furthermore, in CA systems, weed seed is not incorporated into the soil profile but remains on the soil surface where it is more prone to desiccation, consumption by insects and birds, and/or degradation by fungi and bacteria (Baraibar et al., 2009). When conditions are not favourable, seeds germinating on the surface are also more susceptible to desiccation than seed incorporated into the soil profile. In the quest to reduce the soil seed bank and weed populations it is important that farmers ensure that new weed seed is not produced in the field - weeds that escape other control methods should be removed, manually if necessary, throughout the year so that they do not produce seed. The old proverb 'one year's seeding makes seven year's weeding' (University of Oxford, 2008) is very valid for many weed species.

3.2 Crop rotation

Crop rotation is a farming practice that has been recommended since Roman times (White, 1970) and was the basis for successful agriculture prior to the development of inorganic fertilizers and chemical pesticides. Monoculture - the continuous cropping of the same crop on the same land - became more prevalent with the development of industrial agriculture in the twentieth century. Crop rotation is one of the three pillars of CA, but is often the last to be incorporated into the new farming system. Although one of the major benefits of functional crop rotations is improved crop disease management, there are multiple possible benefits from a crop rotation, depending on the species and combination of species that conform the rotation. These benefits, many of which are not well understood, include weed and nematode control; improved nutrient cycling especially when crops with different rooting patterns are included in the rotation; improved soil quality, structure, and infiltration; and, where legumes are included in the rotation, nitrogen supply. Because benefits depend on the particular rotation practised, the effects of the particular species and varieties of crops utilized and the synergies between these, it is impossible to generalize the benefits from crop rotations, and, except for a few well-documented benefits such as nitrogen fixation by rhizobia associated with legume crops, need to be evaluated under local conditions to support the development of optimal rotations. In our experience, however, the benefits of crop rotation are greater in no-tillage agriculture with residue retention than in conventionally tilled agriculture as suggested also by the results of Havlin et al. (1990) on the benefits of rotation on soil properties.

However, unavailability of markets for new crops is often a major hindrance to the adoption of optimal crop rotations. On farms and in regions where monocropping is the common practice, initiation of functional and beneficial crop rotations involves both considerable effort on the part of the farmer in learning how to manage new crops and incorporate them into the farming system, and institutional support to develop and maintain markets for the necessary inputs (including seed) for the new crops, as well as ensuring markets for the produce. In some cases a green manure cover crop can be inserted into the system without major management implications, for instance replacing the 'weedy fallow' in the continuous wheat-fallow system common in North Africa, although even a relatively simple change such as this needs research and extension support to identify the best cover crops and the best varieties of these, and, importantly, institutional support to coordinate seed production and marketing.

3.3 Managerial capacity

Change from one farming system to another often requires new expenditures, inputs and equipment, and capacity to manage the change and the new system. In general terms, CA systems require more precise management than conventionally tilled systems (one cannot cover up mistakes with another pass of the plough or the disk harrow)! This requirement for managerial capacity and the purchase and management of new inputs can, initially, provide an impediment to adoption for resource-poor farmers, especially those managing low external input production systems. Poor management of any system results in poor yields and increased frequency of losses and poor management of CA systems is no different to conventional systems in this regard.

3.4 Poorly drained soils

The benefits of CA in increasing water infiltration into the soil and reducing the loss of water through evaporation become a problem in poorly drained soil, where they lead to increased waterlogging and anoxia. Continual tillage in conventional systems, whether with mechanical, animal, or manual energy, commonly leads to the formation of plough pans that restrict water infiltration and lead to waterlogging above the pan. Where poor drainage is due to a plough pan, if possible this should be broken prior to establishing the CA system. If effective subsoiling is not possible, for instance in animal traction or manual systems, then sowing crops with strong tap roots can, over a space of 2–3 years, help develop drainage channels through the compacted layer, although if wet conditions occur during this initial period, yield reductions are likely. Possible crops for this purpose include safflower (*Carthamus tinctorius*) or green manures such as pigeonpea (*Cajanus cajan*) and oilseed radish (*Raphanus sativus*), which may be sown as a solid stand or even as an intercrop.

Sowing crops onto permanent raised beds is also an option to reduce the effects of poor drainage – the beds help maintain part of the root system of the crop in aerobic conditions even when the rest of the roots are waterlogged. Beds need to be reformed occasionally (the frequency depends largely on the soil type and degree of degradation – sandier and more degraded soils require more frequent reformation) by passing a winged tine in the furrow between the beds. Thus, only the furrow is tilled and the area on top of the beds is left untilled. Permanent raised beds have the added advantages of facilitating more efficient surface irrigation and also aiding in controlling traffic and therefore compaction – tractor wheels are confined to the furrows during all operations (Sayre and Hobbs, 2004; Govaerts et al., 2006; Verhulst et al., 2011b).

3.5 Crop diseases

Plant diseases can be broadly differentiated into those caused by biotrophic organisms that derive their energy from living cells, and generally have a very specific and narrow host range, and necrotrophic organisms that derive their energy from dead tissue and generally have a wide host range (Moore et al., 2011). Necrotrophic diseases can also attack living plants, where they kill the host cells rapidly. CA has little, if any, effect on biotrophic diseases, but can affect the prevalence and severity of necrotrophic diseases that survive on the crop residues. Cultural control of these diseases depends on avoiding seeding a crop into infected residues, and therefore crop rotation is the key to successful management of the diseases. This highlights the need for diversity in the rotations, as a rotation including only crops that are hosts to the same necrotrophic disease, such as a maize-wheat rotation where both crops are host to the *Fusarium graminearum* complex - the cause of fusarium head blight in wheat and maize stalk rot - will result in high disease incidence. Disease intensity on the wheat crop, for instance, depends on varietal resistance level and inoculant level and is, therefore, worst on susceptible wheat varieties sown directly into maize residues with high inoculum levels following weather conditions that promote disease infection (Landschoot et al., 2013). Similar results were found by Santos and Reis (2003) working in southern Brazil on *Bipolaris sorokiniana*, the causal agent of spot blotch and common root rot in wheat. Both root and foliar disease intensity increased with inoculum load on the surface residues. Crop rotation with non-host species reduced inoculum load and infection levels. After an effective break in the rotation disease, infection levels were low but increased with each successive wheat crop. Thus two or three wheat crops could be grown in succession until such time as disease levels reached economically important levels.

3.6 Other factors

There is a cost, monetary and/or in time and effort, involved in any change. As CA is a complex technology involving many changes and adjustments in the farming system, the farmer needs to invest considerable time in learning about the new management requirements if he/she is to be successful with the new system. This is made more difficult where the farmer is not well connected to information sources, as is the case for smallholder farmers in many parts of the world - extension services are generally spread thinly and declining (Ajieh et al., 2008). As stated earlier, managerial capacity is another important factor in the success of CA systems - CA generally requires more precision than is necessary with well-understood conventional systems, and good management is a requirement for success. When the degrading components of farming systems are omitted, there are often changes possible or required in the management of other agronomic components (including seeding dates, fertilizer amounts and management, irrigation amounts and timing) and these need to be adapted to the new cropping system in order to achieve a successful system and reap the potential benefits of the new system. Access to the necessary inputs and equipment is a prerequisite for successful CA, as it is for any new technology. However, again, as CA is a complex of technologies rather than a single component technology, there are often multiple new inputs that the farmer needs to access. This can cause considerable impediments to adoption of a CA system, especially where markets for inputs and services are poor or incipient.

4 Conservation Agriculture adoption

The benefits outlined in the previous section lead to the principal impacts of CA on farm productivity through cost reduction, improved profitability, enhanced agricultural sustainability and, at least after 2–3 years, increased crop yields. The relative importance of the different benefits depends on the local biophysical conditions and the farmers' socio-economic circumstances, as well as the field history, cropping systems, production methodologies, and the level of farmer managerial capacity. As such farmer circumstances define the overall benefits from a CA system and the likelihood of adoption under present market conditions and institutional arrangements.

Although there are no official statistics collected on the extent of CA adoption in the world, most recent estimates show approximately 180 million hectares of zero-tillage annual crops with crop residue cover worldwide (Kassam et al., 2019). Farmers seldom adopt all three CA components from the onset – a normal and logical decision process as they adapt CA principles and systems to their own biophysical and socio-economic conditions. In many situations benefits can be obtained from systems that include only one or two of the three CA principles, and as such these may be a logical starting point for farmers. For instance, in Syria and Iraq, Loss et al. (2015) found that there were appreciable benefits to zero-tillage systems even in the absence of crop residue retention. This is an important starting point for farmers as competition for residues, generally grazed by sheep, is acute. It is likely that additional benefits could be obtained with residue retention, especially as the systems are generally moisture-limited, but it is more important that farmers are able to initiate a more sustainable system with just zero tillage, and then over time adapt residue retention and crop rotation strategies into the system to be able to enjoy all of the benefits of a complete CA system. It is worth noting, however, that in other conditions such as the high-altitude valleys of Bolivia (Wall, 1999) and Mexico (Govaerts et al., 2005), zero tillage without residue retention was not a viable option for farmers, and crops yielded considerably less than with conventional tillage.

CA adoption has not taken place evenly across the globe. The principal areas of zero-tillage adoption are in South America (Brazil, Argentina, Paraguay, Bolivia, Uruguay, Chile, and Venezuela – 70 million ha), North America (principally USA and Canada – 63 million ha), and Oceania (principally Australia – 23 million ha), which together comprise nearly 90% of the area of zero-tillage agriculture in the world (Kassam et al., 2019). There are also extensive areas in China (9 million ha), Russia (5 million ha), Kazakhstan (2.5 million ha,) and South Asia (India and Pakistan – 2 million ha). Most of this area, with the exception of China and South Asia, is on large mechanized farms in rainfed agricultural systems – worldwide there is relatively little CA adoption under irrigated conditions and even less on smallholder farms.

The concentration of CA adoption on mechanized farms in rainfed areas is understandable – the improved crop water balance under CA as a result of the increased water infiltration and reduced evaporation give immediate benefits under water-limited conditions, and rainfed agriculture is often situated in areas with rolling or sloping land where water erosion is a problem, providing important reasons for the adoption of CA. In contrast, under irrigated conditions where benefits in water-use efficiency can be masked by irrigation, and where land is generally more level and without an acute water erosion risk, the incentives to adopt CA are considerably less, especially if the cost of irrigation water is subsidized, thereby negating the need for efficient use of water.

The following paragraphs summarize the major drivers of adoption in different parts of the world:

- Rampant soil erosion drove the initial move to CA in southern Brazil, and once farmers were able to control erosion the benefits in fuel and machinery savings became the drivers of adoption (Bolliger et al., 2006). The exponential increase in adoption rate occurred in the late 1980s after Monsanto's patent on RoundUp expired and cheaper generic glyphosate became available. Today CA is practiced on more than 50% of the annual crop area (de Freitas and Landers, 2014).
- Similarly, in Argentina, erosion as well as structural degradation of the fine-textured Pampean loess soils were the principal initial drivers of CA, as well as glyphosate prices and later the introduction of glyphosate-resistant soybeans that facilitated weed control in the system. By 2010, nearly 80% of annual crops in Argentina were produced under CA (Peiretti and Dumanski, 2014). Adoption in the whole region (Argentina, Paraguay, Bolivia and Uruguay) received considerable support from Brazilian farmers, scientists, and farmers' associations.
- The movement to conservation practices in the USA considerably pre-dates what we know today as CA. The extensive, tillage-induced wind erosion in the grain belt of the US that became known as the Dust Bowl in the 1930s increased awareness of the problems of intensive tillage and soil erosion and led to the definition of Conservation Tillage (CT) – a series of practices that left at least 30% of the soil covered with crop residues after seeding. Farmers on erodible lands had to use CT to access government agricultural support programmes. With the development of herbicides, scientists started trying to develop no-tillage systems, which began to be adopted with the advent of ICI's paraquat herbicide in the late 1960s. Paraquat, however, does not control perennial weeds and populations of these increased, setting back early adoption levels (Duiker and Thomason, 2014). Glyphosate herbicide provided the impetus

for renewed adoption, which increased with the reduction in price of glyphosate and the introduction of glyphosate-resistant varieties of maize, cotton, and soybean.

- Wind, especially, and water erosion in Canada, coupled with the loss of soil organic matter, drove quests to identify and adopt conservation practices, leading to the development and adoption of CA systems once chemical weed control became feasible (Lafond et al., 2014). The reduction in glyphosate prices and interest rates in the early 1990s, coupled with increased fuel prices, stimulated rapid adoption and today 50% of the annual crops of the prairies are produced without tillage (Awada et al., 2014).
- In Australia, the pressure to adopt CA was driven by soil and water conservation and reduced costs (Baker et al., 2001). Today 80–90% of the cropped area of Australia is under CA systems (Bellotti and Rochecouste, 2014).
- Wind erosion was tremendous in the steppes of northern Kazakhstan after these were put under the plough in the 1960s, leading to the development of CT systems that greatly reduced wind erosion. Since 2000, fuel savings and machinery efficiency have driven the adoption of CA on large-scale corporate farms on the steppes and today there are an estimated 2.5 million hectares of zero-tillage agriculture on the steppes (Kassam et al., 2019).
- In contrast, in South Asia, it was savings in turnaround time between the rice and wheat crops that originally drove the move to zero tillage. Later, increases in fuel prices and the emergence of herbicide-resistant canary grass (*Phalaris minor*), which could be effectively controlled by zero tillage through a combination of earlier seeding (when temperatures are above the threshold temperature for *Phalaris* germination) and reduced seed cycling in the soil profile by tillage (Malik and Singh, 1993, 1995), have increased the adoption rate of zero-tillage wheat (Hobbs, Vol. 1, Ch. 6 in this series). More recently a move to more direct seeding and direct transplanting of rice into unpuddled soil (thereby enabling a complete year-round CA system) is largely being driven by a labour shortage and water scarcity for hand transplanting of rice (Hobbs et al., 2019). However, weeds are effectively controlled by standing water in paddy situations and weeds become a serious problem in non-paddy situations.

The reasons for the relative lack of adoption on small farms are complex, and include technological issues such as competition for scarce crop residues, adoption risk, and interactions with the cropping system; social issues such as mindset, knowledge of the CA system, and conflicts with social norms – e.g. communal grazing rights; and institutional issues including input, credit and output markets, policies and subsidies, security of land tenure, equipment availability, and applicable research and extension support (Wall, 2007). In fact, all of these factors affect large, commercial farmers as well as smallholders, but they have been, or are, more easily overcome in regions and areas of large commercial farms. Several recent adoption studies show that the major impediments to smallholder CA adoption are institutional (Bolliger et al., 2005; Affholder et al., 2010; Corbeels et al., 2014a). However, this is often the case for agricultural technologies in general (Ehui and Pender, 2005). Several authors suggest that efforts to promote CA should be confined to groups of farmers who currently have incentives to adopt the system (Giller et al., 2009; Corbeels et al., 2009; Andersson and D'Souza, 2014; Baudron et al., 2012, 2015b). This argument assumes that market and institutional arrangements are static and fixed, whereas we suggest that it rather shows the importance of institutional change to facilitate the adoption of sustainable agricultural practices and stop the current spiral of land degradation evident in many farming areas. Payments to farmers for ecosystem services resulting from CA-based land management, together with increased product prices associated with providence systems informing consumers of sustainable production methods, can change this dynamic rapidly. There are often disincentives to investment in agriculture in general, especially in smallholder agriculture, quite apart from disincentives to CA adoption (Ehui and Pender, 2005). Necessarily, therefore, efforts to limit and revert the effects of extractive and degradative agriculture will involve policies and institutional changes that incentivize agricultural intensification, as suggested by Scherr and McNeely (2008) for the development of ecoagriculture. Fischer et al. (2009) illustrated that a change to more efficient markets (and institutional arrangements) can effect sizeable changes in the economic optimum input and yield levels and therefore change the likelihood of technology adoption. Recently Ward et al. (2018) have shown that directed payments to farmers adopting CA systems can indeed increase the level of adoption of these more sustainable systems.

Research has shown the complexity of farmer adoption and adaptation of agricultural technologies and practices (e.g. Hudson, 1991; Tenge et al., 2004) and a recent review of CA adoption studies concluded that there are few, if any, universal variables that regularly explain its adoption. In contrast to Giller et al. (2009), Corbeels et al. (2014a) and others who suggest that CA is only relevant for particular groups (typologies) of farmers, Knowler and Bradshaw (2007) call for promotional efforts to be tailored to reflect the particular conditions of individual locales. Similarly, Ekboir (2002) found that although the development and adoption of no-till packages by small-scale farmers followed different paths than for large-scale farmers, 'the paths shared one important common feature: all successful programs resulted from networks that worked with participatory research approaches'. He argued for the need to develop local innovation platforms comprising multiple agents representing the principal components of the main farm product value chains to overcome the institutional limitations to smallholder farm productivity and sustainability. An innovation systems approach that fosters dynamic interactions among researchers, extension agents, equipment manufacturers, input suppliers, farmers, traders, and processors, and critically, depends on learning processes, feedback loops, and iterative interactions, offers the opportunity to realistically assess bottlenecks to farm productivity and explore options to overcome them (Spielman et al., 2008; Diaz-Jose et al., 2016; Camacho-Villa et al., 2016).

Smallholder farmers generally manage farms in more marginal areas and have less access to markets for inputs, services (including credit), and outputs than their larger counterparts because of the restricted and variable volume of demand, supply, and equity that they represent. Associative purchasing, selling, and negotiation arrangements are feasible for groups of smallholder farmers but are not easy to organize and maintain. A major role of the innovation platform is to explore and overcome these market limitations. CA equipment for smallholder farmers at different scales and with different energy sources (manual, animal, 2-wheel and 4-wheel tractors) have been developed and can help smallholders increase input-use efficiency (Van Loon et al., 2018), but are not as widely available as larger-scale equipment. Additionally, mechanized equipment with dimensions suitable for smallholders often has a capacity that far exceeds the land holding of smallholders, making the purchase of the equipment not only very expensive but also inefficient. However, this does open an option for some farmers or local entrepreneurs to become service providers, as has happened in both the north-west and eastern India where farmers who were able to purchase seeders were then able to use the excess capacity (time) of their seeders to provide services to nearby farmers creating local pockets of adoption (Hobbs et al., 2019). Support to entrepreneurs to facilitate the development of agricultural contract services is one of the avenues that is likely to pay large dividends. This does not have to begin with services for zero-tillage seeding, but this service can often be incorporated once the business is established. Often, especially in areas where small grains are produced, harvesting costs are extremely high for smallholder farmers, and initiating a business with contract harvesting would likely provide a solid foundation for an agricultural service business that could later expand into custom seeding and spraying.

Access to information has been problematic for smallholder farmers in many regions of the world, although this is rapidly changing with more equitable access to mobile phones and the Internet. While agricultural extension, education and training can help farmers increase production efficiency and productivity; marked reductions in the budgets of public research and extension services in the developing world (Ajieh et al., 2008; Anderson and Van Crowder, 2000) mean that these services are often unavailable to farmers, especially those in marginal environments. New approaches are needed for providing extension services to resource-poor farmers on a more sustainable basis and in a way that supports the emergence of the aforementioned agricultural innovation systems. Bell et al. (2018) found that among smallholder farmers in the Shire Valley of Malawi, the two factors that had most influence on the adoption of zero tillage and residue mulching were whether neighbours were using the practices and access to economic incentives for the use of the technologies. Experience around the world has shown that farmer-to-farmer information exchange is far more effective in changing farmer behaviour than traditional extension messages (Fisher et al., 2018; Holden et al., 2018) – in short farmers believe farmers far more readily than they believe extension agents or researchers (understandably they put more weight on the word of someone who makes his or her living from farming). The role of the extension agent thus changes from one of supplying information to the farmer to a new role of supporting farmer experimentation with new technologies and facilitating farmer-to-farmer information exchange.

Smallholder farmers also manage mixed crop–livestock systems more frequently than their better endowed, mechanized counterparts, and, because of their restricted capital rely more on crop residues for animal feed, albeit of low quality (Valbuena et al., 2012). Leaving residues on the soil surface thus competes directly with the feed requirements of the livestock, and until farmers are convinced that the residues give more benefits remaining on the field than passing through an animal, it is unlikely that surface soil cover with crop residues will be adopted. Wall (2009) proposed several different strategies to facilitate residue retention on small farms, including initial adoption on only a small part of the farm followed by stepwise adoption of CA. Thus, the impact on fodder supply would be reduced, especially when coupled with partial removal of the crop residues and the incorporation of some area of high-quality fodder production into the farming system. In many regions of the world, once the harvest is over, there are communal grazing rights on the stover, and so an individual farmer cannot take a unilateral decision to maintain crop residues on the field. Although overcoming this social impediment to CA adoption will be complicated, there are examples of areas where communities have got together to control grazing patterns and respect farmer rights to maintain crop residues in the field (Wall, 2009). Barriers, whether hedges or fences, are generally not popular among smallholder farmers because of the costs, either monetary or in land, but live barriers using fodder species deserve more attention in some conditions, as does the possibility of intercropping (relay cropping) long-season crops within the main crop - generally social norms prohibit animal entry into fields with growing crops.

Many smallholder farmers depend on family manual labour as the prime source of energy for farming operations. Reducing the labour requirement for land preparation, therefore, has important advantages for these farmers and can be a major driver of adoption (Sorrenson et al., 1998; Ekboir et al., 2002). However, if the farmer relies on manual weed control rather than on green manure cover crops or chemical weed control, or a combination of these, labour requirements for weeding may more than offset the labour savings from tillage, and thus make CA adoption unlikely. Support, therefore, in developing integrated weed control strategies and facilitating access to herbicides, contract herbicide application services (Ekboir et al., 2002), and training (Bellinder et al., 2002) are often important goals for researchers and members of the innovation platform. Giller et al. (2009) referred to weeds as the 'Achilles heel' of CA precisely because weed control is often laborious and costly, with a greater requirement for herbicides than with conventional tillage at least in the first years. There is a huge amount of knowledge embedded in successful chemical weed control, and it is unlikely that the resources will be available to support the acquisition of this knowledge by the majority of smallholder farmers. Again, the support of entrepreneurs to provide weed control services to farmers would appear to be a more reasonable strategy. Pesticides can be dangerous, and their misuse by uninformed personnel is a major hazard in agricultural areas. Concentration of sufficient knowledge in a limited number of well-trained and accountable service providers can help reduce the risk of disasters. Together with efficient service providers, providing smallholder farmers access to reasonably priced herbicides via efficient and effective input supply chains, knowledge on the benefits and management of green manure cover crops, and access to seeds of these will mean that weeds need not be a constraint to the use of CA systems.

5 Benefits of Conservation Agriculture to society

CA offers many benefits to society at a local, national, regional, or even at a global level. Knowler and Bradshaw (2007) and Kassam and Friedrich (2011) summarized many of these benefits and Landers et al. (2001) calculated their importance and impact in Brazil. Societal benefits of CA include:

- Reduced problems with soil erosion, sedimentation of water bodies and waterways, blockage of roads, and (with respect to wind erosion) air pollution.

- Reduced air pollution from not burning crop residues, a common practice, especially in regions of high crop productivity with high levels of residue production that hamper tillage in conventionally tilled systems and seeding in CA systems. Currently, the Government of India is enforcing a ban on residue burning in the north-western States of Punjab and Haryana in order to reduce the choking smoke pollution in the capital, New Delhi, an action that is likely to stimulate considerable adoption of CA practices.
- The increased water infiltration results in less flooding, more stable river flows, aquifer recharge, and the re-emergence of dried wells (e.g. Evers and Agostini, 2001).
- Mitigation of deforestation by the intensification and sustainability of crop/livestock systems on degraded pastures and degraded soils.
- Reduction in the contamination of surface water with soil and toxic compounds due to the reduced water run-off and erosion, and reduced contamination of groundwater as a result of increased organic matter levels in and on the soil.
- Reduction in GHG emissions due to both the reduced fuel use and the sequestration of carbon in soil organic matter (Dendooven et al., 2012) – at least in those conditions where soil organic matter is increased by CA (see section above on SOM).
- Conservation of biodiversity, both on the land and in the soil.
- Greater food security and lower food prices resulting from increased soil fertility.
- Greater rural incomes leading to less rural–urban migration.
- Decreased levels of imports of fuel and fertilizers.

Fuel savings represent a major benefit to society as a whole, not only because of the reduced use of fossil fuels but also because of the reduction in the emissions of carbon dioxide (CO_2) – the principal greenhouse gas (GHG) implicated in the climate change phenomenon. Calegari et al. (2014) calculated that in 2012 alone, Brazil saved 1.34 billion litres of diesel fuel due to the adoption of CA, representing a reduction of CO_2 emissions of 3.5 million MT. However, several studies (summarized by Six et al., 2002) have shown that CA systems can result in increased emissions of nitrous oxide (N_2O) – a more potent GHG than CO_2. This implies the need for more efficient nitrogen fertilizer applications in CA systems – adjustments to the placement and timing of nitrogen fertilizer need to be incorporated into the CA system rather than simply using the same times and methods used in conventionally tilled systems (Six et al., 2002). Tailored crop rotations will also be an important component of nitrogen management in CA systems (Six et al., 2002). The decreased compaction and improved aeration of rice soils that will doubtless be a feature of permanent

CA rice-based systems will undoubtedly reduce methane (CH_4) emissions from rice production adding to the total GHG reductions possible with well-adapted CA systems.

The costs of soil erosion to society can be very large. The costs of dust storms and of clearing and repairing roads blocked by silt, as well as the dredging of rivers, dams, and ports can be high. More common may be the contamination of surface water with nutrients and pesticides from agricultural fields. CA, especially when adopted over most of a watershed, is an effective way to control these effects of soil erosion. For this reason, CA is promoted by the binational company that manages the Itaipu hydroelectric dam between Brazil and Paraguay in an effort to lengthen the useful life of the dam by reducing siltation (Laurent et al., 2011). Problems of pollution effects on the Great Barrier Reef off Australia's north-east coast have resulted in numerous programmes including the Smartcane Best Management Practice (BMP) (https://www.smartcane.com.au/), a voluntary programme through which Queensland cane farmers attain the accreditation of the Johnson River Catchment Management Association. One of the main pillars of the Smartcane BMP is Green Cane Trash Blanketing by sugarcane farmers under which farmers only burn the residues one year in five to reseed the crop, and in the remaining years leave all of the trash from the ratoon crops without burning to maintain complete residue cover.

In 2001, CA was practised on 35% of the agricultural land of Brazil. Landers et al. (2001) estimated that the annual on-farm economic benefits from CA adoption to Brazil were US$ 356 million. At the same time the estimated off-farm, societal benefits from this area of zero tillage were equivalent to US$ 1.03 billion – three times as much as the on-farm benefits. The largest portion of these benefits (786 million) was due to mitigation of deforestation due to the intensification of crop/livestock production on degraded pastures. The remainder of the off-farm societal benefits (US$246 million/year), due principally to reduced erosion, reduced fuel consumption, increased groundwater recharge and carbon sequestration, were equal to 69% of the on-farm benefits. Furthermore, the authors were not able to quantify many off-farm benefits such as reduced chemical pollution in inland and coastal waters; increased terrestrial, soil, and aquatic biodiversity; reduced air pollution by dust and smoke; lower food prices; reduced imports of fuel and fertilizer; greater food security; and reduced emissions of methane and nitrous oxide. Thus, farmers' investment in changing to CA provided more benefits to society as a whole than it did directly to the farmer. The key question, therefore, is how society should recompense farmers for their investment in changing to, and managing, more sustainable production systems, and thereby increase the incentives to farmers to further invest in and adopt these systems.

6 The way ahead: support for Conservation Agriculture

There will be costs to society to achieve widespread adoption of CA, especially investments in research and extension, and in farmer organization. There is also an environmental cost in the increased use of herbicides in many situations, especially in the first years after CA adoption. However, most of the costs implicated in the change from current farming systems that are resulting in severe and widespread land degradation to a more sustainable system based on CA, are currently paid by the farmer, whereas, as discussed previously, there are considerable benefits of this change to society in general. This has led to the discussion of paying the farmer for ecosystem services on the grounds that those who receive the benefits should pay for them. There are many ways that payments for ecosystem services could be paid to farmers through subsidies, preferential access to subsidized services (as was done in the US to benefit farmers who put in place practices to restrict soil erosion), reductions in taxes, increased research and extension services etc. Each of these has their benefits and potential drawbacks. Individual payments to smallholder farmers will be difficult and costly to administer, and communal benefits either in the form of payments to communities or increased services (e.g. research, extension and/or credit facilities) may be better aligned with smallholder circumstances. Scherr and McNeely (2008) have laid out the main institutional changes necessary to successfully promote ecoagriculture. While CA does not qualify as ecoagriculture based on their definition, it is certainly an important component of ecoagricultural systems, and, as such the following measures adapted from the proposals of these authors can equally act as a blueprint for the institutional support necessary for the adoption at scale of CA systems:

- Targeted investment to produce the research and management breakthroughs that will enable farmers to increase farm productivity while conserving and enhancing natural resources.
- The development of niche markets for certified, more environmentally friendly products. In Mexico, major breweries and bread makers have started a move toward the preferential purchase of crops produced using CA. Improved market price and security give added benefits and incentives to CA farmers.
- Payments to farmers and their communities for ecosystem services resulting from the use of environmentally friendly management practices.
- More supportive policy for sustainable agriculture, or simply the removal of major policy barriers. Core policy needs include (1) compatibility and coordination of agricultural development and natural resource conservation policies; (2) environmental legislation that recognizes the rights of farming communities as stewards of the land; and (3) the removal of public subsidies for agricultural systems that degrade natural resources.

Markets, supportive policies, and payments for ecosystem services will help farmers with productive land achieve improved livelihoods for themselves and their families through more productive and sustainable agricultural practices. However, land that is degraded beyond the limits of productive agriculture demands special attention (see chapter on management of degraded lands in this volume).

At the same time, we believe that a civil society conversation on the rights, privileges, and responsibilities of landowners and farmers is warranted, with the aim of developing policies that promote the maintenance of functional ecosystems, account for the effects of our actions on future generations, and develop norms for environmental justice (Guth, 2009).

We end with an excerpt from the Declaration made by the participants of the first World Congress on CA held in Madrid, 2001 (Garcia-Torres et al., 2001) with which we fully concur: 'This Congress calls upon politicians, international institutions, environmentalists, farmers, private industry and society as a whole to recognise that the conservation of natural resources is the co-responsibility, past, present and future, of all sectors of society, in the proportion that they consume products resulting from the exploitation of these resources. Furthermore, it calls upon society, through these stakeholders, to conceive and enact appropriate long-term strategies and to support, further develop and embrace the concepts of CA as the most appropriate means of ensuring the continuity of the land's ongoing capacities to yield food, other agricultural products, water and environmental services in perpetuity. It follows that the environmental services provided by farmers practising CA should be recognised and recompensed by society'.

7 Where to look for further information

Worldwide, interest in CA is growing rapidly and currently there is an explosion in the amount of information being produced and published on different aspects of this more sustainable production system. Recent books on CA, including the following, provide an insight into the theory and practise of CA in different parts of the world:

- *Conservation Agriculture. Global Prospects and Challenges* (2014) edited by Ram Jat, Kanwar Sahrawat, and Amir Kassam
- *Conservation Agriculture* (2015) edited by Muhammed Farooq and Kadambot Siddique
- *Conservation Agriculture for Africa* (2017) edited by Amir Kassam, Saidi Mkomwa, and Theodor Friedrich

The following resource pages also include up-to-date information on CA and CA components:

- http://soilhealth.org of the College of Agriculture and Life Sciences of Cornell University.
- http://www.fao.org/conservation-agriculture of FAO.

8 References

Affholder, F., Jourdain, D., Quang, D. D., Tuong, T. P., Morize, M. and Ricome, A. 2010. Constraints to farmers' adoption of direct-seeding mulch-based cropping systems: a farm scale modelling approach applied to the mountainous slopes of Vietnam. *Agricultural Systems* 103(1), 51–62. doi:10.1016/j.agsy.2009.09.001.

Ajieh, P. C., Agwu, A. E. and Anyanwu, A. C. 2008. Constraints to privatization and commercialization of agricultural extension services as perceived by extension professionals and farmers. *African Journal of Agricultural Research* 3(5), 343–7.

Anderson, W. and Johnson, T. 2016. Evaluating global land degradation using ground-based measurements and remote sensing. In: Nkonya, E., Mirzabaev, A. and Von Braun, J. (Eds), *Economics of Land Degradation and Improvement: a Global Assessment for Sustainable Development*. Springer, New York, pp. 85–116.

Anderson, J. and Van Crowder, L. 2000. The present and future of public sector extension in Africa: contracting out or contracting in? *Public Administration and Development* 20(5), 373–84. doi:10.1002/pad.149.

Andersson, J. A. and D'Souza, S. 2014. From adoption claims to understanding farmers and contexts: a literature review of Conservation Agriculture (CA) adoption among smallholder farmers in southern Africa. *Agriculture, Ecosystems and Environment* 187, 116–32. doi:10.1016/j.agee.2013.08.008.

Awada, L., Lindwall, C. W. and Sonntag, B. 2014. The development and adoption of conservation tillage systems on the Canadian Prairies. *International Soil and Water Conservation Research* 2(1), 47–65. doi:10.1016/S2095-6339(15)30013-7.

Bagayoko, M., Buerkert, A., Lung, G., Bationo, A. and Römheld, V. 2000. Cereal/legume rotation effects on cereal growth in Sudano-Sahelian West Africa: soil mineral nitrogen, mycorrhizae and nematodes. *Plant and Soil* 218(1–2), 103–16.

Baker, C. J., Collins, R. M. and Choudhary, M. A. 2001. Factors affecting the uptake of no-tillage in Australia, Asia and New Zealand. In: Garcia-Torres, L., Benites, J. and Martinez-Vilela, A. (Eds), *Conservation Agriculture: A Worldwide Challenge. I World Congress on Conservation Agriculture* (vol. I). ECAF/FAO, Cordoba, Spain, pp. 35–41.

Baraibar, B., Westerman, P. R., Carrión, E. and Recasens, J. 2009. Effects of tillage and irrigation in cereal fields on weed seed removal by seed predators. *Journal of Applied Ecology* 46(2), 380–7. doi:10.1111/j.1365-2664.2009.01614.x.

Baudron, F., Andersson, J. A., Corbeels, M. and Giller, K. E. 2012. Failing to yield? Ploughs, Conservation Agriculture and the problem of agricultural intensification: an example from the Zambezi Valley, Zimbabwe. *Journal of Development Studies* 48(3), 393–412. doi:10.1080/00220388.2011.587509.

Baudron, F., Sims, B., Justice, S., Kahan, D. G., Rose, R., Mkomwa, S., Kaumbutho, P., Sariah, J., Nazare, R., Moges, G. and Gérard, B. 2015a. Re-examining appropriate mechanization in Eastern and Southern Africa: two-wheel tractors, Conservation Agriculture, and private sector involvement. *Food Security* 7(4), 889–904. doi:10.1007/s12571-015-0476-3.

Baudron, F., Thierfelder, C., Nyagumbo, I. and Gérard, B. 2015b. Where to target Conservation Agriculture for African smallholders? How to overcome challenges associated with its implementation? Experience from eastern and southern Africa. *Environments* 2(4), 338–57. doi:10.3390/environments2030338.

Bell, A. R., Cheek, J. Z., Mataya, F. and Ward, P. S. 2018. Do as they did: peer effects explain adoption of Conservation Agriculture in Malawi. *Water* 10(1), 16pp. doi:10.3390/w10010051.

Bellinder, R. R., Miller, A. J., Malik, R. K., Ranjit, J. D., Hobbs, P. R., Brar, L. S., Singh, G., Singh, S. and Yadev, A. 2002. Improving herbicide application accuracy in South Asia. *Weed Technology* 16(4), 845–50. doi:10.1614/0890-037X(2002)016[0845:IHAAIS]2.0.CO;2.

Bellotti, B. and Rochecouste, J. F. 2014. The development of Conservation Agriculture in Australia–farmers as innovators. *International Soil and Water Conservation Research* 2(1), 21–34. doi:10.1016/S2095-6339(15)30011-3.

Bolliger, A., Damgaard Hansen, K. and Fowler, R. 2005. Constraints limiting smallholder adoption of Conservation Agriculture: some observations based on three South African smallholder-orientated programmes. In: *Proceedings of the III World Congress on Conservation Agriculture*, Nairobi, Kenya, 3–7 October 2005, 6p. On CD.

Bolliger, A., Magid, J., Amado, J. C. T., Skóra Neto, F., Ribeiro, MdFdS., Calegari, A., Ralisch, R. and de Neergaard, A. 2006. Taking stock of the Brazilian 'Zero-till revolution': a review of landmark research and farmers' practice. *Advances in Agronomy* 91, 47–110. doi:10.1016/S0065-2113(06)91002-5.

Borrelli, P., Robinson, D. A., Fleischer, L. R., Lugato, E., Ballabio, C., Alewell, C., Meusberger, K., Modugno, S., Schutt, B., Ferro, V., Bagarello, V., Van Oost, K. V., Montanarella, L. and Panagos, P. 2017. An assessment of the global impact of 21st century land use change on soil erosion. *Nature Communications* 8(1), 2013. doi:10.1038/s41467-017-02142-7.

Bronick, C. J. and Lal, R. 2005. Soil structure and management: a review. *Geoderma* 124(1–2), 3–22. doi:10.1016/j.geoderma.2004.03.005.

Brouder, S. M. and Gomez-Macpherson, H. 2014. The impact of Conservation Agriculture on smallholder agricultural yields: a scoping review of the evidence. *Agriculture, Ecosystems and Environment* 187, 11–32. doi:10.1016/j.agee.2013.08.010.

Byerlee, D., Heisey, P. and Hobbs, P. R. 1989. Diagnosing research priorities for small farmers: experiences from on-farm research in Pakistan. *Quarterly Journal of International Agriculture* 28(3), 254.

Calegari, A., de Araújo, A. G., Costa, A., Fuentes Lanilo, R., Casao Junior, R. and Rheinheimer dos Santos, D. 2014. Conservation Agriculture in Brazil. In: Jat, R. A., Sahrawat, K. L. and Kassam, A. (Eds), *Conservation Agriculture: Global Prospects and Challenges*. CAB International, Wallingford, UK, pp. 54–88.

Camacho-Villa, T. C., Hellin, J., Govaerts, B., Rendon-Medel, R., Martinez, T. E., Almekinders, C., Guevara-Hernández, F. and Beuchelt, T. C. 2016. Lessons and challenges in the implementation of an agricultural innovation network: the case of the MasAgro hubs. *The Journal of Agricultural Education and Extension* 22(5), 455–70.

Cheesman, S., Thierfelder, C., Eash, N. S., Kassie, G. T. and Frossard, E. 2016. Soil carbon stocks in Conservation Agriculture systems of Southern Africa. *Soil and Tillage Research* 156, 99–109. doi:10.1016/j.still.2015.09.018.

Cook, R. J. 1990. Twenty-five years of progress toward biological control. In: Horny, D. (Ed.), *Biological Control of Soil Borne Pathogens*. CAB International, Wallingford, UK, pp. 1–14.

Corbeels, M., Triomphe, B., Giller, K. E. and Kienzle, J. 2009. Tailoring conservation agriculture to local conditions of small holder farmers in Africa. In: *4th World Congress on Conservation Agriculture*, 4–7 February 2009, New Delhi, pp. 238–9.

Corbeels, M., de Graaff, J., Ndah, T. H., Penot, E., Baudron, F., Naudin, K., Andrieu, N., Chirat, G., Schuler, J., Nyagumbo, I., Rusinamhodzi, L., Traore, K., Mzoba, H. D. and Adolwa, I. S. 2014a. Understanding the impact and adoption of Conservation Agriculture in Africa: a multi-scale analysis. *Agriculture, Ecosystems and Environment* 187, 155–70. doi:10.1016/j.agee.2013.10.011.

Corbeels, M., Sakyi, R. K., Kühne, R. F. and Whitbread, A. 2014b. Meta-analysis of crop responses to Conservation Agriculture in sub-Saharan Africa. CCAFS Report No. 12. CGIAR Research Program on Climate Change, Agriculture and Food Security (CCAFS), Copenhagen, 20pp.

Corsi, S., Friedrich, T., Kassam, A., Pisante, M. and de Moraes Sa, J. 2012. *Soil Organic Carbon Accumulation and Greenhouse Gas Emission Reductions from Conservation Agriculture: A Literature Review*. Plant Protection Division Integrated Crop Management Vol. 16. FAO, Rome, 89pp.

de Freitas, P. L. and Landers, J. N. 2014. The transformation of agriculture in Brazil through development and adoption of zero tillage Conservation Agriculture. *International Soil and Water Conservation Research* 2(1), 35–46. doi:10.1016/S2095-6339(15)30012-5.

Dendooven, L., Patiño-Zúñiga, L., Verhulst, N., Luna-Guido, M., Marsch, R. and Govaerts, B. 2012. Global warming potential of agricultural systems with contrasting tillage and residue management in the central highlands of Mexico. *Agriculture, Ecosystems and Environment* 152, 50–8. doi:10.1016/j.agee.2012.02.010.

Derpsch, R., Friedrich, T., Kassam, A. and Li, H. 2010. Current status of adoption of no-till farming in the world and some of its main benefits. *International Journal of Agricultural and Biological Engineering* 3(1), 1–26. doi:10.3965/j.issn.1934-6344.2010.01.001-025.

Derpsch, R., Franzluebbers, A. J., Duiker, S. W., Reicosky, D. C., Koeller, K., Friedrich, T., Sturny, W. G., Sa, J. C. M. and Weiss, K. 2014. Why do we need to standardize no-tillage research? *Soil and Tillage Research* 137, 16–22. doi:10.1016/j.still.2013.10.002.

Derpsch, R., Lange, D., Birbaumer, G. and Moriya, K. 2016. Why do medium- and large-scale farmers succeed practising CA and small-scale farmers often do not? – experiences from Paraguay. *International Journal of Agricultural Sustainability* 14(3), 269–81. doi:10.1080/14735903.2015.1095974.

Diaz-Jose, J., Rendón-Medel, R., Govaerts, B., Aguilar-Ávila, J. and Muñoz-Rodriguez, M. 2016. Innovation diffusion in Conservation Agriculture: a network approach. *The European Journal of Development Research* 28(2), 314–29. doi:10.1057/ejdr.2015.9.

Doran, J. W. and Zeiss, M. R. 2000. Soil health and sustainability: managing the biotic component of soil quality. *Applied Soil Ecology* 15(1), 3–11. doi:10.1016/S0929-1393(00)00067-6.

Duiker, S. and Thomason, W. 2014. Conservation Agriculture in the USA. In: Jat, R. A., Sahrawat, K. L. and Kassam, A. (Eds), *Conservation Agriculture: Global Prospects and Challenges*. CAB International, Wallingford, UK, pp. 26–53.

Ehrlich, P. and Ehrlich, A. 1981. *Extinction: the Causes and Consequences of the Disappearance of Species*. Random House, New York, 305pp.

Ehui, S. and Pender, J. 2005. Resource degradation, low agricultural productivity, and poverty in sub-Saharan Africa: pathways out of the spiral. *Agricultural Economics* 32(s1), 225–42. doi:10.1111/j.0169-5150.2004.00026.x.

Ekboir, J. 2002. Developing no-till packages for small-scale farmers. In: Ekboir, J. (Ed.), *World Wheat Overview and Outlook*. CIMMYT, Mexico City, Mexico, pp. 1–38.

Ekboir, J., Boa, K. and Dankyi, A. A. 2002. Impact of no-till technologies in Ghana. Economics Program Paper 02-01. CIMMYT, Mexico DF, 32pp.

Evers, G. and Agostini, A. 2001. No-tillage agriculture for sustainable land management: lessons learned from the 2000 Brazil study tour. FAO/TCI Occasional Paper no. 12. FAO, Rome.

Farooq, M., Flower, K. C., Jabran, K., Wahid, A. and Siddique, K. H. M. 2011. Crop yield and weed management in rainfed Conservation Agriculture. *Soil and Tillage Research* 117, 172–83. doi:10.1016/j.still.2011.10.001.

Faulkner, E. H. 1943. *Ploughman's Folly*. University of Oklahoma Press, Stillwater, 161pp.

Fischer, R. A., Byerlee, D. and Edmeades, G. O. 2009. Can technology deliver on the yield challenge to 2050 ? In: *Expert Meeting on How to Feed the World in 2050*. FAO, Rome, 46pp.

Fisher, M., Holden, S. T., Thierfelder, C. and Katengeza, S. P. 2018. Awareness and adoption of Conservation Agriculture in Malawi: what difference can farmer-to-farmer extension make? *International Journal of Agricultural Sustainability* 16(3), 310–25. doi:10.1080/14735903.2018.1472411.

Fouzai, A., Smaoui, A., Frija, A. and Dhehibi, B. 2018. Adoption of Conservation Agriculture technologies by smallholder farmers in the semiarid region of Tunisia: resource constraints and partial adoption. *Journal of New Sciences. Sustainable Livestock Management* 6(1), 105–14.

Fryrear, D. W. and Skidmore, E. L. 1985. Methods for controlling wind erosion. In: Follett, R. F. and Stewart, B. A. (Eds), *Soil Erosion and Crop Productivity*. ASA-CSSA-SSSA, Madison, WI, pp. 443–57.

Garcia-Torres, L., Benites, J. and Martinez-Vilela, A. (Eds). 2001. The 1st World Congress on Conservation Agriculture declaration. In: *Conservation Agriculture: A Worldwide Challenge. I World Congress on Conservation Agriculture* (vol. I). ECAF/FAO, Cordoba, Spain, pp. 13–16.

Giller, K. E., Witter, E., Corbeels, M. and Tittonell, P. 2009. Conservation Agriculture and smallholder farming in Africa: the heretics' view. *Field Crops Research* 114(1), 23–34. doi:10.1016/j.fcr.2009.06.017.

Govaerts, B., Sayre, K. D. and Deckers, J. 2005. Stable high yields with zero tillage and permanent bed planting? *Field Crops Research* 94(1), 33–42. doi:10.1016/j.fcr.2004.11.003.

Govaerts, B., Sayre, K. D., Ceballos-Ramirez, J. M., Luna-Guido, M. L., Limon-Ortega, A., Deckers, J. and Dendooven, L. 2006. Conventionally tilled and permanent raised beds with different crop residue management: effects on soil C and N dynamics. *Plant and Soil* 280(1-2), 143–55. doi:10.1007/s11104-005-2854-7.

Govaerts, B., Sayre, K. D., Lichter, K., Dendooven, L. and Deckers, J. 2007a. Influence of permanent raised bed planting and residue management on physical and chemical soil quality in rain fed maize/wheat systems. *Plant and Soil* 291(1-2), 39–54. doi:10.1007/s11104-006-9172-6.

Govaerts, B., Mezzalama, M., Unno, Y., Sayre, K. D., Luna-Guido, M., Vanherck, K., Dendooven, L. and Deckers, J. 2007b. Influence of tillage, residue management, and crop rotation on soil microbial biomass and catabolic diversity. *Applied Soil Ecology* 37(1-2), 18–30. doi:10.1016/j.apsoil.2007.03.006.

Govaerts, B., Mezzalama, M., Sayre, K. D., Crossa, J., Lichter, K., Troch, V., Vanherck, K., De Corte, P. and Deckers, J. 2008. Long-term consequences of tillage, residue management, and crop rotation on selected soil micro-flora groups in the subtropical highlands. *Applied Soil Ecology* 38(3), 197–210. doi:10.1016/j.apsoil.2007.10.009.

Govaerts, B., Verhulst, N., Castellanos-Navarrete, A., Sayre, K. D., Dixon, J. and Dendooven, L. 2009a. Conservation Agriculture and soil carbon sequestration: between myth and farmer reality. *Critical Reviews in Plant Sciences* 28(3), 97–122. doi:10.1080/07352680902776358.

Govaerts, B., Sayre, K. D., Goudeseune, B., De Corte, P., Lichter, K., Dendooven, L. and Deckers, J. 2009b. Conservation Agriculture as a sustainable option for the central Mexican highlands. *Soil and Tillage Research* 103(2), 222–30. doi:10.1016/j.still.2008.05.018.

Guth, J. H. 2009. *A Law to Protect the Earth: the Tort of Ecological Degradation*. Science and Environmental Health Network, 8pp.

Haddad, N., Piggin, C., Haddad, A. and Khalil, Y. 2014. Conservation Agriculture in West Asia. In: Jat, R. A., Sahrawat, L. and Kassam, A. H. (Eds), *Conservation Agriculture. Global Prospects and Challenges*. CAB International, Wallingford, UK, pp. 248–62.

Haggblade, S. and Tembo, G. 2003. Conservation farming in Zambia. EPTD Discussion Paper No. 108. International Food Policy Research Institute, Washington DC, 128pp.

Hatfield, J. L., Sauer, T. J. and Prueger, J. H. 2001. Managing soils to achieve greater water use efficiency: a review. *Agronomy Journal* 93(2), 271–80. doi:10.2134/agronj2001.932271x.

Haque, M. E., Bell, R. W., Islam, A. K. M. S., Sayre, K. and Hossain, M. M. 2011. Versatile multi-crop planter for two-wheel tractors: an innovative option for small-holders. In: Gilkes, R. J. and Nattaporn, P. (Eds), *5th World Congress of Conservation Agriculture Incorporating 3rd Farming Systems Design Conference*, 26–29 September 2011, Brisbane, Australia.

Havlin, J. L., Kissel, D. E., Maddux, L. D., Claassen, M. M. and Long, J. H. 1990. Crop rotation and tillage effects on soil organic carbon and nitrogen. *Soil Science Society of America Journal* 54(2), 448–52. doi:10.2136/sssaj1990.03615995005400020026x.

Hobbs, P. R. 2007. Conservation Agriculture: what is it and why is it important for future sustainable food production? *The Journal of Agricultural Science* 145(2), 127–38. doi:10.1017/S0021859607006892.

Hobbs, P. R. and Gupta, R. K. 2004. Problems and challenges of no-till farming for the rice-wheat systems of the Indo-Gangetic plains in South Asia. In: Lal, R., Hobbs, P. R., Uphoff, N. and Hansen, D. O. (Eds), *Sustainable Agriculture and the Rice-Wheat System*. Ohio State University, Columbus, OH, pp. 101–20.

Hobbs, P. R., Sayre, K. and Gupta, R. 2008. The role of Conservation Agriculture in sustainable agriculture. *Philosophical Transactions of the Royal Society of London. Series B, Biological Sciences* 363(1491), 543–55. doi:10.1098/rstb.2007.2169.

Hobbs, P. R., Gupta, R., Jat, R. K. and Malik, R. K. 2019. Conservation Agriculture in the Indo-Gangetic plains of India: past, present and future. *Experimental Agriculture* 55(2), 339–57. doi:10.1017/S0014479717000424.

Holden, S. T., Fisher, M., Katengeza, S. P. and Thierfelder, C. 2018. Can lead farmers reveal the adoption potential of Conservation Agriculture? The case of Malawi. *Land Use Policy* 76, 113–23. doi:10.1016/j.landusepol.2018.04.048.

Howard, D. D., Chambers, A. Y. and Lessman, G. M. 1998. Rotation and fertilization effects on corn and soybean yields and soybean cyst nematode populations in a no-tillage system. *Agronomy Journal* 90(4), 518–22. doi:10.2134/agronj1998.00021962009000040013x.

Hudson, N. W. 1991. A study of the reasons for success and failure of soil conservation projects. Soils Bulletin 64. Food and Agriculture Organization of the United Nations, Rome, Italy.

Kassam, A. and Friedrich, T. 2011. Conservation Agriculture: global perspectives and developments. In: *Regional Conservation Agriculture Symposium*, 8–10 February 2011, Johannesburg, South Africa, pp. 1–33.

Kassam, A., Friedrich, T. and Derpsch, R. 2019. Global spread of Conservation Agriculture. *International Journal of Environmental Studies* 76(1), 29–51. doi:10.1080/00207233.2018.1494927.

Kladivko, E. J. 2001. Tillage systems and soil ecology. *Soil and Tillage Research* 61(1-2), 61–76. doi:10.1016/S0167-1987(01)00179-9.

Kliewer, I., Casaccia, J., Vallejos, J. and Derpsch, R. 2000. Cost and herbicide reduction in the no-tillage system by using green manure cover crops in Paraguay. In: *Proceedings 15th ISTRO Conference*, Fort Worth, Texas, USA, 2–7 July 2000.

Knowler, D. and Bradshaw, B. 2007. Farmers' adoption of Conservation Agriculture: a review and synthesis of recent research. *Food Policy* 32(1), 25–48. doi:10.1016/j.foodpol.2006.01.003.

Laborte, A. G., De Bie, K. A. J. M., Smaling, E. M. A., Moya, P. F., Boling, A. A. and Van Ittersum, M. K. 2012. Rice yields and yield gaps in Southeast Asia: past trends and future outlook. *European Journal of Agronomy* 36(1), 9–20. doi:10.1016/j.eja.2011.08.005.

Lafond, G. P., Clayton, G. W. and Fowler, D. B. 2014. Conservation Agriculture on the Canadian Prairies. In: Jat, R. A., Sahrawat, K. L. and Kassam, A. (Eds), *Conservation Agriculture: Global Prospects and Challenges*. CAB International, Wallingford, UK, pp. 89–107.

Landers, J. N., Sant'Anna De, C., Barros, G., Rocha, M. T., Manfrinato, W. A. and Weiss, J. 2001. Environmental impacts of zero tillage in Brazil – a first approximation. In: Garcia-Torres, L., Benites, J. and Martinez-Vilela, A. (Eds), *Conservation Agriculture: A Worldwide Challenge. I World Congress on Conservation Agriculture* (vol. I). ECAF/FAO, Cordoba, Spain, pp. 317–26.

Landschoot, S., Audenaert, K., Waegeman, W., De Baets, B. and Haesaert, G. 2013. Influence of maize-wheat rotation systems on Fusarium head blight infection and deoxynivalenol content in wheat under low versus high disease pressure. *Crop Protection* 52, 14–21. doi:10.1016/j.cropro.2013.04.013.

Laurent, F., Leturcq, G., Mello, I., Corbonnois, J. and Verdum, R. 2011. La diffusion du semis direct au Brésil, diversité des pratiques et logiques territoriales: l'exemple de la région d'Itaipu au Paraná. *Confins* (Online), 12(12). Published online 2 July 2011. Consulted 23 October 2018. Available at: http://journals.openedition.org/confins/7143; doi:10.4000/confins.7143.

Lavelle, P. 1997. Faunal activities and soil processes: adaptive strategies that determine ecosystem function. *Advances in Ecological Research* 27, 93–132. doi:10.1016/S0065-2504(08)60007-0.

Loss, S., Haddad, A., Desbiolles, J., Cicek, H., Khalil, Y. and Piggin, C. 2015. *The Practical Implementation of Conservation Agriculture in the Middle East*. Australian Center for International Agricultural Research/international Center for Agricultural Research in the Dry Areas, 98pp.

Lowdermilk, W. C. 1948. Conquest of the land through seven thousand years. S.C.S. MP-32. U.S. Department of Agriculture, Washington DC, 46pp

Malik, R. K. and Singh, S. 1993. Evolving strategies for herbicide use in wheat: resistance and integrated weed management. In: *Proceedings of an International Symposium of the Indian Society of Weed Science* (vol. 1), 18–20 November 1993, Hisar, India., pp. 225–38.

Malik, R. K. and Singh, S. 1995. Littleseed canarygrass (*Phalaris minor* Retz.) resistance to isoproturon in India. *Weed Technology* 9(3), 419–25. doi:10.1017/S0890037X00023629.

Michler, J. D., Baylis, K., Arends-Kuenning, M. and Mazvimavi, K. 2019. Conservation Agriculture and climate resilience. *Journal of Environmental Economics and Management* 93, 148–69. doi:10.1016/j.jeem.2018.11.008.

Montgomery, D. R. 2007a. *Dirt: the Erosion of Civilizations*. The University of California Press, Berkeley.

Montgomery, D. R. 2007b. Soil erosion and agricultural sustainability. *Proceedings of the National Academy of Sciences of the United States of America* 104(33), 98195. doi:10.1073/pnas.0611508104.

Moore, D., Robson, G. D. and Trinci, A. P. J. 2011. *21st Century Guidebook to Fungi*. Cambridge University Press, Cambridge and New York, 627pp.

Mwangi, M. and Kariuki, S. 2015. Factors determining adoption of new agricultural technology by smallholder farmers in developing countries. *Journal of Economic and Sustainable Development* 6(5), 208–16.

Ngwira, A. R., Thierfelder, C., Eash, N. and Lambert, D. M. 2013. Risk and maize-based cropping systems for smallholder Malawi farmers using Conservation Agriculture technologies. *Experimental Agriculture* 49(4), 483–503. doi:10.1017/S0014479713000306.

Nichols, V., Verhulst, N., Cox, R. and Govaerts, B. 2015. Weed dynamics and Conservation Agriculture principles : a review. *Field Crops Research* 183, 56–68. doi:10.1016/j.fcr.2015.07.012.

Nyagumbo, I. 2008. A review of experiences and developments towards Conservation Agriculture and related systems in Zimbabwe. In: Goddard, T., Zoebisch, M. A., Gan, Y. T., Ellis, W., Watson, A. and Sombatpanit, S. (Eds), *No-Till Farming Systems*. Special Publication No. 3. World Association of Soil and Water Conservation, Bangkok, Thailand, pp. 345–72.

Oldeman, L. R. 1994. Global extent of soil degradation. In: Greenland, D. J. and Szabolcs, I. (Eds), *Soil Resilience and Sustainable Land Use*. CAB International, Wallingford, UK, pp. 99–118.

Oldrieve, B. 1993. *Conservation Farming for Communal, Small Scale, Resettlement and Co-Operative Farmers of Zimbabwe: A Farm Management Handbook*. Prestige Business Services (Pvt) Ltd., Harare, 77pp.

Peiretti, R. and Dumanski, J. 2014. The transformation of agriculture in Argentina through soil conservation. *International Soil and Water Conservation Research* 2(1), 14-20. doi:10.1016/S2095-6339(15)30010-1.

Pittelkow, C. M., Linquist, B. A., Lundy, M. E., Liang, X., van Groenigen, K. J., Lee, J., van Gestel, N., Six, J., Venterea, R. T. and van Kessel, C. 2015. When does no-till yield more ? A global meta-analysis. *Field Crops Research* 183, 156-68. doi:10.1016/j.fcr.2015.07.020.

Powlson, D. S., Stirling, C. M., Thierfelder, C., White, R. P. and Jat, M. L. 2016. Does Conservation Agriculture deliver climate change mitigation through soil carbon sequestration in tropical agro-ecosystems? *Agriculture, Ecosystems and Environment* 220, 164-74. doi:10.1016/j.agee.2016.01.005.

Reicosky, D. C. and Lindstrom, M. J. 1993. Fall tillage method: effect on short-term carbon dioxide flux from soil. *Agronomy Journal* 85(6), 1237-43. doi:10.2134/agronj1993.00021962008500060027x.

Rocco, G. 1997. Costo de la maquinaria en la siembra directa. Paper presented in the *Segundas Jornadas Bolivianas de Siembra Directa*, Santa Cruz, Bolivia, 9-11 October 1997.

Romero, A., Causarano, H. and Wall, P. C. 1993. Tillage and rotation effects on aggregate stability in south-eastern Paraguay. In: Tanner, D. G. (Ed.), *Developing Sustainable Wheat Production Systems. 8th Regional Wheat Workshop for Eastern, Central and Southern Africa*, Kampala, Uganda, 7-10 June 1993. CIMMYT, Addis Ababa, Ethiopia, pp. 316-8.

Rusinamhodzi, L., Corbeels, M., Wijk, M. T., Rufino, M. C., Nyamangara, J. and Giller, K. E. 2011. A meta-analysis of long-term effects of Conservation Agriculture on maize grain yield under rain-fed conditions. *Agronomy for Sustainable Development* 31(4), 657-73. doi:10.1007/s13593-011-0040-2.

Santos, H. P. and Reis, E. M. 2003. Efeito de sistemas de rotaçao de culturas sobre controle de doenças. In: Santos, H. P. (Ed.), *Rotaçao de Culturas em Plantio Direto* (2nd edn.). EMBRAPA - Trigo, Passo Fundo, Brazil, pp. 31-9.

Sayre, K. D. and Hobbs, P. R. 2004. The raised-bed system of cultivation for irrigated production conditions. In: Lal, R., Hobbs, P. R., Uphoff, N. and Hansen, D. O. (Eds), *Sustainable Agriculture and the Rice-Wheat System*. Ohio State University, Columbus, OH, pp. 337-55.

Scherr, S. J. and McNeely, J. A. 2008. Biodiversity conservation and agricultural sustainability: towards a new paradigm of 'ecoagriculture' landscapes. *Philosophical Transactions of the Royal Society of London. Series B, Biological Sciences* 363(1491), 477-94. doi:10.1098/rstb.2007.2165.

Schertz, D. L. 1994. Conservation tillage - a national perspective. In: Bauer, P. J. and Busscher, W. J. (Eds), *Proceedings of the 1994 Southern Conservation Tillage Conference for Sustainable Agriculture*. United States Department of Agriculture-ARS, Florence, SC, pp. 1-5.

Siddoway, F. H., Chepil, W. S. and Armbrust, D. V. 1965. Effect of kind, amount, and placement of residue on wind erosion control. *Transactions of the ASAE* 8(3), 327-31.

Sidiras,N. and Pavan, M. A. 1985. Influencia do sistema de manejo do solo no seu nivel de fertilidade. *Revista Brasileira do Ciencias do Solo* 9, 244–54.

Sikka, A. K., Khan, A. R., Singh, S. S. and Subash, N. 2005. Water management imperatives in zero/reduced till. In: Abrol, I. P., Gupta, R. K. and Malik, R. K. (Eds), *Conservation Agriculture - Status and Prospects*. Centre for Advancement of Sustainable Agriculture, New Delhi, pp. 93–8.

Six, J., Elliott, E. T. and Paustian, K. 2000. Soil macroaggregate turnover and microaggregate formation: a mechanism for C sequestration under no-tillage agriculture. *Soil Biology and Biochemistry* 32(14), 2099–103. doi:10.1016/S0038-0717(00)00179-6.

Six, J., Feller, C., Denef, K., Ogle, S. M., de Moraes Sa, J. C. and Albrecht, A. 2002. Soil organic matter, biota and aggregation in temperate and tropical soils - effects of no-tillage. *Agronomie* 22, 755–75. doi:10.1051/agro:2002043.

Siziba, S. 2008. Assessing the adoption of Conservation Agriculture in Zimbabwe's smallholder sector. PhD thesis. University of Hohenheim, 175pp.

Sorrenson, W. J., Duarte, C. and López Portillo, J. 1998. Economics of no-till compared to conventional cultivation systems on small farms in Paraguay, policy and investment implications. Report Soil Conservation Project MAG - GTZ, August 1998.

Spielman, D. J., Ekboir, J., Davis, K. and Ochieng, C. M. O. 2008. An innovation systems perspective on strengthening agricultural education and training in sub-Saharan Africa. *Agricultural Systems* 98(1), 1–9. doi:10.1016/j.agsy.2008.03.004.

Stinner, B. R. and House, G. J. 1990. Arthropods and other invertebrates in conservation-tillage agriculture. *Annual Review of Entomology* 35(1), 299–318. doi:10.1146/annurev.en.35.010190.001503.

Tenge, A. J., De Graaff, J. and Hella, J. P. 2004. Social and economic factors affecting the adoption of soil and water conservation in West Usambara highlands, Tanzania. *Land Degradation and Development* 15(2), 99–114. doi:10.1002/ldr.606.

Thierfelder, C. and Wall, P. C. 2009. Effects of Conservation Agriculture techniques on infiltration and soil water content in Zambia and Zimbabwe. *Soil and Tillage Research* 105(2), 217–27. doi:10.1016/j.still.2009.07.007.

Thierfelder, C. and Wall, P. C. 2010. Rotation in conservation agriculture systems of Zambia: effects on soil quality and water relations. *Experimental Agriculture* 46(3), 309–25. doi:10.1017/S001447971000030X.

Thierfelder, C., Cheesman, S. and Rusinamhodzi, L. 2013a. Benefits and challenges of crop rotations in maize-based Conservation Agriculture (CA) cropping systems of southern Africa. *International Journal of Agricultural Sustainability* 11(2), 108–24. doi:10.1080/14735903.2012.703894.

Thierfelder, C., Mombeyarara, T., Mango, N. and Rusinamhodzi, L. 2013b. Integration of Conservation Agriculture in smallholder farming systems of southern Africa: identification of key entry points. *International Journal of Agricultural Sustainability* 11(4), 317–30. doi:10.1080/14735903.2013.764222.

Thierfelder, C., Bunderson, W. T., Jere, Z. D., Mutenje, M. and Ngwira, A. 2015a. Development of Conservation Agriculture (CA) systems in Malawi: lessons learned from 2005 to 2014. *Experimental Agriculture* 52(4), 579–604. doi:10.1017/S0014479715000265.

Thierfelder, C., Rusinamhodzi, L., Ngwira, A. R., Mupangwa, W., Nyagumbo, I., Kassie, G. T. and Cairns, J. E. 2015b. Conservation Agriculture in Southern Africa: advances in knowledge. *Renewable Agriculture and Food Systems* 30(4), 328–48. doi:10.1017/S1742170513000550.

Thierfelder, C., Matemba-Mutasa, R. and Rusinamhodzi, L. 2015c. Yield response of maize (*Zea mays* L.) to Conservation Agriculture cropping system in Southern Africa. *Soil and Tillage Research* 146, 230–42. doi:10.1016/j.still.2014.10.015.

Thierfelder, C., Baudron, F., Setimela, P., Nyagumbo, I., Mupangwa, W., Mhlanga, B., Lee, N. and Gérard, B. 2018. Complementary practices supporting Conservation Agriculture in southern Africa. A review. *Agronomy for Sustainable Development* 38(2), 16. doi:10.1007/s13593-018-0492-8.

Triplett, G. B. and Dick, W. A. 2008. No-tillage crop production: a revolution in agriculture! *Agronomy Journal* 100 (Suppl. 3), S-153–65. doi:10.2134/agronj2007.0005c.

Umar, B. B., Aune, J. B., Johnsen, F. H. and Lungu, I. O. 2012. Are smallholder Zambian farmers economists? A dual-analysis of farmers' expenditure in conservation and conventional agriculture systems. *Journal of Sustainable Agriculture* 36(8), 908–29. doi:10.1080/10440046.2012.661700.

University of Oxford. 2008. *The Oxford Dictionary of Proverbs* (5th edn.). Jennifer Speake (Ed.). Oxford University Press, Oxford.

Vagen, T. G., Lal, R. and Singh, B. R. 2005. Soil carbon sequestration in sub-Saharan Africa: a review. *Land Degradation and Development* 16(1), 53–71. doi:10.1002/ldr.644.

Valbuena, D., Erenstein, O., Homann-Kee Tui, S., Abdoulaye, T., Claessens, L., Duncan, A. J., Gérard, B., Rufino, M. C., Teufel, N., van Rooyen, A. and van Wijk, M. T. 2012. Conservation Agriculture in mixed crop-livestock systems: scoping crop residue trade-offs in sub-Saharan Africa and South Asia. *Field Crops Research* 132, 175–84. doi:10.1016/j.fcr.2012.02.022.

Van Loon, J., Speratti, A. B. and Govaerts, B. 2018. Precision for smallholder farmers: a small-scale-tailored variable rate fertilizer application kit. *Agriculture* 8(4), 48. doi:10.3390/agriculture8040048.

Verhulst, N., Govaerts, B., Verachtert, E., Mezzalama, M., Wall, P. C., Chocobar, A., Deckers, J. and Sayre, K. D. 2010. Conservation Agriculture, improving soil quality for sustainable production systems ? In: Lal, R. and Stewart, B. A. (Eds), *Advances in Soil Science: Food Security and Soil Quality*. CRC Press, Boca Raton, FL, pp. 137–208.

Verhulst, N., Nelissen, V., Jespers, N., Haven, H., Sayre, K. D., Raes, D., Deckers, J. and Govaerts, B. 2011a. Soil water content, maize yield and its stability as affected by tillage and crop residue management in rainfed semi-arid highlands. *Plant and Soil* 344(1–2), 73–85. doi:10.1007/s11104-011-0728-8.

Verhulst, N., Kienle, F., Sayre, K. D., Deckers, J., Raes, D., Limon-Ortega, A., Tijerina-Chavez, L. and Govaerts, B. 2011b. Soil quality as affected by tillage-residue management in a wheat-maize irrigated bed planting system. *Plant and Soil* 340(1–2), 453–66. doi:10.1007/s11104-010-0618-5.

Wakatsuki, T. and Azwar, R. 1992. Rates of weathering and soil formation. *Geoderma* 52(3–4), 251–63. doi:10.1016/0016-7061(92)90040-E.

Wall, P. C. 1999. Experiences with crop residue cover and direct seeding in the Bolivian highlands. *Mountain Research and Development* 19(4), 313–7.

Wall, P. C. 2007. Tailoring Conservation Agriculture to the needs of small farmers in developing countries: an analysis of issues. *Journal of Crop Improvement* 19(1–2), 137–55. doi:10.1300/J411v19n01_07.

Wall, P. C. 2009. Strategies to overcome the competition for crop residues in Southern Africa: some light at the end of the tunnel. In: *Innovations for Improving Efficiency, Equity and Environment. 4th World Congress on Conservation Agriculture*. Lead Papers. New Dehli, pp. 65–70.

Wall, P. C., Zambrana, L., Gomez, P., Sims, B. and Callisaya, A. 2001. Development of an animal drawn zero tillage seeder for small grains. In: Garcia-Torres, L., Benites, J. and Martinez-Vilela, A. (Eds), *Conservation Agriculture: A Worldwide Challenge. I World Congress on Conservation Agriculture Vol II: Offered Contribution*. ECAF/FAO, Cordoba, Spain, pp. 631–5.

Wall, P. C., Ekboir, J. M. and Hobbs, P. R. 2002. Institutional aspects of Conservation Agriculture. Paper presented at the *International Workshop on Conservation Agriculture for Sustainable Wheat Production in Rotation with Cotton in Limited Water Resource Areas*, Tashkent, Uzbekistan, 13–18 October 2002.

Wall, P. C., Thierfelder, C. T., Ngwira, A., Govaerts, B., Nyagumbo, I. and Baudron, F. 2014. Conservation Agriculture in eastern and Southern Africa. In: Jat, R. A., Sahrawat, L. and Kassam, A. (Eds), *Conservation Agriculture. Global Prospects and Challenges*. CAB International, Wallingford, UK, pp. 263–92.

Ward, P. S., Bell, A. R., Droppelmann, K. and Benton, T. G. 2018. Early adoption of Conservation Agriculture practices: understanding partial compliance in programs with multiple adoption decisions. *Land Use Policy* 70, 27–37. doi:10.1016/j.landusepol.2017.10.001.

White, K. D. 1970. Fallowing, crop rotation, and crop yields in Roman Times. *Agricultural History* 44(3), 281–90. Available at: http://www.jstor.org/stable/3741455.

Chapter 12

Social benefits of Conservation Agriculture systems

Rafael Fuentes Llanillo, Tiago Santos Telles and Dimas Soares Junior, Agricultural Research Institute of Paraná State (IAPAR), Brazil; Sara Kaweesa, University of Natural Resources and Life Sciences (BOKU), Austria; and Anne-Marie B. Mayer, Independent Nutrition and Agriculture Consultant, UK

1 Introduction

Conservation Agriculture (CA), more than a soil conservation method, is a complete set of farming systems that follow the three principles: minimum soil disturbance, soil permanently covered with straw and/or live plants, and crop rotation/association and diversity of species. CA is not only the no-till system of annual crops, but also permanent pastures sown directly and without overgrazing, fruit tree crops and vegetables with permanent soil cover, agroforestry and forest planting without soil disturbance and the protection of water sources. The ideal condition should be the entire planning and management of watersheds and catchments with CA farming systems. This concept was developed by FAO based on the successful experiences of no-tillage (NT) or zero-tillage (ZT) farming in Brazil and Latin America (FAO, 2001). The main driver of CA development in these areas was the struggle against soil erosion after the 1970s by pioneer farmers and pioneer research. Environmental and economic impacts were largely described in literature and in the previous chapters of this book. Despite the small number of specific

http://dx.doi.org/10.19103/AS.2019.0049.12

articles and reports on the subject, social benefits of CA systems cover several areas such as farmers' organization, soil governance and water conservation, social capital and community development, nutritional aspects, among others.

The successful adoption of CA in South America had a strong effect on farmers' organization creating a synergic and supportive process of communication farmer-to-farmer. This process was conducted for the creation of CA farmers' organizations, initially in that region, and after that worldwide. This may be the strongest social benefit of CA.

CA takes time to be adopted at landscape scale and as social effects at the community level can only be expected after a wider uptake, only in regions with massive adoption do social benefits become more apparent. With massive adoption soil and water governance may be expected. This is the case of Itaipu Hydroelectric Plant in the border between Brazil and Paraguay where CA is the key factor to keep and increase the operating life time of the reservoir.

CA is also responsible for varied socioeconomic benefits that represent an important contribution to the sustainability of agricultural production, bringing with it considerable environmental, social and economic benefits that are described in detail ahead.

Improvement in social capital and promotion of community development are also social benefits of CA mainly in Africa where the three principles can function as drivers of mindset changes that are described through some examples in this chapter.

Finally the improvement of nutritional aspects through the effects of diversification and extension of production seasons provoked by CA in Africa is also discussed.

2 Farmers' organization

Perhaps one of the major social benefits of CA is the effect on how farmers are getting organized. In the first days of CA in Brazil the farmer Herbert Bartz in 1971 was inspired by two diligent, generous and technically well-prepared persons that he visited in the United States, Harry Young Jr. and Shirley Phillips, that were doing NT in Christian, Kentucky, since 1962. By their side, they were inspired by the book 'Ploughman's Folly' of Edward Faulkner, published in 1945, that was the first to warn against the harmful effects of using ploughs to prepare agricultural land. Harry Young Jr. was a pioneer farmer with graduation in Agronomy and also teacher at Kentucky University and Shirley Phillips was an associate agricultural agent, state specialist in field crops, associate director for extension of Kentucky University and also he was a defender of the system. Both were very open and showed all the steps and necessary equipment to implement the new system (Santin, 2018). The adoption of NT in Latin America began in 1972 when Herbert Bartz did the first soybean crop under NT in commercial scale in Rolândia, Northern Paraná State, Brazil. Since the very first moment Herbert used the same solidarity he received in the United States to share and communicate his findings and the new knowledge that had arisen with the new system. In 1976, after visiting and knowing the ideas of the 'crazy German', Franke Dijkstra and Manoel Henrique (Nonô) Pereira began the adoption in the Natural Fields Region in Southern Paraná (Campos Gerais). Few research data and information about NT were available in Brazil at that time, so the farmers decided in 1979 to create the Earthworm Club (Clube da Minhoca) in Ponta Grossa, Paraná State, to discuss the practical problems and solutions regarding NT among farmers, extensionists and interested scientists and professionals. The capacity of the system to stop erosion in addition to the enormous economic advantage of reducing costs of tillage launched a strong farmer-to-farmer communication process.

Based on the example of the Earthworm Club other farmers' organizations were created in the State of Rio Grande do Sul, but these were called Friends of the Land Clubs (Clube Amigos da Terra). The objectives were to promote awareness, training and technical development for adoption and refinement of ZT/CA (Borges, 1993). After the successful experience in Rio Grande do Sul, the concept of Friends of the Land Clubs was extended to other Brazilian regions to help farmers promote and develop new technologies related to herbicide management, spraying techniques, fertilization and liming, erosion control and rotation of annual crops with pasture. These farmers' organizations propelled Brazil to becoming a world leader in tropical and subtropical ZT/CA systems (Freitas, 2001). In the Brazilian Savannah, the Farmer Association for the Cerrado Region (Associação de Plantio Direto no Cerrado – APDC) was organized in 1992, headquarters in Goiânia (State of Goiás), with the objective to promote ZT/CA following two successful short courses organized by John Landers (with agribusiness support). It later expanded its network to 47 'Clube Amigos da Terra', foundations and other farmer organizations (Freitas and Landers, 2014). In the same year of 1992 a national CA farmers' organization was created, the Brazilian No-Till Farmers Federation (Federação Brasileira de Plantio Direto na Palha – FEBRAPDP), in order to articulate the action of the former Friends of the Land Clubs created to support NTS expansion. The main founders of FEBRAPDP were Herbert Bartz, Manoel Nonô Pereira and Franke Dijkstra, the three pioneer farmers who started this synergic and supportive process of farmer-to-farmer communication. Since the 1970s they travelled voluntarily, receiving nothing in financial terms, all over Brazil and other South American countries to disseminate No-Till, talking with farmers and professionals in meetings and with students in classes, but also opened their farms for visits of people from all over the world.

The social benefits were huge and changed agriculture in Brazil, South America and many other countries in the world through the action of organized farmers. The willingness of innovative farmers to share their lessons and experience with other farmers contributed enormously to the acceptance of NT agriculture systems (Dumanski et al., 2014). This social movement of farmers' organizations to promote No-Till and CA was much wider and reached other countries and continents. According to Peiretti (2012), based on the need to support each other to cope with the changes of magnitude that were to be carried out, producers from different regions of the world spontaneously began to gather and enhance their interaction. The foundation and consolidation of national civil and non-profit institutions, created by farmers, had a great impact, both economic and social, through the facilitation and progress of agriculture on its path of paradigm shift towards CA. The phenomenon first took place informally and mainly at local levels but then was consolidated and grew to form national associations and confederations of international scope.

There is a large number of examples of this type of organization around the world. In Latin America there was the creation at the national level of the Argentinian No-Till Farmers' Association (AAPRESID, n.d.) in 1989, of the Uruguayan Association of No-Till Farmers (AUSID, n.d.) in 1991, of the Brazilian No-Till Farmers' Federation (FEBRAPDP, n.d.) in 1992, and finalizing this first period there was the creation of a continental organization, the American Confederation of Farmers Organizations Toward a Sustainable Agriculture (CAAPAS, n.d.). Later, the CAAPAS included the Paraguayan No-Till Farmers' Federation (FEPASIDIAS, n.d.) created in 1993, the Soil Conservation Society of Chile – SOCOSCHI then transformed into the Center for Development of Zero Tillage CEDECELA of Chile, the Mexican Association of Conservation Tillage – AMLC, the Oil Seeds and Wheat Farmers Association – ANAPO from Bolivia, the Conservation Technological Information Center – CTIC, West Lafayette, Indiana, USA and the Soil Conservation Council of Canada – SCCC. Even though farmers started organizing No-Till associations at the local level at the end of 1970s, the end of the 1980s until the mid-1990s was the ripening period for creating national and continental institutions in America, mainly Latin America (Peiretti, 2003). In Oceania, the Western Australian No-Tillage Farmers Association – WANTFA was created in 1992 and the Conservation Agriculture Alliance of Australia and New Zealand – CAAANZ. A little bit later, by global expansion of the concept, similar institutions were created in other continents, such as the African Conservation Tillage Network (ACT, n.d.) in Africa created in 1998 and the European Conservation Agriculture Federation (ECAF, n.d.) in Europe created in 1999. In South Asia, Bangladeshi farmers first established Conservation Agriculture Service Providers Association (CASPA) in 2014.

In 2009, a Global Conservation Agriculture Community of Practice (CA-CoP) platform was launched by the FAO to help the spread of CA knowledge under the auspices of the FAO, and this platform has remained very active all these years in increasing support, understanding and adoption of CA globally. In

2015, an attempt was made to form a Global Conservation Agriculture Network - GCAN - as an unofficial assemblage of CA experts and farmers organizations from around the world. It has no physical building or headquarters. It was formed by a group of CA supporters from different countries attracted to the same need of increasing understanding and adoption of CA around the world.

This social process did not stop with the creation of regional, national and continental farmers' organization and the promotion of regional events for promotion of CA adoption and spreading. A worldwide Conservation Agriculture Community was formed bringing together farmers, policy makers, scientists, international organizations and financing organisms. The first World Congress on Conservation Agriculture was organized in Madrid (2001) and a first CA Declaration was released (1st WCCA Declaration, 2001). After that, the second WCCA was held in Brazil (2003), third WCCA in Kenya (2005), fourth WCCA in India (2009), fifth WCCA in Australia (2011), sixth WCCA in Canada (2014) and seventh WCCA in Argentina (2017). The eighth WCCA will be held in Switzerland in 2020. World Congresses are important tools to enhance the adoption and spread of CA in all continents. Other than world congresses, there are a large number of events related to CA and NT that are regularly organized at sub-national, national and regional levels.

With the development of CA around the world, this knowledge, spread and conducted mainly by farmers' organizations, helped to promote a social and cultural change of mindset. Today there is a new social valuation about the increasing importance of straw and stubble making the permanent soil cover in opposition to former values of nude-milled soils in conventional tillage when straw was trash or at maximum waste for animal fodder. Presently, permanent soil cover and undisturbed root systems are of considerable worth. Higher levels of organic matter are synonymous of soil health, and carbon sequestration to mitigate climate change, are very important social values presently.

3 Social governance and water conservation

Another good example of social benefit is soil governance and water conservation obtained with the wide adoption of CA at landscape and regional scale. This is the situation of Parana 3 basin at the border between Brazil and Paraguay around the reservoir of Itaipu Binational, the world biggest hydroelectric operating power plant. On the Brazilian side this basin represents 785 thousand ha and in the Paraguayan side other 720 thousand ha. It is a very powerful case study.

Itaipu started functioning in 5 May 1984 together with a strong intensification of Brazilian and Paraguayan agriculture of annual crops as soybeans, maize and wheat based on intensive soil tillage with ploughs and harrows. After a five-year period, erosion was considered a nightmare that could limit the initial expected lifetime of 172 years (Mendes, 2017). In 1992 after eight years of operation a prediction study of the lifetime of the reservoir showed a life expectancy of 131 years (Mendes, 2017). It showed the disastrous impact of conventional tillage agriculture in the area. A strong policy of protection of riparian areas with restoration of natural vegetation strips, varying from 50 to 300 m, was adopted to prevent the access of sediments to water. Also there was support for terracing and other mechanical soil conservation practices, as well as rural roads maintenance and wells to supply water for sprayers without pollution. But the most important change was the growing adoption of the NT System in the area in the early 1990s.

From 1997 to 2006 an agreement between Itaipu and IAPAR – Agricultural Research Institute of Parana State – studied the quality standard of NT System in the area and proposed certain parameters to be incentivized and supported at farm level. With these parameters, between 2009 and 2012 the Brazilian No-Till Farmers' Federation and Itaipu Binational developed a Participative Quality Index (IQP) of NT System (Roloff et al., 2012).

Since 2003 Itaipu Binational has established a comprehensive programme called Cultivating Good Water. CA together with other soil and water conservation measures is the core of the programme. It is a set of 20 projects, 65 actions and more than 2000 partners involving support to NT System quality improvement, to soil conservation practices, to riparian forest planting and building of fences to protect the riparian forests and river sources, to implement construction of wells for water supplies for agricultural use, to adequate network of roads and its maintenance among other social-environmental actions at the watershed level.

Socio-environmental actions management and execution are participative. Actions are developed through partnerships between Itaipu and both public and private institutions, which are social and environmental entities. The organization and execution of these processes are supported by management committees. Itaipu works together with diverse partners among municipalities, cooperatives, professional organizations, farmers, NGOs, government agencies and civil society. Many of these partners are organized into Catchment Management Committees and participate in the actions developed within the territory. It is a case of success in soil governance (ITAIPU, n.d.). With the huge adoption, the Itaipu reservoir lifetime estimates reached 160 years in 2006, 169 years in 2010 and 168 years in 2014 (Mendes, 2017) showing the capacity of CA in preserving not only soils but also water.

4 Varied socioeconomic benefits

CA represents an unprecedented technological advance that promotes economic growth of the agricultural sector without threatening the environment. CA encapsulates a cohesive collection of technological processes that are geared towards conserving, improving and optimizing resource use through natural means, including integrated management of soil, water and biodiversity that is compatible with external inputs. The adoption of CA principles contributes to improvements in biodiversity, soil fertility, erosion control and carbon sequestration by deploying practices focussed on maintaining and increasing soil organic matter content. CA thus represents an important contribution to the sustainability of agricultural production, bringing with it considerable environmental, social and economic benefits.

CA can be considered as one of the most noteworthy contributors to agricultural development in recent decades. Especially in ZT system of annual crops, the harvest-plant process seeks to minimize the interval between harvest and planting, thus maximizing the number of harvests per year. Furthermore, these practices are in line with methods of soil and water conservation, as well as with integrated management of pests, diseases and weeds. With its efficiency confirmed by documented improvements in soil quality and natural resource conservation, CA enables crop productivity gains, and thus enhances the profitability of agricultural production.

The benefits of CA may include: increased productivity and income; reduction of up to 70% in fuel energy or manual work; up to 50% less fertilizer use; 20% or more reduction in the use of pesticides and herbicides; about 30% less need for water and reduced cost in agricultural machinery. With CA, it is possible to improve adaptation to climate change in cropping systems, due to improved soil moisture ratios and at the same time obtaining higher carbon sequestration and lower greenhouse gas emissions, particularly CO_2, N_2O and CH_4 (Friedrich et al., 2009). Due to increased water infiltration and lower runoff and soil erosion, CA also reduces flood risk, increases water quality and quantity of resources, and can reduce maintenance costs of public infrastructures. As a result, farmers have greater profits. Otherwise, they would not be practicing CA on more than 180 million ha of agricultural land worldwide, nor would it extend at an annual rate of over 10 million ha in last ten years (Kassam et al., 2018). Compared to conventional farming, CA not only offers a way to sequester carbon and intensify production in an ecologically sustainable way, it is considerably less costly and economically, environmentally and socially beneficial. CA uses the entire ecosystem and all natural biodiversity including soil microorganisms to improve soil health and productive capacity, thus protecting crops against weeds, insects and pathogens. Root crops, such as white potatoes, sweet potatoes, cassava, peanuts and sugar beets, are also successfully grown in CA as they are harvested with minimal disturbance of the soil using appropriate harvesting equipment or cultivation practices (Kassam et al., 2013).

Adoption of CA increases farm profitability over the long term by reducing costs associated with farm operations, inputs and labour. By reducing the need for mechanical manipulation of the soil, CA improves the longevity of mechanical implements, reduces repair and fuel costs, and lessens depreciation. One of the principal benefits of CA, and what makes it especially popular among farmers, is that it creates savings not only in terms of money, but also in terms of time and quality of life (Jat et al., 2012).

CA has also reduced production costs in the rice and wheat cultivation systems of South Asia by approximately US$60 per ha, largely as a result of lower fuel, labour and water use. These savings occur because, under CA, planting is generally completed within a single operation, freeing workers to complete other tasks. Farmers who adopted the CA package in India and Pakistan enjoyed similar savings and benefits (Hobbs, 2007).

For instance, cover crops planted as part of the CA package reduce the labour costs and fertilizer required for subsequent cash or main crops. Use of legume cover crops in the rotation can reduce the need for nitrogen fertilizers in the next crop cycle and can increase subsequent yields by approximately 21% (Miguez and Bollero, 2005).

Experiences from various locations in the Indo-Gangetic plains have shown that with NT technology, farmers were able to save on land preparation costs by about 2500 rupees (US$41.7) per ha and reduce diesel consumption by 50–60 litres per ha (Sharma et al., 2005).

The adoption of CA is highest in the Northwest of North America, in the Southern cone of South America and in Western Australia with adoption levels above 50%. Between 2008–09 and 2015–16 the area under CA in the North American region had increased 57.9% from 40.0 million ha to 63.2 million ha. A considerable part of this area is cropped in Canada, where CA was practiced at 19.9 million ha in 2015–16. In the long term the adoption of CA, mainly in the western provinces, has resulted in benefits, including the absence of dust storms and greater biodiversity. The environmental services provided through CA are increasingly recognized, for example, through carbon payment schemes such as in Alberta. In the United States, CA adoption in 2008–09 was 26.5 million ha (21.5% of cultivated area) and in 2015–16, it was 43.2 million ha (35.1% of planted area), despite the long experience with no-till farming (Kassam et al., 2015, 2018; Derpsch et al., 2010; Duiker, 2017). Although NT in the United States is still mainly applied to certain rotational practices, awareness of crop rotation and cover crop rotation as well as the additional benefits of permanent no-till systems is growing. In the United States, much of CA farming has been using maize, soybeans, wheat and canola crops, but more recently cotton systems (Duiker, 2017).

CA systems help Africa's resource-poor farmers maintain livelihoods with sustainability. There, the under CA is still relatively small, but there is a trend of adoption and a CA movement of about two million small farmers, including medium and large-scale farmers, with a total area of about 1.5 million ha in

2015–16. Since 2008–09, CA has spread further in countries such as Kenya, Madagascar, Malawi, Mozambique, Tanzania, Zambia and Zimbabwe. The total area of CA in Africa in 2015–16 was over 1.5 million ha, an increase of about 211% compared to 2008–09 (0.48 million ha). CA is recognized to increase the production of food with less negative effects on the environment and energy costs, and to result in the development of locally adapted technologies in accordance with the principles of CA (Thierfelder et al., 2018; Kassam et al., 2018).

Since 2008–09 the area of CA for annual crops in Europe has increased from 1.6 million ha to 2.0 million ha in 2013–14 and by 30% to 3.5 million ha in 2015–16, an increase of 127.4%. In 2008–09, the area under CA was known for 11 countries, but in 2013–14, it increased to 15 countries and in 2015–16 to 29 countries. Since 1999, ECAF and its members of the national associations, comprising many farmers, have promoted systems in Europe, with significant adoption in Spain, Italy, Finland, France, Romania, Poland, United Kingdom, Switzerland and Germany (Kassam et al., 2018).

The agricultural decision-making process includes an important ethical dimension, given that individuals in the present must consider the intertemporal consequences of their actions on future generations. CA satisfactorily addresses these ethical considerations by appropriately balancing the capacity to produce foods, fibres and energy with the well-being of both current and future inhabitants of the earth. The production of these valuable goods represents one of the main social benefits of CA, which has the potential to satisfy the consumption requirements of a growing population while simultaneously improving incomes for rural producers. In this sense, the social advantages of this conservation practice are felt both as improvements in collective (societal) and individual (producer) welfare. Adequate, effective natural resource management of this kind guarantees both high levels of productivity and profitability, which are indispensable ingredients of economic development and important incentives to agricultural production. In other words, the benefits of CA satisfy the needs of society, as well as those of rural producers.

Farmers, for the most part, are conscious of the need to adopt CA systems and practices, especially sustainable soil and water management practices, to preserve and increase the value of their greatest asset: the land. Even though CA is widely recognized within the area of sustainable agriculture, the environmental services provided by farmers who adopt these practices often are underappreciated. It is fundamental that producers and society at large recognize and value the true contributions of sustainable agriculture, especially considering that decisions concerning the use, management, and conservation of soils and waters are based on expectations, and are taken within a context of biophysical, institutional and economic risks and uncertainties. For all of these reasons, appropriately valuing the positive contributions of CA is essential.

If farmers were obliged to deliver all the benefits achievable through CA, it is likely that adoption of this technological package would approach 100%. Inertia, which slows the diffusion of CA practices, is largely due to the costs involved in purchasing the necessary equipment and to the insecurity involved in shifting to new and unknown practices. Out of these there are no other obstacles standing in the way of the diffusion of this promising set of techniques.

5 Social capital and community development

CA is achievable by all kind and size of farmers (smallholders, family farmers and big commercial farmers), especially improving the quality of life of smallholders due to less labour intensity, higher productions and more profits. This aspect of less labour and time requirement, through the removal of operations of land preparation and the reduction of weeding operations, release time for men and women to take more care of their families, diversify production with high-income activities and even add value transforming their own products. Any of these strategies permit higher life standard levels. However, low adoption rates, particularly in Africa, do not seem to reflect this success on a wide scale as yet.

The study of Kaweesa et al. (2018) explored the extent of practice of CA principles and related them to the socioeconomic status of the population in Mid-Northern Uganda. Other social benefits of CA were documented as labour reduction and improved food security. The challenges faced when practicing CA were the lack of follow-up for tracking progress, little interaction with extension officers, for instance, when they needed to ask questions, the need for further training and little equipment. Most respondents were influenced by profit expectations and their available cash, and hardly relied on loans or even donations. The major reason for adoption of CA in this region was attributed to the information gained through training provided by Rural Enterprise Development Services (REDS), a nongovernmental organization among promoting agencies that spread CA information through given funding cycles. It is evident that access to information played a key role in enabling uptake of CA; for instance some women farmers adopted the technique from their peers and others copied their neighbours. There is empirical evidence from the focus group discussion showing that relationships between women farmers were strong as they offered support to the members in the group and in fact those that were disadvantaged such as the widows and elderly women in the community. There were reports of women being able to support their families from incomes accrued from CA particularly when their spouses were not in position to do so for instance when they lost jobs or were altogether not supportive of their family roles. The knowledge that farmers gained about CA enabled them to understand why and how to practice CA, unlike other programmes that simply required them to follow instructions. Farmers had the opportunity to dialogue with the private extension workers during the programme. Evidence shows that farmers still practiced CA three years after the project had long ended, which was an exciting experience throughout the region.

In Africa, CA adoption is generally dependent on a given context as areas even within the country are unique and may greatly differ; the factors influencing non-adoption are needed to be studied bearing this and other factors in mind. These include complexities within African smallholder-farming systems, unfavourable institutional policy approaches (Andersson and D'Souza, 2014), lack of appropriate extension (Nicol et al., 2015), limited access to credit and underdeveloped input and output markets (Baudron et al., 2015; Thierfelder et al., 2016a,b), competition for crop residues for use as animal fodder (Corbeels et al., 2014; Giller et al., 2009, 2015), the approach of CA promotion as a package (Rusinamhodzi et al., 2011; Rusinamhodzi, 2015; Giller et al., 2011) and the inappropriateness of the technology to the target group (Baudron et al., 2012; Nkala, 2012; Brouder and Gomez-Macpherson, 2014). On the other hand, there are factors that have enabled farmers to take up CA, for instance, peer influence (Kassie et al., 2015; Brown et al., 2018; Bell et al., 2018) and information availability (Ngwira et al., 2014).

There are success CA stories in several African countries. Overall though, the slow rate of CA adoption could also merely reflect the old sectorial problems rather than CA being a misfit for the African farmers or even the notion that the methodology of CA adoption studies is flawed as some studies seem to suggest. However, some of these old problems that continue affecting the agricultural sectors in general, unfortunately apply to most agricultural innovations including CA. These include little up-to-date information technology and thus forcing farmers to rely on ancient farming knowledge, defective agricultural extension system, low access to genuine quality seeds and other inputs like storage facilities needed for post-harvest handling, inaccessibility of financial credit and lack of basic services for rural small holder farmers, low infrastructure such as a poor road network and lack of affordability of appropriate small CA machinery among others.

6 Food security and nutritional aspects

Nyanga (2012) reported improvements in household food security specifically related to growing and consumption of pulses in Zambia. Comparing pulse production of CA adopters and non-adopters, there was higher production and also consumption of legumes among adopters. An early green harvest also contributed to reported food security among women farmers. The legumes grown were groundnuts, cowpeas, soy beans and other beans with those legumes that are part of the diet being prioritized. Women were also selling pulses to provide income.

Wagstaff and Harty (2010) in Zimbabwe reported improvements to household food security assessed using the Household Economy Approach. CA contributed to meeting annual food needs better than in the comparison group and also the diversity of food production for own needs was improved with more emphasis on legumes.

Mayer et al. (2016) conducted research on the potential of CA to improve nutrition among CA adopters in Zambia. It has shown that CA has the potential to improve nutrition of children and also for other demographic groups but evidence of this is scarce. Several pathways from agriculture to nutrition could have a role to play, specially related to women's workload, diversification of diets related with income improvement and child caring practices. Women reported that time saved from a switch to CA enabled them to breastfeed more frequently and prepare special foods for young children, thus offering potential for improvement to infant and young child feeding. Nutrition education is also needed to fill a gap in nutrition knowledge, particularly infant and young child feeding practices and utilization of the foods that are now locally available through CA.

CA projects of the Worldwide Organisation in Zambia and Malawi have been evaluated for impact on gender dynamics in the households. A project evaluation in Malawi and Zambia concluded that 'the benefits of CA are more significant for women as CA has enabled them to prepare their land early and plant on time, which in turn has led to increased yields, an earlier harvest, a shorter hunger gap, and higher household food security'. Women reported reduced labour and labour intensity, especially for weeding, the time saved was then used on other crops, vegetable gardens or other income generating activities, extra income to pay school fees, buy school clothes, school books, household items and improve their children's nutrition especially with the inclusion of groundnuts and soybeans in the diet and enable them to buy productive assets such as livestock or fertilizers. The women are now more self-reliant and have learned new skills and gained knowledge from the CA training which empowers them to improve their own food production and decision making (Reid and Chikarate, 2013).

7 Conclusion

Pioneer farmers that initially developed CA were inspired by the struggle against erosion and soil depletion, thinking together in individual and collective environmental, economic and social effects. Presently in scaling up the adoption at farmer level, the strongest benefits that conduct people to CA are the cash or labour savings related to tillage and weeding operations among others. It is influenced mainly by improving profit and/or life standard expectations more than by environmental issues.

However large-scale adoption provoked and provokes a lot of social benefits such as farmers' organization, soil governance and water conservation, social capital and community development, food security and nutritional aspects. Many additional benefits, mainly linked to the preservation of the human race on Earth, could be addressed. This is the case of mitigating the effects of climate change, conserving the natural capital of soils, forests and water and even the potential of CA to meet growing world demand for food. These themes remain for further reflection.

8 Where to look for further information

The Food and Agriculture Organization of United Nations – FAO provide on its site a complete section on the matter and specific papers on socioeconomic impacts of CA available at www.fao.org/conservation-agriculture/en/.

The main author of this book Dr Amir Kassam keeps the moderation of Global CA-CoP CONSERVATION AGRICULTURE COMMUNITY OF PRACTICE that can be accessed at his e-mail: amirkassam786@googlemail.com or at http://www.fao.org/conservation-agriculture.

All authors of this chapter are available to provide more information as Rafael Fuentes Llanillo (rfuentes@iapar.br), Tiago Santos Telles (telles@iapar.br), Dimas Soares Junior (dimasjr@iapar.br), Sara Kaweesa (sara.kaweesa@boku.ac.at) and Anne-Marie B. Mayer (abmayer17@gmail.com) in your own email addresses.

All CA Farmers Organizations can be accessed on their sites.

9 References

1st World Congress on Conservation Agriculture Declaration. 2001. Available at: https://febrapdp.org.br/download/publicacoes/Declaracao_do_1_WCCA_versao_ingles.pdf.

AAPRESID. n.d. Available at: http://www.aapresid.org.ar/.

ACT. n.d. Available at: http://www.act-africa.org/.

Andersson, J. A. and D'Souza, S. 2014. From adoption claims to understanding farmers and contexts: a literature review of Conservation Agriculture (CA) adoption among smallholder farmers in southern Africa. *Agriculture, Ecosystems and Environment* 187, 116–32. doi:10.1016/j.agee.2013.08.008.

AUSID. n.d. Available at: http://www.ausid.com.uy/.

Baudron, F., Andersson, J. A., Corbeels, M. and Giller, K. E. 2012. Failing to yield? Ploughs, Conservation Agriculture and the problem of agricultural intensification: an example from the Zambezi Valley, Zimbabwe. *Journal of Development Studies* 48(3), 393–412. doi:10.1080/00220388.2011.587509.

Baudron, F., Sims, B., Justice, S., Kahan, D. G., Rose, R., Mkomwa, S., Kaumbutho, P., Sariah, J., Nazare, R., Moges, G. and Gérard, B. 2015. Re-examining appropriate mechanization in Eastern and Southern Africa: two-wheel tractors, conservation agriculture, and private sector involvement. *Food Security* 7(4), 889–904. doi:10.1007/s12571-015-0476-3.

Bell, A. R., Cheek, J. Z., Mataya, F. and Ward, P. S. 2018. Do as they did: peer effects explain adoption of conservation agriculture in Malawi. *Water* 10(1), 51. doi:10.3390/w10010051.

Borges, G. O. 1993. Resumo histórico do plantio direto no brasil. In: Embrapa, CNPT, FUNDACEP e Fundação ABC. *Plantio Direto no Brasil*. Aldeia Norte/FUNDACEP/FECOTRIGO/FUNDAÇÃO ABC, Passo Fundo, Brasil, pp. 13–8.

Brouder, S. M. and Gomez-Macpherson, H. 2014. The impact of conservation agriculture on smallholder agricultural yields: a scoping review of the evidence. *Agriculture, Ecosystems and Environment* 187, 11–32. doi:10.1016/j.agee.2013.08.010.

Brown, B., Nuberg, I. and Llewellyn, R. 2018. Further participatory adaptation is required for community leaders to champion conservation agriculture in Africa. *International Journal of Agricultural Sustainability* 16(3), 286–96. doi:10.1080/14735903.2018.1472410.

CAAPAS. n.d. Available at: https://www.linkedin.com/company/confederaci%C3%B3n-de-asociaciones-americanas-para-la-agricultura-sustentable-caapas.

Corbeels, M., de Graaff, J., Ndah, T. H., Penot, E., Baudron, F., Naudin, K., Andrieu, N., Chirat, G., Schuler, J., Nyagumbo, I., Rusinamhodzi, L., Traore, K., Mzoba, H. D. and Adolwa, I. S. 2014. Understanding the impact and adoption of conservation agriculture in Africa: a multi-scale analysis. *Agriculture, Ecosystems and Environment* 187, 155–70. doi:10.1016/j.agee.2013.10.011.

Derpsch, R., Friedrich, T., Kassam, A. and Li, H. 2010. Current status of adoption of no-till farming in the world and some of its main benefits. *International Journal of Agricultural and Biological Engineering* 3(2), 1–25.

Duiker, S. W. 2017. Planting green – a new cover crop management technique. *Field Crop News*. Available at: https://extension.psu.edu/planting-green-a-new-cover-crop-management-technique.

Dumanski, J., Reicosky, D. C. and Peiretti, R. A. 2014. Preface. Global pioneers in soil conservation: common elements and lessons learned. *International Soil and Water Conservation Research* 2(1), 1–4. Available at: http://www.ecaf.org/. doi:10.1016/S2095-6339(15)30018-6.

ECAF. n.d. Available at: http://www.ecaf.org/.

FAO. 2001. No-till agriculture for sustainable land management: lessons: learned from the 2000 Brazil study tour. Draft for TCI Occasional Paper Series. FAO Investment Centre Divison, Rome.

FEBRAPDP. n.d. Available at: https://febrapdp.org.br/.

FEPASIDIAS. n.d. Available at: http://www.fepasidias.org.py/.

Freitas, P. L. 2001. Organization of a technology platform for cooperative research and development in zero-till agriculture in Brazil. Available at: www.embrapa.br/plantiodireto/ (accessed on 30 October 2018).

Freitas, P. L. and Landers, J. N. 2014. The transformation of agriculture in Brazil through development and adoption of zero tillage Conservation Agriculture. *International Soil and Water Conservation Research* 2(1), 35–46. doi:10.1016/S2095-6339(15)30012-5.

Friedrich, T., Kassam, A. H. and Shaxson, F. T. 2009. Conservation Agriculture. In: *Agriculture for Developing Countries. Science and Technology Options Assessment (STOA) Project*. European Technology Assessment Group, Karlsruhe, Germany.

Giller, K. E., Witter, E., Corbeels, M. and Tittonell, P. 2009. Conservation agriculture and smallholder farming in Africa: the heretics' view. *Field Crops Research* 114(1), 23–34. doi:10.1016/j.fcr.2009.06.017.

Giller, K. E., Corbeels, M., Nyamangara, J., Triomphe, B., Affholder, F., Scopel, E. and Tittonell, P. 2011. A research agenda to explore the role of conservation agriculture in African smallholder farming systems. *Field Crops Research* 124(3), 468–72. doi:10.1016/j.fcr.2011.04.010.

Giller, K. E., Andersson, J. A., Corbeels, M., Kirkegaard, J., Mortensen, D., Erenstein, O. and Vanlauwe, B. 2015. Beyond conservation agriculture. *Frontiers in Plant Science* 6, 870. doi:10.3389/fpls.2015.00870.

Hobbs, P. R. 2007. Conservation agriculture: what is it and why is it important for future sustainable food production? *The Journal of Agricultural Science* 145(2), 127–37. doi:10.1017/S0021859607006892.

ITAIPU. n.d. Available at: https://www.itaipu.gov.br/en/the-environment/governance.

Jat, R. A., Wani, S. P. and Sahrawat, K. L. 2012. Conservation agriculture in the semi-arid tropics: prospects and problems. *Advances in Agronomy* 117, 191–273.

Kassam, A., Basch, G., Friedrich, T., Shaxson, T. F., Goddard, T., Amado, T., Crabtree, B., Hongwen, L., Mello, I., Pisante, M. and Mkomwa, S. 2013. Sustainable soil management is more than what and how crops are grown. In: Lal, R. and Stewart, R. A. (Eds), *Principles of Soil Management in Agro-ecosystems*. Advances in Soil Science. CRC Press, Boca Raton, pp. 338–87.

Kassam, A., Friedrich, T., Derpsch, R. and Kienzle, J. 2015. Overview of the worldwide spread of conservation agriculture. *Field Actions Science Reports* 8.

Kassam, A., Friedrich, T. and Derpsch, R. 2018. Global spread of Conservation Agriculture. *International Journal of Environmental Studies* 76(1), 29–51. doi:10.1080/00207233.2018.1494927.

Kassie, M., Teklewold, H., Jaleta, M., Marenya, P. and Erenstein, O. 2015. Understanding the adoption of a portfolio of sustainable intensification practices in eastern and southern Africa. *Land Use Policy* 42, 400–11. doi:10.1016/j.landusepol.2014.08.016.

Kaweesa, S., Mkomwa, S. and Loiskandl, W. 2018. Adoption of conservation agriculture in Uganda: a case study of the Lango subregion. *Sustainability* 10(10), 3375, doi:10.3390/su10103375.

Mayer, A. B., Smit-Mwanamwenge, M. and Whal, C. 2016. The potential of nutrition-sensitive Conservation Agriculture: lessons from Zambia. *Field Exchange* 51(January). Available https://www.ennonline.net/fex/51/nutsensitiveconservationagrzambia.

Mendes, A. B. 2017. Monitoramento sedimentométrico nos principais afluentes ao Reservatório de Itaipu. In: Coli, A. and Dias, P. (Orgs.), *O Setor Elétrico e o Meio Ambiente*. Synergia: FMASE, Rio de Janeiro, pp. 316–29. Chapter 19. ISBN 978-85-68483-62-6.

Miguez, F. E. and Bollero, G. A. 2005. Review of corn yield response under winter cover cropping systems using meta-analytic methods. *Crop Science* 45(6), 2318. doi:10.2135/cropsci2005.0014.

Nicol, A., Langan, S., Victor, M. and Gonsalves, J. 2015. *Water-Smart Agriculture in East Africa*. IWMI, Colombo, Sri Lanka.

Ngwira, A., Johnsen, F. H., Aune, J. B., Mekuria, M. and Thierfelder, C. 2014. Adoption and extent of conservation agriculture practices among smallholder farmers in Malawi. *Journal of Soil and Water Conservation* 69(2), 107–19. doi:10.2489/jswc.69.2.107.

Nkala, P. 2012. *Assessing the Impacts of Conservation Agriculture on Farmer Livelihoods in Three Selected Communities in Central Mozambique*. University of Natural Resources and Life Science, Vienna, Austria.

Nyanga, P. H. 2012. Food security conservation agriculture and pulses; evidence from smallholder farmers in Zambia. *Journal of Food Research* 1(2), 120–38. doi:10.5539/jfr.v1n2p120.

Peiretti, R. A. 2003. The CAAPAS actions and the development of the MOSHPA. In: *Proceedings of the II World Congress on Conservation Agriculture* (vol. I). Foz de Iguazú, Brazil, 11–15th August 2003, pp. 127–8.

Peiretti, R. A. 2012. La Siembra Directa y las Instituciones de Productores. Sus roles frente al desafío de aumentar la producción agrícola global en forma sustentable. *20th AAPRESID Congress*, Rosario, Argentina, 08–10 August 2012.

Reid, J. and Chikarate, J. 2013. Final evaluation of the Accenture funded project: effecting improvements in livelihoods through Conservation Agriculture Malawi and Zambia.

Roloff, G., Lutz, R. A. T. and Mello, I. 2012. *Metodologia Participativa para avaliação do Sistema de Plantio Direto implantado na Bacia Hidrográfica Paraná 3*. FEDERAÇÃO BRASILEIRA DE PLANTIO DIRETO & ITAIPU BINACIONAL, Foz do Iguaçu, Brasil, 36p. Available at: febrapdp.org.br/download/publicacoes/BOLETIM_TCNICO_02.pdf.

Rusinamhodzi, L. 2015. Tinkering on the periphery: labour burden not crop productivity increased under no-till planting basins on smallholder farms in Murehwa district, Zimbabwe. *Field Crops Research* 170, 66–75. doi:10.1016/j.fcr.2014.10.006.

Rusinamhodzi, L., Corbeels, M., Wijk, M. T., Rufino, M. C., Nyamangara, J. and Giller, K. E. 2011. A meta-analysis of long-term effects of conservation agriculture on maize grain yield under rain-fed conditions. *Agronomy for Sustainable Development* 31(4), 657–73. doi:10.1007/s13593-011-0040-2.

Santin, W. 2018. *O Brasil possível: a biografia de Herbert Bartz* (1st edn.). Author Edition, Londrina, Brazil, 240p. ISBN 978-85-922249-1-2.

Sharma, A. R., Singh, R. and Dhyani, S. K. 2005. Conservation tillage and mulching for optimizing productivity in maize-wheat cropping system in the outer western Himalaya region – a review. *Indian Journal of Soil Conservation* 33(1), 35–41.

Thierfelder, C., Matemba-Mutasa, R., Bunderson, W. T., Mutenje, M., Nyagumbo, I. and Mupangwa, W. 2016a. Evaluating manual conservation agriculture systems in southern Africa. *Agriculture, Ecosystems and Environment* 222, 112–24. doi:10.1016/j.agee.2016.02.009.

Thierfelder, C., Bunderson, W. T., Jere, Z. D., Mutenje, M. and Ngwira, A. 2016b. Development of conservation agriculture (CA) systems in Malawi: lessons learned from 2005 to 2014. *Experimental Agriculture* 52(4), 579–604. doi:10.1017/S0014479715000265.

Thierfelder, C., Baudron, F., Setimela, P., Nyagumbo, I., Mupangwa, W., Mhlanga, B., Lee, N. and Gérard, B. 2018. Complementary practices supporting conservation agriculture in southern Africa. A review. *Agronomy for Sustainable Development* 38(2). doi:10.1007/s13593-018-0492-8.

Wagstaff, P. and Harty, M. 2010. The impact of Conservation Agriculture on Food Security in 3 low veldt districts of Zimbabwe. *Trocaire Development Review*.

Chapter 13

Harnessing ecosystem services with Conservation Agriculture

Amir Kassam, University of Reading, UK; Emilio J. Gonzalez Sanchez, Universidad de Córdoba, Spain, European Conservation Agriculture Federation (ECAF), Belgium and Asociación Española Agricultura de Conservación. Suelos Vivos (AEAC.SV), Spain; Tom Goddard, Alberta Agriculture and Forestry, Canada; Li Hongwen, Conservation Tillage Research Centre, China Agriculture University, China; Ivo Mello, Instituto Rio Grandense do Arroz, Brazil; Saidi Mkomwa, African Conservation Tillage Network, Kenya; Francis Shaxson, Land Husbandry Group, Tropical Agricultural Association, UK; and Theodor Friedrich, Food and Agriculture Organization of the United Nations (FAO), Italy

1 Introduction

Societies everywhere benefit from the many resources and processes supplied by nature. According to the Millennium Ecosystem Assessment (MEA, 2005), these are collectively referred to as ecosystem services. They include the *provisioning* services of clean drinking water, edible and inedible biological products, processes that decompose and transform organic matter, carbon sequestration, biologically fixed nitrogen; *regulating* services that control air quality, soil erosion, pest and diseases, natural hazards, pollination; *supporting* services of soil formation, nutrient, carbon and water cycling, biodiversity, primary production, atmospheric circulation; and *cultural* services related to recreation, sacred and spiritual relationships with nature, particular sites or phenomenon (Fig. 1, MEA, 2005).

As the term implies, processes underlying the supporting services are part of the life-support systems and processes which operate in the natural world without the need of any intervention from human beings. If human beings did not exist on the planet, these supporting services would continue

http://dx.doi.org/10.19103/AS.2019.0049.13

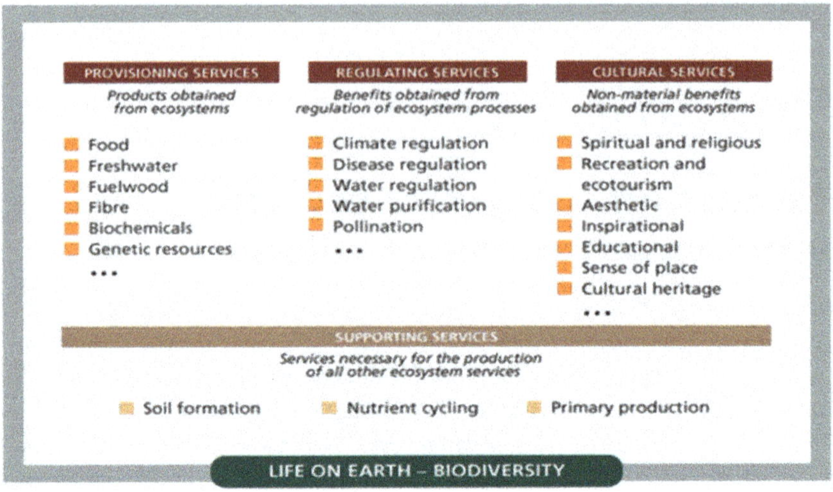

Figure 1 Ecosystem services from nature. Source: Millennium Ecosystem Assessment (2005).

as part of ecosystem functions and processes within the biosphere based on the incoming solar energy, the Earth's rotation, the 23° inclination of its axis, the Earth's elliptical orbit around the sun, land-sea configuration, general atmospheric circulation, local and long-distance ocean currents and daily tides. The regulatory and provisioning service processes are nested within the supporting ecosystem processes and help regulate flows, cycles and stocks of materials upon which all life on Earth depends for its survival, reproduction, evolution and development, and quality and comfort.

Humankind, by its very existence as part of the natural world and the need for it to survive and develop, has been intervening, with its knowledge, technology, action and ignorance, within the natural world in different ways. Consequently, it has been altering and utilizing ecosystem functions and natural resources to generate ecosystem provisioning services, including tangible products such as food, water, nutrients and so forth to meet its large and growing array of needs of biological and mineral products. Humankind has also been taking advantage of the supporting, regulatory and cultural services and benefitting from them in all sorts of ways to meet its social, environmental and spiritual needs. However, the relationship between humankind and the natural world has been largely that of exploitation, disconnection, destruction and degradation (Oldeman et al., 1991; Pretty, 2002; MEA, 2005; Vlek et al., 2010; FAO, 2011a; Juniper, 2013; Nkonya et al., 2016).

Interventions from humankind have resulted in altered ecosystems and, in many cases, in ecosystem degradation, as noted by the Millennium Ecosystem Assessment (MEA, 2005). Some 89% of our ecosystems are degraded or severely degraded. Only about 11% of ecosystems are functioning adequately. Humankind's interventions have led to loss in the natural land resource base (including soil, water, biodiversity, atmosphere etc.).

In the case of agricultural land use, until the mid-1990s and the beginning of this century, relatively little direct attention was paid by governments, farmers and service providers, including research, education and development community, to the need to sustain ecosystem functions and processes to meet the needs of society. Little attention was also paid to the necessity of harnessing ecosystem services from agricultural lands, except for yields of biological products from plants and animals. Ecosystem functions and processes in agricultural landscapes operate at various nested and linked levels, from (i) the level of the production field, affecting soil health and functions as well as crop and agricultural productivity and outputs; (ii) to the farm level, affecting overall farm productivity potential and profitability; (iii) to the landscape and watershed level, affecting large-scale services to society such as clean water resources and control land degradation; (iv) to the regional level and beyond, affecting landscape and regional-level ecosystem functioning and services to societies and the natural world.

Only in recent years, it has become clear that in agricultural landscapes, provisioning ecosystem services to meet the demands of the society and the natural world can only be delivered efficiently and without risks of harming ecosystem functions and processes, including natural balances, when the related regulatory and supporting services are also allowed to operate effectively and efficiently. It is also now clear that this is not possible to achieve in mechanically tilled and disturbed agricultural landscapes. This is because mechanical tillage and tillage-based agriculture debilitates all the soil- and land-mediated supporting and regulatory functions and processes and hence the related provisioning services to meet societal and environmental needs (Montgomery, 2007; Kassam et al., 2013). For example, to maintain the quality of water resources at a particular site or within a watershed, water cycling must function normally. This would include the maximum capture of rainfall by the land surface upon which the drops fall, maximum infiltration, absorption and retention of the water in the soil profile, with the rest or any excess being drained to lower layers and into the groundwater and deeper aquifers, from where it would flow into streams and rivers in a regulated manner. The space-time relationship of the hydrological processes in a landscape or a watershed would depend on the rainfall patterns, on the land use and ground cover, on the soil type and depth, on the topography and landform and on the geomorphology.

However, in tillage-based agriculture because of the regular mechanical disturbance from tillage and of bare and exposed soils, covered with poor crop diversity, soon leads to loss of soil health and functions, soil surface sealing up, soil losing its porosity and structure and unable to maintain adequate infiltration rate and retention and drainage characteristics. With time, there is increasing run-off and erosion of top soil, and run-off water loaded with sediment and agrochemicals enters the streams and rivers within the watershed at a much rapid rate, sometimes, or eventually often, causing floods, clogging stream and river channels and causing water and landscape pollution. Sediment and chemical load in river water affect the quality of water in inland water bodies and reservoirs as well as seas and oceans into which rivers flow, eventually creating dead zones.

This chapter illustrates two types of ecosystem services that are harnessed by Conservation Agriculture (CA) systems, namely: (i) those that are mobilized and operate in the crop fields; and (ii) those that are harnessed and delivered across landscapes and watersheds. The former services relate to soil health and functions and positively affect crop performance and biological products; and the latter relate to large-scale ecosystem services such as water-resource quality, carbon sequestration and soil conservation which represent positive externalities which sustainable agriculture should provide to local communities and to society at large.

2 Ecosystem services in production fields

CA practices that related to three CA principles are now widely used in a range of farming systems in all continents on some 12.5% (180 million ha) of the global cropland (Kassam et al., 2019). They add to sustainability of production and soil systems and generate a range of ecosystem services in the field which relate to soil health and productivity (Table 1). They also improve soil conditions and result in beneficial outcomes for production, ecosystem services and socio-economic conditions. However, to achieve the sustainable *output intensification* desired by producers, these CA practices need to be complemented by additional best practices related to integrated crop, soil, nutrient, water, pest, time, labour and energy management. This whole production system enables the sustainable *intensification of output* to occur, rather than the *intensification of inputs,* and the optimization of the need and utilization of production inputs of seeds, agrochemicals, water, energy, time and labour. Such sustainable production systems are biologically dynamic and ecologically alive, offering farmers many possible combinations of practices to choose from and adapt, according to their local production conditions and constraints (Kassam et al., 2009, 2011a, 2013, 2017; Godfray et al., 2010; FAO, 2011b, 2016; Pretty et al., 2011) (see Chapter 1, Volume 2).

Thus, sustainable crop production intensification is the combination of all of these improved practices applied in a timely and efficient manner within the ecological underpinning created by the application of the three interlinked CA principles adapted to local conditions and constraints. For this, the

Table 1 Effects of production system components fully applied together on sustainability and ecosystem services

System component ► To achieve services ▼ Initially: seek, identify and alleviate any subsurface soil-compaction or pans before planting.	Mulch cover (crop residues cover-crops, green manures)	No-tillage (minimal or no soil disturbance)	Legumes (as crops for fixing nitrogen and supplying plant nutrients)	Crop rotation/association (for several beneficial purposes)
Simulate optimum 'forest-floor' conditions	√	√		
Reduce evaporative loss of moisture from soil surface	√			
Reduce evaporative loss from upper soil layers	√	√		
Minimize oxidation of soil organic matter, CO_2 loss		√		
Minimize compactive impacts by intense rainfall, passage of feet, machinery	√	√		
Minimize temperature fluctuations at soil surface	√			
Provide a regular supply of organic matter as the substrate for soil organisms' activity	√			
Increase and maintain nitrogen levels in root zone	√	√	√	√
Increase CEC of root zone	√	√	√	√
Maximize rain water infiltration, minimize run-off	√	√		

(Continued)

Table 1 *(Continued)*

System component ► To achieve services ▼ Initially: seek, identify and alleviate any subsurface soil-compaction or pans before planting.	Mulch cover (crop residues cover-crops, green manures)	No-tillage (minimal or no soil disturbance)	Legumes (as crops for fixing nitrogen and supplying plant nutrients)	Crop rotation/association (for several beneficial purposes)
Minimize soil loss in run-off, and by wind	√	√		
Permit and maintain natural layering of soil horizons by actions of soil biota	√	√		
Minimize weeds	√	√		√
Increase the rate of biomass production	√	√	√	√
Speed soil-porosity's recuperation by soil biota	√	√	√	√
Reduce labour input		√		
Reduce fuel-energy input		√	√	√
Recycle nutrients	√	√	√	√
Reduce pest pressure of pathogens				√
Re-build damaged soil conditions and dynamics	√	√	√	√
Pollination services	√	√	√	√

Source: based on Friedrich et al. (2009).

ensuring of soil stability, and the favouring of self-recuperation of appropriate soil structural and functional conditions, are essential (see Table 1). Thus, sustainable soil management depends on more than how and what crops are grown. It depends on the dynamic ecological and biological foundation that is created by the production systems themselves through their interaction with the self-protecting and self-regenerating capacity of all the interacting parts of the agroecology (FAO, 2008; Kassam et al., 2013).

In agricultural crop fields, a number of ecosystem services are mediated in the soil root zone and locally that relate to soil life, health and functions, as well as crop health and protection, and therefore to crop growth and productivity, and to system resilience. The contribution of practices that implement the technical principles of CA – no-tillage, mulch cover, legume crops and crop rotations/associations – to important ecosystem services is shown in Table 1 (Kassam et al., 2011a, 2013; FAO, 2012). The first column in Table 1 lists the ecosystem services that can operate in healthy agricultural soils. However, all such conditions and services are debilitated and made suboptimal or dysfunctional in agricultural soils that are managed under conventional tillage (CT) agriculture with bare soil and monocropping. This is because of regular mechanical soil disturbance, that destroys soil life, health and structure and promotes erosion and biological, physical, hydrological and chemical degradation; exposed bare soils that lack surface protection and suffer from inadequate supply of biomass substrate to support soil microorganisms, soil mesofauna and food webs below and above the ground surface; and lack of habitats for natural enemies of insect pests and pathogens to control pest populations.

On the other hand, agricultural land that is managed using the CA approach facilitates ecosystem functions and services, particularly those services related to provisioning, regulating and supporting and derived as a result of improved conditions in the soil volume used by plant roots (Kassam et al., 2009, 2013). The improvement in the porosity of the soil is affected by the actions of the soil biota, which are present in greater abundance in the soil under CA. The mulch on the soil surface in CA systems protects against the compacting and erosive effects of heavy rain, buffers temperature fluctuations and provides energy and nutrients to the organisms below the soil surface. Table 1 shows which of the field-based ecosystem services are facilitated by each of the core practices of CA systems.

Legumes in CA rotations provide increased availability of biologically fixed nitrogen in the soil, thus diminishing the need for large amounts of applied nitrogenous fertilizers. Society gains from CA regardless of farm size by diminished erosion and run-off, less downstream sedimentation and flood damage to infrastructure, better recharge of groundwater, more regular stream flow throughout the year, resulting in a more assured community water supply and quality. Cleaner civic water supplies also reduce costs of treatment for urban/domestic use. CA also contributes to increased stability of food supplies due to greater resilience of crops in the face of drought. The rural community ultimately benefits from better nutrition, overall health and less pressure on curative health services.

In CA systems, rotations and associations of crops also encourage agrobiodiversity as each crop will attract different overlapping spectra of microorganisms. The optimization of populations, range of species and effects of the soil-inhabiting biota are encouraged by the recycling of crop biomass, stubbles and other organic matter that provide the substrate for their metabolism. Rotations and associations of crops inhibit the build-up of weeds, insect pests and pathogens by interrupting their life cycles, making them more vulnerable to natural predator species, and contributing development-inhibiting allelochemicals. The same crop mixtures, sequences and rotations provide above-ground mixed habitats for insects, mammals and birds. Under CA systems, it is possible to harness many of the above ecosystem services from nature mainly because the ecosystem functions that generate these services are enhanced and protected, so that agriculture is not in competition with nature but works in harmony with it.

As has been described in some detail in Chapter 1 of this volume, CA systems, offering full productivity, economic, environmental and social benefits, cannot be established overnight. Depending on the severity of the degradation status of the soil, the CT systems must undergo a transformation into a CA system in space and time in order to harness the benefits and changes that improve productivity (and hence profit), efficiency, resilience and ecosystem services. Often, it is necessary to deal with the detrimental subsurface compaction and hardpans at the outset, otherwise they have been demonstrated to have significant hindering effects on plant production and yield as well as several other soil-mediated ecosystem services, due to increased physical impedance of roots and loss of pore space for retaining water at plant-available water potentials, or decreased water percolation to deeper soil layers and into groundwater or aquifers.

Equally, if soils are hindered from performing some of their basic functions, soil-mediated ecosystem services at the larger-scale landscape, watershed and regional levels, illustrated in the following section, such as water resource quality and quantity, water, nutrient and carbon cycling, carbon sequestration and erosion control, cannot be harnessed optimally. Hence, it is important to physically engender a more permeable *sub*surface before anything else can improve significantly. This has the dual effects noted in Table 1 of '*Maximize rainwater infiltration, minimize run-off*'. In practice, this means that the ground may need to be physically broken at the beginning of the process of transformation to CA, with pickaxes, chisels or rippers, subsoilers and para-ploughs. The process then has to continue biologically and ecologically to turn the soil into a healthy 'carbon' sponge with the help of soil microorganisms including bacteria, mycorrhiza and earthworms, as the practice of no or minimum soil disturbance, return of crop root and shoot biomass into the soil and diversified cropping systems involving legumes and crops of different root systems, all continue to assist the soil enhance its health and ecosystem functions.

3 Large-scale ecosystem services from agricultural landscapes and watersheds

When the field-level ecosystem services (Table 1) are reproduced across farms in a contiguous micro-catchment within a landscape or a region, the ecosystem services provided at the landscape and watershed level - such as clean water, sequestration of carbon, avoidance of erosion and run-off - become more apparent (Lindwall and Sonntag, 2010; Kassam et al., 2011b, 2013). The co-benefits of more water infiltrating into the ground beyond the depth of plant roots is perceptible in terms of more regular stream flow from groundwater through the year, and/or more reliable yields of water from wells and boreholes. The benefits of carbon capture become apparent in terms of the darkening of soil colour and crumbly 'feel' of the soil, accompanied by improvements in crop growth, plus much less erosion and hence less deposition of sediment in adjacent waterways and into reservoirs.

Society and nature gain from CA regardless of farm size by diminished erosion and run-off, less downstream sedimentation and flood damage to infrastructure, better recharge of groundwater, more regular stream flow throughout the year, resulting in a more assured community water supply and quality. Cleaner civic water supplies also reduce costs of treatment for urban/domestic use. CA also contributes to increased stability of food supplies due to greater resilience of crops in the face of drought. The rural community ultimately benefits from better nutrition, overall health and less pressure on curative health services. In CA systems, the sequences and rotations of crops also encourage agrobiodiversity as each crop will attract different overlapping spectra of microorganisms.

The interlinked CA practices of no or minimal soil disturbance, organic soil cover and diversified crop associations and rotations facilitate the harnessing of ecosystem services on agricultural landscapes and watershed. During the last couple of decades or so, harnessing CA-based ecosystem services are operating or being tested in different parts of the world (Kassam et al., 2013). Several good examples include (1) the agricultural carbon offset scheme in Alberta, Canada, that allows regulated companies to offset their emissions by purchasing verified tonnes from a range of approved sources including agriculture projects (Goddard et al., 2009); (2) the hydrological services from Paraná III Basin in Brazil located in the western part of Paraná State on the

Paraguay's border in which the Itaipú Dam Programa *Cultivando* Água Boa (cultivating good water) has established a partnership with farmers to achieve the sustainable use of soil and water in the watershed for efficient electricity generation (Itaipú, 2017); (3) control of erosion and restoration of soil and landscape health in olive orchards in Andalucia, Spain (Gonzalez-Sanchez et al., 2015); and (4) control of soil conservation and wind erosion in China (Li et al., 2017). These examples are elaborated in the following sections.

3.1 Canada: carbon offset scheme in Alberta

Canada has the fifth largest area under CA systems with some 19.9 M ha (Kassam et al., 2019). That constitutes about 53% of all cropland and 29% of all agriculture land (Statistics Canada, 2016). Most of the agriculture land is in the prairie region and Alberta has about 32% of the agriculture land in Canada (Alberta, 2017).

The large amount of energy resources and industry in Alberta is the result of it being the largest greenhouse gas (GHG)-producing province in the country, comprising 37% of national emissions. It is also a large agriculture province and has the largest amount of agriculture GHG emissions of any province (18 MT per year), even though it still compromises only 7% of the total provincial emissions (Environment Canada, 2016). Alberta's large emission profile has required early and ongoing policy attention by governments. Swallow and Goddard (2016) have tracked the interconnectivity of international, national and regional nested networks in formulating policy. Alberta has interacted with other jurisdictions in examining policy, sharing experiences and has included a few characteristics of adaptive policy such as annual meetings with stakeholders to review issues and prospective changes in regulations.

In 2007, the Albertan government made amendments to their earlier climate change policy that required large final emitters to reduce their annual GHG output. In order to reach their compliance targets, facilities had options one of which was to purchase offset credits from Alberta sources using government-approved protocols. One of the agriculture opportunities at the time was the development and application of a no-till annual cropping protocol that recognized the increase in soil carbon sequestration with the adoption of no-till cropping systems (Goddard et al., 2009). The updated version now named the Conservation Cropping Protocol (CPP) also includes a provision for reduction of summer fallow. In 2007, it was the first soil sequestration protocol in the world if not the first agriculture protocol for a GHG emissions compliance system. Other examples did not exist, and guidance literature was sparse.

Unique features of the CPP include the use of pooled coefficients for soil carbon sequestration derived from application of the national emissions inventory to soil map units combined with crops and years to average variance and risks across large areas thus avoided the prohibitive transaction costs of measuring GHG state or flux changes at the site level. Another feature is the use of private aggregator companies to assemble fields and farms into larger projects that attract buyers. They also are responsible for third-party verification of projects and ensuring the project is correctly represented on the registry where tonnes are serialized and monitored by the Canadian Standards Association. The aggregator companies have been either existing crop consulting companies that have expanded their services or new entities that specialized in the one area of business. Government does not take on the costs of delivering the offset system, only of setting the boundaries and guidance then, letting industry find the efficiencies of delivery and administration.

The government set a compliance price in 2007 of Cdn$15 per tonne CO2e which increased to Cdn$20 in 2015 and Cdn$30 in 2016. Facilities that could not comply with the reduction target could purchase offsets or pay into a 'Tech Fund' which invested into GHG mitigation research projects. Regulated facilities had no limits on the amount of offset utilization; however, they ranged from 20% to 50% of the total compliance needs (Swallow and Goddard, 2016). Offset projects did not have to disclose transaction prices (only tonnes were certified and tracked) so the sale price needed to account for aggregation and verification costs. In the decade following 2007, there were slightly more than 12 million tonnes from CPP projects, or that many tonnes of verified soil carbon sequestration. At a price of Cdn$15 per tonne that represents Cdn$180 000 that flowed from large emitters to farmers and the agribusiness sector.

Alberta learned that well-intended policies do not always meet their target or meet associated expectations. The policy was to reduce emissions using a top-down, regulatory compliance policy. As such it could be shown that facilities reduced their emissions. The associated offset system that is more aligned with a market policy was constrained by high transaction costs. Unlike most other offset systems, this one was subjected to government auditors that were unfamiliar with non-metered biologic systems and wanted to impose detailed accounting and information tracking upon farms. Tarnoczi (2011) found that the verification and audit desires to be conservative and accurate and reduce uncertainty to zero went beyond the ISO guidance of 'practicality'. Governments need to take pragmatic approaches tailored to market-based principles in order to foster GHG offsets beyond industrial facilities. This need will be even more pronounced if governments desire to go beyond the relatively simple environment of GHGs into the more complex world of ecosystem services.

On the positive side, the policy was found to be accepted by the regulated industry and the suppliers of offsets. The role of aggregators was key to improving tracking systems and lowering costs in assembling projects and in working with verifiers. Experienced aggregators have provided valuable insight into protocol revisions and 'desk testing' new protocols in the final design stages. Farmers have become more carbon conscious and aware of the records needed to verify their change in practices. The comfort level with protocols, partial life cycle accounting and offset projects has increased knowledge and skills of specialists such that they have moved into more integrated life cycle analysis and sustainable certifications with confidence and understanding beyond just talking about the topics.

Baig and Gamache (2011) have reviewed some of the many other co-benefits of CA on the Canadian prairies. Soil carbon imparts many properties of soils and biologic function and is often utilized as a key indicator for ecosystem services value. Biologic carbon sequestration represents so much more than mitigation of GHGs, and public policy needs not only to incentivize behaviours such as CA adoption for mitigation of emissions but to reward farmers for private costs of delivering a public good.

3.2 Brazil: watershed services in the Paraná Basin

Brazil has long experience in CA, and now has more than 32.0 M ha under various CA cropping systems. It has set important precedents for the engagement of farmers as principal actors in the development and adaptation of new technologies and practices, including the integration of pasture, trees and livestock with arable crops. CA attracted Brazilian farmers because it increased crop yields (at least 10–25%), greatly reduced surface run-off and soil erosion and cut production costs.

As part of a strategy for improvement, conservation and sustainable use of natural resources, the Itaipú Dam *Programa Cultivando Água Boa* ('cultivating good water programme') established a partnership with farmers to achieve their goals in the Paraná III Basin located in the western part of Paraná State on the Paraguay border (Mello and van Raij, 2006; Laurent et al., 2011; ITAIPU, 2019). The Itaipú Dam is one of the largest operational hydroelectric energy producers in the world. The dam works to integrate energy, environment and biodiversity within to achieve sustainable development. The Itaipú Dam is situated in one of the most productive agriculture regions in Brazil and Paraguay. Therefore, the water bodies that connect the territory with the reservoir are subject to environmental threats arising from agriculture such as use of pesticides and agrochemicals, waste from livestock production and soil erosion (ITAIPU, 2017). Thus, the dam's reservoir depends on the sustainable use and management of soil and water in the watershed/catchment for efficient electricity generation. Sediments, plant nutrients and animal manure entering the reservoir result from inappropriate land use pollute the water used by the turbines to generate electricity. This phenomenon shortens the reservoir's useful life and increases the maintenance costs of power generating turbines, increasing thereby electricity generation costs. Thus, in principle, payments could be made through a programme to improve the conditions of electricity generation. The spatial unit covered by this programme is the whole watershed. Functioning as a community, joining many farmers in the watershed, they reached a scale where environmental impact can be monitored with suitable indicators to establish a system for payment for environmental services.

One of the partnerships established in the *Cultivando Água Boa* programme and developed through an agreement with the Brazilian no-till federation (FEBRAPDP) is the Participatory Methodology for Conservation Agriculture Quality Assessment (Laurent et al., 2011; ITAIPU, 2019) based upon former positive experiences with catchment-development in Brazil. The first phase in the programme was the partners plan to measure the impacts of farm management through a scoring system indicating how much each farm is contributing to the improvements of the water conditions (https://www.itaipu.gov.br/en/the-environment/cultivating-good-water).

In this regard, a scoring index model for rating the quality of no-till systems was being devised. The model relies on expert knowledge and is being applied to identify soil erosion and land degradation risks arising from any weakness in the adopted CA practices, and possible action needed to address the weakness (Roloff et al., 2011). Consolidating this phase and adapting the principles established for the 'water producer' by the National Water Agency, the partners will assign values to ecosystem services generated from farms participating in the programme (ANA, 2011). Considering the polluter/payer and provider/receiver principles set in the Brazilian Water Resources Policy, farmers with good scores are paid for their proactive action to deliver watershed services within the Paraná Watershed Plan.

The scheme also allows animal manure to be used for generating biogas that is converted into electricity fed into the grid for distribution and for which farmers also derive an additional income from the Itaipú Dam Authority. All this is part of a new community-based land and agricultural management framework for services provided by farmers as compensation for their proactive approach to improving the reservoir water quality and reducing the costs of electricity generation by the Itaipú Dam Authority. The energy generation programme cuts across several ministries such as energy, water, agriculture and community development, all of whom have an important stake in the development programme. But what is most significant is the fact that because of the change in land-use practice to CA, and the resultant significant improvement in watershed hydrology, leading to improved water resource quality due to significant reduction in run-off and soil erosion and in sediment load and water pollutants, the operating life of the Itaipú Dam has increased by at least fourfold.

The *Cultivando Água Boa* programme was awarded the 'Water for Life' UN-Water best practices Award 2015 as the world's best water management practice (category 1), because the programme is a permanent community participation movement in which Itaipú works together with the society to change its values in addition to mitigating and correcting environmental liabilities (ITAIPÙ, 2019).

3.3 China: soil conservation and erosion control services

China is paying more attention to tilled soils being more susceptible to water and wind erosion. The best solution to control water and soil erosion is to eliminate tillage. Practices which improve water-use efficiency and natural resource management by reducing run-off and erosion are of great importance. Therefore, the adoption of CA practices, providing more residue cover and less soil disturbance, has received considerable attention. Since their beginnings in response to issues from the American 'Dust Bowl' era, several decades of development have demonstrated that CA systems are a valuable means of reducing erosion by both water and wind (Uri et al., 1998) because of low soil disturbance and soil surface protection with crop residue. China already has some 9 M ha of agricultural land that is being managed with CA systems (Li et al., 2016).

CT in dryland farming areas of northern China includes mouldboard ploughing to a depth of about 20 cm, followed by harrowing, hoeing, rolling and levelling. All the residues in the fields are removed for animals or as fuel before ploughing. Long-term mouldboard ploughing and residue removal/ burning have increased the risks of wind and water erosion and the formation of the hardpan in the deep soil layer. It has also resulted in poor soil physical and chemical properties, high inputs of energy and labour which apparently leads to low farmer incomes. In some parts of northern China, particularly in the North China Plain, burning crop residue has increased during the last decades, which caused high levels of air pollution. Therefore, the government is pushing to accelerate the development and spread of various kinds of CA systems and related practices, such as controlled traffic and permanent raised beds (PRB), to protect and improve soil, water and air quality as ecosystem societal services (Bai et al., 2008; Gao et al., 2008; Chen et al., 2008; Kassam et al., 2009, 2013; He et al., 2012; Kautz et al., 2013; Ding et al., 2013; Lal, 2014).

Effect of Conservation Agriculture on soil fertility: Accumulated evidences suggest that CT makes the soil vulnerable to losses of organic matter and other nutrients, and potentially leads to a reduction in soil fertility. In long-term CA systems, as compared to CT soils, minimum soil disturbance decreases the exposure of SOM in inter- and intra-aggregates to microbial processes and thus decreases the mineralization of SOM. Furthermore, long-term retention of crop residue can increase organic matter input. Therefore, CA practices can improve soil C sequestration and nutrition, so as to increase root zone conditions and crop productivity. As indicated in Table 2, the results from long-term (>5 years) field experiments showed CA significantly improved SOM (>7%) and NPK near the soil surface as compared to CT in different regions in China. Moreover, accumulated SOM had a strong relationship with the implementation time of CA. For example, Zhang et al. (2019) founded that after 26 years of experiments, SOM under CA practices (NT+S, NT+S+ST) were >20% and >60% greater than that under CT treatment, respectively. In addition, low aggregate stability of CT soils resulting from frequent soil disturbance via intensive tillage affects NPK accumulation. Under CA the physical protection provided by organic matter is increased leading to significant NPK increase. The SOM, N, P and K in soil are critically needed nutrients for crop production and to a large extent, hold the land productivity and crop yields. Increased soil fertility is helpful for the improvement of crop yields in long-term CA systems. Li et al. (2016), based on meta-analysis, showed that the mean increase in crops yields with CA practised for ≥10 years was >7.0% relative to CT, indicating that productivity continues to increase as the natural capital of the system is enhanced over time.

Effect of Conservation Agriculture on water erosion: In a CA system, soil cover provides protection from raindrop/irrigated water impact, slows run-off and facilitates the deposition of sediments. Further, improved soil property in a CA system is helpful for the holding and infiltration of rainfall or irrigation water. Under natural rainfall conditions, Wang et al. (2014a) and Tan (2015) demonstrated that CA treatments significantly reduced run-off and soil erosion by >25% and >12%, respectively (Table 4). Furthermore, a meta-analysis of CA in northern China (Li et al., 2016) showed that adoption of CA could reduce water run-off and erosion by >40%, increase soil water-holding capacity by >3% and water-use efficiency by >10% for most crops and agro-ecoregions. Rainfall simulation experiment was also carried out to study the effect of CA on water erosion, and the results showed that under the rainfall intensity from 10 to 80 mm/h, CA delayed run-off start time and reduced run-off by 17.9–38.7% (Table 3).

Effect of Conservation Agriculture on wind erosion: CA, which combines no-tillage with soil mulch cover, can increase the roughness of surface, reduce soil surface wind momentum and retain the corrodible particles under straw cover, so as to reduce soil wind erosion. As indicated in Table 4, soil erosion under CA was significantly lower than that under CT. Wind tunnel simulation experiments by Yang et al. (2019) and Lin et al. (2017) showed that CA reduced mean wind erosion rate by 22.5% (wind speed: 6–22 m/s) in Wuwei of Northwest China, and soil loss by 72.6% (wind speed: 14 m/s) in Siping of Northeast China, respectively. In natural wind erosion condition, Wang et al. (2014b) indicated that CA reduced >50% soil loss as compared to CT during an 11-h dust storm event. The reduction of soil loss will be different in a CA system depending upon the measurement duration, ambient conditions and erosive winds.

Table 2 Effects of CA on soil fertility in China

Region	Experimental site	Treatment	Years of CA	Increase compared with CT (%)	
				SOM	NPK
Loess Plateau	Linfen, Shanxi province (Zhang et al., 2019)	NT+S	26 years	21.61	TN:40.26 AN:56.1 AP:26.93 AK:67.48
		NT+S+ST		61.52	TN:48.05 AN:48.01 AP:51.99 AK:48.21
	Weinan, Shaanxi province (Wang et al., 2014a)	ST	4 years	0.27–15.60	AN:3.14%–8.61 AP:3.76%–24.32 AK:15.62%–25.17
Northwest China	Dingxi, Gansu province (Zhang et al., 2016)	NT	12 years	11.99	TN:11.11
		NT+S		7.78	TN:3.33
Northeast China	Shenyang, Liaoning province (Gong, 2014)	NT+S	9 years	23.01	AN:31.79 AP:15.77 AK:31.73
	Tieling, Liaoning province (Zhang et al., 2015a)	NT+S	5 years	62.85	AN:27.69 AP:31.31 AK:22.85
North China Plain	Luoyang, Henan province (Lal, 2014)	RT+S	3 years	32.48	AN:38.78 AP:9.96 AK:32.44

Region	Experimental site	Treatment	Years of CA	Increase compared with CT (%)	
				SOM	NPK
		ST+S		26.49	AN:43.62 AP:12.38 AK:69
	Zhengzhou, Henan province (Wang et al., 2015)	ST	3 years	2.21–3.44	AN:11.19 AP:15.25 AK:11.40

Note: NT: no tillage, ST: subsoiling, RT: rotary tillage, +S: straw mulching; SOM: soil organic matter, TN: total nitrogen, AN: available nitrogen, TP: total phosphorus, AP: available phosphorus, TK: total potassium, AK: available potassium.

Table 3 Effect of CA on water erosion in China

Region	Testing site	Decrease compared with CT (%)
Loess Plateau	Shouyang, Shanxi province (He et al., 2010)	42.5 (The annual run-off in CT was 153 mm, and in CA was 88 mm)
	Shenmu, Shaanxi province (Tan, 2015)	12.2 decrease in soil loss (The annual soil loss in CT was 1234.98 t/ha, and in CA was 1084.58 t/ha)
Northeast China	Shenyang, Liaoning province (Wang et al., 2014b)	25.1 (The monthly run-off in CT was 16.00 $g \bullet m^{-3}$, and in CA was 11.98 $g \bullet m^{-3}$)
Northwest China	Lanzhou, Gansu province (Li et al., 2011)	34.7 (The cumulative run-off in CT was 60.32 $mm \bullet h^{-1}$, and in CA was 44.78 $mm \bullet h^{-1}$)
North China Plain	Zhuozhou, Hebei province[a] (Zhang et al., 2015b)	17.9 (80 mm/h rainfall intensity)-38.7 (60 mm/h rainfall intensity)

Note: [a] A rainfall simulation experiment under 1400 kg/ha straw mulching.

Table 4 Effects of CA on wind erosion in China

Region	Testing site	Decrease compared with CT (%)
Northwest China	Wuwei, Gansu province (Yang et al., 2019)	22.5 (The mean wind erosion rate in CT was 15.47 $g \cdot m^{-2} \cdot min^{-1}$, and in CA was 11.99 $g \cdot m^{-2} \cdot min^{-1}$ by wind tunnel simulation with five different wind speeds of 6–22 m/s)
Northeast China	Siping, Jilin province (Lin et al., 2017)	72.6 (The soil loss in CT was 17.53 $g \cdot m^{-2} \cdot s^{-1}$, and in CA was 4.81 $g \cdot m^{-2} \cdot s^{-1}$ by wind tunnel simulation with 14 m/s wind speed)
	Fuxin, Liaoning province (Chen et al., 2016)	89.3 (wind erosion rate under natural wind erosion condition)
North China	Zhangjiakou, Hebei province (Wang et al., 2014b)	53.1 (The soil loss in CT was 1770 $g \cdot m^{-2}$, and in CA was 830 $g \cdot m^{-2}$ in an 11-h dust storm event)

3.4 Spain: soil conservation services in olive groves

Groundcovers in woody crops are considered an essential part of CA-based perennial systems in Spain (Gonzalez Sanchez et al., 2015). Among the woody perennial crops, olive orchards are an important agro-ecosystem in Spain, where 2.7 M ha or 16% of the surface area is covered with olives groves. In some regions, olive groves are essential for the economy and the development of the society, that is in Andalusia, where it accounts for 60% of the Spanish olive-growing area. The value of the production of the olive oil sector was around €1886 million on average from 2007 to 2012 (MAPA, 2019).

Historically, olive cropping has been concentrated on hilly lands, where soil erosion happens to be a very severe and widespread problem. Locally, historical soil loss rates have been reported to reach up to 184 Mg $ha^{-1}yr^{-1}$ (Vanwalleghem et al., 2010). Erratic but high-intensity rainfall, especially during winter, but also the management of the orchards, underlies the origin of soil erosion. Commercial olive orchards, mainly grown under rainfed conditions, are characterized by extremely adverse management conditions as farmers tend to till intensively to avoid competition of weeds with tree water and nutrient uptake. Therefore, effective conservation strategies, such as the establishment of groundcovers, are not easily adopted by farmers. Indeed, CT has been the dominant management system in olive orchards over the last decades. The combination of this human-induced low vegetation cover with the steep slope gradients on which these orchards are located, together with the high-intensity rainfall events that characterize the Mediterranean climate, explains why high soil erosion rates have been associated with olive oil production (Beaufoy, 2001).

As a result of these alarming erosion rates that have been reported in olive groves on sloping and hilly land, many researchers have started to question the severity and the extent of water erosion in olive orchards in southern Spain (Fleskens and Stroosnijder, 2007). Others, however, insist on soil erosion being a widespread threat to sustainable land use through olive production (Gomez et al., 2008; Vanwalleghem et al., 2010, 2011). Moreover, land use change and the abandonment of the terraced slopes, functioning as anthropic hydrological infrastructures, which protected the soil and preserved the natural vegetation in the recent past, have been progressively collapsing, mainly due to the rapid removal of the soil, causing important land degradation problems (Dunjó et al., 2003).

The spread of CA in woody crops in Spain was facilitated by a first CAP agri-environmental scheme. It was introduced in Andalusia in the late 1990s and early 2000s aiming to fight soil erosion in olive orchards mainly by groundcovers between trees and natural vegetation on the land borders. This is considered a key milestone in the development of CA systems in woody crops, as it helped double the CA area of olive crops over a period of a decade or so from 434 828 ha in 2006 to 825 516 ha in 2018 (Table 5), without any substantial economic support to producers. During the same period, the area of perennial CA systems (all woody crops), including olives, increased by 56% from 832 729 ha to 1 301 174 ha. The proportion of CA perennial area to total perennial area increased from 16.5% in 2006 to 25.1% in 2018. The shift from tillage system to CA system was based on the confidence and the feasibility of the economic side of the system. It was proven that CA was a win-win option not only for annual crops but also for woody crops (Garcia-Torres, 2001). Besides, other soil erosion control practices were also promoted such as soil tillage along contour lines and the maintenance of pruning residues on the ground surface in the interrow space (Franco and Calatrava, 2006).

Table 5 Evolution of area (hectares) under groundcover soil management in perennial CA systems in all woody crops in Spain

CA groundcover area by type of woody crop	2006	%	2010	%	2014	%	2018	%
Citrus	153956	18.5	156012	13.7	135293	10.7	132464	10.2
Fruit	198019	23.8	257733	22.7	273928	21.8	282743	21.7
Olives	434828	52.2	659330	58.1	787938	62.6	825516	63.4
Vineyard	36035	4.3	49633	4.4	51028	4.1	52327	4.0
Others	9891	1.2	12928	1.1	10890	0.9	8124	0.6
Total CA groundcover area	*832729*	*100*	*1135636*	*100*	*1259077*	*100*	*1301174*	*100*
Total perennial crop area	*5039440*		*4986046*		*4925681*		*5191115*	
CA groundcover area/total perennial crop area	*16.5%*		*22.8%*		*25.6%*		*25.1%*	

Source: adapted from ESYRCE (2006–2018).

Despite this notable progress in terms of adoption of soil conservation measures in the case of perennial crop production in Spain, there are still regions where the adoption of soil conservation practices is still very low and, in general, there exists much room for the extension of policy measures to mitigate and reverse soil erosion and degradation (Calatrava et al., 2011). In some cases, farmers do not till the soil between tree lines, using herbicides to control weeds instead. However, this leads to bare soil, with no protection or improvement in soil health and function. Findings of Gomez et al. (2009) showed that bare soil, even though untilled, is capable of more run-off and sediment load in olive groves under certain conditions, and this should be seriously taken into account when designing soil and water conservation strategies. These have to be driven by their real delivery of ecosystem services and not just by cost-effective minimal conservation approaches.

The faster adoption of CA-based groundcovers rather than no-till with bare soil as soil and water conservation measures in perennial woody crops, and especially in olives (Table 5), can therefore be considered an important step towards the mitigation of soil erosion and degradation. These are significant ecosystem services mobilized with CA systems by producers for society at large.

4 Conclusion and future trends

The Rio Earth Summit in 1992 was a milestone in the expression of humankind's awareness of the destructive power of development and the need to address the global challenges resulting from climate change, destruction of biodiversity and land degradation. The Summit established Agenda 21 for development and three commissions to respond to the challenge posed by the need to move forward with sustainable development. However, the Earth Summit in Rio had little to say about the unsustainable nature of tillage-based agriculture and the land degradation caused by tillage and what should be done about it. While the Rio Summit established the three conventions to address climate change, land degradation and biodiversity loss, none of these initiatives have yet come up with a definitive statement regarding the destructive nature of the CT-based agricultural paradigm and its inability to adapt to and mitigate climate change. The Earth Summit in 2005 in Rome also had little to say about the unsustainable nature of tillage-based agriculture.

Yet, CA has offered large-scale empirical and scientific evidence worldwide that it can address all the ecological weaknesses in the CT-based agriculture and that CA-based land use is able to offer maximum and sustainable productivity while at the same time permit the harnessing of societal ecosystem service. Additionally, because of its regenerative powers, CA is able to help rehabilitate degraded agricultural landscapes and restore landscape ecosystem service functions. These benefits from CA land uses are elaborated throughout this volume, and an overview of benefits is provided in chapter 1.

A number of ecosystem services operate at the field level which help enhance soil health and productivity, system resilience, production efficiency and ecological sustainability. They also provide productivity-enhancing support to crop plants that enables the optimization of integrated crop management leading to relatively stable and greater yields. They also help minimize the application of purchased inputs of fertilizer, pesticides, water, energy and labour, and provide protection against abiotic and biotic stresses, and protect the soil from erosion and degradation.

Given that field-based and landscape-based ecosystem services contribute positively to all aspects of sustainability – economic, social and environmental – in theory it is in the interest of farmers themselves to move away from CT agriculture and adopt CA systems. Given the enormous productivity and environmental benefits offered by CA, time-bound extension and financial support to farmers to help them to transform and adopt CA is justifiable in the early years. Equally, the landscape-level ecosystem services to society are of a public goods character, and they deserve recognition and appropriate public sector support. These include services such as carbon sequestration, availability of clean water, control of erosion, enhancement of biodiversity, reduction in chemical use and pollution, promotion of pollination services and climate change mitigation.

Stewardship schemes based on ecosystem services from agricultural land use will become increasingly important in the future as enabling policy and institutional support mechanisms to farmers become popular. Already, there are strong signs in different parts of the world that civil society is in favour of extending 'public support and money for public goods' delivered by farmers and land managers.

5 Where to look for further information

The subject area of ecosystem services and their relationship with rural and urban land use is of relatively recent origin. In agricultural landscapes and watersheds, land use management by farmers and land managers is increasingly expected to ensure that land-mediated ecosystem functions and services to society and nature are protected and/or enhanced. Given the severe historical land degradation caused by CT agriculture, there is a great deal of rehabilitation and regeneration work that needs to be done in the future to repair the damage done to different ecosystems. CA is considered to be a regenerative approach to agricultural land use and therefore will play an increasingly important role in the future in land use and agricultural development programmes aiming to reverse land, water, biodiversity and environmental degradation, and mitigate future degradation.

A major breakthrough in thinking regarding ecosystem functions and services in different ecosystems was made by the UN Millennium Ecosystem Assessment published in 2005 (MEA, 2005). The MEA reports are a significant source of information on concepts, frameworks, principles and relationships between ecosystem management and ecosystem functions as well as between ecosystem and human well-being. These reports are available at: https://www.millenniumassessment.org/en/Reports.html.

In 2013, the European Environment Agency published a review of recent literature on mapping ecosystem services and analysis of methods used (Gotzl et al., 2013). This report is available at: https://pdfs.semanticscholar.org/b10c/5ba623c8fbe59fb8b668fe9c78b23da4af06.pdf.

In 2015, Bioversity International published a literature review on the integration of ecosystem services in agricultural economic models (Rojas Lara et al., 2015). The report is available at: https://www.bioversityinternational.org/e-library/publications/detail/literature-review-on-the-integration-of-ecosystem-services-in-agricultural-economic-models/.

In 2018, a literature review on the ecosystem services and biodiversity of novel ecosystems was published by Evers et al. (2018). This is available at: https://doi.org/10.1016/j.gecco.2017.e00362.

Those interested in understanding biodiversity-ecosystem service relationships in urban areas would find more information in Schwartz et al. (2017) available at: https://doi.org/10.1016/j.ecoser.2017.08.014.

A literature review on cultural heritage and ecosystem services is provided by Holleland et al. (2017) and the review is available at: https://doi.org/10.1080/13505033.2017.1342069.

6 References

Alberta. 2017. *Agriculture Statistics Yearbook 2016*, Table 106. Alberta Agriculture and Forestry, Edmonton, Canada.

ANA. 2011. *Programme for Water Producers* (*Programa Produtor de Água*). Available at: http://www2.ana.gov.br/Paginas/imprensa/noticia.aspx?id_noticia=9304.

Bai, Y. H., Chen, F., Li, H. W., Chen, H., He, J., Wang, Q. J., Tullberg, J. N. and Gong, Y. S. 2008. Traffic and tillage effects on wheat production on the Loess Plateau of China: 2. Soil physical properties. *Soil Research* 46(8), 652–8. doi:10.1071/SR07193.

Baig, M. N. and Gamache, P. M. 2011. *The Economic, Agronomic and Environmental Impact of No-Till on the Canadian Prairies* (vol. 2). Alberta Reduced Tillage Linkages, Canada.

Beaufoy, G. 2001. The environmental impact of olive oil production in the European Union: practical options for improving the environmental impact. *European Forum on Nature Conservation and Pastoralism*. Available at: http://ec.europa.eu/environment/agriculture/pdf/oliveoil.pdf.

Calatrava, J., Barbera, G. G. and Castillo, V. M. 2011. Farming practices and policy measures for agricultural soil conservation in semi-arid Mediterranean areas: the case of the guadalentin basin in southeast Spain. *Land Degradation and Development* 22(1), 58–69. doi:10.1002/ldr.1013.

Chen, H., Bai, Y. H., Wang, Q. J., Chen, F., Li, H. W., Tullberg, J. N., Murray, J. R., Gao, H. W. and Gong, Y. S. 2008. Traffic and tillage effects on wheat production on the Loess Plateau of China: 1. Crop yield and SOM. *Soil Research* 46(8), 645–51. doi:10.1071/SR07106.

Chen, M. F., Wang, H. X. and Yu, G. Q. 2016. Study on residual stubble erosion control effect of peanut-maize intercropping model in wind-sand semi-arid region. *Liaoning Agricultural Sciences* 1, 73–5.

Ding, F., Hu, Y. L., Li, L. J., Li, A., Shi, S., Lian, P. Y. and Zeng, D. 2013. Changes in soil organic carbon and total nitrogen stocks after conversion of meadow to cropland in Northeast China. *Plant and Soil* 373(1–2), 659–72. doi:10.1007/s11104-013-1827-5.

Dunjo, G., Pardini, G. and Gispert, M. 2003. Land use change effects on abandoned terraced soils in a Mediterranean catchment, NE Spain. *Catena* 52(1), 23–37. doi:10.1016/S0341-8162(02)00148-0.

Environment Canada. 2016. *National Inventory Report 1990–2016: Greenhouse Gas Sources and Sinks in Canada - Part 3*. Environment and Climate Change Canada, Gatineau, Canada. Available at: http://www.publications.gc.ca/site/eng/9.506002/publication.html.

ESYRCE. 2006–2018. *Análisis de las técnicas de mantenimiento del suelo y métodos de siembra en España*. Ministerio de Agricultura de España (Official statistics of the Spanish Ministry for Agriculture of the soil management systems). Available at: https://www.mapa.gob.es/es/estadistica/temas/estadisticas-agrarias/agricultura/esyrce/informes-sectoriales/ (in Spanish).

Evers, C. R., Wardropper, C. B., Branoff, B., Granek, E. F., Hirsch, S. L., Link, T. E., Olivera-Lora, S. and Wilson, C. 2018. The ecosystem services and biodiversity of novel ecosystems: a literature review. *Global Ecology and Conservation* 13. doi:10.1016/j.gecco.2017.e00362.

FAO. 2008. *Investing in Sustainable Crop Intensification: The Case for Soil Health*. Report of the International Technical Workshop, FAO, Rome, July. Integrated Crop Management (vol. 6). FAO, Rome. Available at: http://www.fao.org/ag/ca/.

FAO. 2011a. *The State of the World's Land and Water Resources for Food and Agriculture (SOLAW): Managing Systems at Risks*. Summary report. FAO, Rome, 47pp.

FAO. 2011b. *Save and Grow. A policymakers' Guide to the Sustainable Intensification of Smallholder Crop Production*. FAO, Rome, 98 pp. Available at: www.fao.org/ag/save-and-grow/.

FAO. 2012. *Soil Organic Carbon Accumulation and Greenhouse Gas Emission Reductions from Conservation Agriculture: A Literature Review*. Integrated Crop Management (vol. 16). FAO, Rome.

FAO. 2016. *Save and Grow in Practice: Maize, Rice and Wheat*. FAO, Rome.

Fleskens, L. and Stroosnijder, L. 2007. Is soil erosion in olive groves as bad as often claimed? *Geoderma* 141(3–4), 260–71. doi:10.1016/j.geoderma.2007.06.009.

Franco, J. A. and Calatrava, J. 2006. Adoption of soil erosion control practices in southern Spanish olive groves. *Proceedings of the International Association of Agricultural Economists*, Gold Coast, Australia, 12–18 August, 16pp.

Friedrich, T., Kassam, A. and Shaxson, F. 2009. Conservation Agriculture. In: *Agriculture for Developing Countries, Science and Technology Options Assessment (STOA) Project*. European Technology Assessment Group, Karlsruhe, Germany.

Gao, H. W., Li, H. W. and Li, W. Y. 2008. Development of conservation tillage. *Transactions of the Chinese Society for Agricultural Machinery* 9, 43–8. (in Chinese).

García-Torres, L. 2001. *Agricultura de Conservación en el Olivar: Cubiertas Vegetales*. Asociación Española Agricultura de Conservación/Suelos Vivos (AEAC/SV), Córdoba, Spain (in Spanish).

Goddard, T., Haugen-Kozyra, K. and Ridge, A. 2009. Conservation agriculture protocols for green house gas offsets in a working carbon market. Paper presented at the *IV World Congress on Conservation Agriculture*, 3–7 February 2009, New Delhi, India.

Godfray, H. C., Beddington, J. R., Crute, I. R., Haddad, L., Lawrence, D., Muir, J. F., Pretty, J., Robinson, S., Thomas, S. M. and Toulmin, C. 2010. Food security: the challenge of feeding 9 billion people. *Science* 327(5967), 812–8. doi:10.1126/science.1185383.

Gomez, J. A., Giraldez, J. V. and Vanwalleghem, T. 2008. Comments on "Is soil erosion in olive groves as bad as often claimed?" by L. Fleskens and L. Stroosnijder. *Geoderma* 147(1–2), 93–5. doi:10.1016/j.geoderma.2008.07.006.

Gomez, J. A., Sobrinho, T. A., Giraldez, J. V. and Fereres, E. 2009. Soil management effects on runoff, erosion and soil properties in an olive grove of Southern Spain. *Soil and Tillage Research* 102(1), 5–13. doi:10.1016/j.still.2008.05.005.

Gong, L. 2014. Effects of straw returning on physical & chemical characteristics of brown soil and growth of corn. Ph.D. Dissertation. Chinese Academy of Agricultural Science, Beijing, China (in Chinese with English abstract).

Gonzalez-Sanchez, E. J., Veroz-Gonzalez, O., Blanco-Roldan, G. L., Marquez-Garcia, F. and Carbonell-Bojollo, R. 2015. A renewed view of conservation agriculture and its evolution over the last decade in Spain. *Soil and Tillage Research* 146, 204–12. doi:10.1016/j.still.2014.10.016.

Gotzl, M., Tiefenbach, M., Tramberend, P. and Condé, S. 2013. Review of recent literature on mapping ecosystem services and analysis of methods used. Working Paper No. B/2013. The European Topic Centre on Biological Diversity, European Environment Agency.

He, J., Li, H. W., Wang, Q. J., Gao, H. W., Li, W. Y., Zhang, X. M. and McGiffen, M. 2010. The adoption of conservation tillage in China. *Annals of the New York Academy of Sciences* 1195, E96–106. doi:10.1111/j.1749-6632.2009.05402.x.

He, J., McHugh, A. D., Li, H. W., Wang, Q. J., Li, W. Y., Rasaily, R. G. and Li, H. 2012. Permanent raised beds improved soil structure and yield of spring wheat in arid north-western China. *Soil Use and Management* 28(4), 536–43. doi:10.1111/j.1475-2743.2012.00445.x.

Hølleland, H., Skrede, J. and Holmgaard, S. B. 2017. Cultural heritage and ecosystem services: a literature review. *Conservation and Management of Archaeological Sites* 19(3), 210–37. doi:10.1080/13505033.2017.1342069.

ITAIPU. 2017. Itaipu dam works to combine energy, environment and biodiversity. Itaipú Dam statement on commitment at COP 23, External Statement on 8 November 2017, Budapest, Hungary. Available at: https://unfccc.int/news/itaipu-dam-works-to-combine-energy-environment-and-biodiversity.

ITAIPU. 2019. *Cultivando Agua Boa* (*Growing Good Water*). Available at: https://www.itaipu.gov.br/en/the-environment/cultivating-good-water (accessed on 4 July 2019).

Juniper, T. 2013. *What Has Nature Ever Done for Us? How Money Really Does Grow on Trees*. Profile Books, UK, 336pp.

Kassam, A. H., Friedrich, T., Shaxson, F. and Pretty, J. N. 2009. The spread of Conservation Agriculture: justification, sustainability and uptake. *International Journal of Agricultural Sustainability* 7(4), 292–320. doi:10.3763/ijas.2009.0477.

Kassam, A., Friedrich, T., Shaxson, T. F., Reeves, T., Pretty, J. and Sà, J. C. de M. 2011a. Production systems for sustainable intensification: integrated productivity with ecosystem services. *Technikfolgenabschatzung - Theorie und Praxis* 2, 39–45.

Kassam, A. H., Mello, I., Goddard, T., Friedrich, T., Laurent, F., Reeves, T. and Hansmann, B. 2011b. Harnessing ecosystem services with conservation agriculture in Canada and Brazil. *5th World Congress of Conservation Agriculture Incorporating 3rd Farming Systems Design Conference*, September 2011, Brisbane, Australia.

Kassam, A. H., Basch, G., Friedrich, T., Shaxson, F., Goddard, T., Amado, T., Crabtree, B., Hongwen, L., Mello, I., Pisante, M. and Mkomwa, S. 2013. Sustainable soil management is more than what and how crops are grown. In: Lal, R. and Stewart, B. A. (Eds), *Principles of Soil Management in Agro-Ecosystems. Advances in Soil Science*. CRC Press, Taylor & Francis Group, Boca Raton, FL.

Kassam, A. H., Mkomwa, S. and Friedrich, T. 2017. *Conservation Agriculture for Africa. Building Resilient Farming Systems in a Changing Climate*. CABI Publishing, Wallingford, 289pp.

Kassam, A. H., Friedrich, T. and Derpsch, R. 2019. Global spread of Conservation Agriculture. *International Journal of Environmental Studies* 76(1), 29–51. doi:10.1080/00207233.2018.1494927.

Kautz, T., Amelung, W., Ewert, F., Gaiser, T., Horn, R., Jahn, R., Javaux, M., Kemna, A., Kuzyakov, Y. and Munch, J. 2013. Nutrient acquisition from arable subsoils in temperate climates: a review. *Soil Biology and Biochemistry* 57, 1003–22. doi:10.1016/j.soilbio.2012.09.014.

Lal, R. 2014. Societal value of soil carbon. *Journal of Soil and Water Conservation* 69(6), 186A–92A. doi:10.2489/jswc.69.6.186A.

Laurent, F., Leturcq, G., Mello, I., Corbonnois, J. and Verdum, R. 2011. La diffusion du semis direct au Brésil, diversité des pratiques et logiques territoriales: l'exemple de la región d'Itaipu au Paraná. *Confins* 12(12). doi:10.4000/confins.7143. Available at: http://confins.revues.org/7143.

Li, L. L., Huang, G. B., Zhang, R. Z., Bill, B., Guangdi, L. and Kwong, Y. C. 2011. Benefits of conservation agriculture on soil and water conservation and its progress in china. *Agricultural Sciences in China* 10(6), 850–9 (in Chinese with English abstract). doi:10.1016/S1671-2927(11)60071-0.

Li, H. W., He, J., Bharuchab, Z. P., Lal, R. and Pretty, J. 2016. Improving China's food and environmental security with conservation agriculture. *International Journal of Agricultural Sustainability* 14(4), 377–91. doi:10.1080/14735903.2016.1170330.

Lin, Y., Li, H. P. and Xiao, B. 2017. Influencing factors of wind erosion and their quantitative relationships with erosion rate in black soil area of Northeast China. *Journal of Soil and Water Conservation* 4 (in Chinese with English abstract).

Lindwall, C. W. and Sonntag, B. (Eds). 2010. *Landscape Transformed: The History of Conservation Tillage and Direct Seeding, Knowledge Impact in Society*. University of Saskatchewan, Saskatoon, Canada.

MAPA. 2019. Encuesta sobre Superficies y Rendimientos de Cultivos 2018 (Ministry for Agriculture of Spain, Survey on surfaces and crop yields). Available at: https://www.mapa.gob.es/es/estadistica/temas/estadisticas-agrarias/boletin2018_tcm30-504212.pdf (in Spanish).

MEA. 2005. *Ecosystems and Human Well-Being: Synthesis*. Millennium Ecosystem Assessment, Island Press, Washington DC.

Mello, I. and van Raij, B. 2006. No-till for sustainable agriculture in Brazil. *Proceedings of the World Association of Soil and Water Conservation* P1, 49–57.

Montgomery, D. R. 2007. *Dirt: the Erosion of Civilizations*. University of California Press, Berkeley and Los Angeles, 285pp.

Nkonya, E., Mirzabaev, A. and von Braun, J. (Eds). 2016. *Economics of Land Degradation and Improvement - A Global Assessment for Sustainable Development*. IFPRI and ZEF. Springer Open, 695pp (Online). Available at: http://link.springer.com/book/10.1007/978-3-319-19168-3/page/1.

Oldeman, L. R., Hakkeling, R. T. A. and Sombroek, W. G. 1991. *World Map of the Human-Induced Soil Degradation: an Explanatory Note*. International Soil Reference and Information Centre (ISRIC), Wageningen, the Netherlands.

Pretty, J. N. 2002. *Agri-Culture: Reconnecting People, Land and Nature*. Earthscan Publications, London.

Pretty, J. N., Toulmin, C. and Williams, S. 2011. Sustainable intensification in African agriculture. *International Journal of Agricultural Sustainability* 9(1), 5–24. doi:10.3763/ijas.2010.0583.

Rojas Lara, T., Regmi, A. and Kleinwechter, U. 2015. Literature review on the integration of ecosystem services in agricultural economic models. A background discussion paper prepared for the *Workshop on Integrating Biodiversity and Ecosystem Services in to Foresight Models*, 7–8 May 2015, Rome, Italy.

Roloff, G., Lutz, R. A. T. and Mello, I. 2011. An index to rate the quality of no-till systems: a conceptual framework. *5th World Congress of Conservation Agriculture incorporating 3rd Farming Systems Design Conference*, September 2011, Brisbane, Australia.

Schwartz, N., Moretti, M., Bugalho, M. N., Davies, Z. G., Haase, D., Hack, J., Hof, A., Melero, Y., Pett, T. J. and Knapp, S. 2017. Understanding biodiversity-ecosystem service relationships in urban areas: a comprehensive literature review. *Ecosystem Services* 27(Part a), 161–71. doi:10.1016/j.ecoser.2017.08.014.

Statistics Canada. 2016. *National Agriculture Census,* Table 32-10-0406-01 - *Land Use*. Available at: https://www150.statcan.gc.ca/t1/tbl1/en/tv.action?pid=3210040601.

Swallow, B. M. and Goddard, T. W. 2016. Development of Alberta's greenhouse gas offset system within Canadian and international climate policy contexts. *International Journal of Climate Change Strategies and Management* 8(3), 318–37. doi:10.1108/IJCCSM-04-2015-0040.

Tan, C. J. 2015. Effects of conservation tillage on the soil nutrients maintenance and control of water erosion of sloping farmland. Master Dissertation. Northwest A&F University, Yangling, China (in Chinese with English abstract).

Tarnoczi, T. J. 2011. Energy efficiency offset projects under the Alberta-based offset credit system: a protocol and programme-level policy assessment. *Greenhouse Gas Measurement and Management* 1(3–4), 204–13. doi:10.1080/20430779.2011.647013.

Uri, N. D., Atwood, J. D. and Sanabria, J. 1998. An evaluation of the environmental costs and benefits of conservation tillage. *Environmental Impact Assessment Review* 18(6), 521–50. doi:10.1016/S0195-9255(98)00021-3.

Vanwalleghem, T., Laguna, A., Giraldez, J. V. and Jimenez-Hornero, F. J. 2010. Applying a simple methodology to assess historical soil erosion in olive orchards. *Geomorphology* 114(3), 294–302. doi:10.1016/j.geomorph.2009.07.010.

Vanwalleghem, T., Amate, J. I., de Molina, M. G., Fernández, D. S. and Gómez, J. A. 2011. Quantifying the effect of historical soil management on soil erosion rates in Mediterranean olive orchards. *Agriculture, Ecosystems and Environment* 142(3-4), 341–51. doi:10.1016/j.agee.2011.06.003.

Vlek, P. L. G., Le, Q. B. and Tamene, L. 2010. Assessment of land degradation, its possible causes and threat to food security in sub-Saharan Africa. In: Lal, R. and Stewart, B. A. (Eds), *Advances in Soil Science – Food Security and Soil Quality*. CRC Press, Boca Raton, FL, pp. 57–86.

Wang, R. D., Xiao, D. P., Chang, C. P. and Guo, Z. 2014a. Estimation on farmland soil loss by single wind erosion using improved particle-size distribution comparison method. *Transactions of the CSAE* 30, 278–85 (in Chinese with English abstract).

Wang, L. X., Jiang, X., Li, Y. and Ji, H. Y. 2014b. Effects of conservation tillage on soil water erosion and soil compaction degree. *Journal of Irrigation and Drainage* 33, 83–5.

Wang, Q., Wang, J. and Zhang, X. L. 2015. Change characteristics of wheat and maize anniversary production and soil nutrient content under different rotation tillage patterns. *Journal of Henan Agricultural University* 49(04), 429–37 (in Chinese with English abstract).

Yang, C. H., Feng, F. X. and Cai, Q. 2019. Soil wind erosion in wheat/maize intercropping with zero tillage and straw retention. *Journal of Desert Research* 39, 9–15 (in Chinese with English abstract).

Zhang, S. X., Chen, X. W., Jia, S. X., Liang, A. Z., Zhang, X. P., Yang, X. M., Wei, S., Sun, B., Huang, D. and Zhou, G. 2015a. The potential mechanism of long-term conservation tillage effects on maize yield in the black soil of Northeast China. *Soil and Tillage Research* 154, 84–90. doi:10.1016/j.still.2015.06.002.

Zhang, Y. F., Li, H. W., He, J., Wang, Q. J., Li, W. Y., Chen, W. Z. and Zhang, X. Y. 2015b. Effects of maize straw mulching on runoff and sediment process of slope. *Transactions of the CSAE* 31, 118–24 (in Chinese with English abstract).

Zhang, J., Zhang, R. Z. and Zuo, X. A. 2016. Effect of conservation tillage and physical and chemical characteristics under a pea-wheat rotation system in the loess plateau. *Journal of Desert Research* 36, 137–43 (in Chinese with English abstract).

Zhang, G. Y., Lyu, B. B., Zhang, L. P., Liu, Z., Fan, Q. L., Wei, M. F., Yao, Z., Yuan, J. W. and Chai, Y. J. 2019. Effect of long-term no-tillage with stubble on soil fertility and diversity of prokaryotic microbiome in dryland wheat soils on the Loess Plateau, China. *Chinese Journal of Eco-Agriculture* 27, 358–68 (in Chinese with English abstract).

Chapter 14

Rehabilitating degraded and abandoned agricultural lands with Conservation Agriculture systems

Telmo Jorge Carneiro Amado, Federal University of Santa Maria, Brazil; Carlos Alexandre Costa Crusciol, São Paulo State University (UNESP), Brazil; Claudio Hideo Martins da Costa, Universidade Federal de Goiás, Brazil; Otávio dos Anjos Leal, Catarinense Federal Institute, Brazil; and Luan Pierre Pott, Federal University of Santa Maria, Brazil

1 Introduction

Land degradation may be defined as the long-term loss of ecosystem services and productivity caused by disturbances from which the soil and vegetation cannot recover unaided, and which frequently results in site abandonment. A productive agricultural soil is a living system in which biological processes are drivers of the maintenance and enhancement of soil health and productive capacity (Kassam et al., 2013). In this context, soil health represents the physical, chemical and biological status of soil and its ability to respond to agricultural production inputs (Kassam et al., 2013).

In Brazil, agricultural areas currently under degradation represent 22% of the total and account for 5% of global degrading areas. Degraded pasture area in Brazil is estimated to be around 50 M ha (FAO and ITPS, 2015). Adding cropland degraded area to this value takes the total to over 75 M ha. Degradation may be assessed by net primary productivity (NPP) (Bai et al., 2008). The Brazilian

http://dx.doi.org/10.19103/AS.2019.0049.14

NPP loss between 1981–2003, was estimated to be 63 M tons of carbon (C), the third greatest value in the world ranking of land degradation, preceded by Canada and Indonesia, with 94 and 68 M tons of C, respectively (Bai et al., 2008).

This chapter discusses the extent of degradation of croplands and pasturelands in Brazil and how integrated strategies under Conservation Agriculture (CA) can restore soil productivity in degraded agricultural lands.

1.1 Degradation of croplands

Cropland degradation is usually linked to bare fallow periods, continuous monocropping and frequent mechanical soil disturbance (Kassam et al., 2013). These processes, combined with wet and warm climates, rapidly deplete soil organic matter with detrimental effects on soil fertility, structure and biological activity, accentuating erosion and degradation. Cropping systems which do not implement CA principles (ongoing or minimal soil disturbance through no-till, permanent soil mulch cover and crop species diversification by means of rotation or association) usually cause a stratification of soil fertility (a shallow enriched layer with poor adjacent subsoil layers), subsoil compaction, weed and pest pressure, low biological activity, lack of increase of soil organic matter content, change in root architecture and the concentration of roots in shallow layers, causing low precipitation and nutrient use efficiency by plants (Dalla Nora and Amado, 2013; Dalla Nora et al., 2014) and temporal yield instability. Cropping systems which fail to provide permanent soil biomass cover by means of crop rotation/pasture/cover crops, will require soil mechanical disturbance interventions in the short-term as a temporary remediation action for soil decompaction.

Alternation between conventional tillage (CT), mainly by heavy disc or chisel tillage, and no-tillage (NT) in Cerrado and other tropical Brazilian agricultural regions is not in line with the CA concept. Kassam et al. (2013) defined CA as an approach to managing agroecosystems with the practical application of three linked principles (previously cited), along with other complementary agricultural practices of production management. When the three individual principles are applied concomitantly, the ecological foundation of CA systems is established (Kassam et al., 2019). Most of the ecological advantages of these systems: C sequestration, nutrient cycling, microbial activity and diversity, pest and disease suppression, plant stress mitigation, high infiltration and soil water storage, high soil resistance to erosion and temporal yield stability, result from biological processes which are more pronounced in mature systems (> 10 years without interruption) (Amado et al., 2006; Ferreira et al., 2018a; Kassam et al., 2019; Sá et al., 2017, 2018).

The levels of soil degradation and susceptibility to erosion have an inverse relationship with food security. Therefore, regions with large areas of degraded soils are more prone to gradually shrinking and unstable crop yields, even when external interventions are adopted. The loss of topsoil through erosion is one the most severe types of soil degradation. It occurs in a large variety of climatic and land use conditions worldwide and is associated with surface run-off (FAO and ITPS, 2015). This process is normally triggered by tillage and a decrease in soil cover. It is more pronounced in coarse-textured soils and in dry seasons (wind erosion) or in heavy tropical rain events (water erosion) as commonly observed in the Brazilian Cerrado region (Oldemann et al., 1991; Borreli et al., 2017).

Due to the diversity of soil and vegetation types, climate conditions and soil use management in Brazil, soil erosion varies widely across the country. In general, the highest erosion rates occur where vegetation has been cleared without posterior adoption of conservation practices (Guerra et al., 2014). Estimates of the average erosion rates range between 15 and 27 Mg ha^{-1} $year^{-1}$ (Embrapa, 2002; Borreli et al., 2017). These values are far greater than typical soil formation rates of 0.15–2 Mg ha^{-1} $year^{-1}$ (Debarba and Amado, 1997; FAO and ITPS, 2015) and erosion rates which do not impair soil fertility, ~2 Mg ha^{-1} $year^{-1}$ (Guerra et al., 2014).

Between 2001 and 2012, soil erosion increased by around 8% in South America, mostly driven by deforestation and a large expansion of cropland over native grassland areas in Argentina (42%), Brazil (20%), Bolivia (38%) and Peru (6%) as reported by Borreli et al. (2017). The authors compiled data on the land surface of 202 countries (corresponding to 84% of global land surface), and observed that cropland covered only 11% of the studied surface in 2001 and 2012 but was responsible for 50 to 51% of the total predicted erosion in those years. The average erosion rate in croplands (12.7 Mg ha^{-1} $year^{-1}$) was 77-fold higher than that in forests (0.16 Mg ha^{-1} $year^{-1}$) and sevenfold higher compared to that in other natural vegetation sites (1.84 Mg ha^{-1} $year^{-1}$). These figures show the necessity of efforts to accelerate the expansion rate of CA in order to arrest soil degradation and improve soil health of agricultural soils that already present some level of degradation.

1.2 Degradation of pasturelands

Due to the expansion of Brazilian agricultural frontiers, especially in the Cerrado (Midwest) and Amazon biomes, the native vegetation of these areas has been replaced by pasture as well as by crops. Approximately 62% of the deforested Amazon area (Almeida et al., 2016) and 42% or more of the Cerrado area have been shifted to pasture (Klink and Machado, 2005; Beuchle et al., 2015). In the short-term, the conversion of native vegetation to pasture or agriculture promotes a soil-fertilising effect due to the mineralisation of native biomass and soil organic matter which results in a rapid flush of nutrients to the soil. However, if no further soil amendment is carried out, the nutrient concentrations in the soil tend to decrease rapidly. Consequently, these areas quickly become degraded and are abandoned by farmers seeking new and more productive sites. This triggers a further pasture and soil degradation cycle and contributes to the territorial expansion of degraded soil and native biomes (Luizão et al., 2009; Fearnside, 2013).

In general, pasture degradation is a result of poor fertilisation, overgrazing, low biomass input to the soil, exhaustion of natural soil fertility, organic matter depletion, weeds, pest and disease dissemination, loss of plant vigour and soil compaction (Asner et al., 2004; Müller et al., 2004; Santos et al., 2015). This process leads to a gradual reduction of pasture productivity and quality and facilitates the invasion of an area by plants of low nutritional quality. The consequent increase of grazing intensity and the almost complete removal of above-ground biomass limits the capacity of a pasture to regrow. High-intensity movement of livestock in search of sufficient grazing causes soil compaction (especially under conditions of high moisture and bare soil), an increase in surface run-off (Braz et al., 2013; Merten and Minella, 2013) and depletion of soil organic matter and nutrients (Silva et al., 2014).

Agricultural land degraded by overgrazing must be rehabilitated (Kassam et al., 2013). Integrated pasture-livestock systems under CA can increase C sequestration rates in the soil, especially under low grazing intensity conditions, as observed by Cecagno et al. (2018). In a 13-year crop-livestock system experiment in Southern Brazil, the authors reported soil C accumulation rates of 0.308 and 0.097 Mg ha^{-1} $year^{-1}$ with grazing heights of 0.10 and 0.40 m, respectively. Improved grazing management strategies known as 'rotatinuous stocking' have emerged in CA pasture-livestock systems in recent years. This system is based on animal behaviour and was developed to identify the ideal sward heights for different forage species. This has resulted in higher animal nutrient intake and lower methane emission and grazing intensities when compared to traditional rotational stocking with pre- and post-grazing target sward heights of 25 and 5 cm, respectively (Savian et al., 2018). The authors estimated that rotatinuous stocking reduced methane emissions by 170% per unit of animal product compared to the traditional system.

Soil health is under improving in only 10% of global agricultural land (FAO, 2011). Strategies for the rehabilitation of degraded soils are therefore key in sustaining future global food security, since advances in smart-machines, genetics and plant protection can be impaired by soil degradation, frustrating crop yield increase projections for the next decades.

1.3 Rehabilitation of degraded agricultural lands with Conservation Agriculture

Overall, the soil degradation data reported in Sections 1.1 and 1.2 highlight the urgent need for including CA in the global agenda, especially in agricultural countries such as Brazil. It is evident that increasing crop yields by rehabilitating degraded areas is preferable to expanding agricultural lands at the expense of degrading native environments (FAO and ITPS, 2015; CGEE, 2016). CA is estimated to have reduced soil erosion by 70–90% or more over a large area of Brazilian croplands since the 1980s (FAO and ITPS, 2015). In Brazil, CA has also been the means of supporting increases in rice, maize, soybean and wheat yields by 44, 72, 48 and 64%, respectively (FAO and ITPS, 2015). However, crop diversification and the use of cover crops in cropping systems can be further improved to increase soil organic matter content and biological activity.

The rehabilitation of agricultural lands for optimum and competitive yields relies on mitigating degradation processes. Initially, this is done by controlling soil erosion through avoiding tillage and by planting soil cover crops. These protect the soil surface from wind and the impact of direct rainfall. They also improve water infiltration into the soil. This is followed by correcting soil acidity, restoring fertility and improving soil physical attributes in order to facilitate deep root growth. Long-term CA experiments have shown that the adoption of cropping systems which promote continuous soil cover and highly diverse biomass input with well-developed root systems are efficient to enhance nutrient cycling. The use of biological decompaction processes are preferable to mechanical strategies, since they promote increase in water-use efficiency with minimum soil mobilisation, what is critical for maintaining C protection and avoiding the soil aggregates disruption (Bayer et al., 2009; Ferreira et al., 2013, 2018b; Silva et al., 2016). Organic amendments such as cattle manure and swine slurry may further increase C sequestration in soils under CA (Grave et al., 2015; Nicoloso et al., 2016). This may occasionally promote C saturation in the superficial soil layers, and in time, cause C translocation and deposition in the deeper layers,, as reported by Nicoloso et al. (2018) in a 25-year NT experiment with organic fertilisation carried out in Kansas, USA.

Soil C is well-known to be one of the best soil quality indicators due to its close relationship with various soil functions (Amado et al., 2007) and therefore, to soil productivity potential and soil health (USDA, 2015). However, investigative studies on the effects of land use changes on soil C are usually limited to sampling the shallow layers, with less attention being paid to depths of > 0.30 m, which may contain most of the C in the soil profile (Reis et al., 2014; Leal et al., 2019). As mentioned above, the accumulation of C at deeper soil levels tends to increase over time with CA adoption, either via downwards C translocation or by deep root-derived local C deposition, due to improvements in the subsurface of acid tropical soils. Based on a 21-year CA experiment, Albuquerque et al. (2015) concluded that the effects of these processes are magnified if legume cover crop-based rotation systems are adopted instead of grass-based systems.

Investigating six pioneer and consolidated (>20 years) CA areas in Southern Brazil, Ferreira et al. (2016) reported higher C stock recovery levels in high-cropping intensity systems (85–117%) than in areas of medium-cropping intensity (80–85%), taking adjacent native vegetation sites as reference. However, the authors concluded that a more frequent use of soybean in the rotation system led to lower soil C recovery (61%). A range of soil texture and mineralogy characteristics were represented in this work and also played an important role in the soil C recovery levels. Soils with higher clay texture, lower aluminium content and higher calcium, magnesium and potassium had higher soil C recovery. These data reinforce the fact that intensive cropping can contribute to re-establishing soil C stocks, even having the potential to surpass the original values, and that plant diversification is essential to reach CA long-term benefits (Kassam et al., 2019). Comparing CT and NT with or without legume cover crops in a long-term experiment in Southern Brazil, Veloso et al. (2018) found that combining NT and legume crops produced the highest rate of C accumulation in the soil (1.15 Mg ha^{-1} $year^{-1}$).

2 Conservation Agriculture adoption in Brazil, mainly in Brazilian Southern region, as a tool to prevent and reverse soil degradation

The CA concept was initially adopted to control wind erosion in the United States and Canada, water erosion in Brazil, drought in Australia and for increasing cropland area in Canada (Jat et al., 2014). Between 1960–1970 in Brazil, agriculture and soil disturbance were intensified, especially in Southern Brazil (Rio Grande do Sul, Santa Catarina and Paraná States) where the agricultural area increased by 400% between 1969 and 1977 (Calegari et al., 2014), mainly due to soybean and wheat cultivation. After areas of native vegetation had been cleared and the biomass residues burned, ploughing and harrowing would be performed to prepare soil for cropping. Following harvesting, the crop residues were burned to facilitate seeding operations. At that time, mechanical aggregate disruption, direct raindrop energy, soil sealing and bare soil all contributed to severely increased water erosion rates (causing deep gullies) and soil transportation to water bodies. Soil losses by erosion reached up to 10 Mg ha^{-1} per ton of grain harvested in Southern Brazil (Calegari et al., 2014). In this area, crop yields stagnated for a decade despite the advances in plant genetics, machinery, fertilisers and agrochemical products (Amado and Eltz, 2003). Soil conservation strategies were sought to remediate degradation and change this unsustainable production system (Amado et al., 2006).

In the 1970s, NT was introduced in Brazil, particularly in Paraná State (Southern Brazil) as a local initiative of pioneer farmers. It was followed by support from governmental and research institutions to control erosion. NT expansion in Brazil occurred in three periods: (i) up to 1979; (ii) 1980–1991 and (iii) 1992–present day (Calegari et al., 2014). The first stage was characterised by the introduction and regional dissemination of the concept of minimum or no soil disturbance, in contrast to the usual practice of intensive disturbance. By the end of this period, the NT area in Brazil was 0.05 M ha (Fig. 1). In the second stage, the experimental results supported the efficacy of this practice in soil erosion control (Bernoux et al., 2006; Denardin et al., 2008). Within this period, efficient NT planters and knife-rollers or crimpers were developed. New options for the chemical control of weeds were discovered and both the practical knowledge of NT farmers and scientific research supported its regional dissemination. In 1991, the NT area in Brazil covered 1 M ha (Fig. 1). The third expansion period, on a national scale, was boosted by the maturing of NT knowledge which was moving in the direction of CA principles (Kassam et al., 2013). This involved innovations in machinery and agrochemicals during the mid-1990s, qualified assistance and a further consolidation of CA benefits

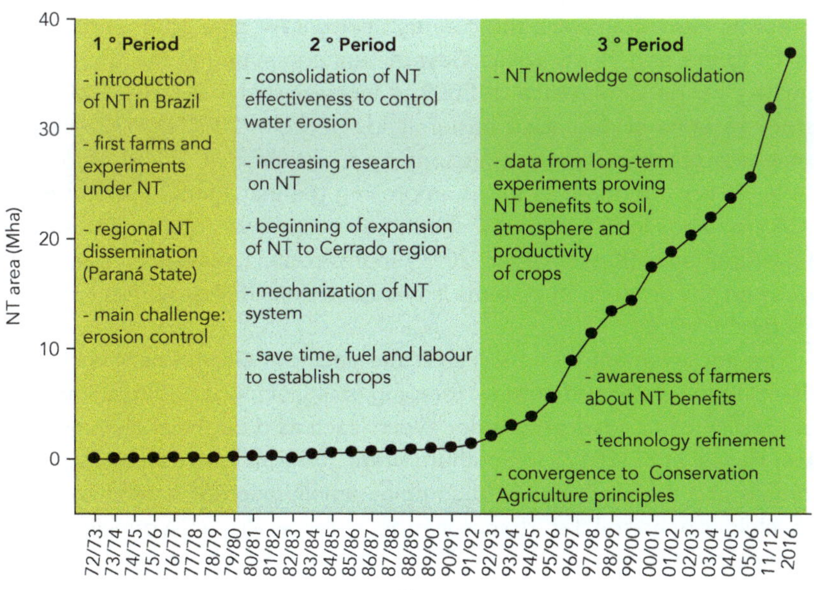

Figure 1 Area and main periods of no-tillage (NT) system evolution and spread in Brazil. Source: adapted from Calegari et al. (2014).

evidenced by medium and long-term experimental data. In 2016, the NT area in Brazil reached almost 37 M ha (Fig. 1).

Five decades after the introduction of NT in Brazil, the challenges have broadened beyond the original focus of controlling soil erosion. Although the NT area has continued to increase (Sá and Ferreira, 2018), there is currently concern over the limited diversification of crop rotation systems due to the frequent cultivation of soybean by farmers and a low diversification of cover crops which is caused by the double-cash crop system and reduces the window for soil restoration. It may prove challenging to continue transformation of the large area of Brazilian NT to the CA concept (Kassam et al., 2019), but it also presents an opportunity to meet the demands for increased crop yield.

CA involves the management of a set of soil-crop-nutrient-water-landscape integrated systems which promote the intensification of sustainable production, reduce greenhouse gas emissions, increase C sequestration into the soil, crop productivity and the enhancing of biodiversity in crop production systems below and above ground (Amado et al., 1998; Friedrich and Kassam, 2012). The 2015/16 global estimate of CA cropland was reported to be around 180 M ha (Kassam et al., 2019).

The optimisation of agricultural lands through CA leads indirectly to the preservation of native ecosystems. According to Motter and Alemida (2015), Brazil has approximately 100 M ha which could be rehabilitated by CA implementation before other native areas are converted to agriculture. This represents more than twice the country's current NT area of around 37 M ha (Fig 1). To this end, the Brazilian Government committed at the Copenhagen Climate Change Conference of 2009 to increase its NT area by 8 M ha and restore 15 M ha of degraded pasturelands between 2011 and 2020, aiming to mitigate greenhouse gas emissions by approximately 111 M Mg eq CO_2 (MAPA, 2012). Important as this is, improving the management of current NT or CA croplands to fully integrate the three interlinked CA principles is also of equal importance (Kassam et al., 2019). In this context, research into means of improving CA production systems is necessary to address current and future societal challenges.

The improvement of subsoil structure and root environment is crucial in enhancing the water and nutrient foraging ability of crops and thus improving plant resilience to adverse weather events such as drought (Dalla Nora et al., 2014; Hansel et al., 2017). In Southern Brazil, drought has caused severe crop shortfalls and the challenge is to combine amelioration of subsoil conditions with minimal or no soil disturbance in order to follow all the CA principles (Dalla Nora et al., 2017).

In the next section, complementary agricultural practices in CA systems are listed in order to address the challenges of sustainable production intensification.

2.1 Complementary agricultural practices

1. Attenuation of chemical properties gradient along tropical soil profiles. This process aims to correct subsoil acidity and decrease high Al^{+3} activity. Farmers are generally advised to incorporate high doses of limestone to the soil at CA implementation and then to maintain CA without soil disturbance in the long-term. Time, fuel, labour and costs are frequently limiting factors for scaling up this strategy. Recently, farmers have been remediating poor chemical subsoil quality by using lime combined with gypsum applied to the soil surface, thus preserving its structure. Related references: Dalla Nora and Amado (2013), Tiecher et al. (2018).
2. Integration of crop-livestock systems to allow greater exploration of soil volume. *Brachiaria* roots can reach considerable depths, improving the cycling of mobile nutrients as N, sulphur (S), potassium (K) and Boron, and increasing C allocation and sequestration, biological activity and deep soil aggregation. Additionally, *Brachiaria* can add 6–10 Mg ha^{-1} $year^{-1}$ of biomass to the soil, promoting biological activity, erosion control, pore continuity, water-use efficiency and interrupting pest and disease cycles. Related research: Balbinot Junior et al. (2017).
3. Fertilisation management. The fertilisation of cash crops can be partially shifted towards cover crops and pasture, promoting higher biomass production and faster plant growth enhancing nutrient cycling. Related research: Fageria et al. (2005).
4. Overseeding cover crops into standing cash crops at maturity or joint sowing of main crops and pasture (e.g. maize plus *Brachiaria*). Related research: Correia and Gomes (2015), Teague et al. (2019).
5. Utilisation of biological soil conditioners such as *Trichoderma*, *Azospirillum*, *Bacillus subtilis* and others (inoculation of biota into the soil or organic fertilisation to multiply native biota) to promote plant growth and resistance. Related research: El-Katatny and Idres (2014), Azcón-Aguilar and Barea (2015), Lehman et al. (2015), Singh et al. (2018).
6. Polyculture of cover crops within cropping windows as a key to enhance biological diversity and ecosystem resilience where soybean is economically predominant. Two strategies have been developed. The first is to implement a summer crop with fast biomass production straight after soybean cultivation, interrupt its cycle after 60–80 days, and sow a winter cover crop preceding the next soybean crop season (double cover crop before a cash crop). The second option is to intercrop 5–8 species sowed simultaneously ('cover crops cocktail'), which rapidly enhances plant and soil biota diversity and improves the overall system quality. Related research: Algeri et al. (2018), Büchi et al. (2018), Forte et al. (2018).

7 Integration of chemical, physical and biological methods of subsoil decompaction to rehabilitate degraded soils. Innovative farmers have combined soil chemical amelioration using lime and gypsum with mechanical chiselling and the cultivation of cover crops to promote soil decompaction and restore the productive potential of the system. Related research: Nicoloso et al. (2008).

Two case studies are presented to illustrate how CA can be part of the solution in reversing soil degradation and promoting sustainable agriculture intensification based on soil health improvement.

3 Case study 1: Integrated strategies for restoration of compacted and low productive soils under no-tillage system in Southern Brazil

3.1 Contextualisation

Large areas of land in Brazil have limited productivity due to poor soil and plant management. Water erosion, subsoil acidity, organic matter depletion and soil compaction are among the main problems to be overcomed in order to rehabilitate these soils, increase economic profit from agricultural activities and support the country's role in meeting increasing global demands for food, energy and the production of fibre.

Compaction increases soil penetration resistance (PR), affects the balance of soil gases and the efficiency of plants in taking up water and nutrients (Batey and Mckenzie, 2006; Hösl and Strauss, 2016). According to Stefanoski et al. (2013) it is the main factor limiting productivity in intensively mechanised croplands. It is linked to intensive tillage, low crop residue input to the soil, soil organic matter depletion and lack of crop diversity. Improving understanding on the remediation of soil compaction and restoration of soil productivity in these poorly managed croplands can have a positive impact on grain production and economic return.

Compaction levels vary due to external and internal soil factors. The external factors are type, intensity and frequency of applied load. The internal factors are soil moisture, texture, structure, C content, initial soil density, biological activity and history of stress. The interaction between these factors is also significant (Richart et al., 2005; Collares et al., 2008; Nawaz et al., 2013). The depth of compaction will vary according to the soil management system adopted and is generally observed to be between 0.08 and 0.15 m (Reichert et al., 2007; Sivarajan et al., 2018).

Mechanical, biological and chemical interventions, in single or combined operations, may be carried out to mitigate soil compaction. In general, the mechanical alleviation of compaction is efficient in the short-term (Nicoloso et al., 2008), although the medium and long-term effectiveness of these interventions have been controversial (Nunes et al., 2014). The cultivation of cover crops, singly or in combination, with different root system architectures and abilities of root penetration, induce (biological) disruption of compacted layers (Reichert et al., 2007; Nicoloso et al., 2008). They may also improve soil surface protection and C allocation at surface and subsurface soil depths (Nicoloso et al., 2018). The alleviation of soil compaction by cover crops is more pronounced over time and is linked to the frequency with which they are grown (Nicoloso et al., 2008). Where the level of compaction is high, the efficiency of cover crops may be limited and in this case, knowledge of the level of physical impedance is critical. Cubilla et al. (2002) reported that soil bulk density > 1.85 Mg m^{-3} was critical to root growth of cover crops, as shown in Fig. 2.

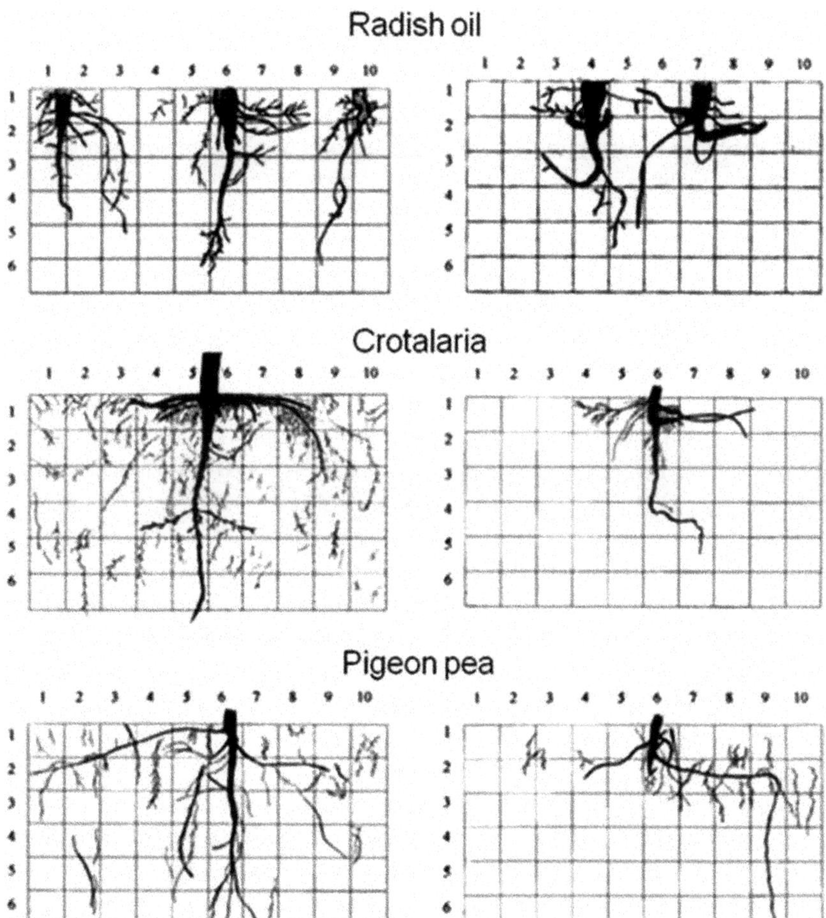

Figure 2 Forage radish (radish oil) (*Raphanus sativus*), crotalaria (*Crotalaria juncea*) and pigeon pea (*Cajanus cajan*) root systems under two soil bulk density conditions: left column = density < 1.75 Mg m^{-3} and right column = density > 1.85 Mg m^{-3}. Source: adapted from Cubilla et al. (2002).

The integration of mechanical and biological chiselling methods may alleviate soil compaction more efficiently than single operations. Nicoloso et al. (2008) reported high soybean yields when both methods (chiselling and a combination of black oat and forage radish) were integrated. These systems had lower soil PR, higher macroporosity and water infiltration rates in relation to chisel tillage alone or single cover crop (black oat) treatments. The integration of mechanical and biological chiselling promotes short-term alleviation through mechanical intervention which enables the deepening of cover crop roots, thus affecting the subsurface structure and extending decompaction effects over the medium to long-term.

The main advantages of integrating chemical (lime and gypsum) and biological (cover crops) methods for soil decompaction are fuel economy, reduced labour demands and conservation of soil organic matter and structure (Ferreira et al., 2013).

This case study selected a NT cropland area with low productive records (yield map) and investigated integrated methods to alleviate soil compaction, restore soil quality and improve soybean grain yield.

3.2 Experimental procedures

The experimental area is located at Não-Me-Toque city, Rio Grande do Sul State, Southern Brazil, geographic coordinates 28° 28' 00'' S, 52° 47' 00'' W, 500 m above sea level. The climate is classified as Cfa (humid subtropical) (Köppen and Geiger, 1928). The average annual temperature and precipitation are 18.7°C and 1680 mm, respectively. The soil is classified as clayey, kaolinitic, thermic Rhodic Hapludox (*Latossolo Vermelho Distrófico típico*) according to Santos et al. (2013), defined in this text as Oxisol.

The crop succession adopted is mainly soybean (*Glycine max* (L.) Merr.) in the summer, followed by black oat (*Avena sativa* L.) as the winter cover crop. This area has been monitored over the last decade in order to define the management zones and particularly to identify low-yielding sites. For this purpose, precision tools were used: cumulative yield maps, electrical conductivity and soil chemical and physical analyses. Soil compaction was identified as a driver of poor crop performance in low-yielding zones, with high PR values, mainly at the 0.07–0.20 m layer.

After identifying low-yielding zones with soil compaction restrictions, three experimental blocks were installed in June 2016. The treatments consisted of (i) control (no intervention); (ii) physical intervention - mechanical soil disturbance (chiselling 0–0.30 m depth); (iii) chemical intervention - improved subsoil quality (phosphogypsum, $CaSO_4$ after lime input in previous year); (iv) biological intervention - polyculture of cover crops; (v) integrated physical and chemical interventions - mechanical chiselling and phosphogypsum; (vi) integrated biological and chemical interventions - polyculture of cover crops and phosphogypsum; (vii) integrated physical and biological interventions - mechanical chiselling and polyculture of cover crops; (viii) fully integrated physical, chemical and biological interventions - mechanical chiselling and polyculture of cover crops and phosphogypsum. The experimental units are 30 × 60 m, totalling 14 400 m^2 per experimental block.

Mechanical chiselling was performed using a chisel with seven rippers (Stara FOX KS®), working at a depth up to 0.30 m, when the soil moisture content was close to the point of friability. The polyculture of cover crops consisted of: garden pea (*Pisum sativum* L.), forage radish (*Raphanus sativus* L.), black oat (*Avena strigosa* Schreb.) and rye (*Secale cereale* L.). The seeding of cover plants was performed simultaneously with the chiselling tillage, using a seed container coupled to the chisel. Broadcast application of phosphogypsum was carried out over the total area prior to the seed operation with a spreader HERCULES 15000® (Stara).

In mid-October, the cover crop was knocked down with herbicide, followed by a knife-roller operation and soybean was sown on 1 November 2017. The cultivar used was Ativa® (GDM Genetics, maturity group = 5.6). The mean plant population was 330 thousand plants per hectare. Basal row fertilisation was performed applying 125 kg ha^{-1} of 10-45-00 (N-P-K) formula. The K was applied by broadcasting as KCl (140 kg ha^{-1}).

The PR was measured (10 replicates per treatment) up to a depth of 0.60 m before soybean was sown using a SoloStar® penetrometer with readings of 0.01m. Cone type 2 with size of 12.83 mm was used with a nominal insertion velocity of the rod of 35 mm sec^{-1}. Prior to the soybean sowing, soil samples were collected at 0–0.05, 0.05–0.10, 0.10–0.20, 0.20–0.30 and 0.30–0.40 m depth using steel cylinders (0.05 m height and 0.05 m diameter) to evaluate soil density and macroporosity (Embrapa, 2017). Soybean grain yield was estimated using an infrared harvesting sensor (Stara APS®), which was installed in an automated harvester (Case 8120).

The data was subjected to analysis of variance and the means of the treatments were compared by Tukey test at 5% probability. All statistical analyses were carried out using R software (R Core Team, 2018).

3.3 Main results and discussion

3.3.1 Soil penetration resistance

In general, the treatments affected soil PR (Fig. 3), density and macroporosity. (Fig. 4). Restrictive soil compaction conditions were observed in the control treatment with PR values around 2000 kPa below a depth of 0.07 m (Fig. 3).

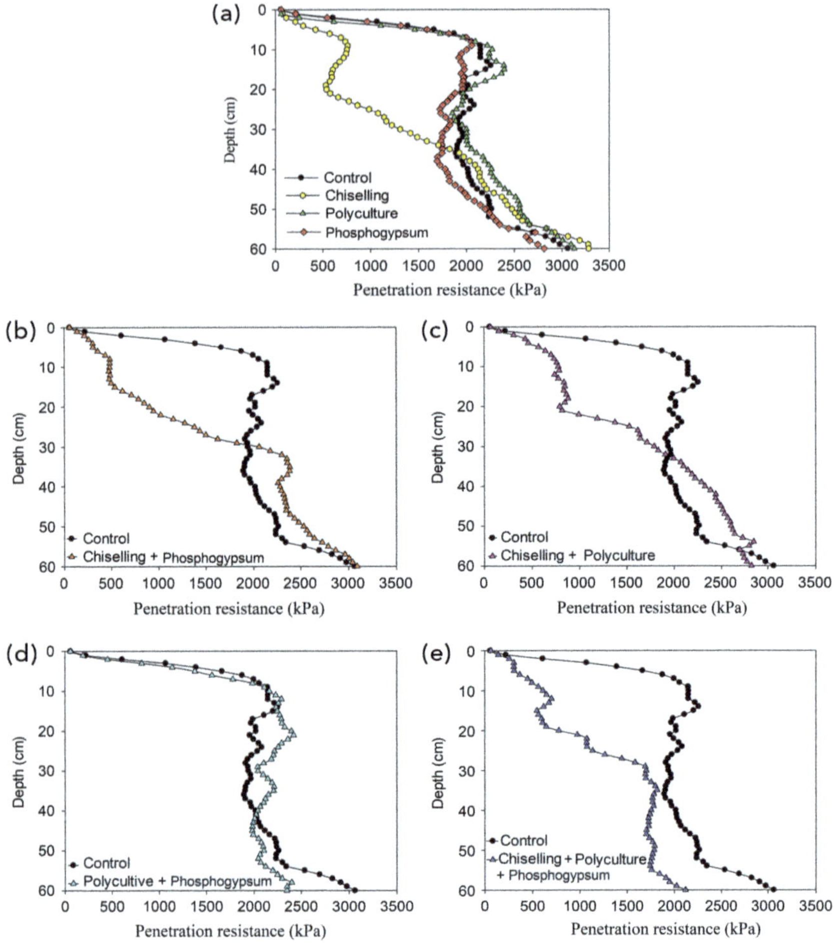

Figure 3 Soil penetration resistance according to single and integrated methods of decompaction: (a) mechanical chiselling, phosphogypsum or polyculture of cover crops in single operation, (b) chiselling and phosphogypsum, (c) chiselling and polyculture of cover crops, (d) polyculture of cover crops and phosphogypsum and (e) chiselling + polyculture of cover crops + phosphogypsum. The control treatment (no intervention) data is presented in all figures to aid visualisation of the treatment effects.

Restrictive PR values for soybean growth in Oxisols have been reported by many authors. Secco (2003) observed critical PR ranging from 2650 to 3260 kPa, with a soybean yield penalty of 24%. These data agree with the findings of Beutler (2003) and Hamza and Anderson (2005). The authors reported that PR values of 2240 kPa or higher decreased soybean yields by 32%. Similar PR values were observed in the control treatment of the present study (Fig. 3), indicating that interventions are needed to rehabilitate the soil and increase its productivity.

Mechanical chiselling as a single intervention decreased PR to values lower than 2000 kPa at a depth of 0–0.30 m (Fig. 3a). However, PR increased progressively below a chiselling operation depth of 0.30 m, reaching PR values around 3400 kPa at a depth of 0.60 m (Fig. 3a). These results confirm that the mechanical chiselling effect is restricted to the depth of the shank operation. However, when mechanical chiselling was combined with phosphogypsum, PR values were maintained at around 1000 kPa up to 0.20 m and 2000 kPa at deeper levels (0.30–0.60 m) (Fig. 3b), significantly increasing soybean yields when compared to mechanical chiselling alone (Fig. 5). Gypsum increases Ca and Mg contents and decreases Al activity at subsurface depths, due to the downward movement of SO_4^{-2}, improving chemical attributes in the rooting zone and stimulating deeper root growth to disrupt compacted layers (Dalla Nora et al., 2017).

Polyculture cover crops as a single intervention did not decrease PR values significantly compared to the control treatment (Fig. 3a), possibly due to the short period of time between implementation and PR measurement which was only four months after seeding. According to Calonego et al. (2017) the effect of cover plants on PR is not immediate as it is for mechanical chiselling and is generally more pronounced in the medium and long-term. Nevertheless, the authors observed that decompaction by mechanical chiselling did not last up to the second year, while cover crops led to soybean yields equal or greater than those obtained by chiselling and also improved subsurface soil structure, providing longer soil decompaction benefits. The combination of mechanical chiselling and polyculture of cover crops efficiently decreased PR values up to a depth of 0.30 m compared to the control (Fig. 3c). Below 0.30 m, PR values in these treatments were similar, most probably due to the chiselling depth (limited to 0.30 m) and to the fact that there was no chemical improvement in the subsoil layers. In this case, phosphogypsum (chemical intervention) can improve soil physics by decreasing clay dispersion, therefore improving aggregation and structure (Dalla Nora et al., 2014, 2017). Only the fully integrated intervention was able to reduce PR values from 0.00 to 0.60 m depth compared to the control (Fig. 3e). It is likely that these results are associated with the alleviation of chemical and physical barriers (up to 0.30 m) which allow the roots of the cover crops to extend to a greater depth, creating biopores where soybean roots can grow more easily.

3.3.2 Macroporosity and bulk density

As expected, mechanical chiselling alone increased macroporosity near the soil surface (0–0.05 m depth) when compared to other single interventions and to the control (Fig. 4a). In general, mechanical chiselling combined with phosphogypsum (Fig. 4b) and/or a polyculture of cover crops (Fig. 4c and 4e) increased macroporosity along the soil profile when compared to the

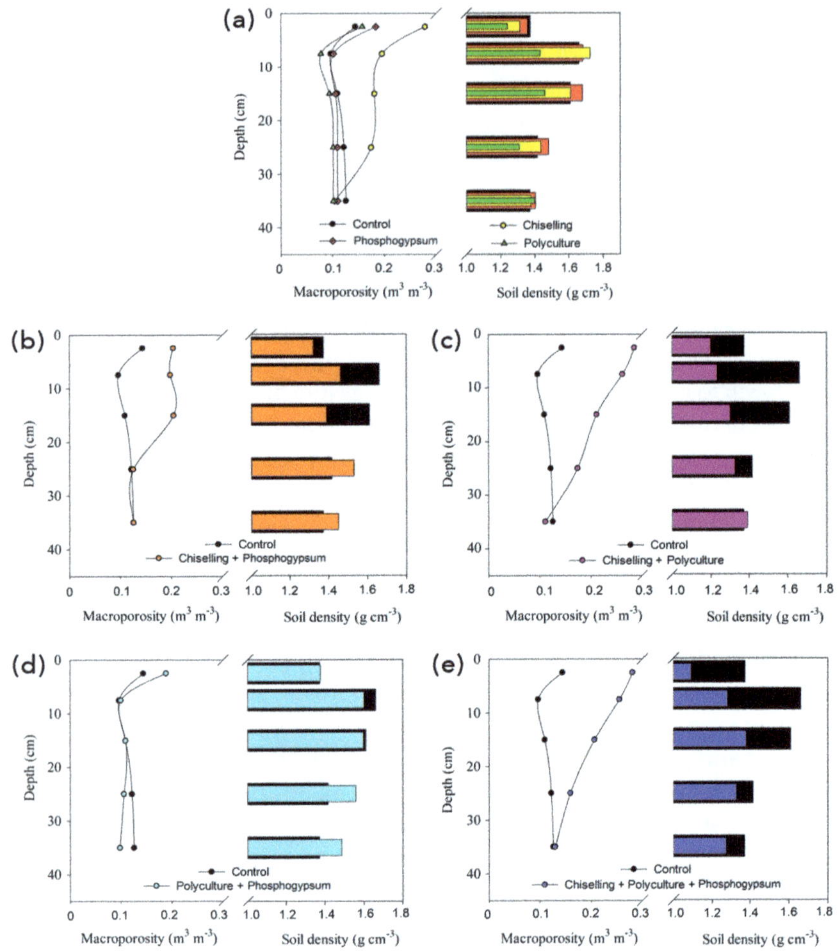

Figure 4 Soil macroporosity and bulk density according to single and integrated methods of soil decompaction: (a) chiselling, phosphogypsum or polyculture of cover crops, (b) chiselling and phosphogypsum, (c) chiselling and polyculture of cover crops, (d) polyculture of cover crops and phosphogypsum and (e) chiselling, polyculture of cover crops and phosphogypsum. The control treatment (no intervention) data is presented in all figures to aid visualisation of treatment effects.

control. However, it is important to note that the effect may be short-lived and that the association of this method with other interventions, both chemical (phosphogypsum) and biological (cover crops), is essential to prevent reconsolidation which occurs frequently in clayey soils (Abreu et al., 2004; Nicoloso et al., 2008; Calonego et al., 2017).

A polyculture of cover crops improved the soil structure at both the surface and subsurface depths: an effect probably associated with the diversity of the cover crop root systems. Plants reciprocally benefit from a better oxygen supply (mechanical decompaction), chemical amelioration by phosphogypsum (increase in Ca and decrease in Al contents) and an increase in water infiltration into the soil (Castro et al., 2011; Garcia et al., 2013). The rooting zone conditions are improved, favouring both root growth and symbiotic associations such as nodulation in soybean (Siczek and Lipiec, 2011).

Soil bulk density values in the control treatment were around 1.6 g cm^{-3} or higher in layers of 0.05 to 0.15 m depth (Fig. 4a). A comparison of the effect of single interventions on soil bulk density showed that only mechanical chiselling was able to reduce values up to a depth of 0.25 m compared to the control (Fig. 4a). Integrating mechanical chiselling and phosphogypsum reduced soil bulk density up to a depth of 0.15 m (Fig. 4b), while mechanical chiselling and a polyculture of cover crops extended the reduction up to a depth of 0.25 m (Fig. 4c). In this case, the bulk density was lower than 1.4 g cm^{-3}. A polyculture of cover crops and phosphogypsum without mechanical chiselling did not change the soil bulk density compared to the control (Fig. 4d). The fully integrated intervention was the most efficient method of reducing soil bulk density along the soil profile, especially up to a depth of 0.15 m (Fig. 4e). In general, reduction of soil bulk density was related to the reduction of soil PR.

3.3.3 Soybean yield

According to Reichert et al. (2009), soil density values higher than 1.2 g cm^{-3} can be critical for soybean root growth and values between 1.4 and 1.7 g cm^{-3} are expected to have a negative effect on growth and yield. In the present study, the lowest soybean yields were observed in single intervention treatments (Fig. 5). These results are in line with the low effect of polyculture of cover crops and phosphogypsum in relation to the control regarding PR (Fig. 3d), macroporosity and soil bulk density (Fig. 4d). Yield in the single intervention mechanical chiselling treatment was intermediate between the other single intervention treatments, biological and chemical, and lower when compared to the fully integrated intervention (Fig. 5). In this last treatment, the soil density values were lower than 1.4 g cm^{-3}, not only at the surface but also at the subsurface layers (Fig. 4e), which may have favoured root development along the soil profile, so improving the plants' nutrient and water foraging capacity and increasing yields in comparison to other treatments (Fig. 5). Thus, the combination of chiselling, polyculture of cover crops and phosphogypsum increased soybean yield by 10% (~500 kg ha^{-1}) compared to the control. It seems that integrating mechanical (chiselling), chemical (phosphogypsum) and biological (cover crops) methods promotes a synergic effect for soil decompaction, increasing soybean yield even in the first crop season in comparison to using the methods in isolation (Fig. 5). These findings corroborate those reported by Nicoloso et al. (2008).

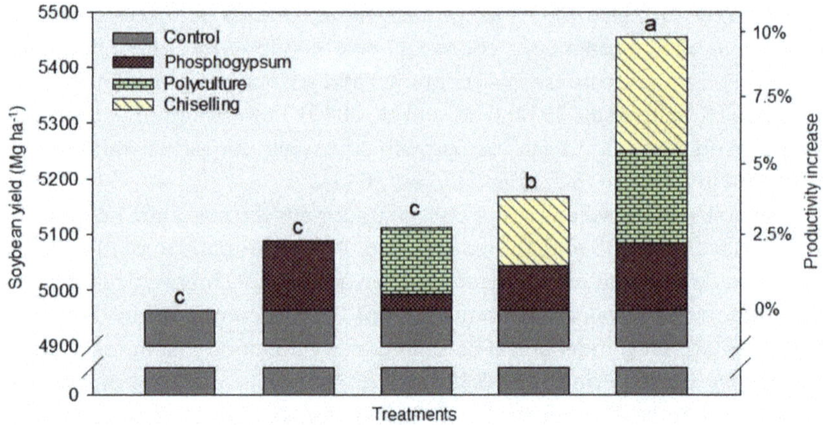

Figure 5 Soybean grain yields and productivity increase according to single and integrated methods of soil decompaction: mechanical chiselling, phosphogypsum, polyculture of cover crops and control treatment (no intervention).

3.4 Main achievements and considerations

Low-yielding zones were associated with poor soil structure and subsoil acidity. The polyculture of cover crops and phosphogypsum application had little effect on soil restoration during the first crop season. Mechanical chiselling as a single intervention was able to reduce soil compaction up to the chiselling depth. However, when integrated with phosphogypsum and polyculture of cover crops, the soil PR, macroporosity and bulk density were improved at both surface and subsurface depths, creating more favourable conditions for root growth.

Overall, the fully integrated intervention was efficient in alleviating soil compaction and resulted in higher soybean yields compared to the control. These results were attributed to the synergistic effect of the fully integrated intervention on the improvement of soil structure and foraging capacity of the plants to acquire water and nutrients stored at subsurface. It was proved to be possible to rehabilitate the degraded soil without the need to go back temporarily to CT.

4 Case study 2: Strategies for soil quality improvement in crop-livestock integration under Conservation Agriculture in acidic tropical soils

4.1 Introduction

In tropical regions, such as Cerrado in Brazil, where dry spells often occur during the rainy season (spring, summer and part of autumn, over six to eight months) and in the dry winter (part of autumn and winter, over three to five months), subsoil acidity and low nutrient availability are factors limiting cash crops and forage productivity (Sumner et al., 1986; Marsh and Grove, 1992). This growth limiting effect has been attributed to soil Ca and P deficiency and to the toxic effects of Al on root growth in the subsoil, inducing water stress and reducing plant nutrient uptake (Caires et al., 2008a).

The use of amendments, such as lime, is one strategy for mitigating soil acidity and restoring its productive potential in agricultural systems. However, in NT systems, lime is applied to the soil surface (without, or with only minimal mechanical incorporation), and due to the low solubility and slow mobility of these materials, alleviation of subsoil acidity is a long-term process (Meda et al., 2002; Caires et al., 2006a; Soratto and Crusciol, 2008a). Because its solubility and mobility are higher than that of limestone, phosphogypsum (lime and gypsum) used as a surface application is an important strategy in ameliorating acidity at lower soil depths, being an effective means of increasing the vertical movement of exchangeable bases in the rooting zone under continuously non-disturbed soils (Dalla Nora and Amado, 2013; Dalla Nora et al., 2017). The alleviation of subsoil acidity can promote greater root development, improving soil structure and increasing the tolerance of plants to water stress during dry spells, thus enabling high and stable yields (Caires et al., 2001).

It is also necessary to improve the quantity and quality of soil organic matter to ensure the productivity of tropical soils. Soil organic matter is considered to be an important factor in reducing free Al levels and increasing nutrient availability. However, tropical soils, such as Oxisols and Ultisols, exhibit a naturally low organic matter content (Carmeis Filho et al., 2017a) and plant straw decomposes relatively rapidly under tropical conditions, reducing the persistence of mulch, with consequent effects on erosion and soil quality (Pariz et al., 2016, 2017a). In CA systems, crop rotation must produce enough biomass to promote soil protection against erosive agents and to cycle nutrients during its biological decomposition (Kassam et al., 2013; Ferrari Neto et al., 2017).

Crop rotation systems have recently evolved through the introduction of tropical perennial grasses, such as palisade grass, intercropped with grains such as corn (Pariz et al., 2016, 2017a,b), sorghum (*Sorghum bicolor* (L.) Moench) (Costa et al., 2016; Mateus et al., 2016) and soybean (*Glycine max* (L.) Merr.) (Crusciol et al., 2012, 2014). This is considered to be an important strategy for producing grain and forage for livestock during the dry season and for increasing the quantity and quality of mulch (Pariz et al., 2011a,b; Borghi et al., 2013a,b; Crusciol et al., 2011, 2012, 2013, 2014, 2015; Mateus et al., 2016).

The following section explores how agricultural production may be intensified through long-term experiments with integrated crop-livestock systems under CA.

4.2 Subsoil acidity alleviation of tropical soils under Conservation Agriculture

In tropical soils such as Oxisols and Ultisols, acidity is commonly observed due to interaction between the material of origin, relief, climate and biological activity during their formation process, which characterise them as highly weathered. In these soils, the charges are mainly formed in kaolinite-type mineral clays, Fe and Al oxyhydroxides and organic matter. The higher the pH of the solution, the greater the magnitude of the negative charges generated, characterising these soils as pH-dependent negative charged. In these soils, the cation exchange capacity (CEC) is mainly attributed to the formation of pH-dependent charges in soil organic matter radicals. This is because the main mineral constituents of these soils (kaolinite, oxy-hydroxides of Fe and Al), in addition to having a reduced total CEC, also have a point of zero charge (PZC) higher than their natural pH, while in organic matter the opposite occurs (Brady and Weil, 2008). Thus, to rehabilitate these soils it is necessary to promote the reduction of acidity, consequently increasing pH, as well as increasing soil organic matter levels. Liming is the most commonly used input for alleviating soil acidity and high P sorption, increasing plant nutrient availability and reducing levels of toxic elements (Caires et al., 2005, 2006a,b, 2011). In NT systems, the neutralisation of subsurface soil acidity with limestone is generally slow, particularly in variably charged soils (Ernani et al., 2004), such as tropical soils. The movement of limestone to deeper layers varies according to the time, dose, form and frequency of application, the soil type, climate, acid fertiliser management and crop system (Blevins et al., 1978; Oliveira and Pavan, 1996; Caires et al., 2008a,b).

Figure 6 shows the correction of soil acidity with limestone applied superficially in Oxisol (Typic Haplorthox). In this study, surface liming was adopted when an NT system was implemented in August 2002. An increase in pH values and a reduction of Al^{3+} and H + Al (total acidity) up to a depth of 0.40 m were observed after 12 months. In October 2004 and 2010, limestone was applied and its effects on soil acidity were observed in the surface and subsurface soil layers. Initially, the level of acidity alleviation, despite some reduction, did not reach the values estimated by the method of elevation of base saturation. When lime was applied superficially the subsoil amelioration, as expected, was a long-term process.

In this context, the use of phosphogypsum associated with limestone has been shown to be efficient in improving soil fertility at depth. Phosphogypsum is composed mostly of calcium sulphate ($CaSO_4 . H_2O$) and is much more soluble (178 times) than limestone; however, its addition does not promote significant changes in subsoil pH. Phosphogypsum applied to the soil surface moves along the profile under the influence of percolating water (Caires et al.,

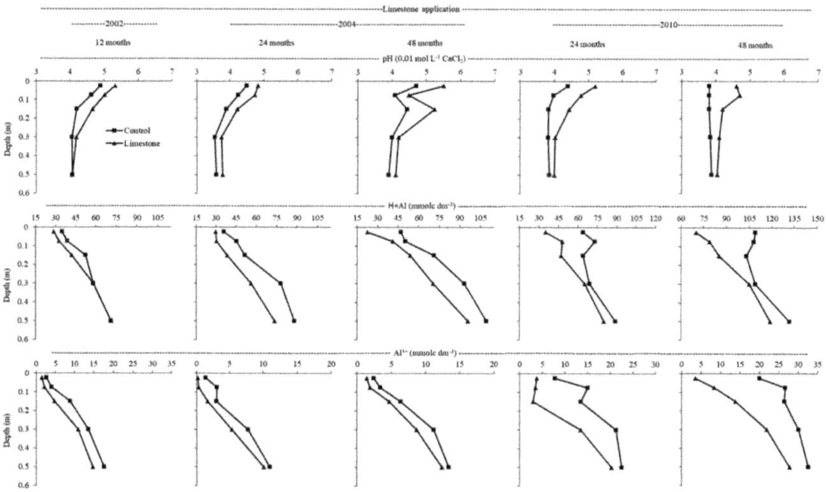

Figure 6 Soil pH, H + Al and Al^{3+} levels 12 months after the first limestone application (in 2002), 24 and 48 months after the second limestone application (in 2004), and 24 and 48 months after the last limestone application (in 2010) in a long-term no-tillage soil. The limestone rates were: 0 and 2700 kg ha^{-1} in 2002, and 0 and 2000 kg ha^{-1} in 2004 and 2010. Botucatu-SP, Brazil. Source: adapted from: Soratto and Crusciol (2008b), Crusciol et al. (2016), Costa et al. (2016), Costa et al. (authors' data) and Carmeis Filho et al. (2017b).

1999). There is a resultant increase in the supply of Ca and a reduction of Al toxicity in the subsoil (Fig. 7).

When improving soil chemical characteristics by reducing acidity, essential nutrients such as Ca^{2+} and Mg^{2+} and the reduction of Al toxicity have a high correlation with the quantity and quality of the soil organic matter. These effects are attributed to the increase in plant growth, above-ground production and deep root growth. Ca input by liming is generally accompanied by an increase in organic C. A greater interdependence of these elements was observed, especially when increased liming rates were applied, playing a fundamental role in soil organic matter stabilisation (Fig. 8). Combining lime and phosphogypsum has a considerable effect on the formation and stabilisation of humin, humic acid and fulvic acid, thus increasing the proportion of C as humic substances (Fig. 9). Because of the large number of charged functional groups on humic substances, with mineral particles through polyvalent cations, such as Ca^{2+}, the associations within soil organic matter have been one of the most important mechanisms for improving the protection and storage of organic matter in highly weathered soils (Briedis et al., 2012).

The physico-chemical protection mechanisms of organic C can stimulate soil aggregation in the long-term, thus improving soil physical quality (Ferreira

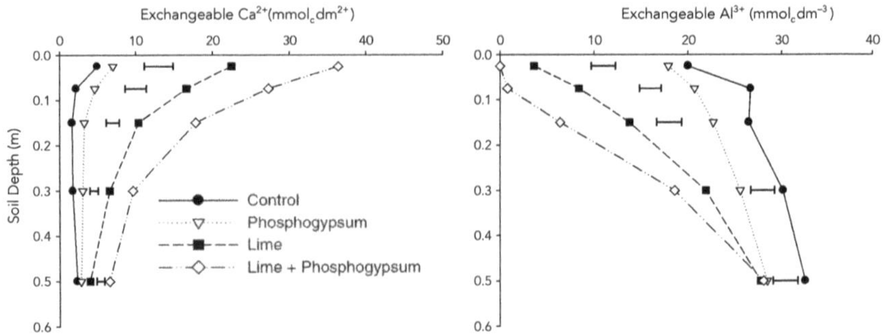

Figure 7 Exchangeable Ca and Al levels in the soil affected by soil amendment application: control (without soil amendment), and application of phosphogypsum (2.1 + 2.1 + 2.1 Mg ha^{-1}), lime (2.7 + 2.0 + 2.0 Mg ha^{-1}), and a combination of lime and phosphogypsum at the given rates, 48 months after the last application. The soil amendments were applied in 2002, 2004 and 2010. Horizontal bars represent the least significant difference at $P = 0.05$ according to the LSD test; ns: non-significant. Source: adapted from Costa et al. (2018).

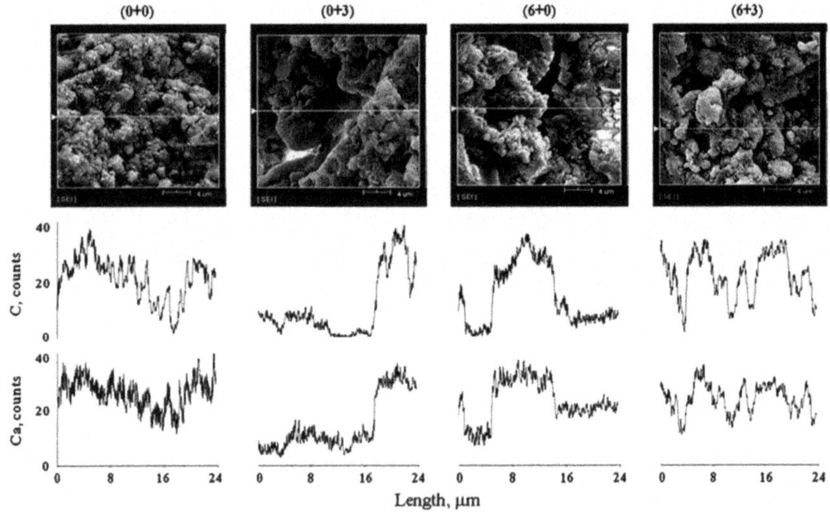

Figure 8 Carbon (C) and calcium (Ca) spectra by energy-dispersive X-ray spectrometer (EDS) attached to a scanning electronic microscope (SEM) in the central region of an 8–19 mm aggregate from the various liming treatments: 0 + 0 (0 ton ha^{-1} in 1993 and 0 ton ha^{-1} in 2000; control), 6 + 0 (6 tons ha^{-1} in 1993 and 0 ton ha^{-1} in 2000), 0 + 3 (0 ton ha^{-1} in 1993 and 3 tons ha^{-1} in 2000) and 6 + 3 (6 tons ha^{-1} in 1993 and 3 tons ha^{-1} in 2000). The white line at the centre of the pictures represents the location of the reading of 512 points by dispersive X-ray spectrometer. Source: adapted from Briedis et al. (2012).

et al., 2018a,b). Frequent lime input is crucial to the formation of water-stable macroaggregates (>2.00 to 8.00 mm), avoiding structural degradation such as a reduction in the stability of the largest aggregates (Fig. 10). Combined lime

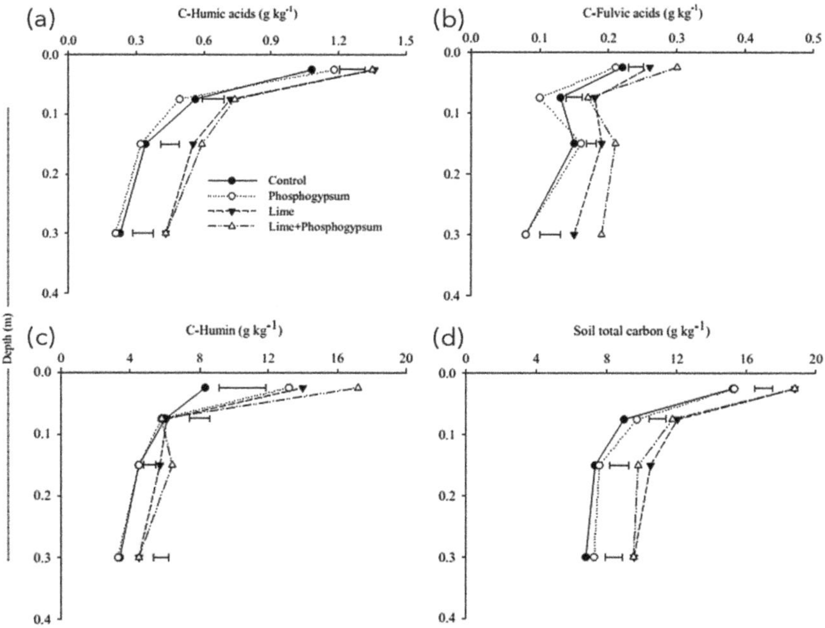

Figure 9 Soil organic carbon in the form of humic acids (a; C-HA), fulvic acids (b; C-FA), humin (c; C-Humin) and soil total carbon (d) as function of soil depth and surface application of lime and phosphogypsum in a tropical no-tillage (NT) system. The horizontal bars indicate the least significant difference between treatments within each soil depth at $P \leq 0.05$. The treatments were: control (without soil amendment), and application of phosphogypsum (2.1 + 2.1 + 2.1 Mg ha^{-1}), lime (2.7 + 2.0 + 2.0 Mg ha^{-1}), and a combination of lime and phosphogypsum at the given rates applied in 2002, 2004 and 2010, respectively. Source: adapted from Carmeis Filho et al. (2017c).

and phosphogypsum applications have a positive effect on the formation and stabilisation of large aggregates, promoting higher mean weight diameter and geometric mean diameter, greater macroporosity in addition to lower soil bulk density and PR even in the deepest layers (0.20–0.40 m) (Figs. 11 and 12, and Table 1).

Soil profile acidity alleviation with combined limestone and phosphogypsum is an important tool in tropical soils for increasing and maintaining basic cations, restoring soil organic matter and improving soil physics under CA. The use of these tools in association with CA principles are considered key strategies for rehabilitating acidic tropical soils.

4.3 Pasture quality and productivity affected by soil amelioration

Crop rotation using tropical forages provides an interesting alternative during the winter-spring period in tropical regions, serving as grazing or for cutting as

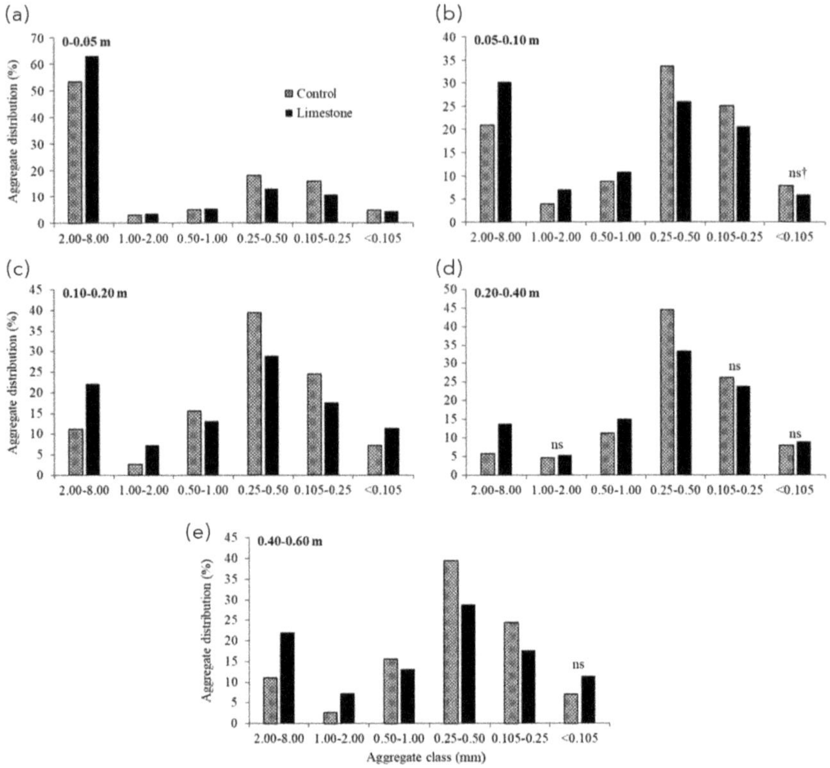

Figure 10 Water-stable aggregate distribution as affected by the surface application of lime in different soil layers in a tropical NT system. The treatments were: control (without soil amendment) and limestone (2.7 + 2.0 + 2.0 Mg ha^{-1}) applied in 2002, 2004 and 2010, respectively. †ns: not significant. Source: adapted from Carmeis Filho et al. (2018).

fodder. It may be intercropped with grain in the summer (Kluthcouski and Aidar, 2003), characterising an integrated crop-livestock system. Maize, sorghum, upland rice and soybean intercropped with tropical forages, such as *Urochloa*, is an excellent alternative for producing grain and forage for livestock during dry seasons and providing an abundant source of straw for NT systems (Pariz et al., 2011a,b; Borghi et al., 2013b; Crusciol et al., 2011, 2012, 2013, 2014, 2015; Mateus et al., 2016). In this context, as grass species are inserted into the crop production system, awareness of the competition between species for water and light and the alleviation of soil acidity become equally important for the success of grain production and adequate forage availability (Pariz et al., 2011b).

Alleviating soil acidity in CA systems with lime input increases forage dry matter yield, crude protein concentrations and meat production in regions with dry winters, such as the Brazilian Cerrado or African Savanna (Fig. 13).

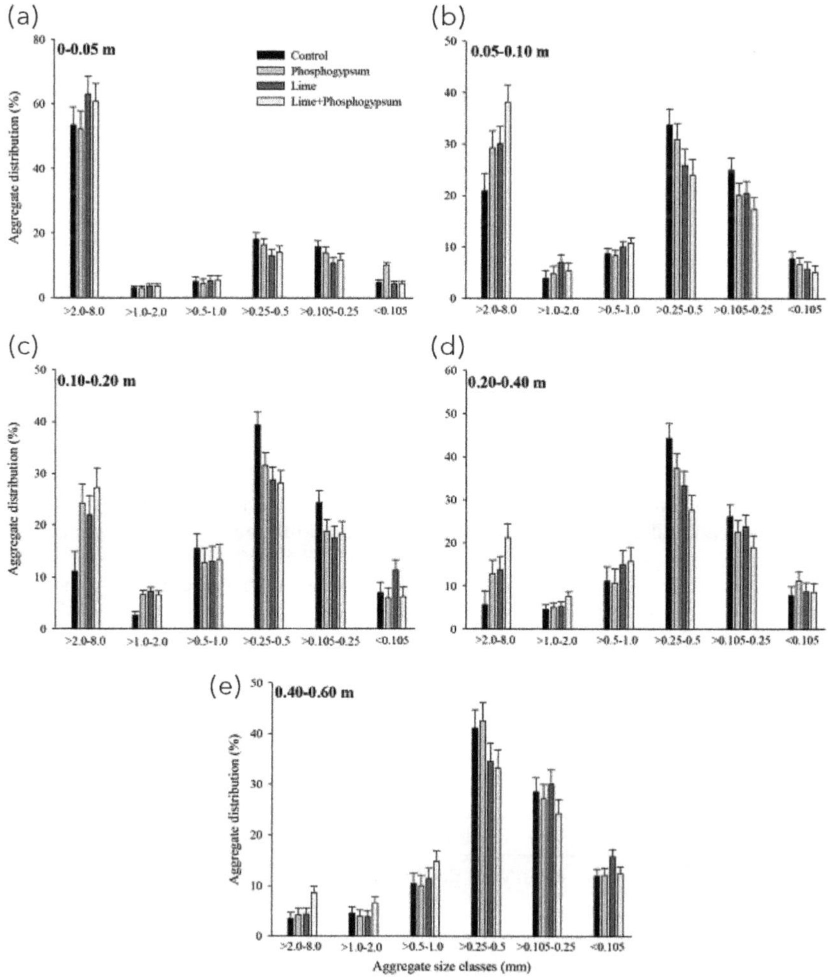

Figure 11 Water-stable aggregate distribution as affected by the surface application of lime and phosphogypsum in different soil layers in a tropical NT system. The vertical bars indicate the least significant difference at $P \leq 0.05$. The treatments were: control (without soil amendment), and application of phosphogypsum (2.1 + 2.1 + 2.1 Mg ha^{-1}), lime (2.7 + 2.0 + 2.0 Mg ha^{-1}), and a combination of lime and phosphogypsum at the given rates applied in 2002, 2004 and 2010, respectively. Source: adapted from Carmeis Filho et al. (2017a).

It is evident that pasture renewal intercropped with grain and low rates of surface limestone application produces a lower yield of forage dry matter in the winter-spring period, consequently reducing meat production. This is because palisade grass has a low tolerance to acidic soils (Pariz et al., 2016, 2017a), particularly where the base saturation in the topsoil is below 60–70%.

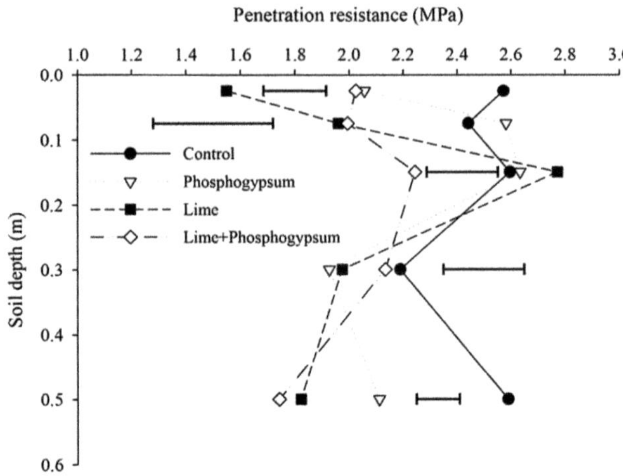

Figure 12 PR as affected by the surface application of lime and phosphogypsum in different soil layers in a tropical NT system. The horizontal bars indicate the least significant difference at $P \leq 0.05$. The treatments were: control (without soil amendment), and application of phosphogypsum (2.1 + 2.1 + 2.1 Mg ha^{-1}), lime (2.7 + 2.0 + 2.0 Mg ha^{-1}), and a combination of lime and phosphogypsum at the given rates applied in 2002, 2004 and 2010, respectively. Source: adapted from Carmeis Filho et al. (2017a).

Table 1 Aggregate stability index, soil bulk density and macroporosity as affected by the surface application of lime and phosphogypsum in different soil layers of a tropical no-tillage system

	Soil layers (m)				
Treatments	0–0.05	0.05–0.10	0.10–0.20	0.20–0.40	0.40–0.60
	Soil Bulk density (kg dm^{-3})				
Control	1.60 a	1.78 a	1.79 a	1.73 a	1.56 a
Phosphogypsum (P)	1.54 a	1.69 ab	1.77 a	1.61 b	1.56 a
Lime (L)	1.66 a	1.68 ab	1.62 b	1.53 c	1.39 b
L + P	1.56 a	1.62 b	1.56 b	1.48 c	1.36 b
F probability	0.4473	0.1531	0.0059	<0.0001	0.0016
	Macroporosity (%)				
Control	12.0 b	6.8 b	5.5 b	6.0 c	10.3 b
Phosphogypsum (P)	16.9 a	10.8 a	8.5 a	14.5 a	11.5 b
Lime (L)	9.0 b	7.0 b	8.7 a	10.5 b	14.3 a
L + P	11.8 b	9.8 a	8.5 a	9.0 b	14.8 a
F probability	0.0023	0.0047	0.0029	0.0001	0.0005

Means with different letters in the column differ statistically according to the LSD test ($P \leq 0.05$).
The treatments were: control (without soil amendment) and application of phosphogypsum (2.1 + 2.1 + 2.1 Mg ha^{-1}), lime (2.7 + 2.0 + 2.0 Mg ha^{-1}) and a combination of lime and phosphogypsum at the given rates applied in 2002, 2004 and 2010, respectively.
Source: Carmeis Filho et al. (2017a).

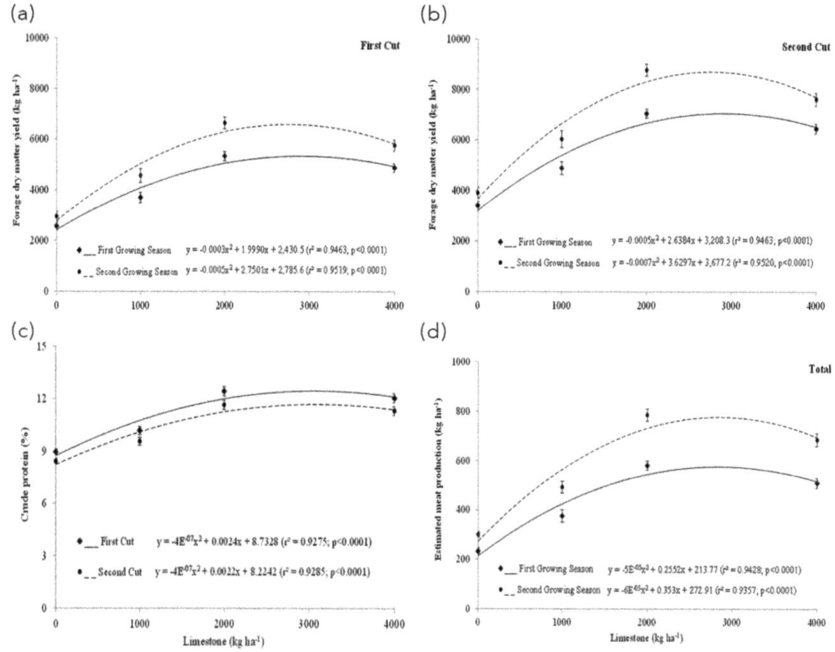

Figure 13 Forage dry matter yield and crude protein content in a pasture of palisade grass over two cuts and two growing seasons, and total meat production over two growing seasons (2007 and 2008) as affected by surface-applied dolomitic limestone rates (0, 1000, 2000 and 4000 kg ha^{-1}) applied in 2002 and 2004 in a long-term no-till soil. Bars indicate the standard error at each lime rate (n = 4; four replicates). Source: adapted from Crusciol et al. (2016).

In Brazilian subtropical regions, the application or reapplication of surface limestone in an integrated crop-livestock system such as soybean/beef cattle also increases the forage dry matter yield (combination of black oat and Italian ryegrass (*Lolium multiflorum*)) and reduces long-term soil acidification. This maintains a higher base saturation and lower Al saturation (Martins et al., 2014a,b, 2016).

Forage dry matter yield and crude protein can also benefit from the combination of phosphogypsum with liming (Table 2). It is important to emphasise that a higher crude protein concentration could lead to a higher average daily weight gain in beef cattle. The higher dry matter forage production and crude protein concentration, together with soil profile acidity alleviation, could therefore give a higher level of cattle meat production per hectare. It is not recommended The application of phosphogypsum alone is not recommended in a tropical NT system annual crop rotation with pasture as this can considerably reduce the yield of forage dry matter after the grain crop is harvested, as well as affecting grain production.

Table 2 Influence of surface-reapplied lime and/or gypsum

Treatments	FDMY (ton ha^{-1})		CP (%)		
	FC[a]	SC[a]	FC[a]	SC[a]	
Control	2.99 d[b]	3.43 d	9.0 d	8.4 d	
Phosphogypsum (P)	3.60 c	4.15 c	9.5 c	9.1 c	
Lime (L)	6.20 b	7.70 b	12.4 b	11.7 b	
L + P	7.56 a	9.33 a	13.3 a	12.3 a	
F probability	<0.01	<0.01	<0.01	<0.01	
Treatments	EASR (AU ha^{-1})[c]		EMP (kg ha^{-1})[d]		
	FC[a]	SC[a]	FC[a]	SC[a]	Total
Control	2.2 d	2.6 d	68.5 d	73.9 d	142.4 d
Phosphogypsum (P)	2.7 c	3.1 c	86.2 c	97.4 c	183.6 c
Lime (L)	4.7 b	5.8 b	163.5 b	204.9 b	368.4 b
L + P	5.7 a	7.0 a	206.3 a	256.9 a	463.2 a
F probability	<0.01	<0.01	<0.01	<0.01	<0.01

[a] First cut (FC) and second cut (SC) in June and August, respectively.
[b] Means with different letters in the column differ statistically according to the LSD test ($P \leq 0.05$).
[c] 1 AU (animal unit) = 450 kg of body weight.
[d] Estimated meat production = kg of body weight gain (cattle) per ha (estimated) × 52% of carcass yield.
Control (without soil amendment), application of phosphogypsum (2.1 + 2.1 Mg ha^{-1}), lime (2.7 + 2.0 Mg ha^{-1}) and a combination of lime and phosphogypsum at the given rates applied between 2002 and 2004 on the forage dry matter yield (FDMY), forage crude protein concentration (CP), estimated animal stocking rate (EASR) and estimated meat production (EMP) in pasture of palisade grass after intercropping with corn in two growing seasons (2007 and 2008) in a long-term no-tillage system.
Source: Crusciol et al. (authors' data).

4.4 Tropical pasture role for improving soil attributes and grain yield

The inclusion of forage grass species through intercropping with grain in annual crop rotation systems associated with soil amelioration can promote the recovery of forage yields in degraded pastures which stand at approximately only 80 M ha in Brazil (Kluthcouski et al., 2003; Crusciol and Soratto, 2007; Crusciol et al., 2012, 2014; Borghi et al., 2013a,b), thus increasing food production without requiring new cropping areas (Sulc and Tracy, 2007; Maughan et al., 2009; Pacheco et al., 2011). The adoption of perennial forage grasses in annual crop areas also increases the yield of the subsequent cash crop by breaking pest cycles and improving soil quality (Calonego et al., 2011; Castro et al., 2015; Crusciol et al., 2015; Nascente et al., 2013).

In order to understand these effects, the maturity cycle of grain crops and the sowing time of forage and forage species must be taken into account. The early maturity cycle of soybean and corn has a positive effect on grain crops so pasture production palisade grass seeds are mixed with fertiliser and sowed in the same operation as the grain, positioned below the grain crop seeds (Fig. 14). For soybean, it is not necessary to use sub-lethal doses of herbicide to control palisade grass growth in the earlier maturity cycle, because the positioning of pasture seeds below the grain crop seeds (to a depth of 8 cm) reduces the speed at which the seedlings emerge. For this reason, the initial growth of soybean is more rapid and effective, and when the growth rate of the palisade grass overtakes it (Cobucci, 2001; Kluthcouski et al., 2003), the soybean yield components are already defined in the cultivars. Occasionally, where corn is intercropped with forage, it will be necessary to use sub-lethal doses of herbicides (nicosulfuron [2-[(4,6-d imethoxypyrimidin-2-ylcarbamoyl)sulfamoyl]-N, N-dimethylnicotinamide]) to delay forage development (palisade grass or guinea grass) and avoid losses in corn grain yields (Freitas et al., 2005, 2008; Jakelaitis et al., 2005, 2006).

As mentioned above, the forage species and the time of intercropping are important factors in avoiding corn yield losses. Pasture species with vigorous development, such as guinea grass, could inhibit corn growth and

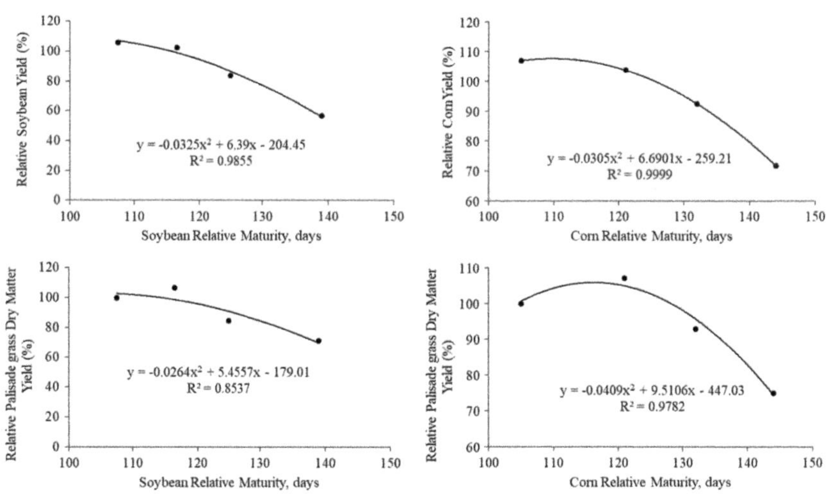

Figure 14 Soybean (left side) and corn yields (right side) as affected by relative maturity (108, 117, 125 and 139 days for soybean; and 105, 121, 132 and 144 days for corn), cropping system (grain crop alone or intercropped with palisade grass), and dry matter production in palisade grass as affected by soybean cultivars and corn hybrid relative maturity. The averages are derived over two years. In the intercropped treatments, palisade grass was sown mixed with fertiliser and sown in the same operation as the grain crops, planted at a depth of 8 cm. Source: adapted from Crusciol et al. (2012, 2013).

reduce yield if sown at the same time. It is recommended that this species should be intercropped with corn at the time of N topdressing fertilisation (V4 corn stage). However, pasture species with a slower rate of development, such as palisade grass, may be simultaneously intercropped with corn without affecting the yield by mixing the seed with fertiliser and sowing in the same operation as the grain crops, or by sowing the corn first and then adding N topdressing fertilisation. However, forage production is higher when intercropping is carried out by seeding grain crops and pasture species simultaneously, largely due to the longer time taken for forage to establish. (Table 3).

Making correct choices in crop-livestock integration systems and soil amelioration will promote the improvement of soil fertility as described, increasing the basic cations (P, K^+, Ca^{2+}, Mg^{2+}) through nutrient cycling, recovery of soil organic matter content and improvement of soil pH conditions, reflecting a higher CEC in the topsoil and subsoil when compared to areas growing only the annual grain crops (Fig. 15). This greater nutrient availability in the soil has significant effects on crop yields in the following growing seasons as a result of a higher concentration of nutrients in the leaves (Table 4).

Another alternative in integrated CA production systems is the intensive use of the soil for animal feed production, that is, the production of silage and forage. In this case, intercropping maize with palisade grass harvested for

Table 3 Corn grain yield in sole stands or intercropped with palisade grass or guinea grass, and forage dry matter production in palisade grass or guinea grass pastures at 60 days (first cut) and 150 days (second cut) after corn harvesting as affected by intercropping treatments

Treatments	Corn grain yield (kg ha^{-1})	Forage dry matter (kg ha^{-1})	
		FC	SC
Sole corn	10 048[a] bc[b]	-	-
Corn plus palisade grass sowed at corn topdressing time[c]	10 212 bc	5369 c	6 980 c
Corn plus palisade grass sowed simultaneously	10 924 b	7923 b	10 299 b
Corn plus guinea grass sowed at corn topdressing time	12 077 a	5988 c	7 785 c
Corn plus guinea grass sowed simultaneously	9 801 c	9045 a	11 759 a

[a] The averages are taken over two years.
[b] Means with different letters in the column differ statistically according to the LSD test ($P \leq 0.05$).
[c] pasture seeds were mixed with the N fertiliser and sown at a 5-cm depth using a fertiliser machine. pasture seeds were mixed with the fertiliser and sown in the same operation as the grain crops, planted at a depth of 8-cm.
Source: Borghi et al. (2013b).

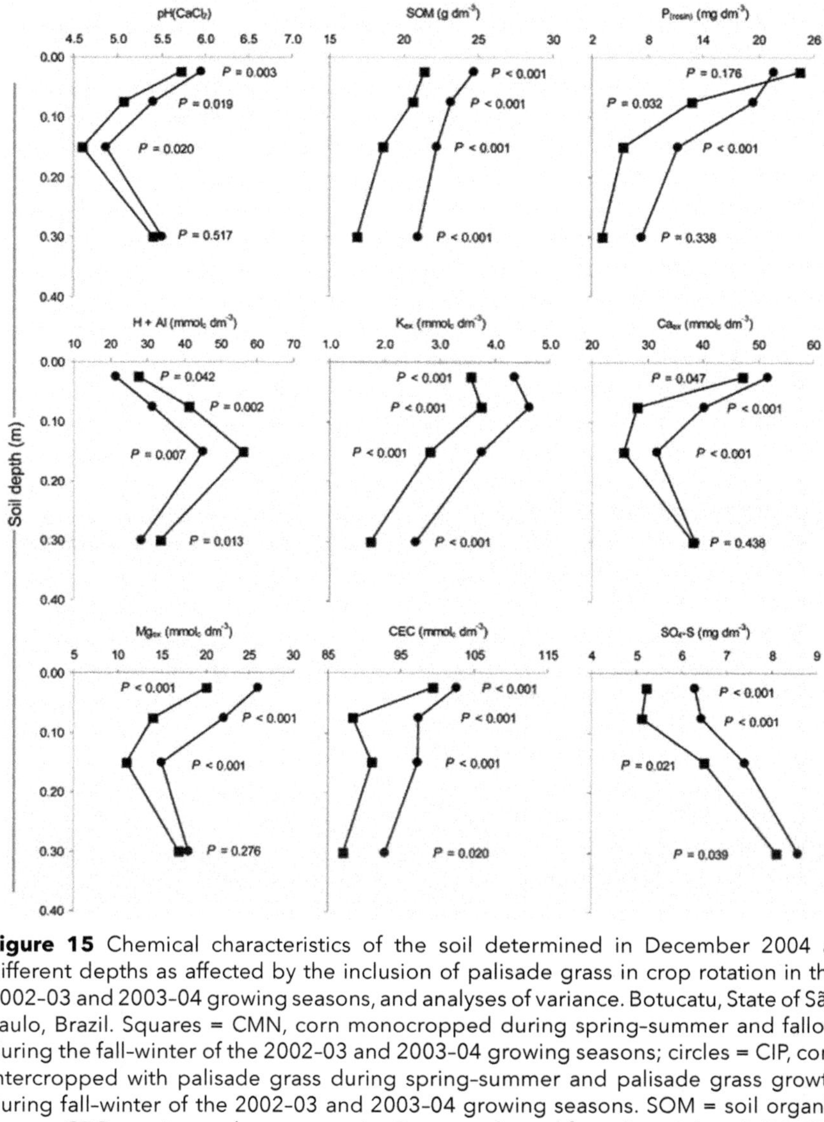

Figure 15 Chemical characteristics of the soil determined in December 2004 at different depths as affected by the inclusion of palisade grass in crop rotation in the 2002-03 and 2003-04 growing seasons, and analyses of variance. Botucatu, State of São Paulo, Brazil. Squares = CMN, corn monocropped during spring-summer and fallow during the fall-winter of the 2002-03 and 2003-04 growing seasons; circles = CIP, corn intercropped with palisade grass during spring-summer and palisade grass growth during fall-winter of the 2002-03 and 2003-04 growing seasons. SOM = soil organic matter; CEC = cation exchange capacity. Source: adapted from Crusciol et al. (2015).

silage is most advantageous when the corn for silage is cut at a height of 0.45 m. This has been the most effective option for silage production and for the improvement of the overall productivity of the system, favouring greater soil fertility, leaf nutrient concentration in the following annual crops and forage mass available for grazing (Table 5). The straw produced and left on the soil surface also reduces soil compaction (Fig. 16).

Table 4 Nutrient (N, P, K, Ca, Mg and S) concentrations in leaves and grain yield of soybean and white oat crops in the 2004-06 growing seasons as affected by the inclusion of palisade grass in crop rotation during the 2002-03 and 2003-04 growing seasons, and analyses of variance at Botucatu, São Paulo State, Brazil

	Treatment			Treatment		
Variable	CMN[a]	CIP[b]	*P* > *F*	CMN[a]	CIP[b]	*P* > *F*
	Soybean: 2004-2005 spring-summer			*White oat: 2005 fall-winter*		
N in leaves (g kg^{-1})	43	52	<0.001	23	28	<0.001
P in leaves (g kg^{-1})	3.0	4.1	<0.001	2.7	3.7	<0.001
K in leaves (g kg^{-1})	20	26	<0.001	20	27	<0.001
Ca in leaves (g kg^{-1})	10	15	<0.001	3.2	3.4	<0.001
Mg in leaves (g kg^{-1})	5.4	7.4	<0.001	2.6	2.8	<0.001
S in leaves (g kg^{-1})	2.4	3.2	<0.001	2.1	3.2	<0.001
Grain yield (kg ha^{-1})	3,373	3,844	<0.001	1540	1910	<0.001
	Soybean: 2005-2006 spring-summer			*White oat: 2006 fall-winter*		
N in leaves (g kg^{-1})	44	52	<0.001	23	29	<0.001
P in leaves (g kg^{-1})	2.8	3.8	<0.001	2.5	3.7	<0.001
K in leaves (g kg^{-1})	20	26	<0.001	22	28	<0.001
Ca in leaves (g kg^{-1})	10	15	<0.001	3.2	3.7	<0.001
Mg in leaves (g kg^{-1})	5.8	7.2	<0.001	2.3	2.8	<0.001
S in leaves (g kg^{-1})	2.5	3.2	<0.001	2.2	2.7	<0.001
Grain yield (kg ha^{-1})	3328	3669	<0.001	2000	2290	<0.001

[a] CMN, corn monocropped during spring-summer and fallow during the fall-winter of the 2002-03 and 2003-04 growing seasons.
[b] CIP, corn intercropped with palisade grass during spring-summer and palisade grass growth during fall-winter of the 2002-03 and 2003-04 growing seasons.
Source: Crusciol et al. (2015).

Table 5 Soil organic matter (SOM), phosphorus (P), potassium (K) and magnesium (Mg) at 0–0.20 m depth after initiating the experiment and P, K and Mg concentration in corn leaves, corn forage mass of silage, and grass forage mass of pasture (GFMP), considering corn intercropped with palisade grass and corn only (signal grass weedy regrowth) after harvest for silage at two cutting heights

Crop systems	Cutting height (m)	SOM ($g\ dm^{-3}$)	P ($mg\ dm^{-3}$)	K	Mg	
				($mmol_c\ dm^{-3}$)		
Corn + signal grass weedy regrowth	0.20	40.3 b[a]	11.8 b	1.0 b	15.3 b	
Corn + signal grass weedy regrowth	0.40	40.1 b	12.0 b	1.0 b	15.3 b	
Corn + palisade grass	0.20	39.3 b	11.8 b	1.0 b	15.6 b	
Corn + palisade grass	0.40	45.7 a	17.1 a	1.5 a	17.5 a	
Crop systems		P	K	Mg	CFMS	GFMP
	Cutting height (m)	($g\ kg^{-1}$)			($Mg\ ha^{-1}$)	
Corn + signal grass weedy regrowth	0.20	1.9 b	16 b	2.5 b	14.1 b	4.1 c
Corn + signal grass weedy regrowth	0.40	2.0 b	15 b	2.4 b	12.0 c	4.5 c
Corn + palisade grass	0.20	1.9 b	15 b	2.5 b	16.2 a	6.3 b
Corn + palisade grass	0.40	2.3 a	18 a	3.0 a	15.9 a	8.2 a

[a] Means within a column of the same category (i.e. crop systems and growing seasons) followed by the same letter are not significantly different at $P \leq 0.05$. Mean of first two growing seasons (2010–11 and 2011–12).
Source: Pariz et al. (2017a).

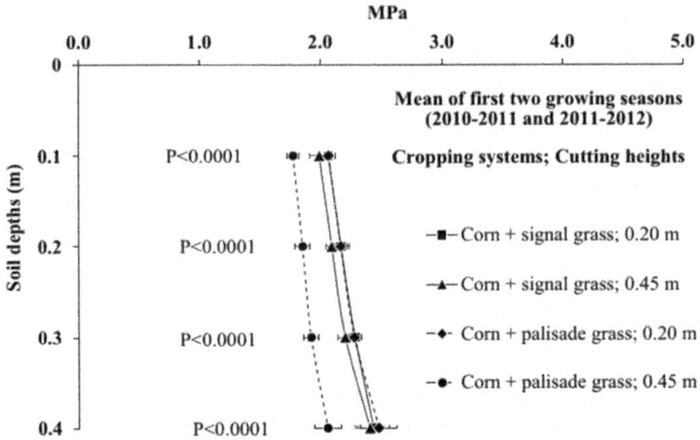

Figure 16 Soil penetration resistance at different soil depths in corn intercropped with palisade grass, and corn only (signal grass weedy regrowth) after harvest for silage in two cuttings during first two growing seasons. Values are mean of twenty-four replicates, and associated error bar is ±1 SE. Source: adapted from Pariz et al. (2017a).

5 Conclusion

Brazil has one of the largest areas of degraded pasture in the world, caused mainly by high grazing pressure, weed infestation and poor pasture fertilisation and management. Croplands under intensified tillage suffer from worsening soil compaction, poor water infiltration, increased water run-off and high levels of soil erosion. The degradation of Brazilian soils, such as tropical Oxisols under tillage agriculture is characterised by organic matter depletion, soil compaction (associated with monocropping), unbalanced fertilisation, lack of amelioration of subsoil acidity, high weed and pest infestations and crop yield instability. CA is the key to reversing degradation and rehabilitating soil health and productivity, although complementary agricultural practices for crop systems and soil management may be required according to the severity of soil degradation. For this reason, CA systems should not be established in degraded soils without a prior rehabilitation process.

The rehabilitation of degraded soils relies on amelioration of surface and subsurface chemical soil attributes, alleviation of soil compaction, biomass input to the soil by redesigning crop systems, recovery of soil organic matter content, nutrient cycling and improvement of soil physical attributes. Integrated crop-livestock systems under CA can improve soil structure, increase C sequestration and decrease pest pressures while providing pasture and increasing cash crop yields, thus enabling the intensification of sustainable agriculture.

Pasture fertilisation does not only focus on the cash crop. It is essential for increasing the vigour of biomass production and nutrient cycling which improves soil protection and the overall quality of the CA system.

Modern precision agricultural tools were used in the first case study to delineate low soil productivity zones which are generally associated with soil compaction and poor physical quality resulting from intensive machinery traffic. In these zones, single interventions (mechanical chiselling, polyculture of cover crops and subsoil acidity alleviation via phosphogypsum application), and integrated interventions were investigated as means of soil rehabilitation. Single interventions had a slight short-term effect in improving soil physical attributes and soybean grain yield. Fully integrated interventions (physical, chemical and biological) had a synergic effect, promoting decompaction of the soil, improving its structure and increasing soybean yield by 10% compared to the control (no intervention) in the first crop season. The second case study showed that in tropical pasture (palisade grass and guinea grass) with non-limiting plant growth factors in the subsoil (i.e. low P and Ca plant availability and high Al concentration), the input of phosphogypsum was effective in restoring soil organic matter levels, decreasing PR and increasing macroporosity, aggregates and cash crop grain yield. The input of lime and phosphogypsum increased forage dry matter production and crude protein, improving the productivity of the system. This contributes to preservation of the environment by enabling a higher level of food production without extending the area under cultivation. It increases profits which helps farmers to diversify activities. Taken together, the two case studies showed that improvements in cropping systems by reducing factors which limit the growth of tropical pasture and cover crops during the cash crop window: decompaction, subsoil chemical amelioration and redesigning crop systems, is critical for soil rehabilitation. Respecting the principles of CA and introducing integrated management strategies which explore the synergistic effects of plant-soil-biology activities are necessary if societal demands are to be met in the coming years.

6 Acknowledgements

The authors thank CNPq and CAPES for research support and scholarships of first and second authors; Dr. Leal thanks the Alexander von Humboldt Foundation for the Return Fellowship and CAPES for the postdoctoral fellowship.

7 References

Abreu, S. L., Reichert, J. M. and Reinert, D. J. 2004. Mechanical and biological chiseling to reduce compaction of a sandy loam Alfisol under no-tillage. *Rev. Bras. Ciênc. Solo* 28(3), 519–31. doi:10.1590/S0100-06832004000300013.

Albuquerque, M. A., Dieckow, J., Sordi, A., Piva, J. T., Bayer, C., Molin, R., Pergher, M. and Ribeiro-Junior, P. J. 2015. Carbon and nitrogen in a ferralsol under zero-tillage rotations based on cover, cash or hay crops. *Soil Use Manag.* 31(1), 1–9. doi:10.1111/sum.12173.

Algeri, A., Vilar, C. C., Ushiwata, S. Y. and Reis, R. G. E. 2018. Produção de biomassa e cobertura do solo por milheto, braquiária e crotalaria cultivados em cultura pura e consorciados. *Glob. Sci. Technol.* 11(2), 112-25.

Almeida, C. Ad., Coutinho, A. C., Esquerdo, J. C. D. M., Adami, M., Venturieri, A., Diniz, C. G., Dessay, N., Durieux, L. and Gomes, A. R. 2016. High spatial resolution land use and land cover mapping of the Brazilian Legal Amazon in 2008 using Landsat-5/TM and MODIS data. *Acta Amaz.* 46(3), 291-302. doi:10.1590/1809-4392201505504.

Amado, T. J. C. and Eltz, F. L. F. 2003. Plantio direto na palha - rumo à sustentabilidade agrícola nos trópicos. *Rev. Ci. e Amb.* 27, 49-66.

Amado, T. J. C., Mielniczuk, J., Fernandez, S. B. and Bayer, C. 1998. Nitrogen availability as affected by ten years of cover crops and tillage systems in Southern Brazil. *J. Soil Water Conserv.* 53(3), 268-71.

Amado, T. J. C., Bayer, C., Conceição, P. C., Spagnollo, E., de Campos, B. H. and da Veiga, M. 2006. Potential of carbon accumulation in no-till soils with intensive use and cover crops in Southern Brazil. *J. Environ. Qual.* 35(4), 1599-607. doi:10.2134/jeq2005.0233.

Amado, T. J. C., Conceição, P. C., Bayer, C. and Eltz, F. L. F. 2007. Qualidade do solo avaliada pelo 'Soil Quality Kit Test' em dois experimentos de longa duração no Rio Grande do Sul. *Rev. Bras. Ciênc. Solo* 31(1), 109-21.

Asner, G. P., Townsend, A. R., Bustamantes, M. M. C., Nardoto, G. B. and Olander, L. P. 2004. Pasture degradation in the central Amazon: linking changes in carbon and nutrient cycling with remote sensing. *Glob. Change Biol.* 10(5), 844-62. doi:10.1111/j.1529-8817.2003.00766.x.

Azcón-Aguilar, C. and Barea, J. M. 2015. Nutrient cycling in the mycorrhizosphere. *J. Soil Sci. Plant Nutr.* 15(2), 372-96. doi:10.4067/S0718-95162015005000035.

Bai, Z. G., Dent, D. L., Olsson, L. and Schaepman, M. E. 2008. Global assessment of land degradation and improvement. 1. Identification by remote sensing. Report 2008/01. ISRIC -World Soil Information, Wageningen.

Balbinot Junior, A. A., Santos, J. C. F. D., Debiasi, H. and Yokoyama, A. H. 2017. Contribution of roots and shoots of *Brachiaria* species to soybean performance in succession. *Pesq. Agropec. Bras.* 52(8), 592-8. doi:10.1590/s0100-204x2017000800004.

Batey, T. and McKenzie, D. C. 2006. Soil compaction: identification directly in the field. *Soil Use Manag.* 22(2), 123-31. doi:10.1111/j.1475-2743.2006.00017.x.

Bayer, C., Dieckow, J., Amado, T. J. C., Eltz, F. L. F. and Vieira, F. C. B. 2009. Cover crop effects increasing carbon storage in a subtropical no-till sandy Acrisol. *Commun. Soil Sci. Plant Anal.* 40(9-10), 1499-511. doi:10.1080/00103620902820365.

Bernoux, M., Cerri, C. C., Cerri, C. E. P., Neto, M. S., Metay, A., Perrin, A. S., Scopel, E., Razafimbelo, T., Blavet, D., Piccolo, MdC., Pavei, M. and Milne, E. 2006. Cropping systems, carbon sequestration and erosion in Brazil: a review. *Agron. Sustain. Dev.* 26(1), 1-8. doi:10.1051/agro:2005055.

Beuchle, R., Grecchi, R. C., Shimabukuro, Y. E., Seliger, R., Eva, H. D., Sano, E. and Achard, F. 2015. Land cover changes in the Brazilian Cerrado and Caatinga biomes from 1990 to 2010 based on a systematic remote sensing sampling approach. *Appl. Geogr.* 58, 116-27. doi:10.1016/j.apgeog.2015.01.017.

Beutler, A. N. 2003. Efeito da compactação do solo e do conteúdo de água nas características agronômicas de soja e arroz de sequeiro. PhD thesis. Faculdade de Ciências Agrárias e Veterinárias, Universidade Estadual Paulista, São Paulo.

Blevins, R. L., Murdock, L. W. and Thomas, G. W. 1978. Effect of lime application on no-tillage and conventionally tilled corn. *Agron. J.* 70(2), 322–6. doi:10.2134/agronj1978.00021962007000020025x.

Borghi, E., Crusciol, C. A. C., Nascente, A. S., Sousa, V. V., Martins, P. O., Mateus, G. P. and Costa, C. 2013a. Sorghum grain yield, forage biomass production and revenue as affected by intercropping time. *Eur. J. Agron.* 51, 130–9. doi:10.1016/j.eja.2013.08.006.

Borghi, E., Crusciol, C. A. C., Mateus, G. P., Nascente, A. S. and Martins, P. O. 2013b. Intercropping time of corn and palisadegrass or guineagrass affecting grain yield and forage production. *Crop Sci.* 53(2), 629–36. doi:10.2135/cropsci2012.08.0469.

Borreli, P., Robinson, D. A., Fleischer, L. R., Lugato, E., Ballabio, C., Alewell, C., Meusburger, K., Modugno, S., Schütt, B., Ferro, V., Bagarello, V., Oost, K. V., Montanarella, L. and Panagos, P. 2017. An assessment of the global impact of 21st century land use change on soil erosion. *Nat. Commun.* 8(1). doi:10.1038/s41467-017-02142-7.

Brady, N. C. and Weil, R. R. 2008. *The Nature and Properties of Soils*. Prentice Hall.

Braz, S. P., Urquiaga, S., Alves, B. J. R., Jantalia, C. P., Guimarães, A. P. P., Santos, C. A., Santos, S. C., Pinheiro, E. F. M. and Boddey, R. M. 2013. Soil carbon stocks under productive and degraded *Brachiaria* pastures in the Brazilian Cerrado. *Soil Sci. Soc. Am. J.* 77(3), 914–28.

Briedis, C., Sá, J. CdM., Caires, E. F., Navarro, JdF., Inagaki, T. M., Boer, A., Neto, C. Q., Ferreira, AdO., Canalli, L. B. and Santos, J. Bd. 2012. Soil organic matter pools and carbon-protection mechanisms in aggregate classes influenced by surface liming in a no-till system. *Geoderma* 170, 80–8. doi:10.1016/j.geoderma.2011.10.011.

Büchi, L., Wendling, M., Amossé, C., Necpalova, M. and Charles, R. 2018. Importance of cover crops in alleviating negative effects of reduced soil tillage and promoting soil fertility in a winter wheat cropping system. *Agric. Ecosyst. Environ.* 256, 92–104. doi:10.1016/j.agee.2018.01.005.

Caires, E. F., Fonseca, A. F., Mendes, J., Chueiri, W. A. and Madruga, E. F. 1999. Corn, wheat and soybean yields as a function of the changes in soil chemical characteristics due to surface application of lime and gypsum under a no-tillage system. *Rev. Bras. Ciênc. Solo* 23(2), 315–27. doi:10.1590/S0100-06831999000200016.

Caires, E. F., Fonseca, A. F., Feldhaus, I. C. and Blum, J. 2001. Root growth and nutrient uptake by soybean as affected by lime and gypsum, under a no-tillage system. *Rev. Bras. Ciênc. Solo* 25(4), 1029–40. doi:10.1590/S0100-06832001000400025.

Caires, E. F., Alleoni, L. R. F., Cambri, M. A. and Barth, G. 2005. Surface application of lime for crop grain production under a no-till system. *Agron. J.* 97(3), 791-8. doi:10.2134/agronj2004.0207.

Caires, E. F., Barth, G. and Garbuio, F. J. 2006a. Lime application in the establishment of a no-till system for grain crop production in Southern Brazil. *Soil Till. Res.* 89(1), 3-12. doi:10.1016/j.still.2005.06.006.

Caires, E. F., Corrêa, J. C. L., Churka, S., Barth, G. and Garbuio, F. J. 2006b. Surface application of lime ameliorates subsoil acidity and improves root growth and yield of wheat in an acid soil under no-till system. *Sci. Agric. (Piracicaba, Braz.)* 63(5), 502–9. doi:10.1590/S0103-90162006000500013.

Caires, E. F., Garbuio, F. J., Churka, S., Barth, G. and Corrêa, J. C. L. 2008a. Effects of soil acidity amelioration by surface liming on no-till corn, soybean, and wheat root growth and yield. *Eur. J. Agron.* 28(1), 57–64. doi:10.1016/j.eja.2007.05.002.

Caires, E. F., Pereira Filho, P. R. S., Zardo Filho, R. and Feldhaus, I. C. 2008b. Soil acidity and aluminium toxicity as affected by surface liming and cover oat residues under a no-till system. *Soil Use Manag.* 24(3), 302–9. doi:10.1111/j.1475-2743.2008.00166.x.

Caires, E. F., Joris, H. A. W. and Churka, S. 2011. Long-term effects of lime and gypsum additions on no-till corn and soybean yield and soil chemical properties in Southern Brazil. *Soil Use Manag.* 27(1), 45–53. doi:10.1111/j.1475-2743.2010.00310.x.

Calegari, A., Araújo, A. G., Costa, A., Lanillo, R. F., Junior, R. C. and Santos, D. R. 2014. Conservation Agriculture in Brazil. In: Jat, R. A., Sahrawat, K. L. and Kassam, A. H. (Eds), *Conservation Agriculture: Global Aspects and Challenges*. Food and Agriculture Organization, pp. 54–88.

Calonego, J. C., Borghi, E. and Crusciol, C. A. C. 2011. Intervalo hídrico ótimo e compactação do solo com cultivo consorciado de milho e braquiária. *Rev. Bras. Ciênc. Solo* 35(6), 2183–90. doi:10.1590/S0100-06832011000600033.

Calonego, J. C., Raphael, J. P. A., Rigon, J. P. G., Oliveira Neto, Ld. and Rosolem, C. A. 2017. Soil compaction management and soybean yields with cover crops under no-till and occasional chiseling. *Eur. J. Agron.* 85, 31–7. doi:10.1016/j.eja.2017.02.001.

Carmeis Filho, A. C., Crusciol, C. A., Guimarães, T. M., Calonego, J. C. and Mooney, S. J. 2017a. Impact of amendments on the physical properties of soil under tropical long-term no till conditions. *PLoS One* 11(12), e0167564.

Carmeis Filho, A. C. A., Crusciol, C. A. C. and Castilhos, A. M. 2017b. Liming demand and plant growth improvements for an oxisol under long term no-till cropping. *J. Agric. Sci.* 155(7), 1093–112. doi:10.1017/S0021859617000235.

Carmeis Filho, A. C. A., Penn, C. J., Crusciol, C. A. C. and Calonego, J. C. 2017c. Lime and phosphogypsum impacts on soil organic matter pools in a tropical oxisol under long-term no-till conditions. *Agric. Ecosyst. Environ.* 241, 11–23. doi:10.1016/j.agee.2017.02.027.

Carmeis Filho, A. C. A., Crusciol, C. A. C., Guimarães, T. M., Calonego, J. C. and da Costa, C. H. M. 2018. Changes in soil physical properties and carbono protection mechanisms by surface application of lime in a tropical no-tillage system. *Soil Sci. Soc. Am. J.* 82(1), 56–65. doi:10.2136/sssaj2017.04.0120.

Castro, G. S. A., Calonego, J. C. and Crusciol, C. A. C. 2011. Soil physical properties in crop rotation systems as affected by liming materials. *Pesq. Agropec. Bras.* 46(12), 1690–8. doi:10.1590/S0100-204X2011001200015.

Castro, G. S. A., Crusciol, C. A. C., Calonego, J. C. and Rosolem, C. A. 2015. Management impacts on soil organic matter of tropical soils. *Vadose Zone J.* 14(1), 0–8. doi:10.2136/vzj2014.07.0093.

Cecagno, D., Gomes, M. V., Costa, S. E. V. G. A., Martins, A. P., Denardin, L. G. O., Bayer, C., Anghinoni, I. and Carvalho, P. C. F. 2018. Soil organic carbon in an integrated crop-livestock system under different grazing intensities. *Rev. Bras. Ci. Agr.* 13(3), 1–7. doi:10.5039/agraria.v13i3a5553.

CGEE. 2016. *Land Degradation Neutrality: Implications for Brazil*. Centro de Gestão e Estudos Estratégicos, Brasilia, 28pp.

Cobucci, T., Kluthcouski, J. and Aidar, H. 2001. Santa Fé System: forage production on off-season. In: *Workshop internacional programa de integração agricultura e pecuária para o desenvolvimento sustentável das savanas sulamericanas. Santo Antonio de Goiás. Anais*. Embrapa Arroz e Feijão, Santo Antonio de Goiás, pp. 125-35. (Embrapa Arroz e Feijão. Documentos, 123).

Collares, G. L., Reinert, D. J., Reichert, J. M. and Kaiser, D. R. 2008. Compactação de um Latossolo induzida pelo tráfego de máquinas e sua relação com o crescimento e produtividade de feijão e trigo. *Rev. Bras. Ciênc. Solo* 32(3), 933–42. doi:10.1590/S0100-06832008000300003.

Correia, N. M. and Gomes, L. J. P. 2015. Sobressemeadura de soja com *Urochloa ruziziensis* e a cultura do milho em rotação. *Pesq. Agropec. Bras.* 50(11), 1017–26. doi:10.1590/S0100-204X2015001100004.

Costa, C. H. M., Crusciol, C. A. C., Ferrari Neto, J. and Castro, G. S. A. 2016. Residual effects of superficial liming on tropical soil under no-tillage system. *Pesq. Agropec. Bras.* 51(9), 1633–42. doi:10.1590/s0100-204x2016000900063.

Costa, C. H. M., Filho, A. C. A. C., Crusciol, C. A. C., Soratto, R. P. and Guimarães, T. M. 2018. Intensive annual crop production and root development in a tropical acid soil under long-term no-till and soil-amendment management. *Crop Pasture Sci.* 69(5), 488–505. doi:10.1071/CP17233.

Crusciol, C. A. C. and Soratto, R. P. 2007. Peanut crop nutrition and yield in no-tillage system in succession to cover crops growth. *Pesq. Agropec. Bras.* 42(11), 1553–60. doi:10.1590/S0100-204X2007001100006.

Crusciol, C. A. C., Mateus, G. P., Pariz, C. M., Borghi, É., Costa, C. and Silveira, J. P. Fd. 2011. Nutrição e produtividade de híbridos de sorgo granífero de ciclos contrastantes consorciados com capim-marandu. *Pesq. Agropec. Bras.* 46(10), 1234–40. doi:10.1590/S0100-204X2011001000017.

Crusciol, C. A. C., Mateus, G. P., Nascente, A. S., Martins, P. O., Borghi, E. and Pariz, C. M. 2012. An innovative crop-forage intercrop system: early cycle soybean cultivars and palisadegrass. *Agron. J.* 104(4), 1085–95. doi:10.2134/agronj2012.0002.

Crusciol, C. A. C., Nascente, A. S., Mateus, G. P., Borghi, E., Leles, E. P. and Santos, N. C. B. 2013. Effect of intercropping on yields of corn with different relative maturities and palisadegrass. *Agron. J.* 105(3), 599–606. doi:10.2134/agronj2012.0426.

Crusciol, C. A. C., Nascente, A. S., Mateus, G. P., Pariz, C. M., Martins, P. O. and Borghi, E. 2014. Intercropping soybean and palisadegrass for enhanced land use efficiency and revenue in a no till system. *Eur. J. Agron.* 58, 53–62. doi:10.1016/j.eja.2014.05.001.

Crusciol, C. A. C., Nascente, A. S., Borghi, E., Soratto, R. P. and Martins, P. O. 2015. Improving soil fertility and crop yield in a tropical region with palisadegrass cover crops. *Agron. J.* 107(6), 2271–80. doi:10.2134/agronj14.0603.

Crusciol, C. A. C., Marques, R. R., Filho, A. C. A. C., Soratto, R. P., Costa, C. H. M., Neto, J. F., Castro, G. S. A., Pariz, C. M. and de Castilhos, A. M. 2016. Annual crop rotation of tropical pastures with no-till soil as affected by lime surface application. *Eur. J. Agron.* 80, 88–104. doi:10.1016/j.eja.2016.07.002.

Cubilla, M., Reinert, D. J., Aita, C. and Reichert, J. M. 2002. Plantas de cobertura do solo: uma alternativa para aliviar a compactação em sistema plantio direto. *R. Plantio Direto* 71, 29–34.

Dalla Nora, D. and Amado, T. J. C. 2013. Improvement in chemical attributes of oxisol Subsoil and crop yields under no-till. *Agron. J.* 105(5), 1393–403.

Dalla Nora, D., Amado, J. C. A., Bortolotto, R. P., Ferreira, A. O., Reichardt, K. and Santi, A. L. 2014. Subsoil chemical amelioration and crop yields under continuos long-term no-till in a subtropical oxisol. *Afr. J. Agric. Res.* 9(45), 3348–9.

Dalla Nora, D., Amado, T. J. C., Nicoloso, R. S., Mazuco, A. C. B. and Piccin, M. 2017. Mitigation of the gradient of chemical properties in the rooting zone of dystrophic Oxisols by gypsum and lime inputs under a no-till system. *Rev. Bras. Ciênc. Solo* 41, 1–22.

Debarba, L. and Amado, T. J. C. 1997. Desenvolvimento de sistemas de produção de milho no sul do Brasil com características de sustentabilidade. *Rev. Bras. Ciênc. Solo* 21(3), 473–80. doi:10.1590/S0100-06831997000300017.

Denardin, J. E., Kochhann, R. A., Bacaltchuk, B., Sattler, A., Denardin, N. D., Faganello, A. and Wiethölter, S. 2008. Sistema plantio direto: fator de potencialidade da agricultura tropical Brasileira. In: Albuquerque, A. C. S. and Silva, A. G. (Eds), *Agricultura tropical: quatro décadas de inovações tecnológicas, institucionais e políticas*. Embrapa Informação Tecnológica, Brasília, DF, vol. 1, ch. 1, pp. 1251–73.

El-Katatny, M. H. and Idres, M. M. 2014. Effects of single and combined inoculation with *Azospirillum brasilense* and *Trichoderma harzianum* on seedling growth or yield parameters of wheat (*Triticum vulgaris l.*, giza 168) and corn (*Zea mays l.*, hybrid 310). *J. Plant Nut.* 37(12), 1913–36. doi:10.1080/01904167.2014.911322.

Embrapa. 2002. *Uso agrícola dos solos brasileiros*. Embrapa Solos.

Embrapa. 2017. *Manual de métodos de análise de solo*. Embrapa Solos.

Ernani, P. R., Ribeiro, M. F. S. and Bayer, C. 2004. Chemical modifications caused by liming below the limed layer in a predominantly variable charge acid soil. *Commun. Soil Sci. Plant Anal.* 35(5–6), 889–901. doi:10.1081/CSS-120030365.

Fageria, N. K., Baligar, V. C. and Bailey, B. A. 2005. Role of cover crops in improving soil and row crop productivity. *Commun. Soil Sci. Plant Anal.* 36(19–20), 2733–57. doi:10.1080/00103620500303939.

FAO. 2011. *The State of the World's Land and Water Resources for Food and Agriculture (SOLAW): Managing Systems at Risks*. Food and Agriculture Organization.

FAO and ITPS. 2015. *Status of the World's Soil Resources (SWSR) – Main Report*. Food and Agriculture Organization of the United Nations and Intergovernmental Technical Panel on Soils.

Fearnside, P. M. 2013. Serviços ambientais: provenientes de florestas intactas,degradadas e secundárias na Amazônia Brasileira. In: Peres, C. A., Gardner, T. A., Barlow, J. and Vieira, I. C. G. (Eds), *Conservação da biodiversidade em paisagens antropizadas do Brasil* (1st edn.). Editora da Universidade Federal do Paraná, Curitiba, pp. 26–57.

Ferrari Neto, J., Crusciol, C. A. C., Soratto, R. P., Costa, C. H. Md and Castro, G. S. A. 2017. Persistence and release of macronutrients and silicon of pigeonpea as a function of fragmentation. *Colloq. Agrariae* 13(1), 43–56. doi:10.5747/ca.2017.v13.n1.a148.

Ferreira, A. O., Jorge Carneiro Amado, T., da Silveira Nicoloso, R., de Moraes Sá, J. C., Ernani Fiorin, J., Santos Hansel, D. S. and Menefee, D. 2013. Soil carbon stratification affected by long-term tillage and cropping systems in Southern Brazil. *Soil Till. Res.* 133, 65–74. doi:10.1016/j.still.2013.05.011.

Ferreira, A. O., Amado, T. J. C., Rice, C. W., Diaz, D. A. R., Keller, C. and Inagaki, T. M. 2016. Can no-till grain production restore soil organic carbon to levels natural grass in a subtropical Oxisol? *Agric. Ecosyst. Environ.* 229, 13–20. doi:10.1016/j.agee.2016.05.016.

Ferreira, A. O., Amado, T. J. C., Rice, C. W., Ruiz Diaz, D. A., Briedis, C., Inagaki, T. M. and Gonçalves, D. R. P. 2018a. Driving factors of soil carbon accumulation in Oxisols in long-term no-till systems of South Brazil. *Sci. Total Environ.* 622–623, 735–42. doi:10.1016/j.scitotenv.2017.12.019.

Ferreira, A. O., de Moraes Sá, J. C., Lal, R., Tivet, F., Briedis, C., Inagaki, T. M., Gonçalves, D. R. P. and Romaniw, J. 2018b. Macroaggregation and soil organic carbon restoration in a highly weathered Brazilian Oxisol after two decades under no-till. *Sci. Total Environ.* 621, 1559-67. doi:10.1016/j.scitotenv.2017.10.072.

Forte, C. T., Galon, L., Beutler, A. N., Perin, G. F., Pauletti, E. S. S., Basso, F. J. M., Holz, C. M. and Santin, C. O. 2018. Coberturas vegetais do solo e manejo de cultivo e suas contribuições para as culturas agrícolas. *Rev. Bras. Ci. Agr.* 13(1), 1-10. doi:10.5039/agraria.v13i1a5504.

Freitas, F. C. L., Ferreira, L. R., Ferreira, F. A., Santos, M. V., Agnes, E. L., Cardoso, A. A. and Jakelaitis, A. 2005. Implantation of pastures via consortium of *Brachiaria brizantha* with corn for silage under no-tillage system. *Planta Daninha* 23(1), 49-58. doi:10.1590/S0100-83582005000100007.

Freitas, F. C. L., Santos, M. V., Machado, A. F. L., Ferreira, L. R., Freitas, M. A. M. and Silva, M. G. O. 2008. Behavior of corn cultivars intercropped with *Brachiaria brizantha* in the presence and absence of sprayed foramsulfuron + iodosulfuron-methyl. *Planta Daninha* 26(1), 215-21. doi:10.1590/S0100-83582008000100022.

Friedrich, T. and Kassam, A. 2012. No-till farming and the environment: do no-till systems require more chemicals? *Outlooks Pest Manag.* 23(4), 153-7. doi:10.1564/23aug02.

Garcia, R. A., Li, Y. and Rosolem, C. A. 2013. Soil organic matter and physical attributes affected by crop rotation under no-till. *Soil Sci. Soc. Am. J.* 77(5), 1724-31. doi:10.2136/sssaj2012.0310.

Grave, R. A., Nicoloso, RdS., Cassol, P. C., Aita, C., Corrêa, J. C., Costa, M. D. and Fritz, D. D. 2015. Short-term carbon dioxide emission under contrasting soil disturbance levels and organic amendments. *Soil Till. Res.* 146, 184-92. doi:10.1016/j.still.2014.10.010.

Guerra, A. J. T., Fullen, M. A., Jorge, M. C. O. and Alexandre, S. T. 2014. Soil erosion and conservation in Brazil. *Anu. Inst. Geoci.* 37, 81-91.

Hamza, M. A. and Anderson, W. K. 2005. Soil compaction in cropping systems: a review of the nature, causes and possible solutions. *Soil Till. Res.* 82(2), 121-45. doi:10.1016/j.still.2004.08.009.

Hansel, F. D., Amado, T. J. C., Diaz, D. A. R., Rosso, L. H. M., Nicoloso, F. T. and Schorr, M. 2017. Phosphorus fertilizer placement and tillage affect soybean root growth and drought tolerance. *Agron. J.* 109(6), 2936-44.

Hösl, R. and Strauss, P. 2016. Conservation tillage practices in the alpine forelands of Austria - are they effective? *Catena* 137, 44-51. doi:10.1016/j.catena.2015.08.009.

Jakelaitis, A., Silva, A. F., Silva, A. A., Ferreira, L. R., Freitas, F. C. L. and Viana, R. G. 2005. Influência de herbicidas e de sistemas de semeadura de *Brachiaria brizantha* consorciada com milho. *Planta Daninha* 23(1), 59-67. doi:10.1590/S0100-83582005000100008.

Jakelaitis, A., Silva, A. A., Silva, A. F., Silva, L. P., Ferreira, L. R. and Vivian, R. 2006. Effects of herbicides on intercropped maize and *Brachiaria brizantha* weed control, growth and yield. *Pesq. Agropec. Trop.* 36, 53-60.

Jat, R. A., Sahrawat, K. L., Kassam, A. H. and Friedrich, T. 2014. Conservation Agriculture for sustainable and resilient agriculture: global status, prospects and challenges. In: Jat, R. A., Sahrawat, K. L. and Kassam, A. H. (Eds), *Conservation Agriculture: Global Aspects and Challenges*. Food and Agriculture Organization, pp. 1-25.

Kassam, A., Gottlieb, B., Theodor, F., Shaxson, F., Goddard, T., Amado, T. J. C., Crabtree, B., Hongwen, L., Mellho, I., Pisante, M. and Mkomwa, S. 2013. Sustainable soil management is more than what and how crops are grown. In: Lal, R. and Stewart, B. A. (Eds), *Principles of Sustainable Soil Management in Agroecosystems*. CRC Press, pp. 338-99.

Kassam, A., Friedrich, T. and Derpsch, R. 2019. Global spread of Conservation Agriculture. *Int. J. Environ. Stud.* 76(1), 29–51. doi:10.1080/00207233.2018.1494927.

Klink, C. A. and Machado, R. B. 2005. Conservation of the Brazilian Cerrado. *Conserv. Biol.* 19(3), 707–13. doi:10.1111/j.1523-1739.2005.00702.x.

Kluthcouski, J. and Aidar, H. 2003. Uso da integração lavoura-pecuária na recuperação de pastagens degradadas. In: Kluthcouski, J., Stone, L. F. and Aidar, H. (Eds), *Integração lavoura-pecuária*. Embrapa, Santo Antonio de Goiás, Goiás, pp. 185–223.

Kluthcouski, J., Stone, L. F. and Aidar, H. 2003. *Integração lavoura-pecuária*. Embrapa Arroz e Feijão.

Köppen, W. and Geiger, R. 1928. *Klimate der Erde*. Verlag Justus Perthes, Gotha.

Leal, O. A., Dick, D. P., Costa, F. S., Knicker, H., Carvalho Júnior, J. A. and Santos, J. C. 2019. Carbon in physical fractions and organic matter chemical composition of an Acrisol after Amazon Forest burning and conversion into pasture. *J. Braz. Chem. Soc.* 30(2), 413–24. doi:10.21577/0103-5053.20180191.

Lehman, R. M., Cambardella, C. A., Stott, D. E., Acosta-martinez, V., Manter, D. K., Buyer, J. S., Maul, J., Smith, J., Collins, H., Halvorson, J., Kremer, R., Lundgren, J., Ducey, T., Jin, V. and Karlen, D. 2015. Understanding and enhancing soil biological health: the solution for reversing soil degradation. *Sustainability* 7(1), 988–1027. doi:10.3390/su7010988.

Luizão, F. J., Fearnside, P. M., Cerri, C. E. P. and Lehmann, J. 2009. The maintenance of soil fertility in Amazonian managed systems. In: Keller, M., Bustamante, M., Gash, J. and Dias, P. S. (Eds), *Amazonia and Global Change*. American Geophysical Union, Washington, pp. 311–36.

MAPA (Ministry of Agriculture, Livestock and Food Supply- Brazil). 2012. *Plano setorial de mitigação e de adaptação às mudanças climáticas para a consolidação de uma economia de baixa emissão de carbono na agricultura*. MAPA/ACS.

Marsh, B. H. and Grove, J. H. 1992. Surface and subsurface soil acidity: soybean root response to sulfate-bearing spent lime. *Soil Sci. Soc. Am. J.* 56(6), 1837–42. doi:10.2136/sssaj1992.03615995005600060031x.

Martins, A. P., Anghinoni, I., Costa, S. E. V. GdA., Carlos, F. S., Nichel, GdH., Silva, R. A. P. and Carvalho, P. CdF. 2014a. Amelioration of soil acidity and soybean yield after surface lime reapplication to a long-term no-till integrated crop-livestock system under varying grazing intensities. *Soil Till. Res.* 144, 141–9. doi:10.1016/j.still.2014.07.019.

Martins, A. P., Costa, S. E. V. Gd, Anghinoni, I., Kunrath, T. R., Balerini, F., Cecagno, D. and Carvalho, P. CdF. 2014b. Soil acidification and basic cation use efficiency in an integrated no-till crop-livestock system under different grazing intensities. *Agric. Ecosyst. Environ.* 195, 18–28. doi:10.1016/j.agee.2014.05.012.

Martins, A. P., Cecagno, D., Borin, J. B. M., Arnuti, F., Lochmann, S. H., Anghinoni, I., Bissani, C. A., Bayer, C. and Carvalho, P. C. F. 2016. Long-, medium- and short-term dynamics of soil acidity in an integrated crop-livestock system under different grazing intensities. *Nutr. Cycl. Agroecosyst.* 104(1), 67–77. doi:10.1007/s10705-015-9759-5.

Mateus, G. P., Crusciol, C. A. C., Pariz, C. M., Borghi, E., Costa, C., Martello, J. M., Franzluebbers, A. J. and Castilhos, A. M. 2016. Sidedress nitrogen application rates to sorghum intercropped with tropical perennial grasses. *Agron. J.* 108(1), 433–47. doi:10.2134/agronj2015.0236.

Maughan, M. W., Flores, J. P. C., Anghinoni, I., Bollero, G., Fernandez, F. G. and Tracy, B. F. 2009. Soil quality and corn yield under crop-livestock integration in Illinois. *Agron. J.* 101(6), 1503–10. doi:10.2134/agronj2009.0068.

Meda, A. R., Pavan, M. A., Cassiolato, M. E. and Miyazawa, M. 2002. Dolomite lime's reaction applied on the surface of a sandy soil of the Northwest Paraná, Brazil. *Braz. Arch. Biol. Technol.* 45(2), 219–22.

Merten, G. H. and Minella, J. P. G. 2013. The expansion of Brazilian agriculture: soil erosion scenarios. *Inter. Soil Water Conserv. Res.* 1(3), 37–48. doi:10.1016/S2095-6339(15)30029-0.

Motter, P. and Almeida, H. G. 2015. *Plantio direto: a tecnologia que revolucionou a agricultura Brasileira*. Parque Itaipu, Foz do Iguaçu.

Müller, M. M. L., Guimarães, M. F., Desardins, T. and Mitja, D. 2004. The relationship between pasture degradation and soil properties in the Brazilian amazon: a case study. *Agric. Ecosyst. Environ.* 103(2), 279–88. doi:10.1016/j.agee.2003.12.003.

Nascente, A. S., Li, Y. C. and Crusciol, C. A. C. 2013. Cover crops and no-till effects on physical fractions of soil organic matter. *Soil Till. Res.* 130, 52–7. doi:10.1016/j.still.2013.02.008.

Nawaz, M. F., Bourrie, G. and Trolard, F. 2013. Soil compaction impact and modelling. A review. *Agron. Sustain. Dev.* 33(2), 291–309.

Nicoloso, R. S., Amado, T. J. C., Schneider, S., Lanzanova, M. E., Girardello, V. C. and Bragagnolo, J. 2008. Eficiência da escarificação mecânica e biológica na melhoria dos atributos físicos de um Latossolo muito argiloso e no incremento do rendimento de soja. *Rev. Bras. Ciênc. Solo* 32(4), 1723–34. doi:10.1590/S0100-06832008000400037.

Nicoloso, R. S., Rice, C. W. and Amado, T. J. C. 2016. Kinetic to saturation model for simulation of soil organic carbon increase to steady state. *Soil Sci. Soc. Am. J.* 80(1), 147–56. doi:10.2136/sssaj2015.04.0163.

Nicoloso, R. S., Rice, C. W., Amado, T. J. C., Costa, C. N. and Akley, E. K. 2018. Carbon saturation and translocation in a no-till soil under organic amendments. *Agric. Ecosyst. Environ.* 264, 73–84. doi:10.1016/j.agee.2018.05.016.

Nunes, M. R., Pauletto, E. A., Denardin, J. E., Faganello, A., Pinto, L. F. S. and Scheunemann, T. 2014. Persistência dos efeitos da escarificação sobre a compactação de Nitossolo sob plantio direto em região subtropical úmida. *Pesq. Agropec. Bras.* 49(7), 531–9. doi:10.1590/S0100-204X2014000700005.

Oldeman, L. R., Hakkeling, R. T. A. and Sombroek, W. G. 1991. *World Map of the Status of Human-induced Soil Degradation (GLASOD): an Explanatory Note*. International Soil Reference and Information Centre, Wageningen.

Oliveira, E. L. and Pavan, M. A. 1996. Control of soil acidity in no-tillage system for soybean production. *Soil Till. Res.* 38(1–2), 47–57. doi:10.1016/0167-1987(96)01021-5.

Pacheco, L. P., Leandro, W. M., Machado, P. L. OdA., Assis, R. Ld, Cobucci, T., Madari, B. E. and Petter, F. A. 2011. Produção de fitomassa e acúmulo e liberação de nutrientes por plantas de cobertura na safrinha. *Pesq. Agropec. Bras.* 46(1), 17–25. doi:10.1590/S0100-204X2011000100003.

Pariz, C. M., Carvalho, M. P., Chioderoli, C. A., Nakayama, F. T., Andreotti, M. and Montanari, R. 2011a. Spatial variability of forage yield and soil physical attributes of a *Brachiaria* decumbens pasture in the Brazilian Cerrado. *Rev. Bras. Zootecn.* 40(10), 2111–20.

Pariz, C. M., Andreotti, M., Buzetti, S., Bergamaschine, A. F., Ulian, N. A., Furlan, L. C., Meirelles, P. R. L. and Cavasano, F. A. 2011b. Straw decomposition of nitrogen-fertilized grasses after intercropping with corn crop in irrigated integrated croplivestock system. *Rev. Bras. Ciênc. Solo* 35(6), 2029–37.

Pariz, C. M., Costa, C., Crusciol, C. A. C., Meirelles, P. R. L., Castilhos, A. M., Andreotti, M., Costa, N. R., Martello, J. M., Souza, D. M., Sarto, J. R. W. and Franzluebbers, A. J.

2016. Production and soil responses to intercropping of forage grasses with corn and soybean silage. *Agron. J.* 108(6), 1–13. doi:10.2134/agronj2016.02.0082.

Pariz, C. M., Costa, C., Crusciol, C. A. C., Meirelles, P. R. L., Castilhos, A. M., Andreotti, M., Costa, N. R., Martello, J. M., Souza, D. M., Potes, V. M., Longhini, V. Z. and Franzluebbers, A. J. 2017a. Production, nutrient cycling and soil compaction to grazing of grass companion cropping with corn and soybean. *Nutr. Cycl. Agroecosyst.* 108(1), 35–54.

Pariz, C. M., Costa, C., Crusciol, C. A. C., Castilhos, A. M., Meirelles, P. R. L., Roça, R. O., Pinheiro, R. S. B., Kuwahara, F. A., Martello, J. M., Cavasano, F. A., Yasuoka, J. I., Sarto, J. R. W., Melo, V. F. P. and Franzluebbers, A. J. 2017b. Lamb production responses to grass grazing in a companion crop system with corn silage and oversowing of yellow oat in a tropical region. *Agric. Syst.* 151, 1–11. doi:10.1016/j.agsy.2016.11.004.

R Core Team. 2018. *R: A Language and Environment for Statistical Computing*. R Foundation for Statistical Computing.

Reichert, J. M., Suzuki, L. E. A. S. and Reinert, D. J. 2007. Compactação do solo em sistemas agropecuários e florestais: identificação, efeitos, limites críticos e mitigação. In: Ceretta, C. A., Silva, L. S. and Reichert, J. M. (Eds), *Tópicos em Ciência do Solo 5*. Sociedade Brasileira de Ciência do Solo, pp. 49–134.

Reichert, J. M., Suzuki, L. E. A. S., Reinert, D. J., Horn, R. and Hakansson, I. 2009. Reference bulk density and critical degree-of-compactness for no-till crop production in subtropical highly weathered soils. *Soil Till. Res.* 102(2), 242–54. doi:10.1016/j.still.2008.07.002.

Reis, C. E. Sd, Dick, D. P., Caldas, JdS. and Bayer, C. 2014. Carbon sequestration in clay and silt fractions of Brazilian soils under conventional and no-tillage systems. *Sci. Agric. (Piracicaba, Braz.)* 71(4), 292–301. doi:10.1590/0103-9016-2013-0234.

Richart, A., Filho, J. T., Brito, O. R., Llanillo, R. F. and Ferreira, R. 2005. Compactação do solo: causas e efeitos. *Semin. Cienc. Agrar.* 26(3), 321–44.

Sá, J. C. M. and Ferreira, A. O. 2018. The soil science in the evolution of no-till system. *SBCS Bulletim*. Available at: https://www.sbcs.org.br/wp-content/uploads/2018/08/boletimsbcs_v44_n2.pdf (accessed on 14 February 2019).

Sá, J. C. M., Lal, R., Cerri, C. C., Lorenz, K., Hungria, M. and Carvalho, P. C. F. 2017. Low-carbon agriculture in South America to mitigate global climate change and advance food security. *Environ. Int.* 98, 102–12. doi:10.1016/j.envint.2016.10.020.

Sá, J. C. M., Potma Gonçalves, D. R., Ferreira, L. A., Mishra, U., Inagaki, T. M., Ferreira Furlan, F. J., Moro, R. S., Floriani, N., Briedis, C. and de Oliveira Ferreira, A. 2018. Soil carbon fractions and biological activity based indices can be used to study the impact of land management and ecological successions. *Ecol. Indic.* 84, 96–105. doi:10.1016/j.ecolind.2017.08.029.

Santos, H. G., Jacomine, P. K. T., Anjos, L. H. C., Oliveira, V. A., Lumbreras, J. F., Coelho, M. R., Lumbreras, J. F. and Cunha, T. J. F. 2013. *Sistema Brasileiro de Classificação de Solos*. Embrapa.

Santos, C. Ad, Krawulski, C. C., Bini, D., Goulart Filho, T., Knob, A., Medina, C. C., Andrade Filho, G. and Nogueira, M. A. 2015. Reclamation status of a degraded pasture based on soil health indicators. *Sci. Agric. (Piracicaba, Braz.)* 72(3), 195–202. doi:10.1590/0103-9016-2013-0274.

Savian, J. V., Schons, R. M. T., Marchi, D. E., Freitas, T. S., Silva Neto, F. S., Mezzalira, C., et al. 2018. Rotatinuous stocking: a grazing management innovation that has high potential to mitigate methane emissions by sheep. *J. Clean. Prod.* 186, 602–8.

Secco, D. 2003. Estados de compactação de dois Latossolos sob plantio direto e suas implicações no comportamento mecânico e na produtividade de culturas. PhD thesis. Faculdade de Agronomia, Universidade Federal de Santa Maria, Rio Grande do Sul.

Siczek, A. and Lipiec, J. 2011. Soybean nodulation and nitrogen fixation in response to soil compaction and surface straw mulching. *Soil Till. Res.* 114(1), 50–6. doi:10.1016/j.still.2011.04.001.

Silva, F. D., Amado, T. J. C., Ferreira, A. O., Assmann, J. M., Anghinoni, I. and Carvalho, P. C. F. 2014. Soil carbon indices as affected by 10 years of integrated crop-livestock production with different pasture grazing intensities in Southern Brazil. *Agric. Ecosyst. Environ.* 190, 60–9.

Silva, F. R., Albuquerque, J. A., Costa, A., Fontoura, S. M. V., Bayer, C. and Warmling, M. I. 2016. Physical properties of a Hapludox after three decades under different soil management systems. *Rev. Bras. Ciênc. Solo* 40, 1–14.

Singh, S., Kumar, R., Yadav, S., Kumar, R., Kumari, P. and Singh, R. K. 2018. Effect of bio-control agents on soil borne pathogens: a review. *J. Pharmacogn. Phytochem.* 7(3), 406–11.

Sivarajan, S., Maharlooei, M., Bajwa, S. G. and Nowatzki, J. 2018. Impact of soil compaction due to wheel traffic on corn and soybean growth, development and yield. *Soil Till. Res.* 175, 234–43. doi:10.1016/j.still.2017.09.001.

Soratto, R. P. and Crusciol, C. A. C. 2008a. Dolomite and phosphogypsum surface application effects on annual crops nutrition and yield. *Agron. J.* 100(2), 261–70.

Soratto, R. P. and Crusciol, C. A. C. 2008b. Nutrição e produtividade de grãos da aveia-preta em função da aplicação de calcário e gesso em superfície na implantação do sistema plantio direto. *Rev. Bras. Ciênc. Solo* 32(2), 715–25.

Stefanoski, D. C., Santos, G. S., Marchão, R. L., Petter, F. A. and Pacheco, L. P. 2013. Uso e manejo do solo e seus impactos sobre a qualidade física. *Rev. Bras. Eng. Agr. Amb.* 17(2), 1301–309.

Sulc, R. M. and Tracy, B. F. 2007. Integrated crop-livestock systems in the U.S. corn belt. *Agron. J.* 99(2), 335–45. doi:10.2134/agronj2006.0086.

Sumner, M. E., Shahandeh, H., Bouton, J. and Hammel, J. 1986. Amelioration of an acid soil profile through deep liming and surface application of gypsum. *Soil Sci. Soc. Am. J.* 50(5), 1254–8. doi:10.2136/sssaj1986.03615995005000050034x.

Teague, R., DeLaune, P. B. and Dowhower, S. L. 2019. Impacts of over-seeding Bermudagrass pasture with multispecies cover crops on soil water availability, microbiology, and nutrient status in North Texas. *Agron. Ecosyst. Environ.* 273, 117–29. doi:10.1016/j.agee.2018.12.013.

Tiecher, T., Pias, O. H. C., Bayer, C., Martins, A. P., Denardin, L. G. O. and Anghinoni, I. 2018. Crop response to gypsum application to subtropical soils under no-till in Brazil: a systematic review. *Rev. Bras. Ciênc. Solo* 42, 1–17.

USDA. 2015. Soil quality indicators – physical, chemical, and biological indicators for soil quality assessment and management. Available at: https://www.nrcs.usda.gov/wps/portal/nrcs/detail/soils/health/assessment/?cid=stelprdb1237387.

Veloso, M. G., Angers, D. A., Tiecher, T., Giacomini, S., Dieckow, J. and Bayer, C. 2018. High carbon storage in a previously degraded subtropical soil under no-tillage with legume cover crops. *Agric. Ecosyst. Environ.* 268, 15–23. doi:10.1016/j.agee.2018.08.024.

Index

www.ingramcontent.com/pod-product-compliance
Ingram Content Group UK Ltd.
Pitfield, Milton Keynes, MK11 3LW, UK
UKHW021824100926
8556IPUK00003B/146

9 781786 762689